NanoScience and Technology

The series NanoScience and Technology is focused on the fascinating nano-world, mesoscopic physics, analysis with atomic resolution, nano and quantum-effect devices, nanomechanics and atomic-scale processes. All the basic aspects and technology-oriented developments in this emerging discipline are covered by comprehensive and timely books. The series constitutes a survey of the relevant special topics, which are presented by leading experts in the field. These books will appeal to researchers, engineers, and advanced students.

More information about this series at http://www.springer.com/series/3705

Roland Wiesendanger
Editor

Atomic- and Nanoscale Magnetism

Springer

Editor
Roland Wiesendanger
Department of Physics
University of Hamburg
Hamburg, Germany

ISSN 1434-4904 ISSN 2197-7127 (electronic)
NanoScience and Technology
ISBN 978-3-319-99557-1 ISBN 978-3-319-99558-8 (eBook)
https://doi.org/10.1007/978-3-319-99558-8

Library of Congress Control Number: 2018951913

This Springer imprint is published by the registered company Springer Nature Switzerland AG
The registered company address is: Gewerbestrasse 11, 6330 Cham, Switzerland

Preface

In the past decade, tremendous progress in the science of magnetism at the nanoscale down to the single-atom limit has been made leading to fundamental insight into spin-dependent phenomena at the atomic level and resulting in numerous unexpected discoveries. These recent fascinating developments in atomic- and nanoscale magnetism have become possible, thanks to novel experimental and theoretical tools which allow for unprecedented insight into static and dynamic spin states from the level of individual spins up to complex spin textures. Novel exciting fields emerged such as the physics of individual magnetic adatoms and single spins, the on-surface synthesis of molecular magnets for spintronic applications, as well as non-collinear spin textures, such as magnetic vortices, chiral domain walls, chiral spin spirals and magnetic skyrmions in ultrathin films and nanostructures.

In this book, we focus on these recent exciting developments in the physics of magnetism from single atoms up to nanoscale structures based on novel experimental and theoretical tools allowing deep insight into spin-dependent phenomena at ultimate spatial and temporal resolution. This includes atomic-resolution imaging of complex spin states by spin-polarized scanning tunnelling microscopy (SPSTM) and magnetic exchange force microscopy, the quantitative determination of magnetic moments and magnetic anisotropies of individual magnetic adatoms by SPSTM-based single-atom magnetometry, as well as studies of dynamic spin states by time-resolved SPSTM, electron microscopy with polarization analysis and X-ray microscopy-based techniques. Moreover, the fundamental magnetic interactions including the Heisenberg exchange interaction, the indirect magnetic exchange interaction as well as the Dzyaloshinskii-Moriya interaction can nowadays be revealed in real space with atomic-scale spatial resolution. Their competition can lead to complex spin textures, such as magnetic skyrmions, which form the basis for novel spintronic devices. The progress in experimental insight into magnetic properties and interactions down to the atomic level is complemented by advances in the theoretical treatment of magnetic ordering at the nanoscale as well as non-equilibrium spin-dependent phenomena.

A significant part of the research reviewed in this book has been conducted in the framework of the Collaborative Research Center (SFB 668) "Magnetism from Single Atoms to Nanostructures" supported by the German Research Foundation (DFG) over a time period of 12 years. We would like to thank all colleagues and particularly our young scientists who contributed to the success of our research programme. We also thank Andrea Beese for her great support in the administration of our Collaborative Research Center.

Hamburg, Germany Roland Wiesendanger

Contents

Contributors

Simon Altenburg Bundesanstalt für Materialforschung und -prüfung, Berlin, Germany

Maximilian W. Aulbach I. Institute of Theoretical Physics, University of Hamburg, Hamburg, Germany

Ivan Baev Institut Für Experimentalphysik, Universität Hamburg, Hamburg, Germany

Matthias Balzer I. Institute of Theoretical Physics, University of Hamburg, Hamburg, Germany

Torben Beeck Institut Für Experimentalphysik, Universität Hamburg, Hamburg, Germany

Richard Berndt Institut für Experimentelle und Angewandte Physik, Christian-Albrechts-Universität zu Kiel, Kiel, Germany

Tabea Buban Institute of Inorganic and Applied Chemistry, University of Hamburg, Hamburg, Germany

Peter Burger Institute of Inorganic and Applied Chemistry, University of Hamburg, Hamburg, Germany

Andreas Burtzlaff Institut für Experimentelle und Angewandte Physik, Christian-Albrechts-Universität zu Kiel, Kiel, Germany

Kai Chen Institut Für Experimentalphysik, Universität Hamburg, Hamburg, Germany

Edna C. Corredor Institut für Nanostruktur- und Festkörperphysik,Universität Hamburg, Hamburg, Germany

Markus Drescher The Hamburg Centre for Ultrafast Imaging (CUI), Hamburg, Germany; Institut Für Experimentalphysik, University of Hamburg, Hamburg, Germany

Stefan Freercks Department of Physics, University of Hamburg, Hamburg, Germany

Robert Frömter Institut für Nanostruktur- und Festkörperphysik,Universität Hamburg, Hamburg, Germany

Leif Glaser Institut Für Experimentalphysik, Universität Hamburg, Hamburg, Germany

Lynn Groß Institute for Inorganic and Applied Chemistry, University of Hamburg, Hamburg, Germany

Manuel Gruber Institut für Experimentelle und Angewandte Physik, Christian-Albrechts-Universität zu Kiel, Kiel, Germany

Sebastian Hankemeier Institut für Nanostruktur- und Festkörperphysik, Universität Hamburg, Hamburg, Germany

Jürgen Heck Institute of Inorganic and Applied Chemistry, University of Hamburg, Hamburg, Germany

Carmen Herrmann Institute for Inorganic and Applied Chemistry, University of Hamburg, Hamburg, Germany

Simon Hesse Department of Physics, University of Hamburg, Hamburg, Germany

Mirek Hänsel I. Institute of Theoretical Physics, University of Hamburg, Hamburg, Germany

Michael Karolak I. Institute of Theoretical Physics, Hamburg University, Hamburg, Germany

Alexander A. Khajetoorians Department of Physics, University of Hamburg, Hamburg, Germany

Fridtjof Kielgast Institut Für Experimentalphysik, Universität Hamburg, Hamburg, Germany

Fabian Kloodt-Twesten Institut für Nanostruktur- und Festkörperphysik, Universität Hamburg, Hamburg, Germany

Thomas Knaak Institut für Experimentelle und Angewandte Physik, Christian-Albrechts-Universität zu Kiel, Kiel, Germany

Stefan Krause Department of Physics, University of Hamburg, Hamburg, Germany

Jörg Kröger Institut für Physik, Technische Universität Ilmenau, Ilmenau, Germany

André Kubetzka Department of Physics, University of Hamburg, Hamburg, Germany

Susanne Kuhrau Institut für Nanostruktur- und Festkörperphysik,Universität Hamburg, Hamburg, Germany

Alexander Lichtenstein I. Institute of Theoretical Physics, Hamburg University, Hamburg, Germany

Laurent Limot Université de Strasbourg, CNRS, IPCMS, UMR 7504, Strasbourg, France

Fabian Lofink Institut für Nanostruktur- und Festkörperphysik,Universität Hamburg, Hamburg, Germany

Michael Martins Institut Für Experimentalphysik, Universität Hamburg, Hamburg, Germany

Guido Meier Max-Planck Institute for the Structure and Dynamics of Matter, Hamburg, Germany

Roberto Mozara I. Institute of Theoretical Physics, Hamburg University, Hamburg, Germany

Alexander Neumann Department of Physics, University of Hamburg, Hamburg, Germany

Nicolas Néel Institut für Physik, Technische Universität Ilmenau, Ilmenau, Germany

Hans Peter Oepen Fachbereich Physik, Institut für Nanostruktur- und Festkörperphysik, Universität Hamburg, Hamburg, Germany

Matthias Peschke I. Institute of Theoretical Physics, University of Hamburg, Hamburg, Germany

Oswald Pietzsch Department of Physics, University of Hamburg, Hamburg, Germany

Michael Potthoff I. Institute of Theoretical Physics, University of Hamburg, Hamburg, Germany

Marc H. Prosenc Institute of Physical Chemistry, TU Kaiserslautern, Kaiserslautern, Germany

Matthias Pues Institut für Angewandte Physik und Zentrum für Mikrostrukturforschung, Hamburg, Germany

Sarah Puhl Institute of Inorganic and Applied Chemistry, University of Hamburg, Hamburg, Germany

Stefan Rößler Institut für Nanostruktur- und Festkörperphysik,Universität Hamburg, Hamburg, Germany

Andrej Schwabe I. Institute of Theoretical Physics, University of Hamburg, Hamburg, Germany

Alexander Schwarz Department of Physics, University of Hamburg, Hamburg, Germany

Johannes Schöneberg Institut für Experimentelle und Angewandte Physik, Christian-Albrechts-Universität zu Kiel, Kiel, Germany

Suranjan Shil Institute for Inorganic and Applied Chemistry, University of Hamburg, Hamburg, Germany

Philipp Staeck Department of Physics, University of Hamburg, Hamburg, Germany

Torben Steenbock Institute for Inorganic and Applied Chemistry, University of Hamburg, Hamburg, Germany

Falk-Ulrich Stein Max-Planck Institute for the Structure and Dynamics of Matter, Hamburg, Germany

Martin Stier I. Institut für Theoretische Physik, Universität Hamburg, Hamburg, Germany

Michael Thorwart I. Institut für Theoretische Physik, Universität Hamburg, Hamburg, Germany

Carsten Thönnissen Department of Physics, University of Hamburg, Hamburg, Germany

Irakli Titvinidze I. Institute of Theoretical Physics, University of Hamburg, Hamburg, Germany

Takashi Uchihashi International Center for Materials Nanoarchitectonics, National Institute for Materials Science, Ibaraki, Japan

Maria Valentyuk I. Institute of Theoretical Physics, Hamburg University, Hamburg, Germany

Elena Vedmedenko Department of Physics, Institute of Nanostructure and Solid-State Physics, University of Hamburg, Hamburg, Germany

Bodo Alexander Voigt Institute for Inorganic and Applied Chemistry, University of Hamburg, Hamburg, Germany

Kirsten von Bergmann Department of Physics, University of Hamburg, Hamburg, Germany

Alexander Weismann Institut für Experimentelle und Angewandte Physik, Christian-Albrechts-Universität zu Kiel, Kiel, Germany

Philipp Wessels The Hamburg Centre for Ultrafast Imaging (CUI), Hamburg, Germany

Jens Wiebe Department of Physics, University of Hamburg, Hamburg, Germany

Roland Wiesendanger Department of Physics, University of Hamburg, Hamburg, Germany

Eva-Sophie Wilhelm Department of Physics, University of Hamburg, Hamburg, Germany

Wilfried Wurth Institut Für Experimentalphysik, Universität Hamburg, Hamburg, Germany

Jianwei Zhang Institut für Experimentelle und Angewandte Physik, Christian-Albrechts-Universität zu Kiel, Kiel, Germany

Part I
From Single Spins to Complex Spin Textures

Chapter 1
Magnetic Spectroscopy of Individual Atoms, Chains and Nanostructures

Jens Wiebe, Alexander A. Khajetoorians and Roland Wiesendanger

Abstract We review the magnetism of tailored bottom-up nanostructures which have been assembled of $3d$-transition metal atoms on nonmagnetic metallic substrates. We introduce the newly developed methodology of single atom magnetometry which combines spin-resolved scanning tunneling spectroscopy (SPSTS) and inelastic STS (ISTS) pushed to the limit of an individual atom. We describe how it can be used to measure the magnetic moment, magnetic anisotropy, and g-factor of individual atoms, as well as their pair-wise Ruderman-Kittel-Kasuya-Yosida (RKKY)-interaction. Finally, we will show that, using these measured quantities in combination with STM-tip induced manipulation of the atoms, nanostructures ranging from antiferromagnetic chains and two-dimensional arrays over all-spin based logic gates to magnetic memories composed of only few atoms can be realized and their magnetic properties characterized.

1.1 Introduction

Magnetic nanostructures which are composed of atom-by-atom assembled arrays of atomic spins on nonmagnetic substrates have attracted a lot of attention in the last ten years as model systems to understand atomic-scale magnetism in the transition region between few interacting spins and macroscopic materials, as well as a platform for the proof of principle of nanospintronic technologies. The pathway into this field was paved by the ability of the scanning tunneling microscope (STM) tip to move individual atoms on a surface [1] and to measure the magnetic properties of single atoms [2, 3]. These advances enabled the study of the magnetic moment [3],

J. Wiebe · A. A. Khajetoorians · R. Wiesendanger (✉)
Department of Physics, University of Hamburg, Hamburg, Germany
e-mail: wiesendanger@physnet.uni-hamburg.de

J. Wiebe
e-mail: jwiebe@physnet.uni-hamburg.de

R. Wiesendanger (ed.), *Atomic- and Nanoscale Magnetism*, NanoScience and Technology, https://doi.org/10.1007/978-3-319-99558-8_1

g-factor [4–6], and magnetic anisotropy [4, 7–9] of individual atoms, the exchange interaction in pairs [10, 11], the properties of bottom-up chains [12–16] and two-dimensional arrays [14], as well as logic gates [17] and magnetic memories [18–20].

The magnetism of such nanostructures not only depends on the atom type used, but also crucially on the interaction of the atomic spin with the substrate conduction electrons which can dramatically modify the magnetic moment and the delocalization of the atomic spin. One strategy has been focused on the use of thin decoupling layers in order to strongly reduce the overlap of the electronic orbitals responsible for the atomic spin from the orbitals of the substrate conduction electrons [2, 7, 12, 15, 19], which typically enhances the "quantumness" of the nanostructures [20]. In this review, we will focus on the other extreme, i.e. in which the atomic spins are adsorbed directly onto a metallic substrate. As we will show, this enables to make use of a large range of substrate-conduction electron mediated Ruderman-Kittel-Kasuya-Yosida (RKKY)-interactions for the coupling between the atomic spins, which offers huge flexibility and tunability.

The review is organized as follows. Section 1.2 introduces the development of the experimental methodology towards characterizing the magnetic properties of single atoms on metallic surfaces. In Sect. 1.3 we review the application of these methods to atoms which are RKKY-coupled to magnetic layers. Furthermore, we consider the RKKY-coupling in pairs of atoms with a particular focus on the non-collinear contribution to the RKKY interaction. Section 1.4 deals with the investigation of tailored dilute chains and two-dimensional arrays of different numbers of atoms. Finally, we show the experimental realization of model systems of logic gates and magnetic memories made from only few atoms in Sect. 1.5.

1.2 Single Atom Magnetometry

For the investigation of the magnetic properties of individual atoms, two complementary scanning tunneling spectroscopy (STS) based techniques have been developed. The first is the spin-resolved STS (SPSTS) based measurement of the magnetization of an atom as a function of an externally applied magnetic field, which is introduced in Sects. 1.2.1 and 1.2.2. The second method is the inelastic STS (ISTS) based measurement of the excitations of the magnetization of an atom, which will be introduced in Sect. 1.2.3.

1.2.1 SPSTS on Individual Atoms

For the application of the technique of SPSTS to individual atoms, we first chose the sample system of cobalt atoms adsorbed on the (111) surface of platinum. This sample system had the following advantages: (i) it was extensively characterized by spatially averaging techniques, (ii) the magnetic moment of the Co atom is large

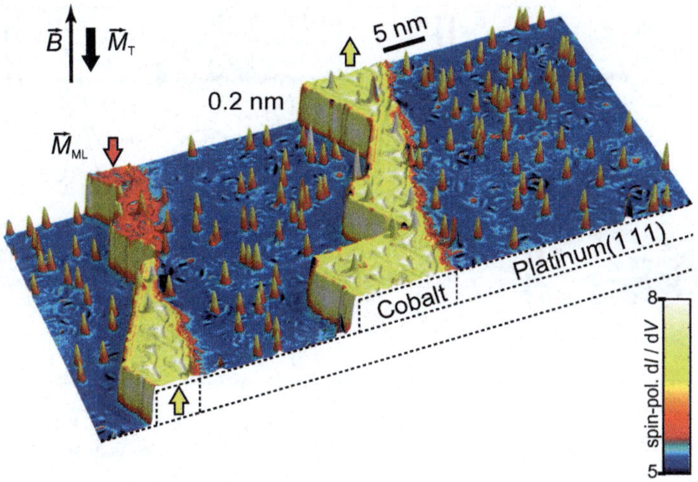

Fig. 1.1 Large-scale 3D rendering of a constant-current SP-STM image of single Co atoms and monolayer (ML) stripes on Pt(111), the first system utilized for the development of single atom magnetometry. The image was acquired with a chromium coated STM tip, magnetized antiparallel to the surface normal. An external magnetic field $\boldsymbol{B}$ can be applied perpendicular to the sample surface to change the magnetization of atoms $\boldsymbol{M}_{\mathrm{A}}$, ML stripes $\boldsymbol{M}_{\mathrm{ML}}$, or tip $\boldsymbol{M}_{\mathrm{T}}$. The ML appears red (yellow) when $\boldsymbol{M}_{\mathrm{ML}}$ is parallel (antiparallel) to $\boldsymbol{M}_{\mathrm{T}}$. Figure reprinted with permission from [3]. Copyright (2008) by AAAS

($m \approx 5\mu_{\mathrm{B}}$) and (iii) it has a large uniaxial magnetic anisotropy of $\mathcal{K} \approx -9\,\mathrm{meV}$ which forces the atomic spin of the Co to point perpendicular to the (111) surface (out-of-plane) [21].

Figure 1.1 shows an overview of the used sample. It consists of individual cobalt atoms on the (111) surface of platinum (blue) and cobalt monolayer (ML) stripes (red and yellow) which are attached to the step edges. The statistical distribution of the Co atoms on this surface results in a variety of different adsorption sites. Isolated Co atoms on Pt(111) can sit either on an fcc or on an hcp hollow site. Co atoms are adsorbed on the hcp or fcc areas of the Co ML. We also find closed-packed Co dimers, as well as pairs, triples or even larger ensembles with different inter-atomic distances (cf. Sect. 1.3.2). The advantage of the additional Co ML stripes is twofold. As will be shown in Sect. 1.3.1 it allows us to measure the magnetic interaction between the stripes and the individual Co atoms. Furthermore, the ML stripes which have a magnetization $\boldsymbol{M}_{\mathrm{ML}}$ perpendicular to the surface serve for the calibration of the orientation of the magnetization of the SPSTM tip. Using out-of-plane oriented (chromium coated) tips the up and down domains exhibit a different spin-resolved $\mathrm{d}I/\mathrm{d}V$ signal as visible in Fig. 1.1. Thereby, it is possible to characterize the spin polarization and magnetization $\boldsymbol{M}_{\mathrm{T}}$ of the foremost tip atom acting as a detector for the magnetization of the atom on the surface $\boldsymbol{M}_{\mathrm{A}}$, as will be described in the following.

In an SPSTS experiment, the spin-resolved differential tunneling conductance $\mathrm{d}I/\mathrm{d}V$ as a function of the applied sample bias voltage V, as long as V is not too

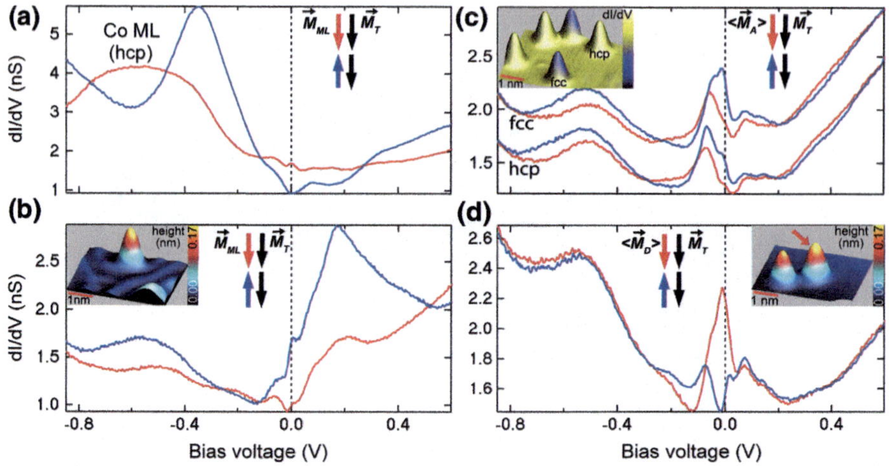

Fig. 1.2 SPSTS curves $\mathrm{d}I^{\uparrow\uparrow}/\mathrm{d}V(V)$ and $\mathrm{d}I^{\uparrow\downarrow}/\mathrm{d}V(V)$ for parallel (red) and antiparallel (blue) orientations of sample and tip magnetization, respectively, taken on **a** an hcp area of the Co ML, **b** a single atom on an fcc region of a ML (see inset), **c** single atoms sitting on fcc and hcp lattice sites on a Pt terrace (see inset), and **d** on the center of a dimer with both atoms sitting on nearest neighboring fcc sites on a Pt terrace (see inset). Figure reprinted with permission from [33]. Copyright (2010) by the American Physical Society

large, is given by

$$
\begin{aligned}
dI/dV(x, y, V) \propto \rho_T(E_\mathrm{F}) \cdot \rho_S(E_\mathrm{F} + \mathrm{eV}, \boldsymbol{R}_\mathrm{T}) \\
\cdot (1 + P_\mathrm{T}(E_\mathrm{F}) \cdot P_\mathrm{S}(E_\mathrm{F} + \mathrm{eV}, \boldsymbol{R}_\mathrm{T}) \cos\theta) .
\end{aligned}
\tag{1.1}
$$

Here $\rho_S(E, \boldsymbol{R}_\mathrm{T})$ is the local electron density of states (LDOS) *above* the sample, $\rho_T(E_\mathrm{F})$ is the LDOS of the tip, P_S and P_T are their spin polarizations given by the difference between the majority and minority LDOSs normalized by their sum, i.e. $P = (\rho^\uparrow - \rho^\downarrow)/(\rho^\uparrow + \rho^\downarrow)$, $\boldsymbol{R}_\mathrm{T}$ is the position of the foremost tip atom and θ is the angle between its magnetization $\boldsymbol{M}_\mathrm{T}$ and that of the sample $\boldsymbol{M}_\mathrm{S}$. If the tip material has a much larger coercivity than the sample, as e.g. Cr, an appropriate external magnetic field $\boldsymbol{B}$ can align tip and sample magnetization parallel ($\uparrow\uparrow$) or antiparallel ($\uparrow\downarrow$). This results in the spin-resolved differential tunneling conductances $\mathrm{d}I^{\uparrow\uparrow}/\mathrm{d}V(V)$ and $\mathrm{d}I^{\uparrow\downarrow}/\mathrm{d}V(V)$. Thereby, the product of tip and sample spin-polarizations can be deduced from the measured magnetic asymmetry, assuming a constant distance between the tip and sample for the two cases ($\uparrow\uparrow$, $\uparrow\downarrow$), i.e.

$$
\begin{aligned}
A_\mathrm{mag}(V) &\equiv \left(\mathrm{d}I^{\uparrow\uparrow}/\mathrm{d}V - \mathrm{d}I^{\uparrow\downarrow}/\mathrm{d}V\right) / \left(\mathrm{d}I^{\uparrow\uparrow}/\mathrm{d}V + \mathrm{d}I^{\uparrow\downarrow}/\mathrm{d}V\right) \\
&= P_\mathrm{T}(E_\mathrm{F}) \cdot P_\mathrm{S}(E_\mathrm{F} + \mathrm{eV}, \boldsymbol{R}_\mathrm{T}).
\end{aligned}
\tag{1.2}
$$

Thus, P_T has to be known in order to extract P_S.

Figure 1.2 illustrates how the sign of the spin-polarization of an atom was determined by measuring $\mathrm{d}I^{\uparrow\uparrow}/\mathrm{d}V(V)$ and $\mathrm{d}I^{\uparrow\downarrow}/\mathrm{d}V(V)$ on the Co ML which has a well-known $P_\mathrm{S}(E_\mathrm{F}+\mathrm{eV}, \boldsymbol{R}_\mathrm{T})$ [22]. Exactly the same tip was then used to characterize the Co atoms with unknown $P_\mathrm{S}(E_\mathrm{F}+\mathrm{eV}, \boldsymbol{R}_\mathrm{T})$ (Fig. 1.2b, c). As seen from Fig. 1.2a, the magnetic asymmetry A_mag defined in (1.2) is positive around E_F, i.e. $P_\mathrm{T}(E_\mathrm{F}) \cdot P_\mathrm{S}^\mathrm{ML}(E_\mathrm{F}, \boldsymbol{R}_\mathrm{T}) > 0$. On the other hand, first-principles calculations of the spin-resolved LDOS above the Co ML on Pt(111) yield $P_\mathrm{S}^\mathrm{ML}(E_\mathrm{F}, \boldsymbol{R}_\mathrm{T}) < 0$ [22]. Therefore, the tip must have a negative spin polarization at E_F, i.e. $P_\mathrm{T}(E_\mathrm{F}) < 0$. By comparison to the spectra measured with the same tip on a Co atom on the ML (Fig. 1.2b) and on a Co atom on the Pt substrate (Fig. 1.2c) we see that the strengths of the $\mathrm{d}I/\mathrm{d}V(V)$ signals at E_F for the parallel and antiparallel alignment of tip and sample (order of red and blue curves) is reversed with respect to the ML. This leads to the conclusion, that the sign of the vacuum spin polarization above the atoms around E_F is reversed with respect to that of the ML. Interestingly, this effect is already reversed back to the normal situation of the ML for a Co dimer, as visible in Fig. 1.2d.

1.2.2 Single-Atom Magnetization Curves

The magnetization of the atoms on Pt(111) in Fig. 1.2c was aligned parallel or antiparallel relative to the tip magnetization by changing the orientation of the external magnetic field $\boldsymbol{B}$. As a consequence, the intensity of the measured $\mathrm{d}I/\mathrm{d}V$ signal changes in a large energy interval around the Fermi energy. This signal change can be used to record the magnetization curves of single atoms as described in the following.

To this end, we use an anti-ferromagnetically coated tip, typically with Cr, whose magnetic moment orientation is not affected by $\boldsymbol{B}$. Then, $\mathrm{d}I/\mathrm{d}V$ at a particular voltage is measured as a function of $\boldsymbol{B}$ on the same atom at the same tip-sample distance (see Fig. 1.3a, b). The time resolution of SPSTS is typically much worse than the time scale of the magnetization switching of an atom which is adsorbed on a metal substrate. Therefore, $P_\mathrm{T}(E_\mathrm{F}) \cdot P_\mathrm{S}(E_\mathrm{F}+\mathrm{eV}, \boldsymbol{R}_\mathrm{T}) \cos\theta$ is proportional to the scalar product of the tip magnetization vector with the time average of the atom magnetization vector ($\langle \boldsymbol{M}_\mathrm{A} \rangle$), and the measured $\mathrm{d}I/\mathrm{d}V$ is given by (cf. 1.1)

$$\mathrm{d}I/\mathrm{d}V \propto (\mathrm{d}I/\mathrm{d}V)_0 + (\mathrm{d}I/\mathrm{d}V)_\mathrm{SP} \boldsymbol{M}_\mathrm{T} \cdot \langle \boldsymbol{M}_\mathrm{A} \rangle (\boldsymbol{B}) \ . \tag{1.3}$$

In words, recording of $\mathrm{d}I/\mathrm{d}V$ as a function of the external magnetic field results in the measurement of the projection of the time-average of the atom magnetization onto the tip magnetization direction.

In practice, a series of $\mathrm{d}I/\mathrm{d}V$ maps is recorded as a function of an external magnetic field $\boldsymbol{B}$ on an area with several atoms as shown in Fig. 1.3a, b. From this data set, the magnetization curve of each atom in this area is received by plotting the corresponding $\mathrm{d}I/\mathrm{d}V$ value averaged on top of each individual atom as a function of B. This is shown in Fig. 1.3c, d for several different atoms (on fcc and hcp stacking

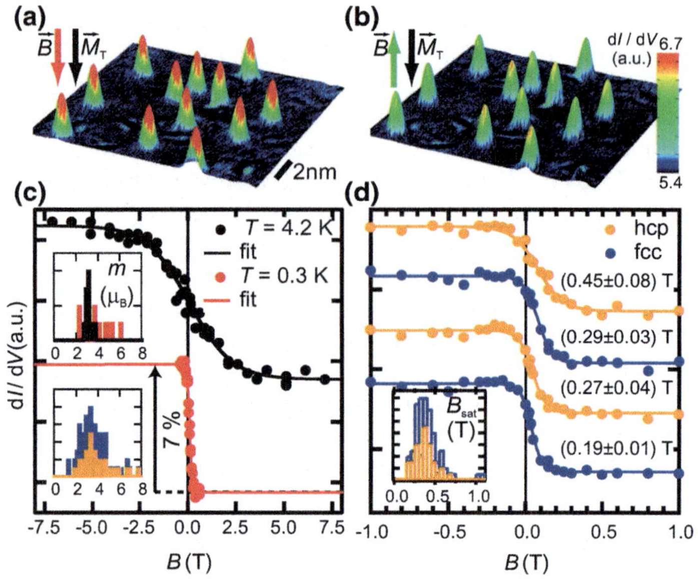

Fig. 1.3 **a**, **b** Spin-polarized dI/dV maps of 12 Co atoms on Pt(111) at $B = -0.5$ T parallel to the tip magnetization $\boldsymbol{M}_{\mathrm{T}}$ (**a**) and $B = +0.5$ T antiparallel to $\boldsymbol{M}_{\mathrm{T}}$ (**c**). The maps have been recorded using a Cr-coated tip that is magnetized perpendicular to the surface. **c** Magnetization curves from one of the atoms taken at different temperatures as indicated (dots). The solid lines are fits to the data (see text). The insets show the resulting histograms of the fitted magnetic moments (in μ_B) for 11 atoms at $T = 4.2$ K (black) and at 0.3 K (red) (upper histogram), and for 38 hcp (orange) and 46 fcc (blue) atoms at 0.3 K (lower histogram, fcc bars stacked on hcp). **d** Magnetization curves of four atoms at 0.3 K with fit curves and resulting B_{sat} of 99% saturation. The inset shows the histogram of B_{sat} (in Tesla) for the same atoms used in the lower histogram in **c** (fcc bars stacked on hcp). Figure reprinted with permission from [3]. Copyright (2008) by AAAS

position) and at two different temperatures $T = 4$ K and $T = 0.3$ K. The resulting s-shaped curves resemble the magnetization curves of paramagnetic atoms.

Such single-atom magnetization curves can be used to determine the magnetic moment of the particular atom, as shown in Fig. 1.3c, d. For this purpose, the curves were fitted to the following classical model:

$$E\,(\theta, B) = -mB\cos\theta + \mathcal{K}(\cos\theta)^2 \tag{1.4}$$

$$\langle M_{\mathrm{A}} \rangle \propto \frac{\int d\theta \sin\theta\, e^{\frac{-E(\theta,B)}{k_{\mathrm{B}}T}}}{\int d\theta\, e^{\frac{-E(\theta,B)}{k_{\mathrm{B}}T}}} \tag{1.5}$$

Here, m is the effective magnetic moment of the atom, and $\mathcal{K}$ is its uniaxial magnetic anisotropy energy in the direction of $\boldsymbol{B}$. Please note that usually, m and $\mathcal{K}$ can only be determined independently from magnetization curves in two perpendicular magnetic field directions. Here, we considered the $\mathcal{K}$-value known from XMCD measurements [21]. The fitted curves which are shown in Fig. 1.3c, d on top of the

Table 1.1 Values of magnetic moments m determined from single atom magnetization curves for different sample systems [3, 5]. The given errors are reflecting the variation from the fitting of the magnetization curves of different atoms on the surface

System	m in μ_B
fcc Co on Pt(111)	3.5 ± 1.5
hcp Co on Pt(111)	3.9 ± 1.5
Fe on Cu(111)	3.5 ± 1.5

measured curves nicely reproduce the data. The resulting magnetic moments are given in the insets of Fig. 1.3c.

A similar measurement and analysis has been done for Fe atoms on Cu(111) and the determined magnetic moments are summarized in Table 1.1. While the values for Co on Pt(111) are considerably smaller than the ones which have been determined by XMCD measurements [21], the values for Fe on Cu(111) fit with values from XMCD [23].

Most importantly, even though the atom has a strong magnetic anisotropy, its magnetization is not stable but switches on a time scale which is much faster than the detection limit of conventional SPSTM. However, we will see in Sect. 1.5 that direct exchange coupling of only three Fe atoms already increases the lifetime of the magnetization to hours. Moreover, there is a strong scattering of m which is a result of the residual RKKY interaction from the background of statistically distributed atoms. We will later see, how the single-atom magnetization curves can be used in order to measure this RKKY interaction in pairs of atoms as a function of their distance (see Sect. 1.3.2).

1.2.3 Magnetic Field Dependent Inelastic STS

A complementary STS based method for the detection of the spin excitations of single atoms is inelastic STS (ISTS). The method was originally applied to magnetic atoms whose spin is decoupled from the conduction electrons of a metal substrate by using thin decoupling layers [2]. Later it was also adapted to the investigation of magnetic atoms adsorbed directly on the surface of a metal [4–6, 8]. The method is illustrated in Fig. 1.4a for an fcc Fe atom on Pt(111). It is based on magnetic field dependent ISTS which reveals steps at positive and negative bias voltages V (symmetrically around zero bias) shifting as a function of B. The steps are located at the energies $E_{ex} = |\pm eV|$ of the spin excitations of the atom (in this case only one). Typically, effective spin Hamiltonians of the form $\hat{\mathcal{H}} = K \cdot \hat{S}_z^2 - g\mu_B \hat{\mathbf{S}} \cdot \boldsymbol{B}$ have been considered for the analysis of such ISTS data. Within this model, the zero field E_{ex} reflects the magnetic anisotropy parameter K of the atomic spin via $K = E_{ex}/(2S - 1)$. E_{ex} is shifting with B due to the Zeeman splitting and the corresponding slope is directly proportional to the g-factor of the atom.

For a transition metal atom which is adsorbed directly on a metal substrate, there are typically strong charge fluctuations within the d-orbitals, such that the spin quan-

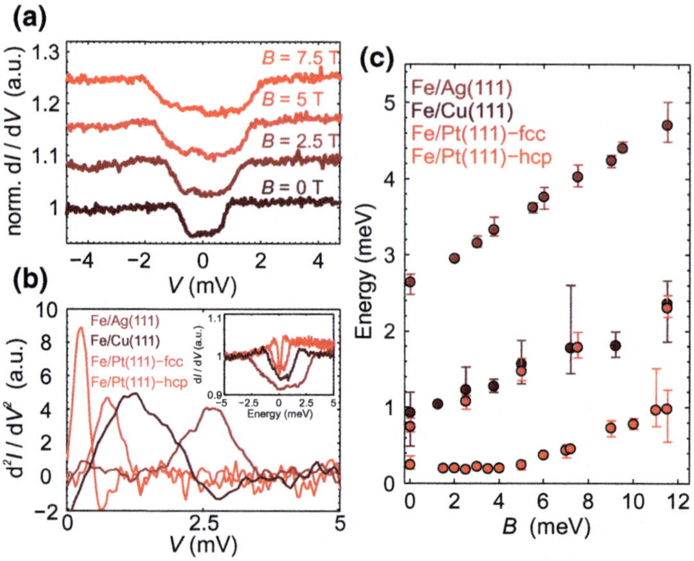

Fig. 1.4 **a** ISTS taken at various out-of-plane magnetic fields, as indicated, on an fcc Fe atom adsorbed on Pt(111) [4]. **b** ISTS taken at zero magnetic field for Fe atoms on various substrates as indicated (second derivative, the corresponding differential conductance data is given in the inset). **c** Spin-excitation energy extracted from the magnetic field dependent ISTS of the various systems as indicated. The data has been taken from [4–6]

Table 1.2 Values of magnetic anisotropy parameter K from the effective spin model and g-factor of Fe on three different substrates determined from magnetic field dependent ISTS [4–6]. The indicated spin quantum numbers S are estimated via $m = g\mu_B S$ from the magnetic moment m measured by single-atom magnetization curves for Cu(111) [5] or calculated from DFT for Pt(111) [4] and Ag(111) (here the calculated spin moment is 3.5 μ_B)

System	S	K in meV	g
hcp Fe on Pt(111)	5/2	0.08	2
fcc Fe on Pt(111)	5/2	−0.19	2.4
Fe on Cu(111)	3/2	−0.5	2.1
Fe on Ag(111)	1	−2.7	3.13

tum number S is no longer well-defined [5, 24]. Surprisingly, even in this case, the excitations can be reasonably reproduced by the effective spin model assuming the magnetic anisotropy and an exchange mechanism for the spin-flip probability, by using an S closest to the magnetic moment of the atom [24]. The latter can be either extracted experimentally from single-atom magnetization curves (Sect. 1.2.2) or determined from DFT calculations.

Figure 1.4b, c illustrate a comparison of magnetic-field dependent ISTS taken on Fe atoms adsorbed on three different substrates. The extracted parameters are shown in Table 1.2. Obviously both, K and g, vary for the different systems, and K even changes from out-of-plane to easy-plane magnetic anisotropy from fcc to hcp for Fe on Pt(111) (see the sign change). Thus, K and g crucially depend on the interaction of the Fe atom with the substrate. Single-atom magnetization curves and ISTS not only were used to reveal the magnetic moment of individual atoms, but also to study their magnetic interactions as will be shown in the following.

1.3 Measurement of the RKKY Interaction

1.3.1 RKKY Interaction Between a Magnetic Layer and an Atom

Figure 1.5a–c illustrate out-of-plane magnetization curves that have been recorded on one of the Co monolayer stripes of the sample of Fig. 1.1 and on three Co atoms with different separations to the stripe. As visible from the square shaped hysteresis, the coercive field of the monolayer stripe is 0.5 T. In stark contrast to the s-shaped magnetization curves of the uncoupled Co atoms (see Fig. 1.3), the curves measured on the three atoms close to the monolayer show hysteresis. This effect can be traced back to the RKKY interaction between the atom and the monolayer. The corresponding exchange bias fields B_{ex} (see arrows in Fig. 1.5a–c) which are given by the magnetic fields at which the RKKY interaction energy J is compensated by the Zeeman energy of the atom can be used to extract the absolute value of J via $|J| = m B_{ex}$ using the magnetic moment of the Co atom of $m \approx 3.7\mu_B$. On the other hand, the sign of J is given by the symmetry of the magnetization curve in Fig. 1.5 [3, 32]. The extracted values are plotted in Fig. 1.5d as a function of distance d of the atom from the monolayer stripe. It shows the typical oscillatory damped behavior of the RKKY interaction. Fits to isotropic models of the asymptotic RKKY interaction $J(d) = J_0 \cdot \cos(2k_F d)/(2k_F d)^D$ with the Fermi wavevector k_F and different assumed dimensionalities D are shown in Fig. 1.5d. D is determined by the dimensionality of the electron system that induces the interaction, which is not known a priori, since it depends on the localization character of the underlying substrate-electron states that induce the interaction. The best fit is found for $D = 1$ which leads to the conclusion that the responsible substrate-electron states are strongly localized in the surface and have a Fermi wavelength of $\lambda_F = 2$–3 nm.

1.3.2 RKKY Interaction Between two Atoms

The RKKY interaction also leads to a measurable coupling between single Co atoms as illustrated in Fig. 1.6. The figure shows single-atom magnetization curves, that have been measured on the two atoms of Co pairs with decreasing separations between 2 and 5 lattice constants. Again, the magnetization curves show clear deviations from the s-shaped magnetization curves of the uncoupled Co atoms (see

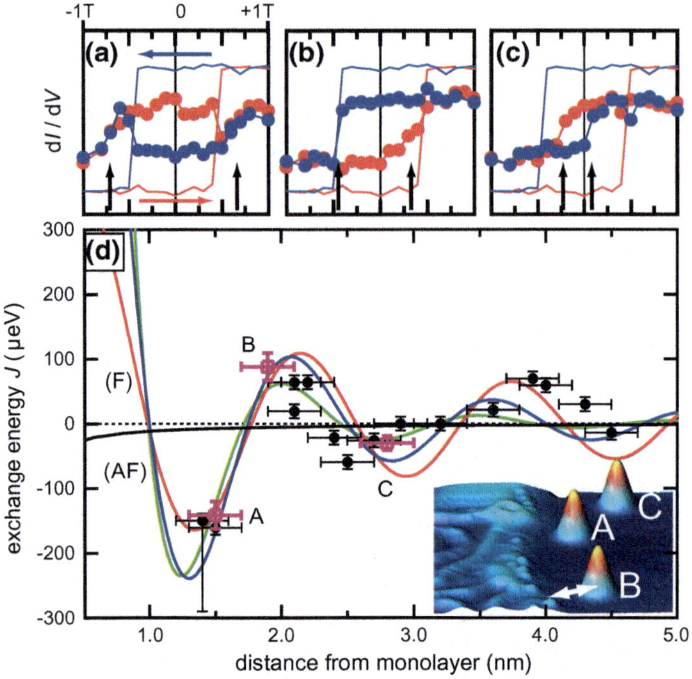

Fig. 1.5 **a**, **b**, **c** Out-of-plane magnetization curves measured on the monolayer (straight lines) and on three atoms (dots) A, B and C with different distances from the monolayer (see inset in (**d**), blue (red) color indicates down (up) sweep). The vertical arrows indicate the magnetic field B_{ex} at which the RKKY interaction between the Co atom and monolayer is compensated by the Zeeman energy. **d** Dots: Extracted RKKY exchange energy as a function of distance between atom and monolayer. The black line is the dipolar interaction. The red, blue and green lines are fits assuming 1D, 2D and 3D itinerant electron systems. Figure reprinted with permission from [3]. Copyright (2008) by AAAS

Fig. 1.3). While for some pairs, the two magnetization curves are still s-shaped, but with a steeper slope around zero magnetic field (Fig. 1.6f, k), other pairs reveal an additional oscillation or a plateau around zero magnetic field (Fig. 1.6g–j). While the former indicates ferromagnetic coupling, the latter is a result of an antiferromagnetic interaction between the two atoms. Note, that there is no hysteresis, indicating that the atoms are coupled, but still fluctuate on a time scale much faster than our measurement. This conclusion is substantiated and quantitatively analysed within the following Ising model:

$$\mathcal{H} = -\frac{1}{2} \sum_{i,j(i \neq j)} J_{ij}\left(\boldsymbol{r}_{ij}\right) \boldsymbol{S}_i \cdot \boldsymbol{S}_j - \sum_i m_i \boldsymbol{S}_i \cdot \boldsymbol{B} \tag{1.6}$$

where $i(j)$ labels the atoms 1 and 2 in the pair, $\boldsymbol{S}_i = \pm \boldsymbol{e}_z$ with the unit vector $\boldsymbol{e}_z$ along the surface normal z, and the absolute values of the magnetic moments m_i

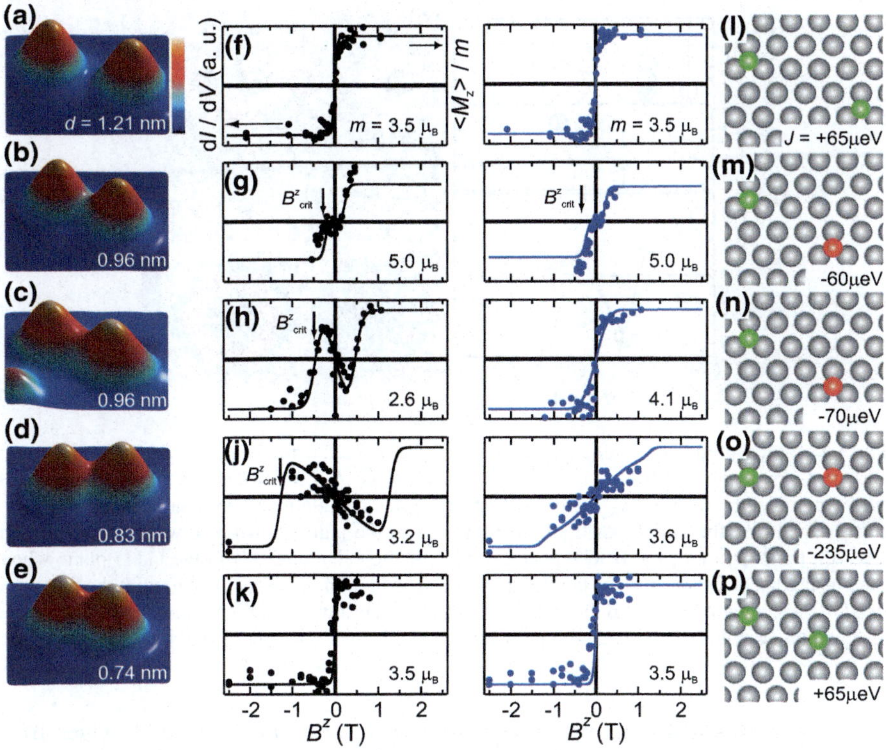

Fig. 1.6 Out-of-plane single-atom magnetization curves of Co pairs on Pt(111) with different distances shown in the STM images in **a–e**, and corresponding ball models of the positions of the two atoms in the pair on the substrate lattice in **l–p**. **f–k** shows the single-atom magnetization curves measured on the left atom (black dots) and on the right atom (blue dots) of each pair. The straight lines are calculated from the Ising model assuming magnetic moments m given in each curve and RKKY interaction energies given in **l–p**. Figure reprinted with permission from [11]. Copyright (2010) by Springer Nature

(in μ_B). While the first term describes the distance dependent exchange interaction, the second term is the Zeeman energy. Note that the Ising limit is justified by the large out-of-plane magnetic anisotropy of $\mathcal{K} = -9.3$ meV of the system of Co atoms on Pt(111) [21]. The results of the fits of the model to the measured single atom magnetization curves by variation of m_1, m_2 and J_{12} are shown in Fig. 1.6f–k as straight lines. They demonstrate an excellent reproduction of the measured data. The corresponding values of the RKKY interaction energy for about 10 pairs with different distances d placed at different locations on the bare Pt(111) substrate are shown in Fig. 1.7a. It reveals the typical oscillation between ferromagnetic ($J > 0$) and antiferromagnetic ($J < 0$) interaction which is reminiscent of the RKKY interaction.

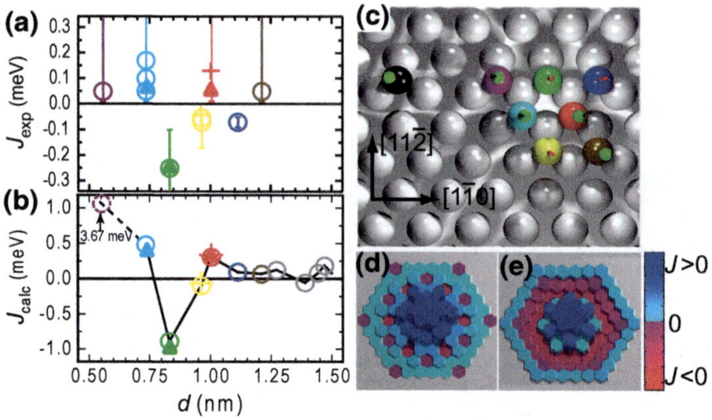

Fig. 1.7 **a** Measured RKKY interaction energy J_{exp} extracted from the single-atom magnetization curves of Co pairs on Pt(111) as a function of the distance d of the two atoms in the pair. **b** shows the corresponding values J_{calc} from the first-principles calculation. The data points are colored corresponding to the lattice positions of the atoms in the pairs shown in **c** where the first atom stays on the position of the black sphere. **d** Map of the calculated J's in the (111) plane where each hexagon corresponds to the lattice site of the second atom in a specific pair with the first atom fixed in the center. **e** Same as **d**, but with J's from the 2D isotropic RKKY model evaluated on the discrete lattice. Figure reprinted with permission from [11]. Copyright (2010) by Springer Nature

The experimental data of J was compared to ab initio calculated values from density functional theory utilizing the KKR method [11] (Fig. 1.6b). While the qualitative behavior of the experimental data is nicely reproduced by the calculation, the calculated values are a factor of about three times larger than the experimental ones. Most interestingly, the RKKY interaction shows a strong directionality, which is revealed by a 3D plot of the calculated J's in Fig. 1.7d in comparison to a similar plot of a 2D isotropic RKKY model given in Fig. 1.7e.

1.3.3 Dzyaloshinskii–Moriya Contribution to the RKKY Interaction

As shown in the preceding section, single-atom magnetization curves of the out-of-plane magnetization of interacting Co atoms on Pt(111) are approximately described within the Ising limit due to their large uniaxial out-of-plane anisotropy. We therefore so far only considered the usual Heisenberg contribution J to the RKKY interaction. However, as theoretically shown by Smith [25] and Fert [26] there is an additional Dzyaloshinskii–Moriya (DM) contribution to the RKKY interaction if the interaction is mediated by a heavy-element substrate featuring strong spin-orbit coupling. The magnetization of the coupled pair of quantum spins with spin operators $\hat{\boldsymbol{S}}_1$ and $\hat{\boldsymbol{S}}_2$ can then be quantified by the following spin Hamiltonian:

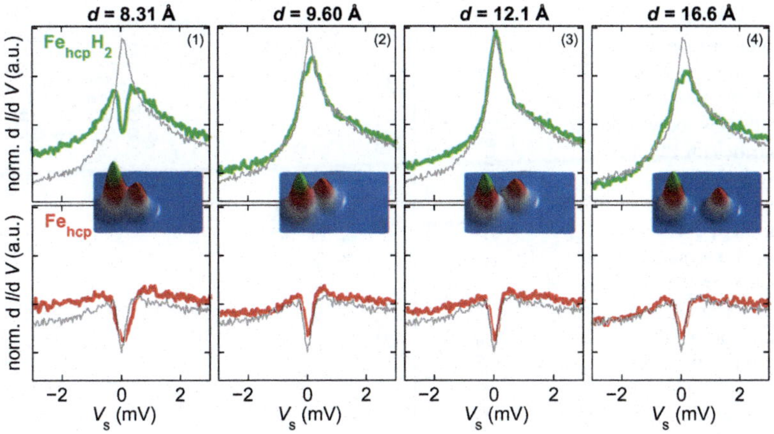

Fig. 1.8 ISTS of $Fe_{hcp}H_2$ (green) and Fe_{hcp} (red) within selected pairs of different distance as shown in the topographs in the insets. In comparison, the spectra measured on the corresponding isolated atoms are shown in gray. Figure reprinted with permission from [27]. Copyright (2016) by Springer Nature

$$\hat{\mathcal{H}} = -J_{12}\left(\hat{\boldsymbol{S}}_1 \cdot \hat{\boldsymbol{S}}_2\right) - \boldsymbol{D}_{12} \cdot \hat{\boldsymbol{S}}_1 \times \hat{\boldsymbol{S}}_2 + \sum_i K_i \left(\hat{S}_i^z\right)^2 - \sum_i g_i \mu_B \hat{\boldsymbol{S}}_i \cdot \boldsymbol{B} \quad (1.7)$$

where g_i are the g-factors of the two atoms. In comparison to (1.7) the additional term with the so called DM-vector $\boldsymbol{D} = \left(D_{||}, D_{\perp}, D_z\right)$ (see the definition of the components in Fig. 1.9d) favors a perpendicular orientation of the two spins.

In order to investigate the non-collinear behavior of the RKKY interaction, we studied pairs of Fe atoms and Fe-hydrogen complexes on Pt(111). Unlike Co atoms, the Fe atoms and complexes can exhibit very small values of magnetic anisotropy, both in-plane and out-of-plane, and Kondo behavior. Atoms with weak easy-plane magnetic anisotropy are no longer correctly described by Ising-like spins and non-collinear interactions need to be considered [24]. By investigating pairs of an Fe-hydrogen complex and an Fe atom on Pt(111) using the method of magnetic-field dependent ISTS in comparison to simulations within a Kondo model based on (1.7) [27], we were indeed able to reveal the DM contribution to the RKKY interaction. Figure 1.8 shows ISTS curves of the two atoms in such pairs of increasing distances. The Fe atom Fe_{hcp} adsorbed on the hcp lattice site has the usual spin-excitation (cf. Fig. 1.4b). The Fe-hydrogen complex $Fe_{hcp}H_2$ consisting of two H atoms and an Fe_{hcp} atom reveals a resonance at zero bias voltage (see the gray curve of the isolated $Fe_{hcp}H_2$) which is due to a multi-orbital Kondo effect. When these two adsorbates are coupled, the interplay of $\boldsymbol{J}$ and $\boldsymbol{D}$ induces a splitting of the Kondo resonance of $Fe_{hcp}H_2$ and a modification of the magnetic excitation of Fe_{hcp} as compared with the isolated atoms, which is oscillating as a function of distance. By

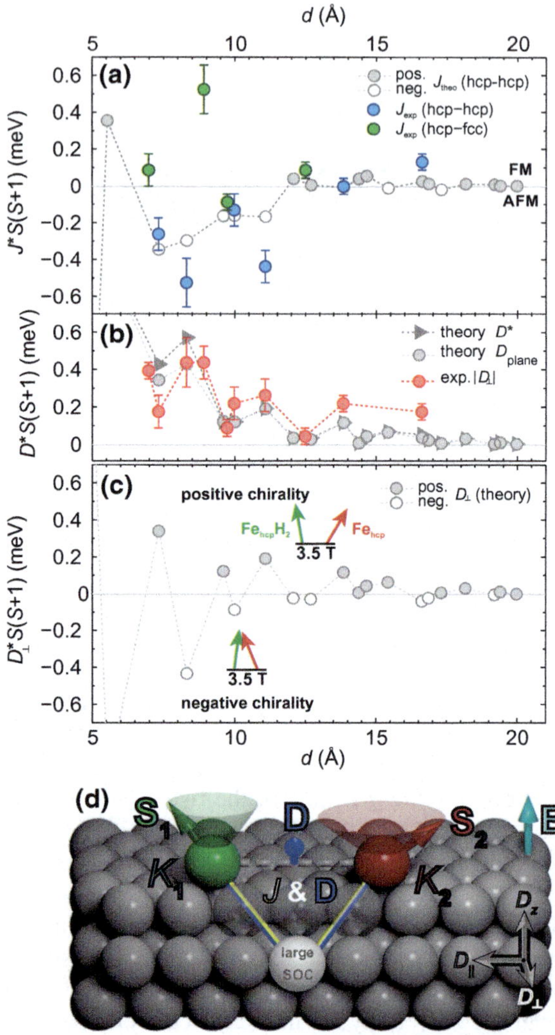

Fig. 1.9 a, b Experimentally determined values of the Heisenberg (J) and the DM interaction (D, colored circles) between two Fe impurities on Pt(111) compared with the theoretical ab initio calculation of J and the different components of the DM vector (gray and white circles and triangles). **c** Calculated perpendicular component of the DM vector resulting in a different rotational sense of the magnetizations of the two impurities for positive and negative sign as indicated by the arrows. **d** Cartoon diagram of the J and DM contributions to indirect conduction electron-mediated exchange interactions between two magnetic atoms with spins S_1 and S_2, and magnetic anisotropies K_1 and K_2. The interaction is mediated by scattering of conduction electrons at a substrate atom (gray) with strong spin-orbit coupling (SOC). The arrows show the orientation of the experimental magnetic field B and components of the DM vector [27]

fitting according magnetic field dependent spectra with simulations, we were able to extract both, J and the largest component of the DM-vector $D_{\perp}$, as a function of separation d of the two magnetic impurities (Fig. 1.9a, b). For most of the distances, the experimentally determined values are nicely reproduced by a theoretical ab initio calculation. Most interestingly, the resulting oscillatory behavior of the sign of $D_{\perp}$ with increasing distance shown in Fig. 1.9c induces a distance dependent chirality of the non-collinear magnetization in the pair as shown in the inset of Fig. 1.9c. The same interaction, which is determined here for pairs of atoms, is also responsible for the formation of complex non-collinear magnetization states as skyrmions in layers of magnetic materials.

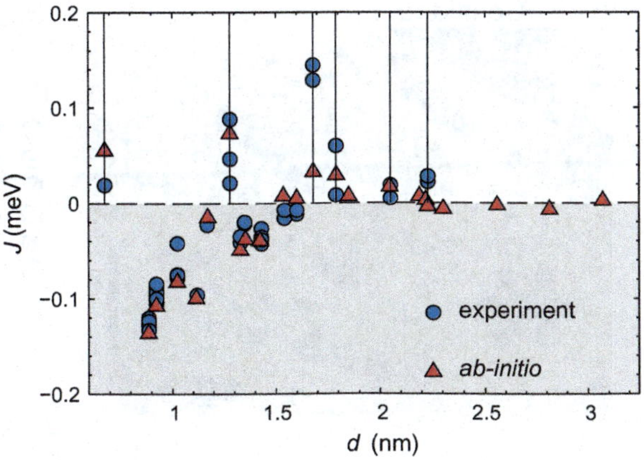

Fig. 1.10 Distance dependency of the RKKY interaction in pairs of Fe atoms on Cu(111). Measured (blue circles) and KKR-calculated (red triangles, renormalized by a factor of 1/2) exchange energy J_{ij} in pairs of Fe atoms on Cu(111) as a function of separation. Data taken from [14]

1.4 Dilute Magnetic Chains and Arrays

In Sect. 1.3, it has been shown how the experimental techniques SPSTS and ISTS can be used to measure maps of the distance dependent RKKY interaction in pairs of atoms adsorbed to metallic substrates. Using such maps it is possible to design artificial nanostructures of a larger number of atoms with tailored interatomic couplings and different topology, e.g. chains or two-dimensional arrays, which can then be built via STM-tip induced atom manipulation. Afterwards, the magnetization curve of each atom within such arrays can be measured by SPSTS and compared to simulations. This methodology has been applied to the system of Fe atoms on Cu(111) as will be shown in the following.

Due to the relatively small spin-orbit interaction of Cu as compared to Pt [28], and the large uniaxial out-of-plane magnetic anisotropy of Fe on Cu(111) of $K = -0.5\,\text{meV}$ the DM contribution to the RKKY interaction can be neglected in this case [29]. The measured distance dependence of the Heisenberg part J of the RKKY interaction is shown in Fig. 1.10. Note, that there is a pronounced minimum at a distance of $d \approx 1\,\text{nm}$, where the RKKY interaction is antiferromagnetic with a large strength of $J \approx -0.1\,\text{meV}$.

Dilute magnetic nanostructures of different topology and number of atoms (chains, 2D arrays) with nearest neighbor distances in the range of this minimum of strongest antiferromagnetic coupling have been assembled using STM-tip induced manipulation as displayed in Fig. 1.11a. Indeed, in the spin-resolved STM image taken in a small magnetic field of $B \approx -0.7\,\text{T}$ they typically show a spin-contrast alternating between dark and bright revealing the trend of an antiferromagnetic alignment of neighboring atoms. The detailed investigation of the underlying magnetization states of all nanostructures is described in [14]. As an example, Fig. 1.11b–m show the investigation of the Fe chains of six and seven atoms, and of a Kagomé of 12 Fe atoms.

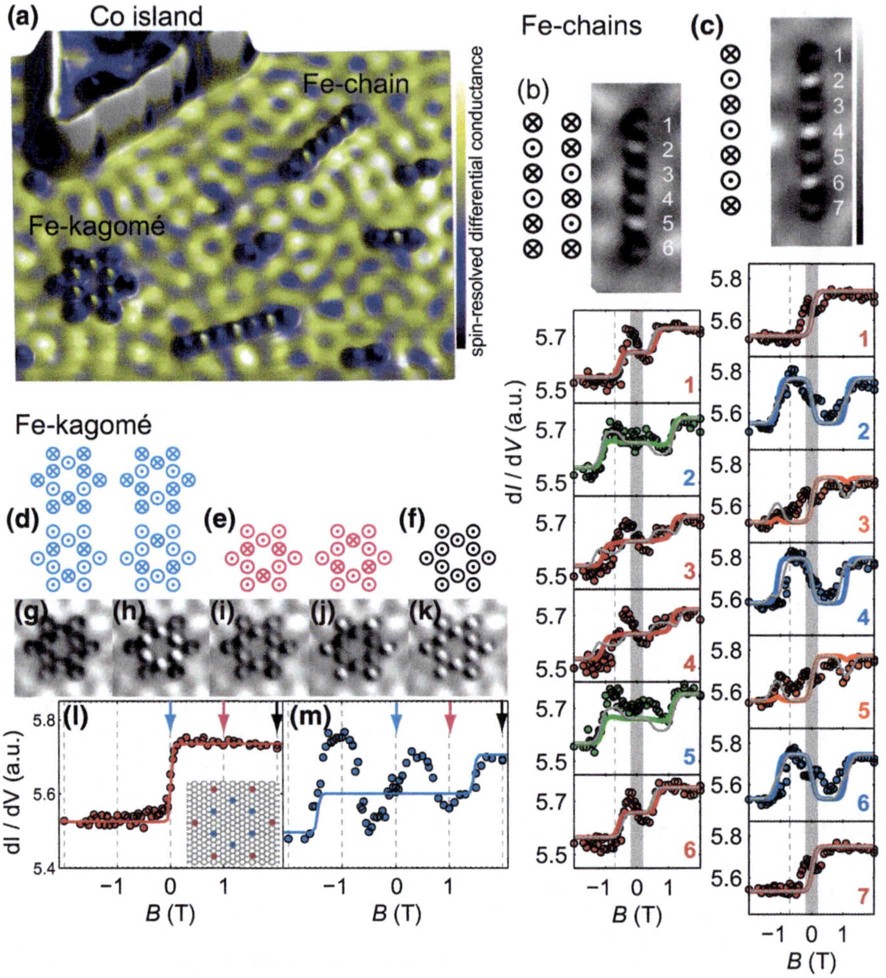

Fig. 1.11 **a** SPSTS image taken at $B \approx -0.7\,\mathrm{T}$ of different manipulated magnetic arrays consisting of RKKY coupled Fe atoms with a neighbor distance of $d \approx 1\,\mathrm{nm}$ on Cu(111) (hillocks). **b**, **c** Top panels: magnetization states from the Ising model (left, partly degenerate) and SPSTS images (right) of chains of anti-ferromagnetically coupled Fe atoms with a length of six (**b**) and seven (**c**) atoms. Bottom panels: magnetization curves measured (circles) on each atom of the chains. The magnetic images have been taken at B indicated by the dashed vertical line. **d–f** degenerate magnetization states of an Ising model for an array of 12 anti-ferromagnetically coupled atoms forming a Kagomé at B as indicated by the arrows in **(l, m)**. **g–k**, SPSTS images of the Kagomé recorded at B as indicated by the dashed lines in **(l, m)**. **l**, **m** magnetization curves measured on the Kagomé atoms (circles). Thick colored and thin gray lines in the magnetization plots show magnetization curves as calculated from different Ising models (see text). The gray shaded areas show the B range, where the experimental magnetization curve deviates from the Ising model. Images partly taken from [14]

The single-atom magnetization curves taken on each atom of the six (Fig. 1.11b) and seven (Fig. 1.11c) atom chains reveal striking differences between the even and odd number chains. For the odd number chain, the magnetization of nearest neighbors alternates between up and down in a magnetic field of $B \approx \pm 0.5\,\mathrm{T}$ indicating the stabilization of an antiferromagnetic Néel state (top part of Fig. 1.11c). In contrast, for the even number chain (Fig. 1.11b), this is not the case for all neighbors (see atom 3 and 4). As shown by simulations within the Ising model (1.7) using the RKKY couplings from the pairs (Fig. 1.10), this can be ascribed to the superposition of multiple degenerate magnetization states for the even number chains (top part of Fig. 1.11b). Interestingly, for the best possible simulation of the magnetization curves within the Ising model, the next-nearest neighbor interaction is crucial. This is shown by a comparison of the simulated curves using the same nearest neighbor J but different next-nearest neighbor interactions with the experimental data (see colored and gray lines in Fig. 1.11b, c). The best agreement was found (gray curves) when the next-nearest neighbor J's from an ab initio calculation of the full chain were used [14].

The investigation of the Kagomé using SPSTS revealed a superposition of four degenerate magnetization states in zero magnetic field (Fig. 1.11d, i). This degeneration is first partly lifted in a small magnetic field where only two degenerate states are remaining (Fig. 1.11e, h, j), and finally all atoms are aligned in a strong magnetic field (Fig. 1.11f, g, k). Surprisingly, there is a strong discrepancy between the measured and calculated magnetization curves for the inner six atoms of the Kagomé in a large magnetic field range between $B = \pm 1.5\,\mathrm{T}$ (Fig. 1.11m). Similarly, there are deviations for some atoms in the chains in a small field window around zero magnetic field (see gray shaded areas). These discrepancies are either due to hidden magnetic moments [14] or effects of a rather slow magnetization dynamics of the arrays.

1.5 Logic Gates and Magnetic Memories

Model systems of magnetic memories and logic gates can be realized using STM-tip induced atom manipulation of the investigated systems of atoms and their interactions.

As shown in the preceding section Sect. 1.4, the antiferromagnetic state of an RKKY coupled chain can be stabilized using a small magnetic field. Moreover, it was shown in Sect. 1.3.1 that the magnetization of an individual atom can be stabilized by RKKY interaction to patches of ferromagnetic monolayers. It is therefore an obvious question to ask, whether the antiferromagnetic state of an RKKY coupled chain could as well be stabilized by RKKY coupling to a ferromagnetic island. Such a stabilization would enable device concepts as illustrated in Fig. 1.12a. Here, the ends of two chains of anti-ferromagnetically coupled atoms are each strongly coupled to a ferromagnetic island (input islands 1 and 2). The chains are intended to transfer the information of the magnetization state of the two input islands towards the actual gate region. The latter consists of the other two end atoms of the two chains (input

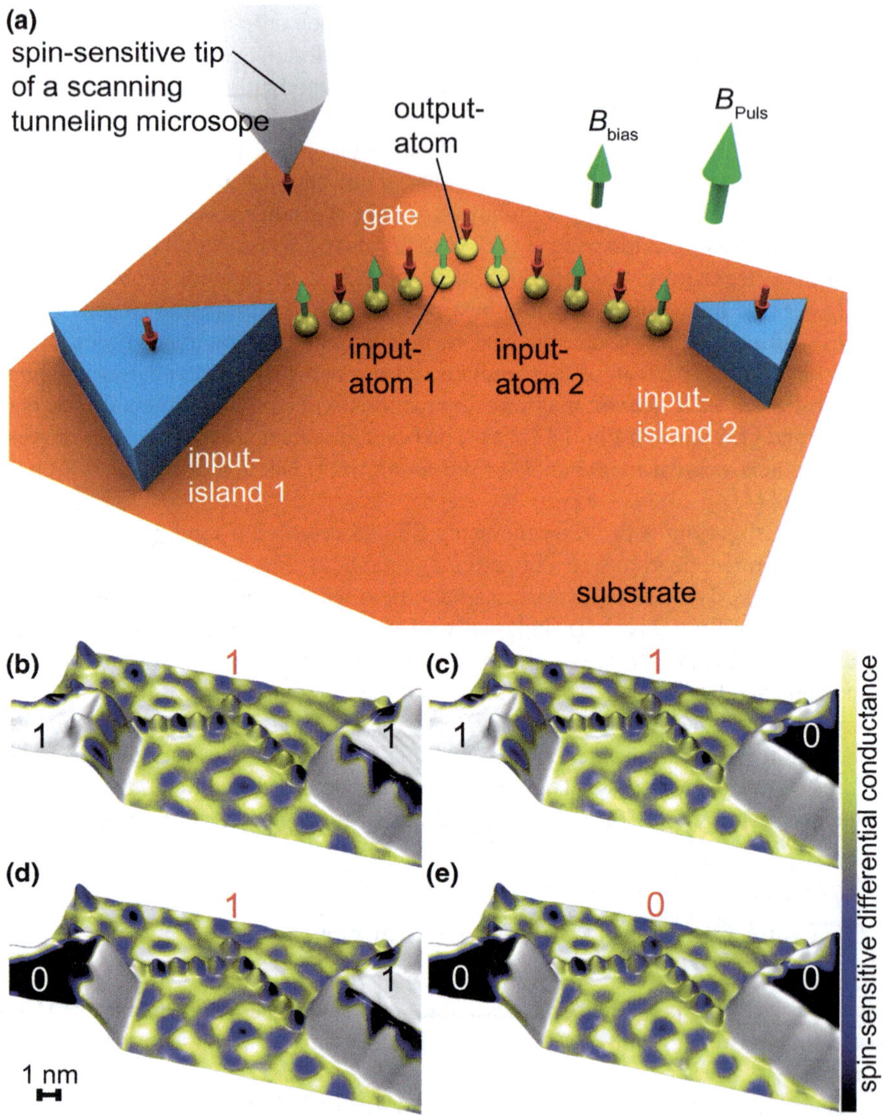

Fig. 1.12 **a** Concept of a logic gate made of two chains of anti-ferromagnetically RKKY coupled magnetic atoms (yellow spheres), which are exchange-coupled to two "input islands" (1, 2) of different size, consisting of patches of ferromagnetic layers. The "input atom" (1, 2) of each spin lead and the final "output atom" form a magnetically frustrated triple with an anti-ferromagnetic coupling which constitutes the logic gate. The field pulse $\boldsymbol{B}_{\text{pulse}}$ is used to switch the inputs. The magnetic tip of an STM is used to construct and characterize the device. **b–e** Side view 3D topographs colored with the simultaneously measured SPSTS image of the constructed OR gate for all four possible input permutations (bright and dark contrasts correspond to states (1) and (0), respectively). By applying out-of-plane magnetic field pulses of different strength and direction, each input island can be controllably switched and the two spin leads transmit the information to their end atoms. The output atom in the gate triple reflects the logical operation of the inputs according to the indicated numbers reflecting the truth table [17]

atoms 1 and 2) and an additional output atom, which together form an equilateral antiferromagnetic triple with a smaller RKKY interaction as inside the chains. Using this equilateral configuration, the output atom will align antiparallel to the two input atoms if these are in the same magnetization state. However, if the two input atoms are in a different magnetization state, the output atom will be in a frustrated state, i.e. the two orientations perpendicular to the surface are degenerate. By using a small bias magnetic field, one of the two orientations will be preferred, which finally determines the logical operation of the gate as a function of the states of the two inputs.

The experimental realization of such a logic gate is shown in Fig. 1.12b. Two chains of 5 Fe atoms with an interatomic distance of $d = 0.923$ nm have been assembled on Cu(111) resulting in strong antiferromagnetic coupling (Fig. 1.10). The chains were assembled in such a way that the atoms on one of the ends of each chain were positioned close to the corner of a ferromagnetic Co island, while the two atoms on the other end of each chain have a mutual distance of $d = 1.35$ nm. Thereby, both chain ends are strongly antiferromagnetically coupled to the islands, but the mutual interaction between the two chains is kept smaller as the interaction within each chain. Finally, the output atom is positioned at the same distance of $d = 1.35$ nm to both chain ends.

The operation of the gate is shown in the spin-resolved images in Fig.1.12b–e. By using magnetic field pulses of appropriate strength, the two Co islands, which have different size and therefore different coercivity, were put into the four different states (11), (10), (01), (00) as revealed by the spin-resolved images. Obviously, the magnetization states of both chains are following the states of their respective input island thereby transmitting the input to the gate region. Here, the output atom is forced into the state (0) if and only if the inputs are in the state (00), which proves the operation of the gate as an OR gate.

An interesting question concerning the down scaling of such logic elements is how small the input islands can be made and still remain stable in either of their magnetization states. This relates to the very fundamental question of how many atoms such an island has to contain in order to behave like a permanent magnet showing remanence [30]. In order to answer these questions, clusters of a small number of direct-exchange coupled Fe atoms have been assembled on Cu(111) and Pt(111), and were investigated by time-resolved SPSTS of the telegraph signal of such clusters.

On the substrate Cu(111), a cluster of five Fe atoms constitutes a stable magnet [31]. An even smaller permanent Fe magnet can be made on the substrate Pt(111) as illustrated in Fig. 1.13 [18]. It consists of only three Fe atoms that have been assembled onto neighboring fcc lattice sites using STM-tip induced atom manipulation (Fig. 1.13e). Figure 1.13a–d show spin-resolved STM images of two of such Fe_3 clusters assembled with a separation of only 2.5 nm. In these images a larger or smaller apparent height of the cluster indicates its spin state up (1) or down (0), respectively. By feeding spin-polarized electrons with sufficient energy from the magnetic STM tip through one of the clusters, which was done between the acquisition of the images, it was possible to write its spin state. Thereby, all four possible spin states (01), (11), (10), and (00) of the two-Fe_3 cluster memory were prepared.

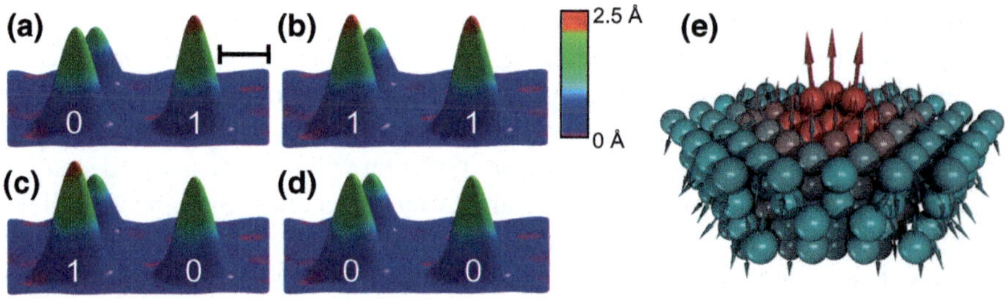

Fig. 1.13 **a–d** Spin-resolved STM images of the four possible spin states (01), (11), (10), and (00) of two Fe_3 clusters (0 and 1 correspond to downwards and upwards pointing magnetization, respectively). An additional Fe atom in the back serves as a marker for the apparent height. The scale bar has a length of 1 nm. **e** ab initio (KKR) calculation of the magnetic moments (arrows) in the Fe_3 cluster and the induced magnetization (color) in the substrate indicating their strong non-collinearity [18]

At the measurement temperature of 0.3 K these spin states were stable for at least 10 h.

The system of Fe_3 on Pt(111) is additionally interesting, as the heavy element Pt supports strong spin-orbit coupling and thereby a considerable DM contribution to the RKKY interaction (see Sect. 1.3.3). Consequently, the induced magnetization in the Pt underneath the cluster is highly non-collinear, as proven by ab initio calculations (Fig. 1.13e). Due to the resulting non-collinear RKKY interaction to neighboring magnetic atoms, the use of such a material combination in spintronic elements as the one shown in Fig. 1.12 might have advantages with respect to other materials, that feature only collinear states.

1.6 Conclusions

As we have shown in this review, the combination of SP(I)STS with STM-tip induced atom manipulation is a powerful experimental methodology to study the magnetic properties of artificial atomic-scale nanostructures. In particular, the magnetic moments, anisotropies, and *g*-factors of different atom/substrate systems, and the RKKY interaction in pairs have been measured directly. It was shown that the RKKY interaction offers a huge flexibility for tailoring the magnetic couplings in assembled nanostructures, ranging from ferromagnetic, over antiferromagnetic to non-collinear interactions. Due to this ultimate control of the atomic composition, positions and magnetic couplings, the results can be directly compared to model and ab initio calculations, in principle without the need to guess any unknown parameters. Finally, the knowledge was applied to build model systems for future atomic spintronic and information storage elements and we have shown that for all-metallic systems, a stable magnet requires only three Fe atoms on a Pt substrate.

Acknowledgements We thank the Deutsche Forschungsgemeinschaft DFG for the financial support of the project. AAK also acknowledges the DFG via the Emmy Noether Program (KH324/1-1). We gratefully acknowledge our colleagues Kirsten von Bergmann, Bruno Chilian, Jan Hermenau, Stefan Krause, Focko Meier, Tobias Schlenk, Andreas Sonntag, Manuel Steinbrecher, Khai Ton That, and Lihui Zhou, who were directly involved in the project. Furthermore, we thank Benjamin Baxevanis, Stefan Blügel, Mohammed Bouhassoune, Antonio T. Costa, Peter H. Dederichs, Manuel dos Santos Dias, Paolo Ferriani, Swantje Heers, Stefan Heinze, Christoph Hübner, Julen Ibañez-Azpiroz, Jindrich Kolorenc, Frank Lechermann, Alexander I. Lichtenstein, Samir Lounis, Phivos Mavropoulos, Doug L. Mills, Daniela Pfannkuche, Benjamin Schweflinghaus, Sergej Schuwalow, Alexander B. Shick, Markus Ternes, Maria Valentyuk, Elena Vedmedenko, and Tim O. Wehling for their theoretical support.

References

1. D.M. Eigler, E.K. Schweizer, Nature **344**(6266), 524 (1990)
2. A.J. Heinrich, J.A. Gupta, C.P. Lutz, D.M. Eigler, Science **306**(5695), 466 (2004)
3. F. Meier, L. Zhou, J. Wiebe, R. Wiesendanger, Science **320**(5872), 82 (2008)
4. A.A. Khajetoorians, T. Schlenk, B. Schweflinghaus, M. dos Santos Dias, M. Steinbrecher, M. Bouhassoune, S. Lounis, J. Wiebe, R. Wiesendanger, Phys. Rev. Lett. **111**, 157204 (2013)
5. A.A. Khajetoorians, S. Lounis, B. Chilian, A.T. Costa, L. Zhou, D.L. Mills, J. Wiebe, R. Wiesendanger, Phys. Rev. Lett. **106**, 037205 (2011)
6. B. Chilian, A.A. Khajetoorians, S. Lounis, A.T. Costa, D.L. Mills, J. Wiebe, R. Wiesendanger, Phys. Rev. B **84**, 212401 (2011)
7. C.F. Hirjibehedin, C.Y. Lin, A.F. Otte, M. Ternes, C.P. Lutz, B.A. Jones, A.J. Heinrich, Science **317**(5842), 1199 (2007)
8. T. Balashov, T. Schuh, A.F. Takács, A. Ernst, S. Ostanin, J. Henk, I. Mertig, P. Bruno, T. Miyamachi, S. Suga, W. Wulfhekel, Phys. Rev. Lett. **102**, 257203 (2009)
9. A.A. Khajetoorians, B. Chilian, J. Wiebe, S. Schuwalow, F. Lechermann, R. Wiesendanger, Nature **467**, 1084 (2010)
10. P. Wahl, L. Diekhöner, M.A. Schneider, L. Vitali, G. Wittich, K. Kern, Phys. Rev. Lett. **93**, 176603 (2004)
11. L. Zhou, J. Wiebe, S. Lounis, E. Vedmedenko, F. Meier, S. Blügel, P.H. Dederichs, R. Wiesendanger, Nat. Phys. **6**(3), 187 (2010)
12. C.F. Hirjibehedin, C.P. Lutz, A.J. Heinrich, Science **312**(5776), 1021 (2006)
13. N. Néel, R. Berndt, J. Kröger, T.O. Wehling, A.I. Lichtenstein, M.I. Katsnelson, Phys. Rev. Lett. **107**, 106804 (2011)
14. A.A. Khajetoorians, J. Wiebe, B. Chilian, S. Lounis, S. Blügel, R. Wiesendanger, Nat. Phys. **8**(6), 497 (2012)
15. A. Spinelli, B. Bryant, F. Delgado, J. Fernndez-Rossier, A.F. Otte, Nat. Mater. **13**(8), 782 (2014)
16. R. Toskovic, R. van den Berg, A. Spinelli, I.S. Eliens, B. van den Toorn, B. Bryant, J.S. Caux, A.F. Otte, Nat. Phys. **12**(7), 656 (2016)
17. A.A. Khajetoorians, J. Wiebe, B. Chilian, R. Wiesendanger, Science **332**(6033), 1062 (2011)
18. J. Hermenau, J. Ibañez-Azpiroz, C. Hübner, A. Sonntag, B. Baxevanis, K.T. Ton, M. Steinbrecher, A.A. Khajetoorians, M. dos Santos Dias, S. Blügel, R. Wiesendanger, S. Lounis, J. Wiebe, Nat. Commun. **8**(1), 642 (2017)
19. S. Loth, S. Baumann, C.P. Lutz, D.M. Eigler, A.J. Heinrich, Science **335**(6065), 196 (2012)
20. F.D. Natterer, K. Yang, W. Paul, P. Willke, T. Choi, T. Greber, A.J. Heinrich, C.P. Lutz, Nature **543**(7644), 226 (2017)
21. P. Gambardella, S. Rusponi, M. Veronese, S.S. Dhesi, C. Grazioli, A. Dallmeyer, I. Cabria, R. Zeller, P.H. Dederichs, K. Kern, C. Carbone, H. Brune, Science **300**(5622), 1130 (2003)

22. F. Meier, K. von Bergmann, P. Ferriani, J. Wiebe, M. Bode, K. Hashimoto, S. Heinze, R. Wiesendanger, Phys. Rev. B **74**, 195411 (2006)
23. G. Pacchioni, L. Gragnaniello, F. Donati, M. Pivetta, G. Autés, O. Yazyev, S. Rusponi, H. Brune, Phys. Rev. B **91**, 235426 (2015)
24. A.A. Khajetoorians, M. Valentyuk, M. Steinbrecher, T. Schlenk, A. Shick, J. Kolorenc, A.I. Lichtenstein, T.O. Wehling, R. Wiesendanger, J. Wiebe, Nat. Nanotechnol. **10**(11), 958 (2015)
25. D. Smith, J. Magn. Magn. Mater. **1**(3), 214 (1976)
26. A. Fert, P.M. Levy, Phys. Rev. Lett. **44**, 1538 (1980)
27. A.A. Khajetoorians, M. Steinbrecher, M. Ternes, M. Bouhassoune, M. dos Santos Dias, S. Lounis, J. Wiebe, R. Wiesendanger, Nat. Commun. **7**, 10620 (2016)
28. O. Šipr, S. Bornemann, H. Ebert, J. Minár, J. Phys. Condens. Matter **26**(19), 196002 (2014)
29. J. Bouaziz, M. dos Santos Dias, A. Ziane, M. Benakki, S. Blügel, S. Lounis, New J. Phys. **19**(2), 023010 (2017)
30. F. Donati, S. Rusponi, S. Stepanow, C. Wäckerlin, A. Singha, L. Persichetti, R. Baltic, K. Diller, F. Patthey, E. Fernandes, J. Dreiser, Ž. Šljivančanin, K. Kummer, C. Nistor, P. Gambardella, H. Brune, Science **352**(6283), 318 (2016)
31. A.A. Khajetoorians, B. Baxevanis, C. Hübner, T. Schlenk, S. Krause, T.O. Wehling, S. Lounis, A. Lichtenstein, D. Pfannkuche, J. Wiebe, R. Wiesendanger, Science **339**(6115), 55 (2013)
32. J. Wiebe, L. Zhou, R. Wiesendanger, J. Phys. D: Appl. Phys. **44**(46), 464009 (2011)
33. L. Zhou, F. Meier, J. Wiebe, R. Wiesendanger, Phys.Rev. B **82**, 012409 (2010)

Chapter 2
Scanning Tunneling Spectroscopies of Magnetic Atoms, Clusters, and Molecules

Jörg Kröger, Alexander Weismann, Richard Berndt, Simon Altenburg, Thomas Knaak, Manuel Gruber, Andreas Burtzlaff, Nicolas Néel, Johannes Schöneberg, Laurent Limot, Takashi Uchihashi and Jianwei Zhang

Abstract The Kondo effect of adatoms on surfaces may to some extent be controlled by manipulating their electronic and geometric environment. Results are presented from artificial structures like quantum well systems, from arrangements of single atoms made with a scanning tunneling microscope, and from custom-made molecules. Spin-orbit coupling at single adatoms is probed via measurements of the anisotropic magnetoresistance, in particular in single atom contacts. Such junctions are also investigated with respect to the current shot noise, which is influenced by the electron spin.

The Kondo effect is one of the key correlation effects in condensed matter physics. Originally it was observed as an increase of the resistivity of metals with dilute magnetic impurities below a characteristic temperature [1]. Kondo explained these observations in terms of conduction electron scattering off the impurity magnetic moment [2]. The scattering processes involve spin flips of the conduction electrons and of the magnetic impurity, which leads to an efficient screening of the magnetic moment. The spectroscopic signature of the resulting many-body nonmagnetic

J. Kröger · N. Néel
Institut für Physik, Technische Universität Ilmenau, 98684 Ilmenau, Germany

A. Weismann · R. Berndt (✉) · T. Knaak · M. Gruber · A. Burtzlaff · J. Schöneberg
J. Zhang
Institut für Experimentelle und Angewandte Physik, Christian-Albrechts-Universität zu Kiel, 24098 Kiel, Germany
e-mail: berndt@physik.uni-kiel.de

S. Altenburg
Bundesanstalt für Materialforschung und -prüfung, 12205 Berlin, Germany

L. Limot
Université de Strasbourg, CNRS, IPCMS, UMR 7504, F-67000 Strasbourg, France

T. Uchihashi
International Center for Materials Nanoarchitectonics, National Institute for Materials Science, 1-1, Namiki, Tsukuba, Ibaraki 305-0044, Japan

R. Wiesendanger (ed.), *Atomic- and Nanoscale Magnetism*, NanoScience and Technology, https://doi.org/10.1007/978-3-319-99558-8_2

singlet ground state is the Abrikosov-Suhl [3–5] or Kondo resonance at the Fermi energy, E_F [6, 7]. The resonance width reflects the Kondo energy scale, $k_B T_K$ (k_B: Boltzmann constant, T_K: Kondo temperature), where T_K is characteristic of the magnetic impurity and the substrate host. Since the first observation of the Kondo effect in scanning tunneling microscope (STM) experiments [8, 9] a wealth of scanning tunneling spectroscopy (STS) data has been published for a variety of adatoms on metal surfaces [10].

Here we focus on experiments that aim at controlling the Kondo effect of adatoms by manipulating their electronic and geometric environment. A number of approaches have been used so far. Electronic quantum well states affect the electronic density of states but have little impact on the adsorption geometry of an adatom. In Sect. 2.1.1 we show that they effectively modulate the Kondo temperature T_K of Co atoms. The addition of adatoms to a magnetic impurity also modifies T_K as demonstrated by experiments on heterogeneous clusters comprised of Co and non-magnetic Cu atoms (Sect. 2.1.2). Chains of Cu atoms may also be used to mediate interactions between two Co atoms (Sect. 2.1.3). Next, the apex atom of an STM tip may be brought closer to a magnetic impurity to controllably increase the interaction with an additional neighbor atom (Sect. 2.1.4). Natural extensions of the approach outlined so far are to use organic molecules to obtain interconnected magnetic impurities (Sect. 2.2) or to tune the substrate-impurity interaction by an interlayer of graphene (Sect. 2.3). The role of spin-orbit coupling (SOC) in transport phenomena is another research frontier in spintronics [11–16]. One of its intriguing consequences is the occurence of anisotropic magnetoresistance (AMR). Section 2.4 focuses on AMR in single-atom contacts, for which particularly large effects have been predicted. Finally, Sect. 2.5 presents measurements of the current shot noise in single atom junctions. Electron correlation tends to reduce the shot noise and noise data consequently are an additional tool to investigate spin physics in nanoscale junctions.

2.1 Tuning the Kondo Effect on the Single-Atom Scale

2.1.1 Co Atoms on a Quantum Well System

A model system was used for tuning the Kondo temperature by the local density of states (LDOS) at E_F [17]. Co atoms adsorbed on layers of Cu and Co on a Cu(100) substrate [Fig. 2.1a] were investigated by spectroscopy of the differential conductance (dI/dV). An excellent lattice match enables epitaxial growth of face-centered cubic Co(100) on Cu(100) [Fig. 2.1b] and subsequently of Cu(100) on Co(100) [Fig. 2.1c]. This layered system is an ideal platform for acquiring spectra of single Co atoms on Cu layers [Fig. 2.1d] of various thicknesses without changing the imaging area. The local thickness of Cu layers (d_{Cu}) was determined from the energies of quantum well states (QWS) vertically confined to the Cu layers by reflection at the vacuum–Cu and Cu–Co interfaces [18–24].

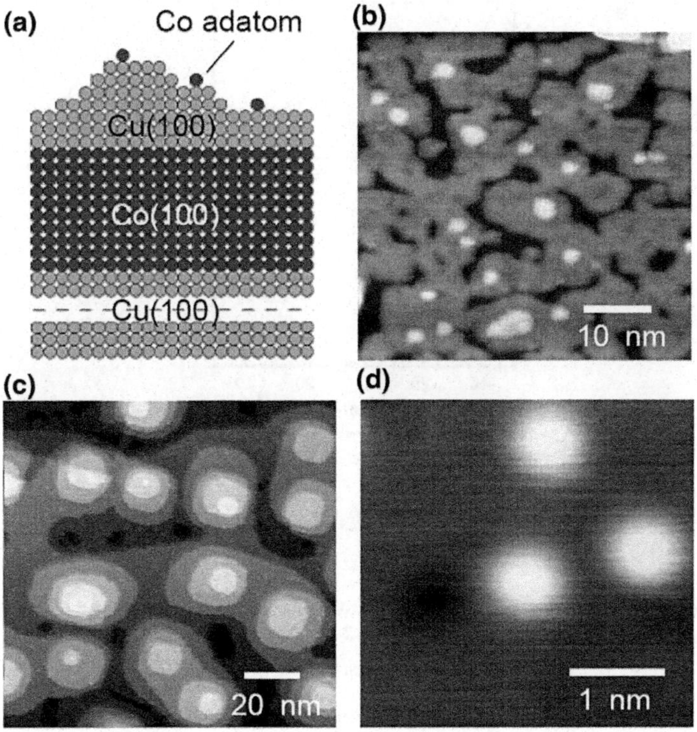

Fig. 2.1 **a** Sketch of Cu/Co/Cu(100) multilayer with adsorbed single Co atoms. STM images of **b** a 10-ML-thick Co overlayer grown on a Cu(100) substrate (current $I = 100\,\text{pA}$, bias voltage $V = 200\,\text{mV}$), **c** a 15-ML-thick Cu overlayer grown on a Co/Cu(100) multilayer (100 pA, 200 mV), and **d** of Co atoms deposited on a Cu/Co/Cu(100) multilayer (100 pA, 300 mV). Reprinted with permission from T. Uchihashi et al., Phys. Rev. B **78**, 033402 (2008) [17]. Copyright 2008 by the American Physical Society

Figure 2.2a shows $\mathrm{d}I/\mathrm{d}V$ spectra of Co atoms on a pristine Cu(100) surface (bottom) and on Cu layers above Co(100) on Cu(100) with thicknesses of 11 (middle) and 14 monolayers (top). The Kondo resonance of Co leads to characteristic asymmetric features around zero bias voltage. In STS the Kondo resonance generally appears with a Fano line shape, which may adopt diplike, peaklike or asymmetric profiles [25, 26]. To quantitatively analyze the $\mathrm{d}I/\mathrm{d}V$ data, the experimental spectra were fit by

$$\frac{\mathrm{d}I}{\mathrm{d}V} = af(V) + b + cV \tag{2.1}$$

with

$$f(V) = \frac{[q + \varepsilon(V)]^2}{1 + \varepsilon(V)^2} \tag{2.2}$$

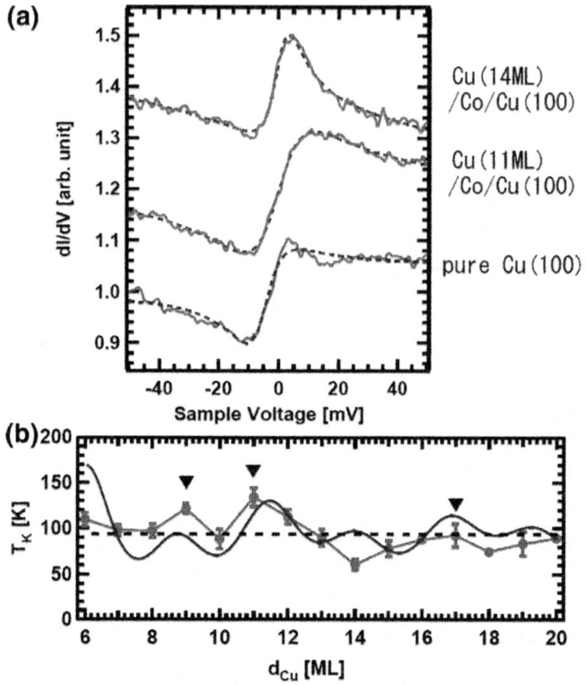

Fig. 2.2 **a** dI/dV spectra measured above Co atoms on Cu/Co/Cu(100) multilayers. Top: 14 ML Cu. Middle: 11 ML Cu. Bottom: bare Cu(100). Spectra are offset for clarity. Fits according to (2.1) are shown as dashed lines. **b** Mean Kondo temperature T_K of Co atoms on Cu/Co/Cu(100) multilayers versus Cu overlayer thickness d_{Cu} (circles). Half the error bar indicates the standard error of T_K for each d_{Cu}. A solid line shows calculated T_K values. The used fit parameters are $A_b = 0.046\,\text{eV}^{-1}\,\text{ML}$, $\Phi_b = 0.26\pi$, $A_n = 0.011\,\text{eV}^{-1}\,\text{ML}$, $\Phi_n = -0.03\pi$. A dotted line indicates the mean $T_K = 94$ K of Co atoms on pure Cu(100). Reprinted with permission from T. Uchihashi et al., Phys. Rev. B **78**, 033402 (2008) [17]. Copyright 2008 by the American Physical Society

the Fano function, $\varepsilon(V) = (eV - \varepsilon_K)/(k_B T_K)$ (e: elementary charge, ε_K: resonance shift from E_F), q the asymmetry parameter of the Fano theory, a the resonance amplitude, and $b + cV$ a linear background. For the spectra shown in Fig. 2.2a the extracted Kondo temperatures are $T_K = 67$ K (14 monolayers) and $T_K = 139$ K (11 monolayers), which significantly deviate from $T_{K,0} = 94 \pm 5$ K for Co on clean Cu(100) [27, 28]. Figure 2.2b shows T_K for Co atoms adsorbed on Cu layers with all investigated thicknesses. The obvious modulation of T_K with d_{Cu} reflects the variation of the LDOS at E_F, ϱ_F, caused by the QWS confined to the Cu slab. Using

$$T_K = T_0 \exp\left(-\frac{1}{2\varrho_F J}\right) \tag{2.3}$$

with T_0 a prefactor, J the exchange coupling constant between conduction electrons and impurity states together with

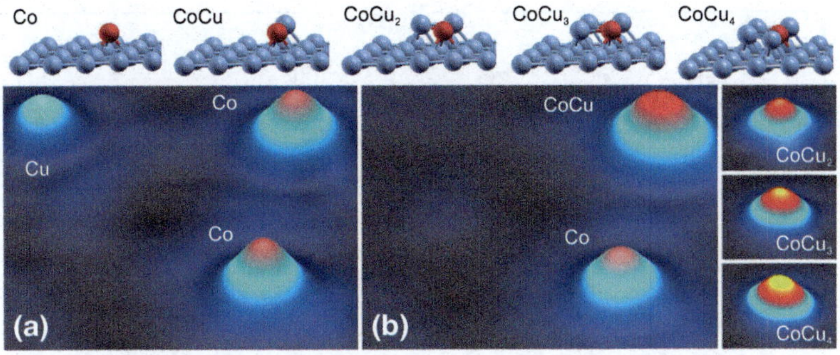

Fig. 2.3 Top: Ball-and-stick models of fully relaxed adsorption structures of CoCu_n $(0 \leq n \leq 4)$ on Cu(111). **a** STM constant current topograph of Cu and Co atoms on Cu(111) prior to the fabrication of CoCu_n clusters. **b** STM topographs showing different CoCu_n clusters. Images were acquired with $V = 100\,\mathrm{mV}$ and $I = 0.1\,\mathrm{nA}$. Reprinted with permission from N. Néel et al., Phys. Rev. Lett. **101**, 266803 (2008) [32]. Copyright 2008 by the American Physical Society

$$\varrho_{\mathrm{F}} = \varrho_{\mathrm{F},0} + \sum_{i=\mathrm{b,n}} \frac{A_i}{d_{\mathrm{Cu}}} \cos\left(\frac{2\pi d_{\mathrm{Cu}}}{\Lambda_i} + \Phi_i\right) \tag{2.4}$$

[A_i, Λ_i, Φ_i: amplitude, spatial periodicity, phase of the quantum oscillation at the belly ($i = b$) and the neck ($i = n$) of the Cu Fermi surface] leads to a reasonable fit [full line in Fig. 2.2b] of the experimentally observed T_{K} variation. The fit was performed for fixed $\Lambda_b = 5.88\,\mathrm{ML}$, $\Lambda_n = 2.67\,\mathrm{ML}$ [18, 21, 29], and free parameters A_i, Φ_i. For Co on pristine Cu(100) the free-electron estimate $\varrho_{\mathrm{F},0} = 0.11\,\mathrm{eV}^{-1}$ was used together with $J \approx 1\,\mathrm{eV}$ [25, 28, 30], $T_{\mathrm{K}} = 94\,\mathrm{K}$.

It is worth mentioning that the exchange interaction between a Co atom and the Co layer through the Cu layers, J_{ex}, is too weak to account for the observed changes in T_{K}. Indeed, $J_{\mathrm{ex}} = 1.5\,\mathrm{meV}$ for $d_{\mathrm{Cu}} = 5\,\mathrm{ML}$ [31] decreases rapidly with increasing d_{Cu}. Only for $d_{\mathrm{Cu}} \leq 4\,\mathrm{ML}$, J_{ex} was found to become comparable and larger than the Kondo energy scale, $k_{\mathrm{B}} T_{\mathrm{K}} \approx 8\,\mathrm{meV}$. In summary, by changing ϱ_{F} through QWS confined to thin metal films the Kondo effect of single adsorbed atoms may be altered.

2.1.2 *Kondo Effect in CoCu_n Clusters*

The key result of this section is the important role of the local and anisotropic electronic structure at the magnetic-impurity site for the Kondo effect [32]. The local environment of a Co atom on Cu(111) was modified by changing its coordination to successively attached Cu atoms. Sketches of the resulting CoCu_n $(0 \leq n \leq 4)$ clusters are shown in the top row of Fig. 2.3, accompanied by corresponding STM images in Fig. 2.3a, b.

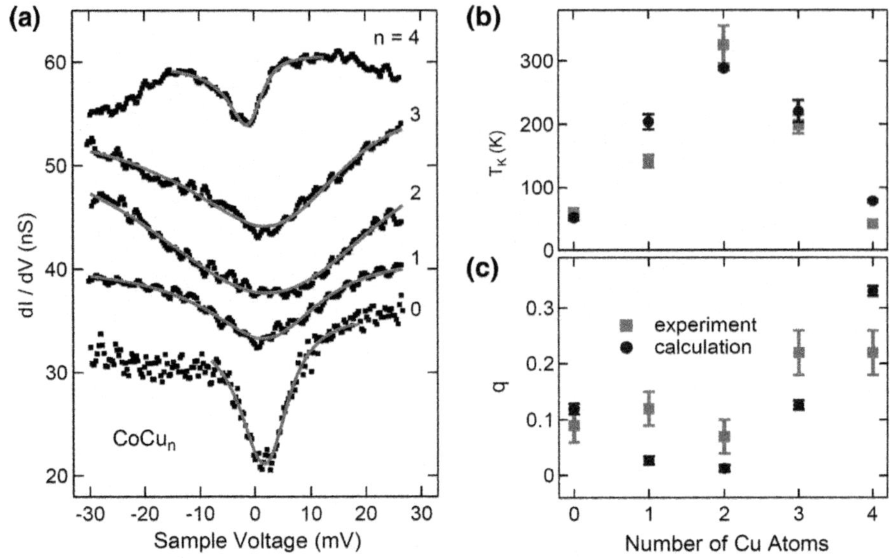

Fig. 2.4 **a** $\mathrm{d}I/\mathrm{d}V$ spectroscopy of $CoCu_n$ clusters. The spectroscopic feature around zero bias is caused by the Kondo effect. Fano line shapes (2.1) were fit to the experimental data and are indicated as lines. Prior to spectroscopy tunneling parameters of $V = 30$ mV and $I = 1$ nA were used. Spectra with $n = 1, \ldots, 4$ are vertically offset by 5, 10, 20, 25 nS, respectively. Kondo temperatures (**b**) and line shape parameters (**c**) versus cluster size obtained from calculations (dots) and experiments (squares). Reprinted with permission from N. Néel et al., Phys. Rev. Lett. **101**, 266803 (2008) [32]. Copyright 2008 by the American Physical Society

Next, $\mathrm{d}I/\mathrm{d}V$ spectra were acquired on top of single Co atoms and of $CoCu_n$ clusters. A representative data set is shown in Fig. 2.4a. For the single Co atom the Kondo resonance appears as a sharp indentation around zero bias voltage. Upon adding Cu atoms, the $\mathrm{d}I/\mathrm{d}V$ signature of the Kondo resonance strongly broadens and sharpens again after attaching a third and fourth Cu atom to Co. In contrast, the energies of the p_z states of the clusters decrease monotonically [32].

To see the changes clearly, the Kondo temperature and the asymmetry parameter were obtained by fitting (2.1) to $\mathrm{d}I/\mathrm{d}V$ data and plotted as a function of the number of Cu atoms in Fig. 2.4b. The nonmonotonic behavior is readily visible. This peculiar variation of the Kondo resonance with the cluster size is at variance with the monotonic behavior of T_{K} with the average hybridization strength expected from bulk models of the Kondo effect.

Calculations of the local electronic structure of the $CoCu_n$ clusters on Cu(111) reproduce the experimental data [32]. Calculated T_{K} and q are shown as dots in Fig. 2.4b and are summarized in Table 2.1 along with experimental data. T_{K} may be estimated as

$$T_{\mathrm{K}} \approx W\sqrt{|\varrho_F J|}\,\exp\left(-\frac{1}{\varrho_{\mathrm{F}} J}\right) \tag{2.5}$$

Table 2.1 Experimental and calculated Kondo temperatures (T_K) and line shape parameters (q) of $CoCu_n$ extracted from fits of Fano lines (2.1) to the spectra of Fig. 2.4a. Experimental uncertainty margins correspond to standard deviations obtained by a statistical analysis of a variety of fits. Calculated T_K and q are the arithmetic mean of values obtained by a Gaussian broadening of the sp LDOS by 100 meV and 50 meV. The uncertainty margins reflect the deviations of the arithmetic mean from data obtained using 100 meV and 50 meV broadening

	Experiment		Calculation	
n	T_K (K)	q	T_K (K)	q
0	61 ± 4	0.09 ± 0.03	52 ± 4	0.12 ± 0.01
1	142 ± 10	0.12 ± 0.03	204 ± 12	0.03 ± 0.01
2	326 ± 30	0.07 ± 0.03	288 ± 3	0.01 ± 0.01
3	200 ± 15	0.22 ± 0.04	221 ± 18	0.13 ± 0.01
4	43 ± 6	0.22 ± 0.04	79 ± 3	0.33 ± 0.01

(W: width of the sp conduction band, J: sp–d exchange interaction energy, ϱ_F: LDOS at E_F at the impurity site). Moreover, the asymmetry parameter reads

$$q = \frac{\gamma + \mathrm{Re}\, G(E_F)}{\mathrm{Im}\, G(E_F)} \tag{2.6}$$

with G the local Green function of the conduction electrons. γ measures the ratio of the coupling of the STM tip to conduction electrons of the substrate and to localized Co d states [33]. Details of the calculations may be found elsewhere [32]. The simulations ($W = 20\,\mathrm{eV}$, $J = 1.3\,\mathrm{eV}$, $\gamma = 0.22\,\mathrm{eV}^{-1}$) show that ϱ_F varies nonmonotonically with the cluster size and thus leads to the variation of T_K and q observed in the experiments.

The evolution of ϱ_F can qualitatively be understood as follows. Co p orbitals have their main spectral weight above E_F, whereas s orbitals are mainly occupied. With increasing cluster size all orbitals shift to lower energies and the p (s) orbital weight increases (decreases) at E_F. These variations do not perfectly cancel and thus lead to a nonmonotonic change of ϱ_F. In summary, atom-by-atom manipulation of the electronic structure at the Kondo impurity site induces a nonmonotonic variation of ϱ_F, the Kondo temperature, and the asymmetry factor. Simple scaling rules that predict a monotonous increase of T_K with the coordination number do not hold.

2.1.3 Two-Site Kondo Effect in Atomic Chains

Next, linear atomic chains are considered that host two Kondo atoms at their ends [34]. $CoCu_nCo$ ($1 \leq n \leq 6$) clusters on Cu(111) are a model system for studying correlation effects. Local correlations are caused by Coulomb interaction at a particular site i. Nonlocal correlations are due to electrons propagating from site i

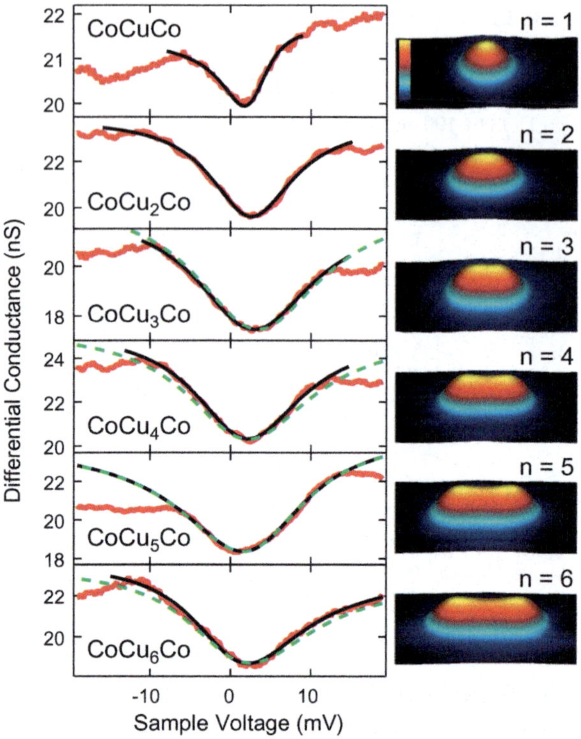

Fig. 2.5 **Left**: dI/dV spectra acquired above Co atoms of $CoCu_nCo$ chains (red dots). Solid lines indicate Fano line shapes fit to the experimental data. The dashed lines are Fano lines for $T_K = 110\,K$, which correspond to a Co adatom attached to a Cu_4 chain. **Right**: Pseudo-three-dimensional representations of STM topographs (38 Å×21 Å) of linear $CoCu_nCo$ clusters ($n = 1 \ldots 6$). Reprinted with permission from N. Néel et al., Phys. Rev. Lett. **107**, 106804 (2011) [34]. Copyright 2011 by the American Physical Society

to another site j with Coulomb scattering at both sites. Therefore, the Ruderman-Kittel-Kasuya-Yosida (RKKY) interaction [35–37] can cause magnetic correlations between distant atoms. When a Kondo effect is involved, the nonlocal interaction may alter the characteristic Kondo energy scale, $k_B T_K$, and may lead to new regimes including ferromagnetically locked spins or an interimpurity singlet [38–40]. Indeed, our model system covers the range from weakly coupled Kondo atoms to interimpurity exchange coupling that is comparable to the Kondo energy scale.

Figure 2.5 presents the main experimental findings [34]. It shows the dI/dV signature of the Kondo resonance of one of the Co atoms in $CoCu_nCo$ ($1 \leq n \leq 6$) chains on Cu(111) together with STM images of the assemblies. dI/dV data were fit using (2.1) (black line in Fig. 2.5). The extracted Kondo temperatures exhibit a remarkable variation with the number of Cu atoms connecting the Co atoms. T_K nearly doubles from $n = 1$ ($T_K \approx 46\,K$) to $n = 2$ ($\approx 79\,K$), increases further to $\approx 108\,K$ for $CoCu_3Co$, and then oscillates (≈ 93, 110, and 91 K for n =4–6, respectively). For clarity these results are summarized in Fig. 2.6. The maxima of the oscillation match the Kondo temperature of a Co atom at the end of Cu_3 and Cu_4 chains ($\tilde{T}_K = 110\,K$), which approximate a $CoCu_\infty$ chain (dashed line in Fig. 2.6).

The calculations of [34] first showed that single-impurity Kondo physics cannot reproduce the experimentally observed variation of T_K. In the absence of magnetic

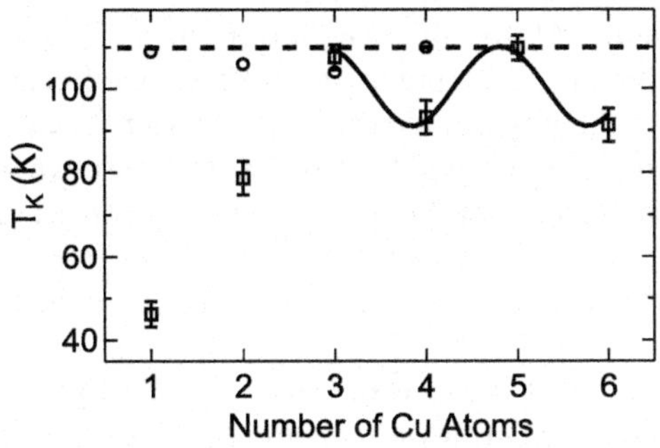

Fig. 2.6 T_K of $CoCu_nCo$ chains versus n (squares). The Kondo temperature $\tilde{T}_K = 110\,K$ of a single Co adatom at the end of a Cu_4 chain is indicated by a dashed line. Calculated *single*-impurity Kondo temperature, that would be expected in the absence of magnetic Co–Co interactions, are indicated as circles. The calculated periodicity in the RKKY interaction regime is shown as a sinusoidal line. Reprinted with permission from N. Néel et al., Phys. Rev. Lett. **107**, 106804 (2011) [34]. Copyright 2011 by the American Physical Society

Co–Co interactions in $CoCu_nCo$ chains the calculated variation of T_K (circles in Fig. 2.6) is much smaller than in the experiments and does not even follow the trend of the experimental data.

In a second step magnetic coupling between the Co atoms was introduced. For the relaxed CoCuCo chain an exchange interaction energy $E_{ex} = E_{\uparrow\downarrow} - E_{\uparrow\uparrow} = 14 \pm 6\,meV$ was found giving rise to ferromagnetic ($\uparrow\uparrow$) coupling of the Co atoms. An antiferromagnetic ($\uparrow\downarrow$) interaction with $E_{ex} = -17 \pm 3\,meV$ couples the Co atoms in $CoCu_2Co$ linear chains. Despite the ferromagnetic Co–Co interaction in CoCuCo chains, a Kondo resonance with $T_K = 46\,K$ is observed. The interplay of the coupling between the impurities, $H = -J\mathbf{S}_1 \cdot \mathbf{S}_2$, and Kondo screening was addressed theoretically in terms of two-site spin-1/2 Kondo models [38–40]. If the exchange interaction is ferromagnetic ($J > 0$) and dominates over the Kondo energy scale of the single-impurity ($k_B\tilde{T}_K \ll J$), the two spins lock to a total spin, $S_1 + S_2 = 1$, and give rise to a spin-1 Kondo effect [38]. This results in a reduced Kondo temperature, $T_K \approx k_B\tilde{T}_K^2/J$, which with $T_K = 46\,K$, $\tilde{T}_K = 110\,K$, $E_{ex} = 14 \pm 6\,meV$ is compatible with the experiments. Consequently, Co atoms in the CoCuCo linear chains are in a crossover regime between two independent and two ferromagnetically locked Kondo atoms, for which a narrower rather than a completely suppressed Kondo resonance is found.

For $CoCu_2Co$ the calculations revealed $|E_{ex}| \approx 2\,k_B\tilde{T}_K$ [34], which is close to the quantum critical point at $J \approx 2.2\,k_B\tilde{T}_K$ predicted for two-site spin-1/2 Kondo impurities in the particle-hole-symmetric case. At this critical point a ground state with an antiferromagnetically locked interimpurity singlet is separated from a ground state of two individually screened Kondo impurities [39, 40]. If particle-hole symmetry is lost, a crossover region replaces the quantum critical point, and the Kondo resonance continuously evolves into a pseudogap feature [41–46]. This crossover region has two characteristic energy scales $T_L < T_H$ which represent spin fluctuations and quasiparticle excitations [41–44, 46]. The lower energy scale, T_L, gives rise to the sharper and more pronounced feature in the spectral function at E_F [44] and characterizes the onset of local Fermi liquid behavior [46]. Hence, $k_B T_L$ should correspond to the experimentally observed width of the Kondo resonance with $T_L = T_K$. As $T_L < \tilde{T}_K$ [41–44, 46], a narrowed Kondo resonance is consistent with the crossover regime and in agreement with the experimental observation. As an important result we obtained that ferromagnetic and antiferromagnetic interactions lead to $T_K < \tilde{T}_K$ [47, 48].

The oscillations of T_K for $n \geq 3$ result from the RKKY interaction between the Co atoms. To show this, the Fermi wave vector, k_F, was determined for Cu_∞ chains on Cu(111) as $k_F \approx 0.37 \cdot (2\pi/a)$ with $a = 2.57$ Å the Cu(111) lattice constant. Therefore, Co–Co RKKY interactions oscillate with $2\,k_F \approx 0.74 \cdot (2\pi/a)$, which by subtracting a reciprocal lattice vector is identical with $-0.26 \cdot (2\pi/a)$ and corresponds to a direct-space periodicity of $\approx 3.8\,a$. While this spatial periodicity can be seen in the oscillatory magnetization density along $CoCu_nCo$ ($n \geq 3$) chains [34], it differs from the periodicity observed for T_K. For weak RKKY interactions, $\tilde{T}_K^2 - T_K^2 \approx E_{ex}^2 / k_B^2$ [34, 47, 48], with $E_{ex}^2 \propto [\sin(2\,k_F n a)]^2$. Therefore, the spatial periodicity is reduced to $\approx 1.9\,a$, which corresponds to the even-odd oscillations of T_K in the experiments. In conclusion, the two-site Kondo effect has been addressed by $CoCu_nCo$ atomic chains. Co–Co magnetic interactions ranging from ferromagnetic coupling ($n = 1$) via an antiferromagnetic singlet ($n = 2$) to RKKY interaction ($n \geq 3$) have been probed using line shape variations of the Kondo resonance.

2.1.4 Spectroscopy of the Kondo Resonance at Contact

Deviating from the preceding paragraphs where the Kondo resonance was probed by dI/dV spectroscopy in the tunneling range, spectroscopies are now performed close to and at contact between the STM tip and the magnetic impurity. In particular, single-atom junctions will be considered where currents on the order of 1 μA flow across the Kondo atom, which corresponds to a junction conductance on the order of the quantum of conductance, $G_0 = 2e^2/h$ (e: elementary charge, h: Planck constant). The break junction technique, in which thin metal wires are ruptured, was used to measure the conductances of metallic nanowires [49]. STM experiments enable imaging of the contact region prior to and after contacting the atom of interest. Modifications of the junction, such as atom transfer [50–52], that could change the electronic properties may be identified and avoided. Moreover, the electrodes and the contacted atom may be chosen independently as different materials, which is particularly appealing for Kondo atoms on nonmagnetic substrate surfaces.

Figure 2.7a shows a typical conductance trace acquired simultaneously with a tip approach towards a single Co atom on Cu(100) [inset to Fig. 2.7a] [27]. As detailed in [53, 54] the exponential part of the conductance curve reflects the tunneling range (region I), which is followed by region II reflecting the transition to the contact range (III). The latter is reached at a junction conductance of $\approx 1\,G_0$.

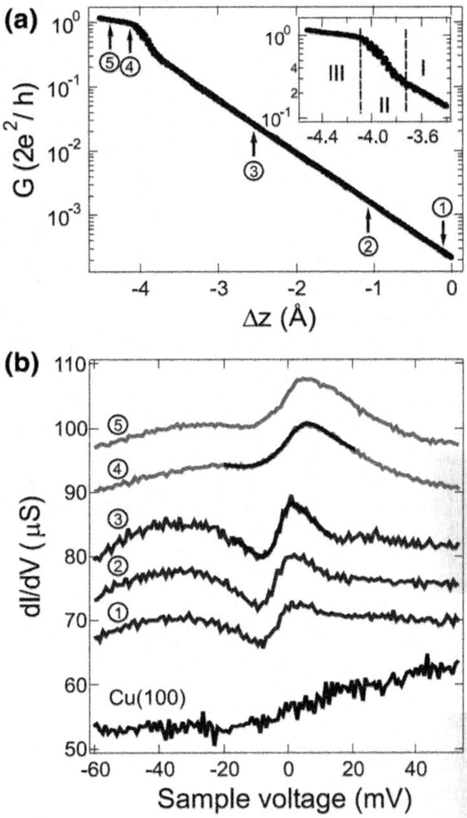

Fig. 2.7 **a** Conductance G as a function of tip displacement Δz. Inset: Transition (II) from the tunneling (I) to the contact (III) range. **b** $\mathrm{d}I/\mathrm{d}V$ spectra taken at tip displacements shown in (**a**). Bottom curve: Spectrum of the bare Cu(100) surface at 5 μA. Curves 1, 2 and 3 show $\mathrm{d}I/\mathrm{d}V$ spectra of a single Co atom in the tunneling range at 1 nA, 10 nA, 100 nA. Spectra 4 and 5 were taken in the contact range at currents of 5.5 μA and 6 μA, respectively. Calculated Fano profiles using the parameters $q = 1.2$, $T_\mathrm{K} = 78\,\mathrm{K}$ (spectrum 3) and $q = 2.1$, $T_\mathrm{K} = 137\,\mathrm{K}$ (spectrum 4) are shown as solid lines. Reprinted with permission from N. Néel et al., Phys. Rev. Lett. **98**, 016801 (2007) [27]. Copyright 2007 by the American Physical Society

In Fig. 2.7b $\mathrm{d}I/\mathrm{d}V$ spectra in the tunneling (1, 2, 3) and the contact range (4, 5) are presented together with a tunneling $\mathrm{d}I/\mathrm{d}V$ spectrum on clean Cu(100) for comparison. The Kondo resonance appears with an asymmetric line shape around zero bias voltage, in agreement with a previous report [55]. Intriguingly, this resonance is likewise observed in the contact range, albeit broadened.

Figure 2.8 summarizes the experimentally obtained Kondo temperatures (triangles) as extracted from fits of (2.1) to $\mathrm{d}I/\mathrm{d}V$ data together with calculated data (circles). Both data sets exhibit an abrupt change of T_K at $\Delta z \approx -4.1\,\text{Å}$. For $\Delta z > -4.1\,\text{Å}$ experimental and calculated T_K vary between 70 and 100 K. In the contact range, $\Delta z < -4.1\,\text{Å}$, experimental values for T_K vary between 140 K and 160 K while calculated data scatter within 200–290 K. The sudden broadening of the Kondo resonance upon contact can thus be related to an abrupt increase of T_K.

For the calculation of the Kondo temperature the single-impurity Anderson model [6, 25, 56] was used. In this model T_K reads

$$T_\mathrm{K} = \frac{1}{\mathrm{k_B}}\sqrt{\frac{\Delta U}{\pi}}\exp\left[-\frac{\pi}{\Delta}\left(\left|\frac{1}{\varepsilon_d}\right| + \left|\frac{1}{\varepsilon_d + U}\right|\right)^{-1}\right] \tag{2.7}$$

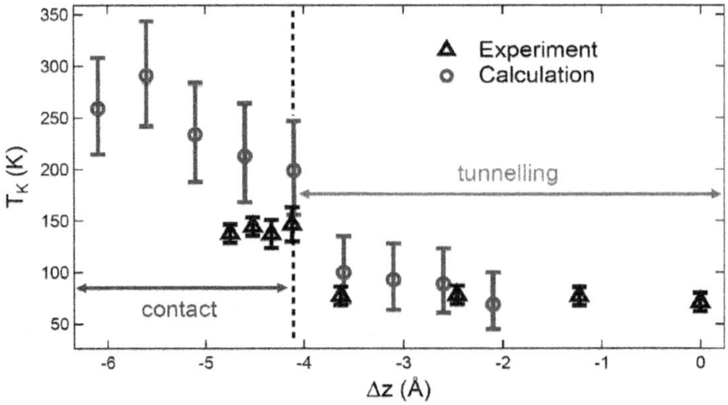

Fig. 2.8 Kondo temperature T_K versus tip displacement Δz. Experimental data are depicted by triangles, theoretical data are presented by circles. The dashed line separates tunneling and contact ranges. Reprinted with permission from N. Néel et al., Phys. Rev. Lett. **98**, 016801 (2007) [27]. Copyright 2007 by the American Physical Society

(Δ: width of the impurity state, ε_d: energy of occupied d band center with respect to E_F, U: on-site Coulomb interaction energy between spin-up and spin-down states). In the case of several impurity levels, as for the Co d states, Δ reflects the broadening due to the crystal field splitting between these states. Using density functional theory Δ, ε_d, U are accessible from ground state calculations. We found that the main changes during tip approach occur in the spin-up states of the Co atom [27]. The closer proximity of the tip induces a shift of these states towards E_F—thus reducing the exchange splitting—and a broadening of the individual peaks. This behavior reflects the additional hybridization of the adsorbed Co atom with the tip. Indeed, the crystal field splitting increases from $\Delta = 0.24$ eV in the tunneling range to $\Delta = 0.40$ eV in the contact range.

While the increase of T_K upon contact is well reproduced by the calculations, the actual values for T_K deviate from experimental data. These deviations may in part be due to the rather simple model that does not do justice to the complex junction. In addition, the surface-atom-tip system considered in the simulations is very rigid due to the limited number of layers on either side of the contacted atom. An increased T_K has likewise been observed in more recent experiments on Co-covered Cu(100) [57].

The above experiments were extended by using a Cu(111) substrate and different tip materials, nonmagnetic Cu-coated W and ferromagnetic bulk Fe tips [58]. The observed broadening of the Kondo resonance is not due to a voltage drop at the transimpedance amplifier, nor do local heating effects play a significant role [52]. However, to some extent, the broadening may be due to a change of the adatom's electrostatic potential. While it is pinned to the sample potential in the tunneling range, this may change at contact [59]. In contrast to results obtained for Co on Cu(100) [27] the width of the Kondo resonance does not increase upon contact with Cu-coated

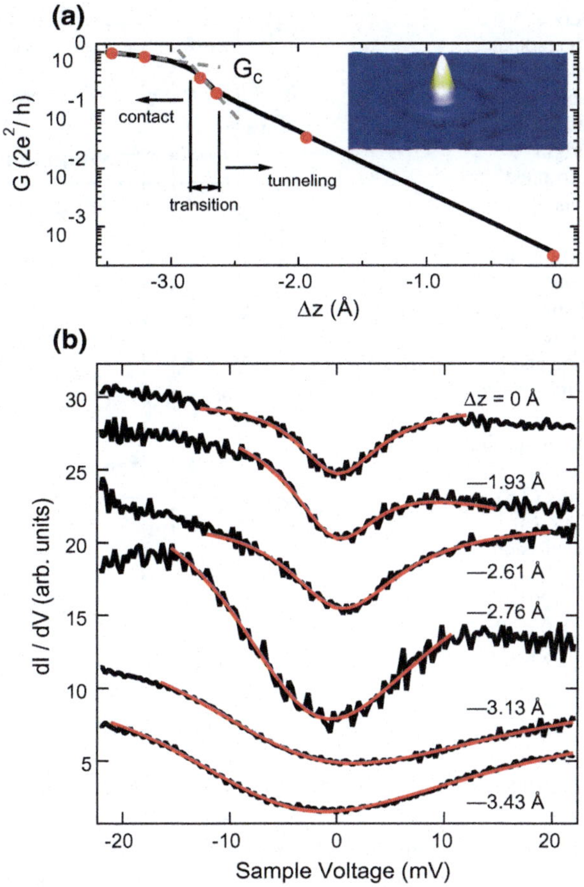

Fig. 2.9 **a** Conductance curves of a Co atom on Cu(111) versus displacement of a Fe tip. Tunneling, transition and contact ranges are indicated. The conductance trace between 0 and ≈ -2 Å is an exponential extrapolation of tunneling conductance in the displacement range from ≈ -2 to -2.5 Å. The contact conductance, G_c, is defined by the intersection of linear fits (dashed lines) to conductance data in the contact and transition ranges. Zero displacement is defined by the parameters of the feedback loop (33 mV and 1 nA). Inset: STM topograph of a single Co atom on Cu (111) (8 Å × 8 Å, 33 mV, 1 nA). **b** dI/dV spectra acquired at conductances, that are indicated on the conductance trace by dots (**a**). Fits of Fano resonance to the experimental spectra are shown as solid lines. Reprinted with permission from N. Néel et al., Phys. Rev. **B** 82, 233401 (2010) [58]. Copyright 2010 by the American Physical Society

W tips, which is in agreement with a previous report [60]. However, hybridization with Fe tips leads to a significant broadening. Figure 2.9a shows the evolution of the single-Co junction conductance depending on the Fe tip displacement. While the transition between tunneling and contact ranges is gradual as observed in contact experiments for Co on Cu(100) with a Cu-coated W tip, the contact conductance is slightly lower, $G_c \approx 0.7\, G_0$. In Fig. 2.9b the evolution of the dI/dV signature of

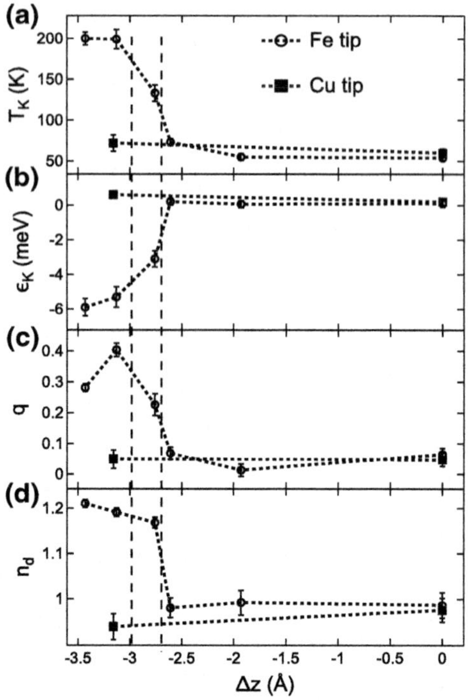

Fig. 2.10 **a** Kondo temperature T_K, **b** resonance energy ε_K **c** asymmetry factor q for a Co atom on Cu(111) contacted with a Cu-coated W (upright triangles) and Fe (rotated triangles) tip versus tip displacement Δz. Uncertainty margins result from fitting multiple dI/dV spectra. Vertical dashed lines separate tunneling, transition and contact ranges as introduced in Fig. 2.9a. **d** The occupation number n_d of the Co d orbitals versus Δz. n_d was calculated using (2.8) with fitted T_K (a) ε_K (b). Reprinted with permission from N. Néel et al., Phys. Rev. **B** 82, 233401 (2010) [58]. Copyright 2010 by the American Physical Society

the single-Co Kondo resonance obtained with a Fe tip from tunneling to contact is presented.

Fits to dI/dV data according to (2.1) appear as solid lines and lead to T_K, q, ε_K as a function of the tip displacement Δz (Fig. 2.10). In the tunneling range ($\Delta z > -2.65$ Å) these parameters are almost equal and constant for both tip materials, which reflects that the interaction between the tip and the sample is negligible in this conductance range. Starting from the transition range, however, Fe and Cu-coated W tips lead to strikingly different results. For Fe tips all parameters start to deviate from their tunneling range values. T_K [Fig. 2.10a] increases from $\approx$60 to $\approx$130 K, the ε_K [Fig. 2.10b] drops from $\approx$0 to -3 meV, and q [Fig. 2.10c] increases from $\approx$0.05 to 0.25. At contact this trend is continued with $T_K \approx 200$ K, $\varepsilon_K \approx -6$ meV, $q \approx 0.3$–0.4. For Cu-coated W tips all parameters are essentially constant throughout tunneling, transition and contact ranges.

The hybridization seems to be determined by chemical identity of the atom at the tip apex. A splitting of the Kondo resonance due to a magnetic stray field from the Fe tip at the adatom site would be too low to explain the observed broadening. The dipole field can be estimated to be $H \approx 1$ T at contact [61], which would result in a splitting of $\approx 2g\mu_B H \approx 0.2$ meV (g: Landé factor, μ_B: Bohr magneton) [62]. This is more than one order of magnitude lower than $k_B T_K$ with $T_K \approx 60$ K.

The line shape parameter q increases significantly upon contact formation between the Co adatom and the Fe tip, which reflects the additional hybridization of Co d levels with Fe states. The d bands of ferromagnetic Fe are located around at E_F [63, 64] and may hybridize with the Co d states. In contrast, the copper d bands are well below E_F [65], which hampers effective hybridization and results in an almost constant q. In the tunneling range, d states decay much faster into the vacuum compared to s states [66], so that s states dominate the hybridization, which results in similar q values for Cu-coated W and Fe tips. The different hybridization results in a change of the Co d state occupation number, n_d, which is related to ε_K and T_K *via* [6, 25]:

$$n_d = 1 - \frac{2}{\pi} \tan^{-1}\left(\frac{\varepsilon_K}{k_B T_K}\right) \tag{2.8}$$

and may be extracted from the fit parameters. For this, we assume that only a single Co d level hybridizes with the tip. This assumption can be justified by a theoretical study of a Co adatom on Au(111) [25] which showed that due to sp–d hybridization four Co d orbitals are completely filled while one partially occupied orbital ($n_d = 0.8$) is responsible for the Kondo effect. Empty, half-filled, and filled d levels correspond to $n_d = 0$, 1, and 2, respectively. Figure 2.10d displays n_d obtained from (2.8) as a function of Δz for Fe and Cu-coated W tips. Clearly, n_d changes upon contact from $\approx$0.98 (average value in the tunneling range) to $\approx$1.2 (contact) while it remains almost constant for the Cu-coated W tip ($\approx$0.98). A value of $n_d \approx 0.98$ is in good agreement with n_d obtained for Co atoms on Cu(111), while $n_d \approx 1.2$ comes close to the value of a Co atom on Cu(100) [30]. Recent work using Ni tips [67] showed that a splitting of the Kondo resonance due to the exchange field is possible and may contribute to the broadened feature observed here for Fe tips. In summary, the hybridization with the tip apex atom changes in the crystal field splitting and the d level occupation, which affects T_K.

2.2 Magnetic Molecules

Magnetic sandwich complexes are of particular interest for investigating interactions between molecular spin centers. Localized magnetic moments also interact with the conduction electrons of non-magnetic substrates, which may lead to a Kondo resonance that serves as a convenient read-out channel.

Benzene-Bridged Cobaltocene-Like Complexes

To explore the above approach, a trinuclear sandwich complex interconnected by benzene linkers (1,3,5-tris-(η^6-borabenzene-η^5-cyclopentadienylcobalt), TCBB, Fig. 2.11a) was synthesized following [68] and [69]. TCBB was deposited on Cu(111) using an electrospray ionization (ESI) setup with mass selection [70].

During the deposition process some fragmentation of TCBB occurs. However, besides η^6-borabenzene-η^5-cyclopentadienylcobalt monomers and dimers intact TCBB molecules are frequently observed. Figure 2.11b displays a typical topograph,

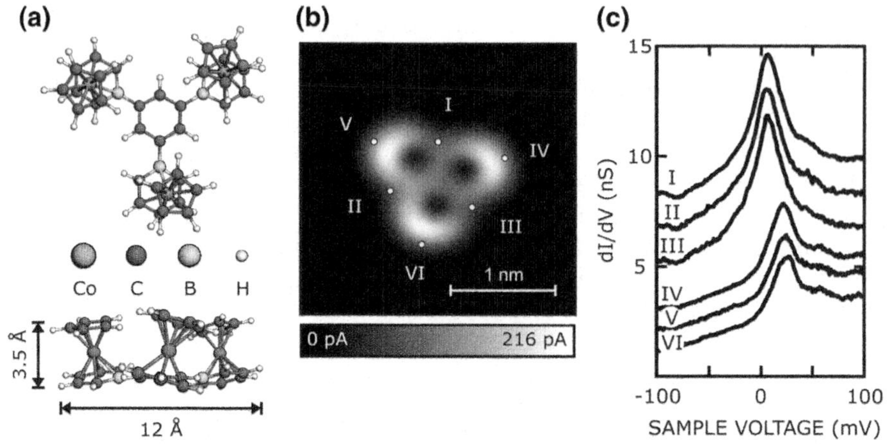

Fig. 2.11 **a** Top and side views of the calculated gas-phase structure of TCBB. **b** Constant-height image ($V = 0.1$ V) of TCBB on Cu(111). Markers indicate positions where $\mathrm{d}I/\mathrm{d}V$ spectra were recorded. **c** Spectra, vertically shifted for clarity. STM images have been processed with WSxM [71]. Reprinted with permission from T. Knaak et al., Nano Lett. **17**, 7146 (2017) [72]. Copyright 2017 American Chemical Society

which clearly shows the clover-like shape of TCBB lying flat on the substrate. Owing to a tilt of the CBB units towards the center of the molecule, the depressions (dark spots) corresponding to the centers of the Cp rings are slightly displaced from the center of the surrounding protrusions [73, 74].

Spectra of the differential conductance acquired at several positions above a TCBB molecule (Fig. 2.11c) show a clear peak close to the Fermi level E_F, which we attribute to a Kondo resonance. Interestingly, the resonance energy and lineshape vary in a systematic fashion between these positions. In particular, the shift of the resonance peak from E_F is $\varepsilon_\mathrm{K} = 5.1$ meV at positions I–III, while $\varepsilon_\mathrm{K} = 22.2$ meV at positions IV–VI. However, this variation does not reflect the interaction of the Co centers, because identical spectroscopic features are found on CBB monomers and dimers as well. Mapping of the position of the Kondo resonance on monomers showed that the areas with $\varepsilon_\mathrm{K} = 5.1$ meV match the calculated spatial distributions of the singly occupied molecular orbital (SOMO, mainly d_{yz} character) whereas the areas with $\varepsilon_\mathrm{K} = 22.2$ meV match the lowest unoccupied molecular orbital (LUMO, mainly d_{xz}). A Kondo resonance occurs for partially filled orbitals, a condition that is obviously met by the SOMO. The d_{xz}, which is empty in the gas-phase molecules, is broadened by the interaction with the substrate giving rise to a possible charge transfer. The occupation number n_d of a localized state involved in the Kondo resonance may be estimated from (2.8). Fits to the experimental spectra yield $n_d \approx 0.9$ and 0.5, which is consistent with the above interpretation. So far, experimental results on multiple Kondo resonances from a single molecule had been rather scarce [75, 76]. In contrast to magnetic adatoms, where the orbitals at the origin of Kondo resonances may be degenerate and spatially overlap, the ligand field in molecular systems can

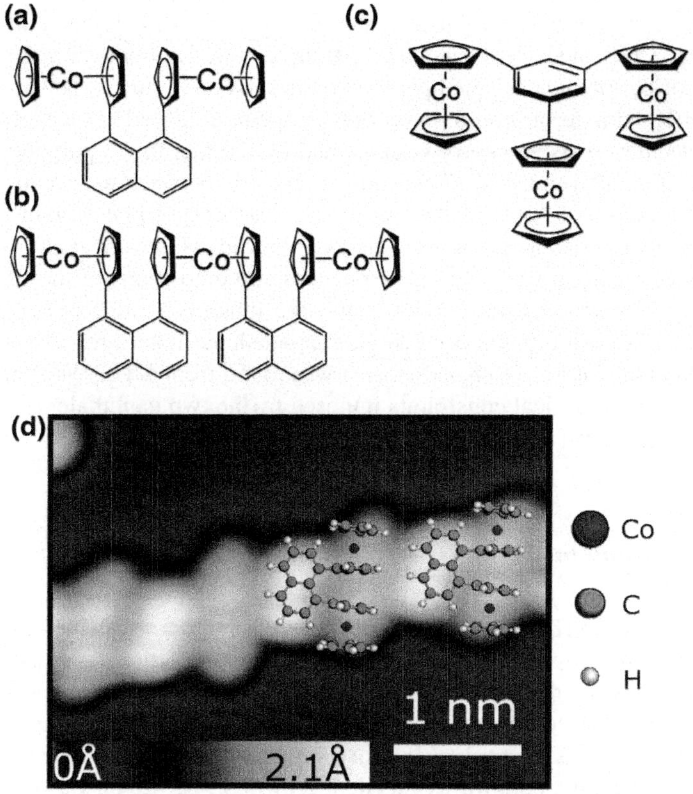

Fig. 2.12 **a** Benzene-bridged triscobaltocene (BTC). **b** Naphthalene-bridged biscobaltocene (NBC). **c** Naphthalene-bridged triscobaltocene (NTC). **d** STM topograph of a NBC chain on Au(111). $I = 30$ pA, $V = 0.2$ V. Scaled models of NBC are superimposed. Reprinted with permission from T. Knaak et al., J. Phys. Chem. C **121**, 26777 (2017) [77]. Copyright 2017 American Chemical Society

strongly influence the degeneracy and the extent of the orbitals relevant for the Kondo resonances.

Benzene- and Naphthalene Bridged Cobaltocenes

Benzene-bridged triscobaltocene (Fig. 2.12a, BTC) seems rather similar to TCBB and other metallocenes [72, 73, 78]. The cobaltocene ($CoCp_2$) subunits should carry a spin $\frac{1}{2}$. However, we did not observe any zero-bias feature reminiscent of a Kondo effect from BTC on Au(111). Topographs of BTC, however, suggest that the $CoCp_2$ subunits are significantly tilted whereas the magnetic units of the TCBB molecules are nearly perpendicular on the substrate. We tentatively suggest that the large tilt reduces the coupling of the localized spins to the substrate, which effectively prevents a Kondo effect.

For further comparison, we investigated naphthalene-bridged bis- and triscobaltocene (Fig. 2.12b and c, NBC and NTC). It turned out that both compounds adsorb with the $CoCp_2$ subunits oriented parallel to the surface. Figure 2.12d shows a topograph of NBC with the proposed adsorption geometry indicated by an overlaid model of the molecules. This geometry is motivated by the fact that a perpendicular orientation is sterically hindered. This leads to an angle between the naphthalene and the Cp-ring planes of $\approx 47°$ in the gas phase. On the substrate, a gain in interaction energy is expected as the angle is further reduced. Moreover, spatially resolved dI/dV spectroscopy reveals Kondo resonances at the positions of the Co atoms in the model. The resonance amplitudes of the two subunits are different, presumably because the Co positions imposed by the naphthalene linker are incommensurate with the Au(111) lattice. No Kondo effect was found from NTC. We speculate that the additional geometrical constraints imposed by the two naphthalene linkers lead to insufficient Co-Au(111) coupling.

2.3 Graphene on Ir(111)

The direct interaction between a magnetic molecule or atom with a metallic substrate often has a drastic effect on the molecular or atomic properties. As a result, various approaches of tuning the degree of interaction with a substrate have been proposed. They include self-decoupling molecules like cyclophanes [79] or spacer layers of the relevant molecules themselves [80] or inorganic insulators [81, 82]. Graphene is particularly appealing, both for its excellent growth properties on various substrates and its unique electronic properties. As a substrate, Ir(111) is remarkable because an almost unchanged graphene band structure has been found [83]. The effect of the graphene layer on the electronic states of the Ir substrate has hardly been explored.

Graphene Islands

To characterize the electronic structure of graphene on Ir(111) close to the Fermi level in nanoscale structures, we prepared nearly circular graphene islands with characteristic diameters of $\approx 5 \ldots 10$ nm [84]. Maps of the differential conductance dI/dV recorded above these islands (Fig. 2.13) reveal confined electronic states. In contrast to earlier reports, that interpreted similar data in terms of Dirac states of graphene, we attribute these states to confinement of the occupied electronic surface resonance of the Ir(111) substrate. As demonstrated by further STS data and DFT calculations, the interaction with the graphene layer shifts this state closer to the Fermi level. In other words, the islands effectively gate the Ir surface resonance. Confinement of the graphene states may also occur, but compared to the Ir resonance, which is located around the $\bar{\Gamma}$ point of the surface Brillouin zone, their contribution to the tunneling current is minor, because they are centered around $\bar{K}$ points. Their high in-plane momentum drastically reduces the tunneling probability.

W–H Complexes on Graphene

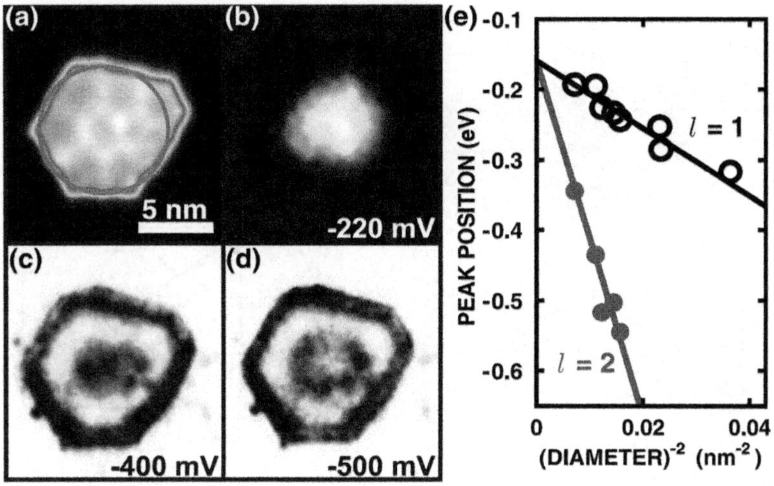

Fig. 2.13 **a** Graphene island on Ir(111) imaged at −220 mV and 1 nA. The effective island diameter of 8.3 nm used in modelling its electronic states is indicated by a light gray circle. **b–d** Normalized $\mathrm{d}I/\mathrm{d}V$ maps of this island reveal LDOS oscillations as expected for the confinement of an electron gas in a quantum dot. **e** Data from various graphene islands show the energies of confined states with a principal quantum number ($n = 0$). States with $l = 1$ (circles) and $l = 2$ (dots) are resolved. Lines show a fit involving the binding energy (160 meV) and the effective mass (0.18 m_e) of an occupied Ir surface resonance. Reprinted with permission from S. Altenburg et al., Phys. Rev. Lett. **108**, 206805 (2012) [84]. Copyright 2012 by the American Physical Society

Single transition metal atoms on graphene may exhibit unusual properties such as large magnetic moments, which are often quenched on metallic substrates [85, 86], and may also exhibit an orbital Kondo effect [87]. We therefore prepared W adatoms on graphene on Ir(111) by sublimation [88]. It turned out that most of the W atoms reacted with hydrogen from the residual gas of the ultra-high vacuum system. These complexes may be dehydrogenated using the electric field of the STM tip and exhibit a number of peculiar effects, such as electric field induced shifts of spectroscopic features as well as reversible switching and charging effects. These observations for W are related to earlier reports of hydrogen-induced spectroscopic features of transition metal–hydrogen complexes [89, 90].

2.4 Ballistic Anisotropic Magnetoresistance of Single Atom Contacts

A fundamental consequence of spin-orbit coupling (SOC) is anisotropic magnetoresistance (AMR), i.e. the dependence of the electrical resistance on the direction of magnetization, which was predicted to be enhanced in ballistic electron transport through atomic scale junctions [91–95]. Although some experimental results [96,

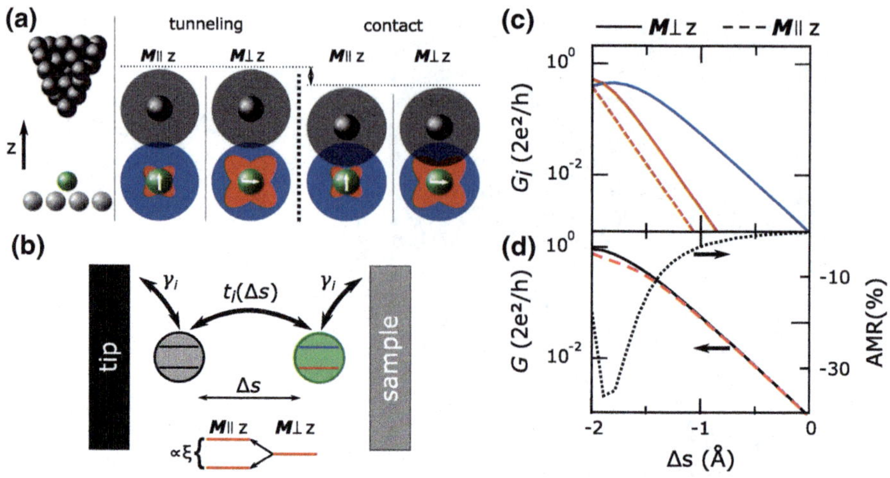

Fig. 2.14 **a** Left: The STM tip (black) is positioned close to a single atom (green) on a substrate (gray). Right: The overlap between the orbitals of tip (black) and adatom (blue and orange) determines the current. The extent of the orange orbital (indicated as an isosurface of spectral electron density) changes with magnetization direction $\boldsymbol{M}$ due to SOC. This orbital does not significantly contribute to the current at large tip-sample separations but becomes relevant in the contact regime. **b** simplified tight-binding scheme. Two localized states representing tip apex atom and adatom are coupled to metallic contacts using self-energy terms $-i\gamma_i$ and mutually via hopping matrix elements $t_i(\Delta s)$ that depend on the tip displacement Δs (defined relative to the distance in the tunneling regime). s (blue) and $d_{xz,yz}$ (orange) states have a different spatial decay of the wavefunctions into the vacuum which affects $t_i(\Delta s)$. The change of the magnetization direction $\boldsymbol{M}$ affects the splitting of $d_{xz,yz}$ orbitals via SOC. **c** Conductance of different transport channels calculated versus separation Δs. **d** total conductance and resulting AMR. Reprinted with permission from J. Schöneberg et al., Nano Lett. **16**, 1450 (2016) [102]. Copyright 2016 American Chemical Society

97] were interpreted as ballistic AMR (BAMR), unambiguous experimental evidence of this effect is difficult to achieve. Large variations of its magnitude, likely due to the unknown atomic geometry of the junction [96, 98], suggested alternative interpretations such as telegraph noise [99], quantum fluctuations [100] and magnetostriction [101]. Here, we study BAMR without the need of an external magnetic field and thus eliminate magnetostriction, which could lead to artifacts by changing the junction geometry. We deposited single Co and Ir adatoms on the ferromagnetic Fe double layer on W(110), which provides different orientations of the magnetization direction due to its domain structure. By exchange interaction with the substrate, the magnetic moments of the adatoms can be oriented out-of-plane and in-plane when positioned on domains and domain walls, respectively. Nonmagnetic STM tips were brought into contact with these adatoms to determine the junction conductances on domains (G_d) and domain walls (G_w). These quantities were measured versus tip-sample displacement from the tunneling to the contact regime, yielding the AMR defined as $\mathrm{AMR} = (G_\mathrm{d} - G_\mathrm{w})/G_\mathrm{d}$.

A distance-dependent AMR is essentially caused by multiple transport channels that contribute differently in the tunnelling and in the contact regimes and that are differently affected by SOC. This may be demonstrated with a simplified microscopic model (Fig. 2.14) comprising a local SOC of the adatom's d-states and a different distance dependence of the tunnelling matrix element across the junction for different orbitals. Figure 2.14a sketches a model of the experimental situation and the orbitals that contribute to the current depending on their spatial overlap for two different tip sample separations and two different magnetization directions $\boldsymbol{M}$. For larger distances, a relevant overlap exists only between the tip and the adatom orbital that extends the furthest into the vacuum and as a consequence the AMR is only sensitive to SOC induced effects of the extended orbital and is not influenced by SOC related effects on a localized state. The latter, however, increasingly contributes to the current and also potentially to the AMR when the tip-sample distance is reduced.

The conductance as a function of tip-adatom separation can be modeled using a tight-binding description (Fig. 2.14b). Two atoms are coupled via hopping terms and attached to contacts via self-energy terms causing a broadening of the atomic states. The adatom is modeled with three orbitals s, d_{xz} and d_{yz}, that each have a different decay length while both d states are assumed to be degenerate without SOC. The effect of SOC is included by a term $H_{\text{SOC}} = \xi \, \mathbf{l}\,\mathbf{s}$ (**l**: orbital momentum operator, **s**: spin momentum operator, ξ: the SOC strength) in the Hamiltonian (details see [102]). The current is dominated by the weakly decaying s orbital in the tunnelling regime, while for smaller separations ($\Delta s < -1.5\,\text{Å}$), the $d_{xz,yz}$ orbitals increasingly contribute (Fig. 2.14c). Their transmissions change under rotation of the magnetization, which can also be observed in the total conductance at small separations (Fig. 2.14d). The resulting AMR is negligible for large separations and becomes negative at contact.

To experimentally check this expectation, we prepared W(110) surfaces by repeated heating cycles in O_2 atmosphere and intermediate short annealing up to a temperature of 2200 K. Fe double layers were grown by Fe evaporation from an electron beam evaporator or a filament at elevated substrate temperatures 500 K. Single Co and Ir atoms were adsorbed on the Fe double layer at sample temperatures of 10 K.

The Fe bilayer on W(110) has ferromagnetic domains with a magnetization direction pointing out-of-plane and an in-plane magnetization within the domain walls. Due to SOC the magnetization direction affects the LDOS and leads to contrast in $\text{d}I/\text{d}V$ maps with non-magnetic tips [103] (Fig. 2.15a). Thus, the position of single atoms on the domain structure may be determined. According to DFT calculations, the magnetic moments of Co and Ir adatoms align parallel to the Fe magnetization due to exchange interaction [104, 105].

The conductance G was measured versus vertical tip displacement Δz. Here $\Delta z = 0\,\text{pm}$ is defined by the tip-sample separation at which the feedback loop was opened. As the feedback parameters (Co: $V = 50\,\text{mV}$, $I = 1.1\,\mu\text{A}$; Ir: $V = 100\,\text{mV}$, $I = 0.5\,\mu\text{A}$) were identical for adatoms on domains and domain walls, this results in AMR $= 0\%$ for $\Delta z = 0\,\text{pm}$.

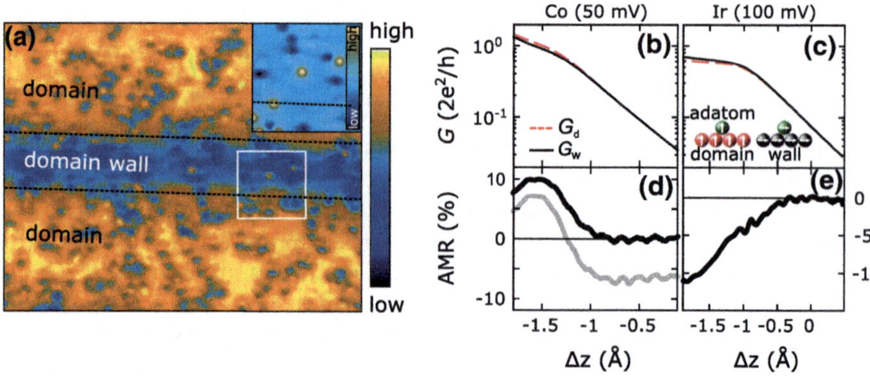

Fig. 2.15 **a** $\mathrm{d}I/\mathrm{d}V$ map ($52 \times 42\,\mathrm{nm}^2$, $V = 70\,\mathrm{mV}$, $I = 0.5\,\mathrm{nA}$) recorded on the double layer Fe on W(110) with single adsorbed Ir atoms. Domains and domain walls can be identified and marked by dashed lines. In the inset a constant current topograph ($10 \times 10\,\mathrm{nm}^2$, $V = 100\,\mathrm{mV}$, $I = 50\,\mathrm{pA}$) is shown, which was recorded in the area marked by a white square. From the comparison of $\mathrm{d}I/\mathrm{d}V$ map and topograph, the atom in the middle can be identified to be positioned on a domain wall. **b** Conductance G versus tip displacement Δz recorded at $V = 50\,\mathrm{mV}$ (Co) and $V = 100\,\mathrm{mV}$ (Ir), respectively. Dashed (solid) curves correspond to data recorded on a domain G_d(domain wall G_w). **d** and **e** show the AMR of Co and Ir adatoms. Reprinted with permission from J. Schöneberg et al., Nano Lett. **16**, 1450 (2016) [102]. Copyright 2016 American Chemical Society

The conductance versus displacement curves of both atoms show smooth transitions form the tunnelling regime to contact (Fig. 2.15b, c). At contact ($\Delta z < -1$ Å) Co adatoms display a larger conductance on domains than on domain walls, in contrast to Ir atoms, for which G_d is smaller than G_w in the contact regime. This directly influences the AMR, which becomes negative for Ir and positive for Co at contact (black curves in Fig. 2.15d, e).

The above results demonstrate a BAMR of single-atom junctions containing either Co or Ir. The magnetoresistance at contact is $\approx$10% but different in sign for Co and Ir. This demonstrates that sign and magnitude of BAMR can be tuned by choosing suitable adatoms.

2.5 Shot Noise Spectroscopy on Single Magnetic Atoms on Au(111)

Compared with magnetoresistance, a less common approach to probe the influence of the electron spin on the electrical current through nanostructures is the analysis of current fluctuations [106–111]. Shot noise results from charge quantization and its power spectral density was derived by Schottky for vacuum diodes as $S_0 = 2eI$, [112]. In a quantum mechanical systems the Pauli principle causes anticorrelations between the electrons with identical spins [113] and decreases the shot noise. By

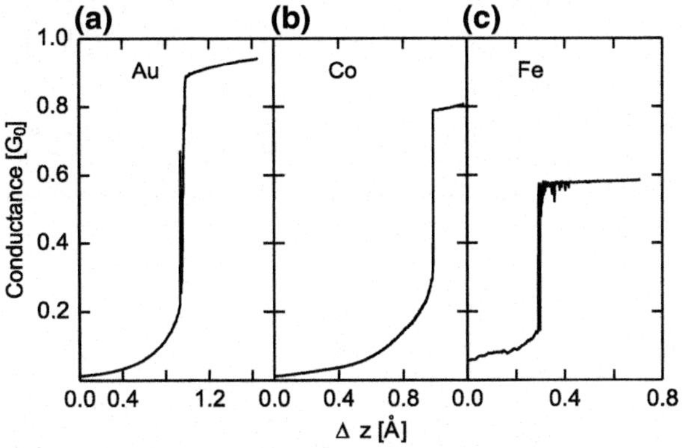

Fig. 2.16 Conductance versus displacement of the tip towards the sample for contacts to single (**a**) Au, (**b**) Co, and (**c**) Fe atoms ($V = 128$ mV). Reprinted with permission from A. Burtzlaff et al., Phys. Rev. Lett. **114**, 016602 (2015) [115]. Copyright 2015 by the American Physical Society

this, spin-polarized electron transport may be identified and a lower boundary of the spin polarization can be extracted from the measurement. We implemented noise spectroscopy in a low temperature STM to probe spin effects in the transport through single atoms and molecules without requiring a magnetic tip, as used in SPSTM [114]. Single Au, Co and Fe atoms were evaporated onto a Au(111) surface and contacted with a Au covered W tip. While single Au atoms show a noise that is comparable to previous studies on mechanically controlled break junctions [109, 110], a significant reduction of current fluctuations below the minimum value for spin-degenerate transport can be observed on both, Fe and Co atoms.

Figure 2.16 shows current versus displacement curves, $I(\Delta z)$, of contacts to Au, Fe and Co atoms. The conductances at contact ($V = 128$ mV) were 0.9–1.0 G_0 for Au (Fig. 2.1a), $\approx$0.8 G_0 for Co (Fig. 2.1b) while for Fe a considerably broader range of conductances between 0.47 and 0.68 G_0 (Fig. 2.16c) was observed.

To measure the noise, we used relays to switch the junction between the STM electronics and a low-noise battery driven power supply, that provides a constant bias current. Two home-built amplifiers measured the voltage fluctuations of the junction in parallel. These signals were cross-correlated to suppress the uncorrelated amplifier noise [116]. The low pass behavior ($f_{-3\text{dB}} = 60$ kHz at $G = 0.5\,G_0$) caused by finite contact impedance and cabling capacitance was numerically compensated and the DC current was recorded by switching a current to voltage converter into the circuit before and after each noise measurement. We recorded STM images before and after each noise measurement, which occasionally showed atom movements, material transfer from tip to sample or a change of the imaging properties of the tip. In these cases the data were discarded and only data recorded on stable contacts were used for further analysis.

Figure 2.17 shows the spectral noise densities of Au (0.96 G_0) and Fe contacts (0.66 G_0) that were biased using identical currents. White noise is observed for

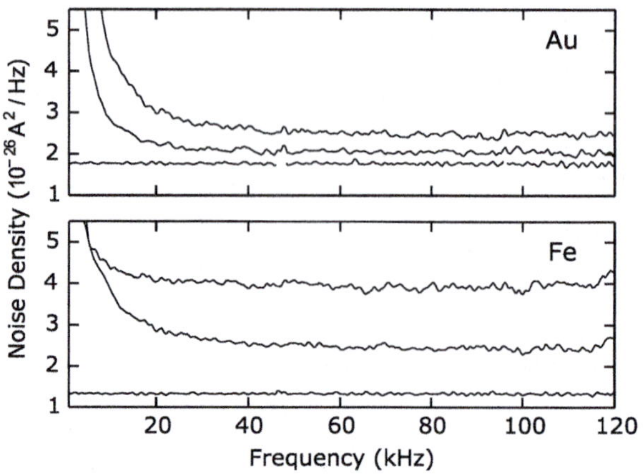

Fig. 2.17 Current noise density of Au and Fe adatoms. The bias currents were 0 (lowest spectra), 0.17 and 0.34 μA. The current noise increases significantly stronger with current for Fe compared with Au. The data has been smoothed using a moving average filter to simplify comparison. Reprinted with permission from A. Burtzlaff et al., Phys. Rev. Lett. **114**, 016602 (2015) [115]. Copyright 2015 by the American Physical Society

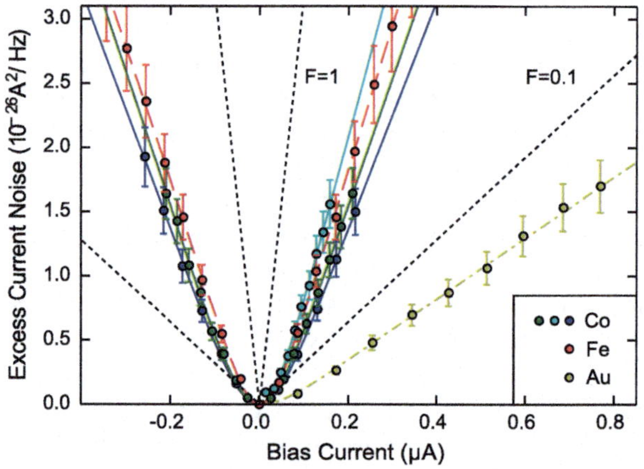

Fig. 2.18 Excess noise power as a function of bias current. Fits of (2.10) to the experimental data (circles) from Fe, Au, and three Co atoms are shown in dashed, dash dotted, and solid lines. The Co data were measured using different tips and on different atoms. Slopes corresponding to Fano factors of $F = 1$ and 0.1 are indicated as dotted lines. Reprinted with permission from A. Burtzlaff et al., Phys. Rev. Lett. **114**, 016602 (2015) [115]. Copyright 2015 by the American Physical Society

$I = 0$, which is caused by thermal current fluctuations. This Johnson-Nyquist noise $S_\Theta = 4\,k_B T G$ is larger for Au compared due to the higher conductance G. For $I > 0$ additional noise contributions appear that increase significantly faster with bias current for Fe than for Au.

For further analysis the excess noise ΔS was calculated by subtracting the $I = 0$ spectrum, which contains S_Θ and correlated amplifier noise (input current noise). The spectral density was averaged within a frequency interval (110–120 kHz) in which white noise was present and displayed as a function of bias current (Fig. 2.18).

In ballistic transport through nanostructures, the current is distributed over a finite number of conduction channels that are characterized by transmission probabilities τ_i giving a total conductance $G = G_0 \sum_i \tau_i$ [49]. Without spin degeneracy, each of these channels contains two spin channels $j = (i, \sigma)$ ($\sigma = \uparrow, \downarrow$) with potentially different transmission probability τ_j. The degree of shot noise reduction due to electron correlations compared to the uncorrelated Poissonian noise S_0 is quantified by the Fano factor, which is a function of τ_j:

$$F = \frac{\sum_j \tau_j (1 - \tau_j)}{\sum_j \tau_j}. \tag{2.9}$$

Thus the Fano factor contains information on the transparencies τ_j of the individual channels and their spin polarizations.

The excess noise as a function of bias current and at finite temperature can be written by the Lesovik expression [117]:

$$\Delta S = S - S_\Theta = F \left[S_0 \coth\left(\frac{S_0}{S_\Theta}\right) - S_\Theta \right] \tag{2.10}$$

This equation was fitted to the experimental excess noise versus current curves of Fig. 2.18 with the temperature T and the Fano factor F as parameters.

Figure 2.19 shows Fano factors and conductances measured on different stable contacts to Fe, Co, and Au atoms. The lower boundary of the Fano factor in a spin-polarized (spin-degenerate) situation is indicated by a solid (dashed) line (cf. 2.9). The data for Au atoms reproduce prior results [109, 110]. Data from noise measurements on Co and Fe atoms lie between the solid and dashed lines indicating spin-polarized transmission. We observed this behavior in all valid measurements on Fe atoms and in all except one measurement on Co atoms.

Although one would expect that the high valency of Co and Fe atoms makes single-channel transport unlikely [118], DFT calculations (details see [115]) confirmed the presence of one dominating transport channel for both Fe and Co. This channel has Γ_1 symmetry (invariant under rotations around the tip axis) as is the case for the s, p_z, and d_{z^2} orbitals. The transmission probabilities of all others channels are lower by at least two orders of magnitude. This is due to the s-symmetry of the Au tip atom, which acts as an orbital filter, as the hybridization of other d- and p-orbitals of the adatom to the s-orbital of the tip is suppressed because of symmetry mismatch.

As the DFT calculations of our systems confirmed transport through one relevant channel (per spin), the spin polarization $P = (\tau_\uparrow - \tau_\downarrow)/(\tau_\uparrow + \tau_\downarrow)$ can be extracted from the Fano factor. Dotted lines in Fig. 2.19 represent the Fano factors resulting from different degrees of spin polarization. The Fano factors measured on Co and Fe

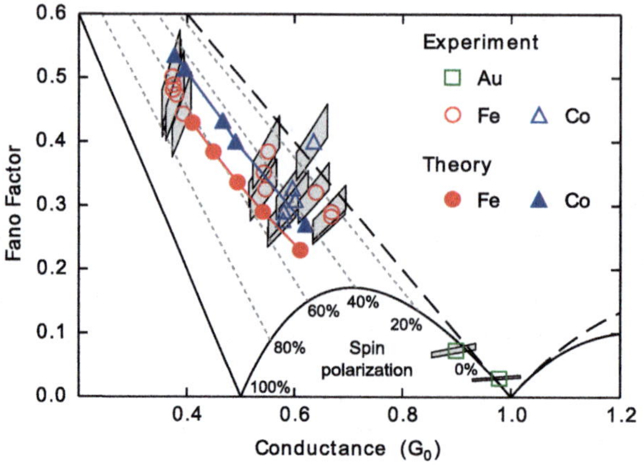

Fig. 2.19 Fano factors F and conductances G of Au, Fe, and Co adatom contacts. The smallest possible Fano factors consistent with spin-polarized ($P = 100\%$ up to 0.5 G_0) and spin degenerate ($P = 0\%$ up to 1 G_0) transport are indicated as solid and dashed lines. For finite spin-polarizations, the Fano factors in a single channel scenario are plotted in gray dotted lines. Uncertainty margins are marked by gray areas. For Au the measured data are consistent with one single spin-unpolarized channel. For Co atoms 4 out of 5 data sets (recorded on different atoms with different tips) indicate a spin-polarized transmission involving a single channel. For Fe all data suggest single channel, spin-polarized transmission. Higher spin polarizations were observed on 5 data sets around 0.38 G_0, that were all measured with the identical tip on different atoms in close vicinity to each other and to a step edge. Theoretical data was calculated for tip-adatom distances (left to right) of 4.10, 4.00, 3.95, 3.85, 3.70 Å (Fe) and 4.25, 4.20, 4.10, 4.05, 3.85 Å (Co). Reprinted with permission from A. Burtzlaff et al., Phys. Rev. Lett. **114**, 016602 (2015) [115]. Copyright 2015 by the American Physical Society

atoms around 0.6–0.7 G_0 correspond to spin-polarizations around 30–50%. Larger values of P ($\approx$60% for data around 0.38 G_0) were measured with a single tip on different Fe adatoms that were placed in close proximity to each other and to a nearby step edge. The role of the tip, a possible inter-atomic magnetic coupling and the electronic structure will be investigated in future experiments.

References

1. W. Meissner, B. Voigt, Annalen der Physik **399**(8), 892 (1930)
2. J. Kondo, Prog. Theor. Phys. **32**(1), 37 (1964)
3. A.A. Abrikosov, Physics **2**, 5 (1965)
4. A.A. Abrikosov, Physics **2**, 61 (1965)
5. W. Marshall (ed.), *Theory of Magnetism in Transition Metals* (Academic, London, 1967), pp. 116–205
6. A.C. Hewson, *The Kondo Problem to Heavy Fermions* (Cambridge University Press, UK, 1993)
7. K. Yoshida, *Theory of Magnetism* (Springer, New York, 1996)
8. J.T. Li, W.D. Schneider, R. Berndt, B. Delley, Phys. Rev. Lett. **80**, 2893 (1998)

9. V. Madhavan, W. Chen, T. Jamneala, M.F. Crommie, N.S. Wingreen, Science **280**, 569 (1998)
10. M. Ternes, A.J. Heinrich, W.D. Schneider, J. Phys.: Condens. Matter **21**, 053001 (2009)
11. N. Nagaosa, J. Sinova, S. Onoda, A.H. MacDonald, N.P. Ong, Rev. Mod. Phys. **82**(2), 1539 (2010)
12. B.G. Park, J. Wunderlich, X. Martí, V. Holý, Y. Kurosaki, M. Yamada, H. Yamamoto, A. Nishide, J. Hayakawa, H. Takahashi et al., Nat. Matter. **10**(5), 347–351 (2011)
13. A. Manchon, H.C. Koo, J. Nitta, S.M. Frolov, R.A. Duine, Nat. Matter. **14**(9), 871 (2015)
14. J. Sinova, S.O. Valenzuela, J. Wunderlich, C.H. Back, T. Jungwirth, Rev. Mod. Phys. **87**(4), 1213 (2015)
15. C. Gould, C. Rüster, T. Jungwirth, E. Girgis, G.M. Schott, R. Giraud, K. Brunner, G. Schmidt, L.W. Molenkamp, Phys. Rev. Lett. **93**, 117203 (2004)
16. S.H. Yang, K.S. Ryu, S. Parkin, Nat. Nanotechnol. **10**(3), 221 (2015)
17. T. Uchihashi, J. Zhang, J. Kröger, R. Berndt, Phys. Rev. B **78**, 033402 (2008)
18. Z.Q. Qiu, N.V. Smith, J. Phys.: Condens. Matter **14**(8), R169 (2002)
19. J.E. Ortega, F.J. Himpsel, Phys. Rev. Lett. **69**, 844 (1992)
20. J.E. Ortega, F.J. Himpsel, G.J. Mankey, R.F. Willis, Phys. Rev. B **47**, 1540 (1993)
21. P. Bruno, J. Phys.: Condens. Matter **11**(48), 9403 (1999)
22. P. van Gelderen, S. Crampin, J.E. Inglesfield, Phys. Rev. B **53**, 9115 (1996)
23. R.K. Kawakami, E. Rotenberg, E.J. Escorcia-Aparicio, H.J. Choi, T.R. Cummins, J.G. Tobin, N.V. Smith, Z.Q. Qiu, Phys. Rev. Lett. **80**, 1754 (1998)
24. R.K. Kawakami, E. Rotenberg, E.J. Escorcia-Aparicio, H.J. Choi, J.H. Wolfe, N.V. Smith, Z.Q. Qiu, Phys. Rev. Lett. **82**, 4098 (1999)
25. O. Újsághy, J. Kroha, L. Szunyogh, A. Zawadowski, Phys. Rev. Lett. **85**, 2557 (2000)
26. M. Plihal, J.W. Gadzuk, Phys. Rev. B **63**, 085404 (2001)
27. N. Néel, J. Kröger, L. Limot, K. Palotas, W.A. Hofer, R. Berndt, Phys. Rev. Lett. **98**, 016801 (2007)
28. P. Wahl, P. Simon, L. Diekhöner, V.S. Stepanyuk, P. Bruno, M.A. Schneider, K. Kern, Phys. Rev. Lett. **98**, 056601 (2007)
29. P. Bruno, Phys. Rev. B **52**, 411 (1995)
30. P. Wahl, L. Diekhöner, M.A. Schneider, L. Vitali, G. Wittich, K. Kern, Phys. Rev. Lett. **93**, 176603 (2004)
31. O.O. Brovko, P.A. Ignatiev, V.S. Stepanyuk, P. Bruno, Phys. Rev. Lett. **101**, 036809 (2008)
32. N. Néel, J. Kröger, R. Berndt, T.O. Wehling, A.I. Lichtenstein, M.I. Katsnelson, Phys. Rev. Lett. **101**, 266803 (2008)
33. V. Madhavan, W. Chen, T. Jamneala, M.F. Crommie, N.S. Wingreen, Phys. Rev. B **64**, 165412 (2001)
34. N. Néel, R. Berndt, J. Kröger, T.O. Wehling, A.I. Lichtenstein, M.I. Katsnelson, Phys. Rev. Lett. **107**, 106804 (2011)
35. M.A. Ruderman, C. Kittel, Phys. Rev. **96**, 99 (1954)
36. T. Kasuya, Prog. Theor. Phys. **16**(1), 45 (1956)
37. K. Yosida, Phys. Rev. **106**, 893 (1957)
38. C. Jayaprakash, H.R. Krishna-murthy, J.W. Wilkins, Phys. Rev. Lett. **47**, 737 (1981)
39. B.A. Jones, C.M. Varma, Phys. Rev. Lett. **58**, 843 (1987)
40. B.A. Jones, C.M. Varma, J.W. Wilkins, Phys. Rev. Lett. **61**, 125 (1988)
41. O. Sakai, Y. Shimizu, T. Kasuya, Solid State Commun. **75**(2), 81 (1990)
42. O. Sakai, Y. Shimizu, J. Phys. Soc. Jpn. **61**(7), 2333 (1992)
43. O. Sakai, Y. Shimizu, J. Phys. Soc. Jpn. **61**(7), 2348 (1992)
44. O. Sakai, Y. Shimizu, Nh Kaneko, Physica **186**, 323 (1993)
45. I. Affleck, A.W.W. Ludwig, B.A. Jones, Phys. Rev. B **52**, 9528 (1995)
46. L. Zhu, J.X. Zhu, Phys. Rev. B **83**, 195103 (2011)
47. V.Y. Irkhin, M.I. Katsnelson, Z. Phys. B **75**(1), 67 (1989)
48. V.Y. Irkhin, M.I. Katsnelson, Phys. Rev. B **56**, 8109 (1997)
49. N. Agraít, A.L. Yeyati, J.M. van Ruitenbeek, Phys. Rep. **377**(2), 81 (2003)

50. L. Limot, J. Kröger, R. Berndt, A. Garcia-Lekue, W.A. Hofer, Phys. Rev. Lett. **94**, 126102 (2005)
51. J. Kröger, H. Jensen, R. Berndt, New J. Phys. **9**(5), 153 (2007)
52. J. Kröger, N. Néel, A. Sperl, Y.F. Wang, R. Berndt, New J. Phys. **11**(12), 125006 (2009)
53. J. Kröger, N. Néel, L. Limot, J. Phys.: Condens. Matter **20**(22), 223001 (2008)
54. R. Berndt, J. Kröger, N. Néel, G. Schull, Phys. Chem. Chem. Phys. **12**, 1022 (2010)
55. N. Knorr, M.A. Schneider, L. Diekhöner, P. Wahl, K. Kern, Phys. Rev. Lett. **88**, 096804 (2002)
56. P.W. Anderson, Phys. Rev. **124**, 41 (1961)
57. D.J. Choi, M.V. Rastei, P. Simon, L. Limot, Phys. Rev. Lett. **108**, 266803 (2012)
58. N. Néel, J. Kröger, R. Berndt, Phys. Rev. B **82**, 233401 (2010)
59. S. Karan, D. Jacob, M. Karolak, C. Hamann, Y. Wang, A. Weismann, A. I. Lichtenstein, R. Berndt, Phys. Rev. Lett. **115**, 016802 (2015). https://www.doi.org/10.1103/PhysRevLett.115.016802
60. L. Vitali, R. Ohmann, S. Stepanow, P. Gambardella, K. Tao, R. Huang, V.S. Stepanyuk, P. Bruno, K. Kern, Phys. Rev. Lett. **101**, 216802 (2008)
61. A. Kubetzka, M. Bode, O. Pietzsch, R. Wiesendanger, Phys. Rev. Lett. **88**, 057201 (2002)
62. T.A. Costi, Phys. Rev. Lett. **85**, 1504 (2000)
63. A.M. Turner, A.W. Donoho, J.L. Erskine, Phys. Rev. B **29**, 2986 (1984)
64. T. Nautiyal, S. Auluck, Phys. Rev. B **34**, 2299 (1986)
65. W.F. Krolikowski, W.E. Spicer, Phys. Rev. **185**, 882 (1969)
66. M. Häfner, J.K. Viljas, D. Frustaglia, F. Pauly, M. Dreher, P. Nielaba, J.C. Cuevas, Phys. Rev. B **77**, 104409 (2008)
67. D.J. Choi, S. Guissart, M. Ormaza, N. Bachellier, O. Bengone, P. Simon, L. Limot, Nano Lett. **16**, 6298 (2016)
68. U. Hagenau, J. Heck, E. Hendrickx, A. Persoons, T. Schuld, H. Wong, Inorg. Chem. **35**, 7863 (1996)
69. C. Lindström, Synthese und Charakterisierung mehrkerniger benzolverbrückter Cobalt-Sandwich-Komplexe. Ph.D. thesis, Universität Hamburg, 2016
70. C. Hamann, R. Woltmann, I.P. Hong, N. Hauptmann, S. Karan, R. Berndt, Rev. Sci. Instrum. **82** (2011)
71. I. Horcas, R. Fernández, J.M. Gómez-Rodríguez, J. Colchero, J. Gómez-Herrero, A.M. Baro, Rev. Sci. Instrum. **78**, 013705 (2007)
72. T. Knaak, M. Gruber, C. Lindström, M.L. Bocquet, J. Heck, R. Berndt, Nano Lett. **17**, 7146 (2017)
73. M. Ormaza, P. Abufager, N. Bachellier, R. Robles, M. Verot, T. Le Bahers, M.L. Bocquet, N. Lorente, L. Limot, J. Phys. Chem. Lett. **6**(3), 395 (2015)
74. N. Bachellier, M. Ormaza, M. Faraggi, B. Verlhac, M. Vérot, T. Le Bahers, M.L. Bocquet, L. Limot, Phys. Rev. B **93**, 195403 (2016)
75. K.J. Franke, G. Schulze, J.I. Pascual, Science **332**, 940 (2011)
76. G.E. Pacchioni, M. Pivetta, L. Gragnaniello, F. Donati, G. Autès, O.V. Yazyev, S. Rusponi, H. Brune, ACS Nano **11**(3), 2675 (2017)
77. T. Knaak, M. Gruber, C. Lindström, J. Heck, R. Berndt, J. Phys. Chem. C **121**, 26777 (2017)
78. M. Ormaza, N. Bachellier, M.N. Faraggi, B. Verlhac, P. Abufager, P. Ohresser, L. Joly, M. Romeo, F. Scheurer, M.L. Bocquet, N. Lorente, L. Limot, Nano Lett. **17**, 1877 (2017)
79. F. Matino, G. Schull, F. Köhler, S. Gabutti, M. Mayor, R. Berndt, PNAS **108**, 961 (2011)
80. X. Ge, C. Manzano, R. Berndt, L. Anger, F. Köhler, R. Herges, J. Am. Chem. Soc. **131**, 6096 (2009)
81. S. Fölsch, A. Helms, S. Zöphel, J. Repp, G. Meyer, K.H. Rieder, Phys. Rev. Lett. **84**, 123 (2000)
82. S. Abbet, U. Heiz, H. Häkkinen, U. Landman, Phys. Rev. Lett. **86**, 5950 (2001)
83. I. Pletikosić, M. Kralj, P. Pervan, R. Brako, J. Coraux, A.T. N'Diaye, C. Busse, T. Michely, Phys. Rev. Lett. **102**, 056808 (2009)
84. S.J. Altenburg, J. Kröger, T.O. Wehling, B. Sachs, A.I. Lichtenstein, R. Berndt, Phys. Rev. Lett. **108**, 206805 (2012)

85. Y. Zhang, W. Ong, I. Arakelyan, J.E. Thomas, Phys. Rev. Lett. **108**, 235302 (2012)
86. S. Frota-Pessôa, Phys. Rev. B **69**, 104401 (2004)
87. T.O. Wehling, A.V. Balatsky, M.I. Katsnelson, A.I. Lichtenstein, A. Rosch, Phys. Rev. B **81**, 115427 (2010)
88. S.J. Altenburg, R. Berndt, New J. Phys. **16**, 093047 (2014)
89. M. Pivetta, M. Ternes, F. Patthey, W.D. Schneider, Phys. Rev. Lett. **99**, 126104 (2007)
90. F. Natterer, F. Patthey, H. Brune, Surf. Sci. **615**, 80 (2013)
91. J. Velev, R.F. Sabirianov, S.S. Jaswal, E.Y. Tsymbal, Phys. Rev. Lett. **94**, 127203 (2005)
92. D. Jacob, J. Fernández-Rossier, J.J. Palacios, Phys. Rev. B **77**, 165412 (2008)
93. G. Autès, C. Barreteau, D. Spanjaard, M.C. Desjonquères, Phys. Rev. B **77**(15), 155437 (2008)
94. A. Smogunov, A. Dal Corso, E. Tosatti, Phys. Rev. B **78**, 014423 (2008)
95. M. Häfner, J.K. Viljas, J.C. Cuevas, Phys. Rev. B **79**, 140410 (2009)
96. M. Viret, M. Gabureac, F. Ott, C. Fermon, C. Barreteau, G. Autes, R. Guirado-Lopez, Eur. Phys. J. B **51**(1), 1 (2006)
97. A. Sokolov, C. Zhang, E.Y. Tsymbal, J. Redepenning, B. Doudin, Nat. Nanotechnol. **2**(3), 171 (2007)
98. S. Egle, C. Bacca, H. Pernau, M. Huefner, D. Hinzke, U. Nowak, E. Scheer, Phys. Rev. B **81**, 134402 (2010)
99. B. Doudin, M. Viret, J. Phys.: Condens. Matter **20**(8), 083201 (2008)
100. K.I. Bolotin, F. Kuemmeth, D.C. Ralph, Phys. Rev. Lett. **97**, 127202 (2006)
101. S.F. Shi, D.C. Ralph, Nat. Nanotechnol. **2**, 522 (2007)
102. J. Schöneberg, F. Otte, N. Nel, A. Weismann, Y. Mokrousov, J. Kröger, R. Berndt, S. Heinze, Nano Lett. **16**(2), 1450 (2016)
103. M. Bode, S. Heinze, A. Kubetzka, O. Pietzsch, X. Nie, G. Bihlmayer, S. Blügel, R. Wiesendanger, Phys. Rev. Lett. **89**, 237205 (2002)
104. N. Néel, S. Schröder, N. Ruppelt, P. Ferriani, J. Kröger, R. Berndt, S. Heinze, Phys. Rev. Lett. **110**, 037202 (2013)
105. S. Schröder, Ph.D. thesis, University of Kiel, 2013
106. H. Birk, M.J.M. de Jong, C. Schönenberger, Phys. Rev. Lett. **75**, 1610 (1995)
107. M. Herz, S. Bouvron, E. Cavar, M. Fonin, W. Belzig, E. Scheer, Nanoscale **5**, 9978 (2013)
108. U. Kemiktarak, T. Ndukum, K.C. Schwab, K.L. Ekinci, Nature **450**(7166), 85 (2007)
109. H.E. van den Brom, J.M. van Ruitenbeek, Phys. Rev. Lett. **82**, 1526 (1999)
110. M. Kumar, R. Avriller, A.L. Yeyati, J.M. van Ruitenbeek, Phys. Rev. Lett. **108**, 146602 (2012)
111. N.L. Schneider, G. Schull, R. Berndt, Phys. Rev. Lett. **105**, 026601 (2010)
112. W. Schottky, Ann. Phys. **362**, 541 (1918)
113. T. Martin, R. Landauer, Phys. Rev. B **45**, 1742 (1992)
114. S. Heinze, M. Bode, A. Kubetzka, O. Pietzsch, X. Nie, S. Blügel, R. Wiesendanger, Science **288**, 1805 (2000)
115. A. Burtzlaff, A. Weismann, M. Brandbyge, R. Berndt, Phys. Rev. Lett. **114**, 016602 (2015)
116. A. Kumar, L. Saminadayar, D.C. Glattli, Y. Jin, B. Etienne, Phys. Rev. Lett. **76**, 2778 (1996)
117. G.B. Lesovik, Sov. Phys. JETP Lett. **49**, 592 (1989)
118. E. Scheer, N. Agraít, J.C. Cuevas, A.L. Yeyati, B. Ludoph, A. Martin-Rodero, G.R. Bollinger, J.M. van Ruitenbeek, C. Urbina, Nature **394**, 154 (1998)

Chapter 3
Electronic Structure and Magnetism of Correlated Nanosystems

Alexander Lichtenstein, Maria Valentyuk, Roberto Mozara and Michael Karolak

Abstract Magnetic nanostructures based on transition metals represent a main building block of standard memory devices. Their unique electronic properties are related to a complex multiplet structure of the partially filled d-shell with strong Coulomb interactions. Starting from a general formulation of the effective multi-orbital impurity problem for a transition metal atom in a fermionic bath of conduction electrons, the exact Quantum Monte Carlo solution is discussed. The concept of Hund's impurities to describe the electronic structure and magnetism of transition metal adatoms becomes very useful for the interpretation of numerous experimental data.

3.1 Electron Correlations in Magnetic Nanosystems

Transition metal atoms and small clusters on metallic substrates represent unique quantum systems to study complex many-body physics beyond standard mean-field electronic theories [1]. Recent progress in solid state theory allows for the analysis of the electronic structure and magnetic properties of correlated systems, while taking into account realistic dynamical many-body effects. These new approaches unify the Stoner theory of itinerant electron magnetism with the Heisenberg model for local spin systems into a unique spin-fluctuation Hubbard approach for real multi-orbital complex materials (see Fig. 3.1). Using the calculated electronic structure of different materials enables one to analyse magnetic properties and effective exchange interactions [2]. Understanding the properties of transition metal ions in different environments is a key ingredient and starting point for the modern theory of magnetism. The tremendous progress over the last years in experimental fabrication of new classes of materials, such as iron-based superconductors, magnetoresistance

A. Lichtenstein (✉) · M. Valentyuk · R. Mozara · M. Karolak
I. Institute of Theoretical Physics, Hamburg University, Hamburg, Germany
e-mail: alichten@physnet.uni-hamburg.de

R. Wiesendanger (ed.), *Atomic- and Nanoscale Magnetism*, NanoScience and Technology, https://doi.org/10.1007/978-3-319-99558-8_3

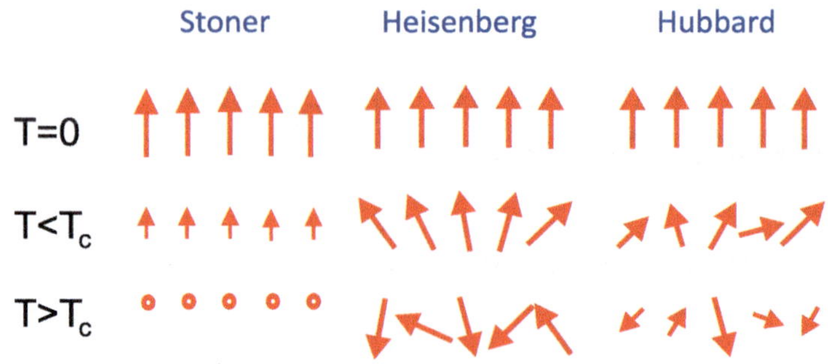

Fig. 3.1 Outline of classical magnetic models for different temperatures: Stoner theory of weakly correlated itinerant electrons, Heisenberg theory of local spin systems and Hubbard theory of spin and charge fluctuations in transition metal systems

systems, and two-dimensional artificial super-lattices put forward new challenges to the theory of transition metal systems. It is well known that the ground-state properties of antiferromagnetic insulators or compounds with orbital ordering cannot be obtained within the standard density functional theory (DFT) [2]. Recent angle-resolved photoemission studies of different cuprate materials [3] pointed out the existence of so-called incoherent peaks in the spectral density, which signals the strong inter-electron correlations in transition metal compounds.

The origin of such complicated features in the spectral properties of correlated materials is connected with the strong excitations to various low-energy electronic configurations, which are represented as a general pattern in Fig. 3.2. Let us discuss one common situation in which the free energy of correlated materials has one well-separated non-magnetic ground state. In this case, it is clear that electron fluctuations will be very small at low temperatures which results in the standard nonmagnetic quasiparticle structure. In opposing situations, when there are few closed local minima corresponding to different spin and orbital structures, as depicted in Fig. 3.2, we can be sure that strong many-body fluctuations will result in a non-quasiparticle structure of the spectral density, originating from Hund's rule behavior [4]. In order to describe systems with such a complicated energy spectrum, one has to use general quantum path-integral methods [5] and investigate different correlation functions using the recently developed continuous-time quantum Monte Carlo schemes [6], which efficiently describe different local minima of the free-energy functional (Fig. 3.2).

The complicated example of ferromagnetic iron with long-range exchange interactions [8] shows important quantum magnetic fluctuations at high temperature and high pressure [7]. So-called half-metallic ferromagnets [9] can be a playground for interesting magnetic correlation effects related to non-quasiparticle states in the minority-spin gap [10] which in principle can be detected in tunneling experiments [11]. Ultrafast dynamics of spin systems [12] and spin-spin correlations in magnetic systems [13] represent future directions of research.

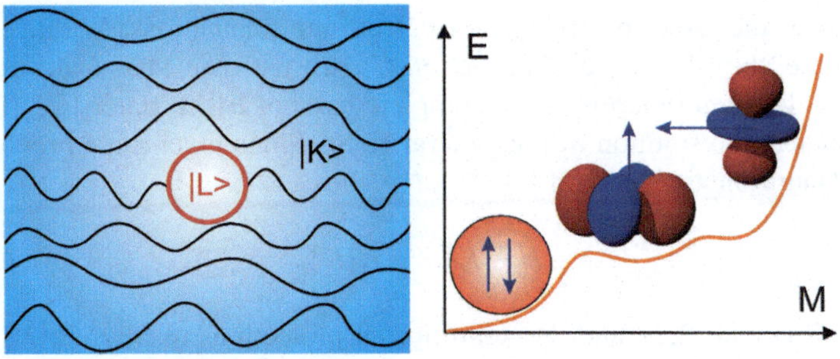

Fig. 3.2 Left: Simple view of the projection technique from itinerant Bloch states $|K\rangle$ to a localised Wannier basis $|L\rangle$ of a correlated subset. Right: Scheme of multiple local minima for nonmagnetic, magnetic and orbital states for strongly interacting electron materials

3.2 Realistic Impurity Models for Correlated Electron Systems

We developed a general scheme for partitioning the local orbital degrees of freedom, which provides practical tools for the investigation of different magnetic adatoms on metallic substrates. In the case of classical Kondo systems, based on cobalt adatoms on the gold surface, these local orbitals are related to the $3d$ electrons of the Co atom. Only a few electronic states from the total basis are needed to be taken into account in the many-body treatment. One of the most useful electronic structure approaches was thus related to the projector scheme which separates the total basis into the subset of Bloch states describing the standard itinerant electrons $|K\rangle$, and these local correlated d orbitals $|L\rangle$ represented by a numerical Wannier basis (see Fig. 3.2 for simple illustration).

In order to use the Monte-Carlo method for correlated subsystems one first needs to calculate the local Green functions and hybridization functions for the five local d orbitals. Standard density functional computer codes use a plane-wave basis set $|K\rangle$. In this case the transformation of the basis set is straightforward, and the convergence properties are easy to control. One of the most precise and efficient plane-wave based approaches is related to the projector augmented wave (PAW) method [14] and was successfully used in the general projection scheme from the Bloch itinerant basis to the local orbital states, seen in Fig. 3.2, using the overlap matrix $\langle K|L\rangle$.

Our universal projection scheme is based on the implementation of a projection operator $\mathcal{P} = \sum_L |L\rangle \langle L|$ within a DFT+DMFT method which is described in detail in [15–17]. Using this projector it is easy to transform the full Kohn–Sham Green function $G^K(\omega)$ into a set of five d orbitals $\{|L\rangle\}$:

$$G^L(\omega) = \mathcal{P}G^K(\omega). \tag{3.1}$$

The subspace $\{|L\rangle\}$ will represent the local correlated d-orbitals. Only these five d orbitals will be used in the many-body investigations which produce the important corrections to the DFT spectrum due to electronic fluctuations. Within the plane-wave scheme, the Bloch Green's function $G^K(\omega)$ with the Matsubara frequencies $i\omega$ can be calculated in terms of the complete basis of Bloch states $|K\rangle$. The Bloch states represent the solution of the general Kohn–Sham eigenvalue problem for an effective Hamiltonian H_K, with

$$H_K\,|K\rangle = \varepsilon_K\,|K\rangle\,. \tag{3.2}$$

Using (3.1) and (3.2), and the definition of projections overlap $\langle L|K\rangle$, one can easily evaluate the Green's function in the local basis $|L\rangle$ with a given chemical potential μ as

$$G^L(i\omega) = \sum_K \frac{\langle L|K\rangle\langle K|L\rangle}{i\omega + \mu - \varepsilon_K}. \tag{3.3}$$

For magnetic transition metal adatoms on different surfaces the set of correlated states are represented by five *d* orbitals. These correlated orbitals are located mostly inside PAW augmentation spheres (Fig. 3.2) which allowed us to use the standard representation of a Bloch state $|K\rangle$ projection onto local five *d* orbitals [14]. If we use only a small number of bands near the Fermi energy for projection onto the local impurity orbitals, it is important to properly orthogonalize the local basis functions [15, 16].

We can define the effective hybridization matrix $\Delta(i\omega)$ for the *d* orbitals impurity model using the following equation for the local Green's function

$$G^{-1}(i\omega) = i\omega - \epsilon_d - \Delta(i\omega). \tag{3.4}$$

The impurity energy ϵ_d describes the crystal field effects from substrates. In general, (3.4) represents a $L \times L$ matrix equation for Δ. In order to separate the static DFT crystal field energy ϵ_d from the frequency dependent hybridization function Δ, one normally evaluates the limit $\omega \to \infty$, where $\Delta(i\omega) \to 0$, and therefore $G^{-1}(i\omega) \to i\omega - \epsilon_d$.

3.3 Multiorbital Quantum Impurity Solvers

The formulation of a numerical solution to the multi-orbital impurity model was a challenge for the quantum many-body problems. During the last decade we developed the novel continuous-time quantum Monte-Carlo (CT-QMC) solver [6] for the general multi-orbital impurity problem. The CT-QMC scheme is based on stochastic Monte-Carlo sampling and consists of two complementary approaches: the interaction and the hybridization expansion. We describe here the most efficient approach for the strongly correlated case, which is the hybridization algorithm (CT-HYB). For simplicity, we discuss the so-called segment scheme, which allows for a fast

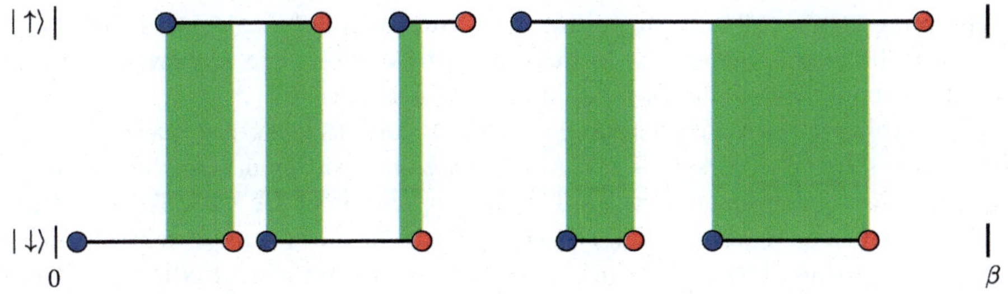

Fig. 3.3 Example of a single-orbital CT-HYB expansion in the segment formalism: The annihilation operators are represented by the blue dots and the creation operators by red ones. The black lines describe hybridization functions $\Delta(\tau_i - \tau_j')$ for two spin projections. The time interval in which two electrons are present on the impurity is marked by a green region with total time l_d, and consequently an energy penalty of U has to be payed here

analytical evaluation of the path-integral trace for d-electrons in the case of diagonal-density types of Coulomb interaction.

The quantum impurity problem at temperature β^{-1} can be represented by the action

$$S_{\mathrm{imp}} = S_{\mathrm{at}} + S_{\Delta} \tag{3.5}$$

with the atomic component

$$S_{\mathrm{at}} = \sum_{\sigma} \int_0^{\beta} \mathrm{d}\tau c^*_{\sigma\tau}[\partial_\tau - \mu]c_{\sigma\tau} + \int_0^{\beta} \mathrm{d}\tau U n_{\uparrow\tau} n_{\downarrow\tau} \tag{3.6}$$

where $c^*_{\sigma\tau}$, $c_{\sigma\tau}$ are Grassmann variables which depend on spin σ, and τ is the imaginary time space. For simplicity we skip orbital indices and $n_{\sigma\tau} = c^*_{\sigma\tau}c_{\sigma\tau}$. The hybridization action S_Δ contains the term $\Delta(\tau)$, and can be written as

$$S_{\Delta} = \sum_{\sigma} \int_0^{\beta} \int_0^{\beta} \mathrm{d}\tau \mathrm{d}\tau' c^*_{\sigma\tau} \Delta(\tau - \tau') c_{\sigma\tau'} \ , \tag{3.7}$$

and is the Fourier transform of the $\Delta(i\omega)$ matrix.

To simplify the notation, we suppress the spin indices and view the proceeding expressions as diagonal matrices in spin and orbital space. We expand the impurity action (3.5) in the hybridization part (3.7) around the atomic limit (3.6). It can then be found, that at a given perturbation order k of the hybridization expansion of the impurity action S_{imp} in power of S_Δ, different terms can be combined into a determinant of hybridization functions. Therefore, the impurity partition function may be written in the following form:

$$\mathcal{Z}/\mathcal{Z}_{\mathrm{at}} = \sum_k \int_0^{\beta} \mathrm{d}\tau_1 \ldots \int_{\tau_{k-1}}^{\beta} \mathrm{d}\tau_k \langle c^*_{\tau_1} \ldots c_{\tau_k} \rangle_{\mathrm{at}} \det \hat{\Delta}^{(k)}. \tag{3.8}$$

The average trace $\langle c^*_{\tau_1} \ldots c_{\tau_k} \rangle_{at}$ in this expression should be calculated over exact states of the atomic action S_{at}. In principle, this can be done numerically for an arbitrarily complicated multi-orbital interaction matrix U [6].

The hybridization matrix determinant det $\hat{\Delta}^{(k)}$ consists of a $k \times k$ matrix in imaginary time space $\hat{\Delta}_{ij} = \Delta(\tau_i - \tau_j)$. Assembling the $k!$ different terms into a single hybridization determinant is crucial for the suppression of the so-called fermionic sign problem in CT-QMC [6]. One should point out that the time interval in imaginary space of equation (3.8) can be considered as a circle with antiperiodic (fermionic) boundary conditions.

In the simplest case of a single-orbital impurity model with Hubbard interaction (3.6), the segment formalism gives a very intuitive picture of CT-QMC insertion in imaginary τ-space in the interval $[0, \beta]$, which is shown in Fig. 3.3. An arbitrary configuration can be represented by two separate world-lines for different spins. In this scheme one can exactly calculate the impurity trace which is related to contributions from the chemical potential μ and the Hubbard-U interaction term for time-intervals of double occupancy on the impurity, and resulting in the simple expression $e^{-Ul_d + \mu(l_\uparrow + l_\downarrow)}$, where l_σ represents the time spent by a spin-σ electron on the impurity, and l_d corresponds to the total time of a doubly occupied impurity state [6].

3.4 Transition Metal Impurities on Metallic Substrates

Using the continuous-time quantum Monte Carlo methods within the interaction (CT-INT) [18] or hybridization (CT-HYB) expansion [19], we investigated accurate low-temperature spectral functions of transition metal impurities on metallic substrates. As an example, we present the electronic structure of cobalt atoms on the Cu(111) surface based on realistic DFT supercell calculations, in combination with the many-body CT-QMC investigation of the multi-orbital local impurity problem [19].

The electronic structure of the cobalt adatom on the Cu(111) surface was first analyzed within a large supercell scheme of twelve atoms in a plane with a thickness of five atomic layers, using the PAW scheme [14] (see Fig. 3.4). Using (3.4) and results of DFT calculations we obtained the hybridization functions shown in Fig. 3.4 for orbitals within the local C_{3v} point group symmetry. The five Co d orbitals split into three subblocks of two doubly degenerate and one non-degenerate representations with corresponding orbitals d_{xz}, d_{yz} for the E_1 representation, $d_{x^2-y^2}$, d_{xy} for E_2, and d_{z^2} for A_1, respectively. A full four-index Coulomb correlation vertex U for the five d orbitals [2] was obtained via screened Slater parameters F^0, F^2, and F^4 corresponding to an effective Hubbard parameter $U = 4$ eV and to a Hund interaction $J = 0.9$ eV.

Since the hybridization of a cobalt atom with the substrate Cu(111) is rather weak (Fig. 3.4), the correlation effects will be strong. The results of the many-body CT-QMC calculations for the density of states (DOS) shows the corresponding new features. There is strong renormalization of the DFT quasiparticle structure near E_F,

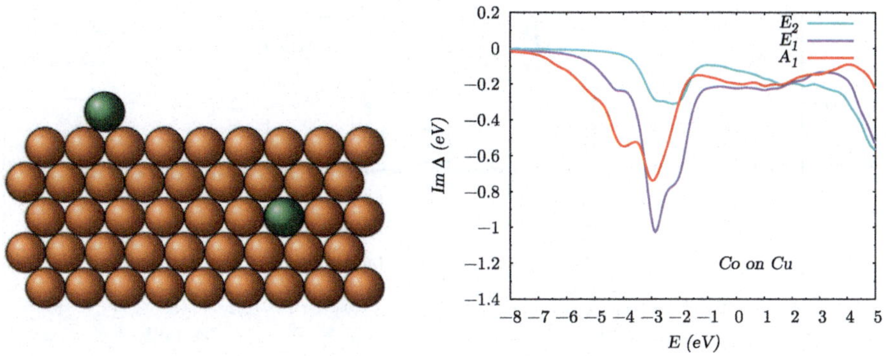

Fig. 3.4 Left: Sketch of cobalt impurities on top of a copper surface and in the bulk. Right: the imaginary parts of the hybridization functions Im Δ for different orbital symmetries

producing sharp Kondo-like peaks, and at higher energies the formation of lower and upper Hubbard bands is clearly visible (Fig. 3.5). One can also see the strong anisotropy of the local DOS of Co on Cu(111) for the different orbitals belonging to the E_1, E_2, and A_1 representations. The smallest hybridization function corresponds to orbitals of E_2 symmetry located in the x-y plane and therefore atomic-like Hubbard bands are well pronounced with the strong suppression of the quasiparticle peak at E_F. We noted that the low-energy quasi-particle peaks appear in all five Co d orbitals, which was not expected within the model for the two-channel Kondo problem for spin $S = 1$, which was assumed to apply for the Co adatom, and where the resonance at E_F would appear within two orbitals only.

The formation of a local Fermi-liquid state of Co on Cu for all five d orbitals at low-temperature indicate a strong hybridization with the metallic substrate. For the estimation of the Kondo temperature T_K one can use the quasiparticle renormalization factor Z calculated from the CT-QMC scheme and a general result from the single impurity Anderson model [20]: $T_K = -\frac{\pi}{4} Z \,\mathrm{Im}\,\Delta(0)$. The corresponding Kondo temperatures within the irreducible representations are: $T_K = 60\,\mathrm{K}$ in E_2, $T_K = 310\,\mathrm{K}$ in E_1, and $T_K = 180\,\mathrm{K}$ in A_1, and the agreement with the experimental result $T_K \approx (54 \pm 5)\,\mathrm{K}$ is reasonable [21]. We were also able to reproduce the large difference of the Kondo scale for the impurity in the bulk and the adatom on the surface [19]. To our knowledge, this is the first successful calculation of the Kondo temperature for a realistic correlated impurity with five orbitals.

3.5 Hund's Impurities on Substrates

The magnetic behavior and electronic structure of d-metal impurities in the fermionic bath of the substrate crucially depends on the multiplet structures and Hund's rule physics [4]. In order to show such Hund's effects we investigated single Mn, Fe,

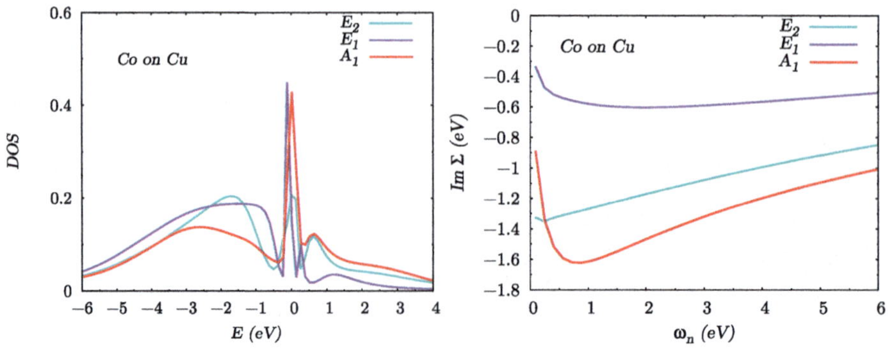

Fig. 3.5 Left: Orbitally resolved DOS of the Co impurities on Cu obtained by analytical continuation of the CT-QMC imaginary time Green's function for $\beta = 40$ eV^{-1}. Right: **a** corresponding orbitally resolved self-energies on the Matsubara axis

Co and Ni adatoms on the metallic Ag(100) surface [22]. The experimental photoemission spectra for the $3d$ series shows a monotonous reduction of high-energy splittings together with non-monotonous features at low-energy peaks. We can now explain this behavior by means of Hund's physics. On the one hand, the high-energy peaks are related to splitting of the ground state energy into multiplets with different spin quantum numbers, to a monotonous decrease of the local magnetic moment m, and to the Hund's splittings Jm which monotonously decrease to the end of the $3d$ series due to filling of the d band. On the other hand, the effective Hubbard energies $U = E_{n+1} + E_{n-1} - 2E_n$ [2, 4], (where E_n is the impurity ground state energy with n particles) has strongly non-monotonous variation in the $3d$ series related to Hund's rules physics.

Using a rotationally invariant Coulomb interaction matrix [2] one can find a strong dependence of the effective Hubbard parameter $\tilde{U}$ as a function of $3d$-occupation n:

$$\tilde{U} \approx \begin{cases} U + 4J & (n = 5) \\ U - 3J/2 & (n = 6, 9) \\ U - J/2 & (n = 7, 8). \end{cases} \tag{3.9}$$

From these results we were able to make a conclusion about the non-monotonous behavior of the charge fluctuations and the renormalization of the DOS at E_{F} which depends on the effective Hubbard parameter $\tilde{U}$ in the $3d$ series. Our results show strong charge fluctuations for Fe ($n = 6$) and Ni ($n = 9$) related to the almost mixed-valence regime due to the small value of $\tilde{U}$. For the case of Mn ($n = 5$) and Co ($n = 7$) the Hubbard parameter U is much larger, which suppresses the charge fluctuations and promotes the multi-orbital Kondo behavior.

Figure 3.6 shows the valence photoemission spectrum for Mn, Fe, Co and Ni adatoms on the silver surface together with theoretical QMC results of the corresponding impurity problem containing first-principle hybridization functions [22]. The Mn ($n = 5$) impurity has the largest effective interaction U (3.9), and the single

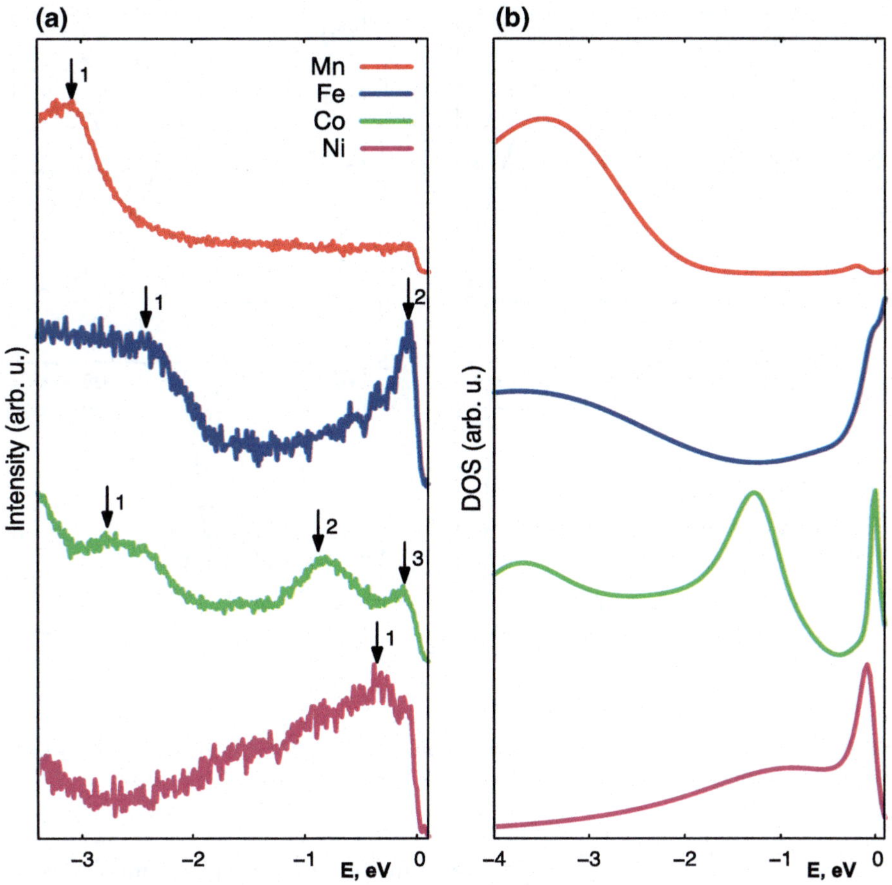

Fig. 3.6 **a** Experimental valence electron photoemission spectra of $3d$ adatoms on Ag(100) surface. **b** Theoretical spectral function from QMC results at $\beta = 20\,\text{eV}^{-1}$

multiplet spectrum for the maximum-spin ground state $S = 5/2$ can explain the single high-energy peak at -3.5 eV in the spectral function. For the Fe ($n = 6$) impurity, the calculations show a broad lower-Hubbard band at -3 eV and a sharp quasiparticle resonance below $E_{\rm F}$, and can be well compared with the experimental peaks 1 and 2 (Fig. 3.6a). This broad Hubbard band at -3 eV is formed by all $3d$ orbitals and can be found in simple atomic exact diagonalization (ED) results, and is related to $d^6 \rightarrow d^5$ excitations (Fig. 3.6b). From the orbitally resolved DOS in Fig. 3.6b one can identify the $d_{x^2-y^2}$ orbital which is responsible for the experimental peak 2 (Fig. 3.6a). The occupation of the $d_{x^2-y^2}$ orbital is equal to $n = 0.8$ due to strong charge fluctuations in the Fe $3d$ shell and indicates that this peak is not related to a spin-Kondo resonance. If one inspects the atomic ED calculations for the Fe impurity with crystal field splitting from the surface hybridization (Fig. 3.6b), this peak also results from

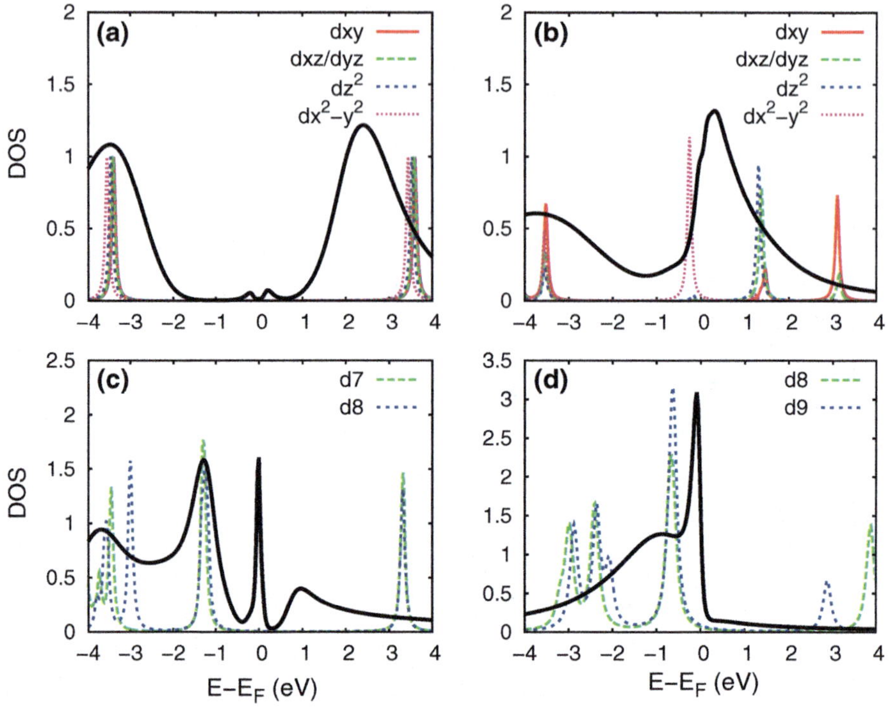

Fig. 3.7 DOS for Mn **a**, Fe **b**, Co **c** and Ni **d** impurities from QMC calculations (thick black lines) and from exact diagonalization (dashed lines) for different occupations of the *d* shell

multiplet $d^6 \rightarrow d^5$ excitations. The energy difference between photoemission peaks 1 and 2 can be understood from the ED results as splitting of the final d^5 multiplets: low-energy peak 2 is related to the $S = 5/2$, $L = 0$ state and high-energy peak 1 with the $S = 3/2$, $L \geq 0$ states. Moreover, we can estimate this energy difference as Jm, which relates to Hund's rule exchange.

Our theoretical QMC calculations for the Co adatom on Ag(100) with occupation $n = 7.8$ describe well the three-peak structure of the experimental photoemission spectrum (Fig. 3.6). The orbital character of the DOS from ED calculations (Fig. 3.7c) shows that the experimental peak 2 at -1 eV comes from excitations within the d_{xz}, d_{yz} and $d_{x^2-y^2}$ orbitals. Moreover, the high-energy experimental peak 1 is related to multi-orbital transitions between $d^8 \rightarrow d^7$ multiplets. Similar to the Fe case, the energy difference between peaks 1 and 2 is related to Hund's rule exchange and becomes smaller by J due to the different magnetic moment of Co.

The experimental and theoretical photoemission spectrum of the Ni adatom with only one broad peak below E_F is very different from other 3*d* impurities (Fig. 3.6). We can understand such a featureless spectrum from the Hund's rule physics related to a strong reduction of exchange splitting in Ni and very small splitting between atomic multiplets (Fig. 3.7d) which are washed out by hybridization with the substrate.

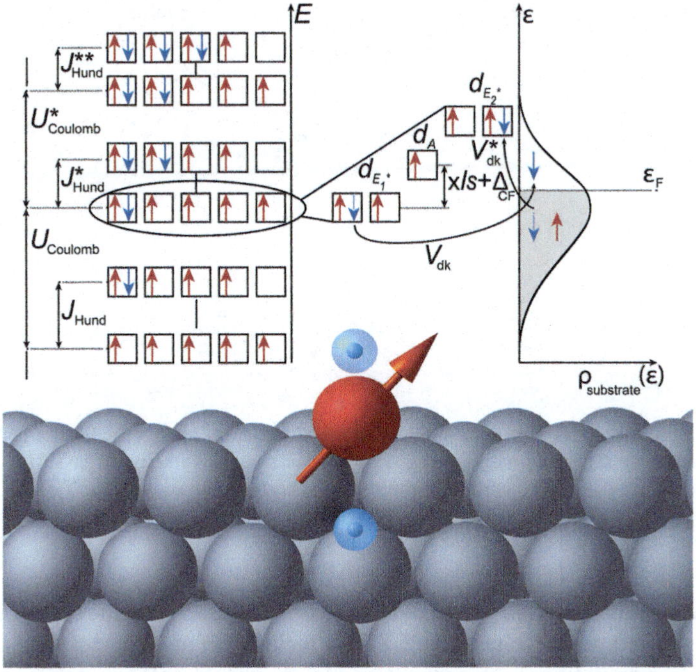

Fig. 3.8 Schematic view of a magnetic transition metal adatom (red) with additional hydrogens on a substrate. Top-left panel: The electronic spectrum with spin-up (red arrows) and spin-down (blue arrows) states of the five adatom 3*d* orbitals is filled up. The Hubbard energy U has to be paid if an additional electron is put into an orbital where there is already one, and Hund's rule energy J is paid if electron spins are flipped. Top-middle panel: Optimal electronic structure for the five orbitals in valance configurations with crystal field splitting Δ_{CF} and spin-orbit coupling ξls. Top-right panel: Hybridization parameter V_{dk} of the adatom orbitals with the substrate DOS $\rho_{\mathrm{substrate}}$ which results in the broadening of impurity states

To understand why the experimental peak 2 for Fe, the peak 3 for Co and the broad peak for Ni are quite close to E_{F}, we investigate effects of valence fluctuations and formation of the Kondo resonance in 3*d* adatoms. One can see from Fig. 3.7b that for the d_{z^2} state of Fe there is only one peak in the DOS, just above the Fermi energy, without any signature of Hubbard bands. Moreover, the occupation of the Fe 3*d* shell of about $n = 6.4$ indicate strong charge fluctuations with mixed valence behavior. The similar situation applies for the Ni impurity where QMC calculations show broad spectra with strong renormalization of the quasiparticle peak towards E_{F}. Opposite to Fe and Ni, for the Co adatom on Ag(100) our results show much smaller charge fluctuations and the formation of Hubbard bands together with a sharp resonance at E_{F} which can be related to the multi-orbital Kondo effect. This conclusion is supported by STM spectroscopy of the Kondo resonance for a Co adatom on Ag(100) [21]. Similar correlation effects can be found for adatoms on insulating surfaces [23–28] and in *f* electron systems [29, 30].

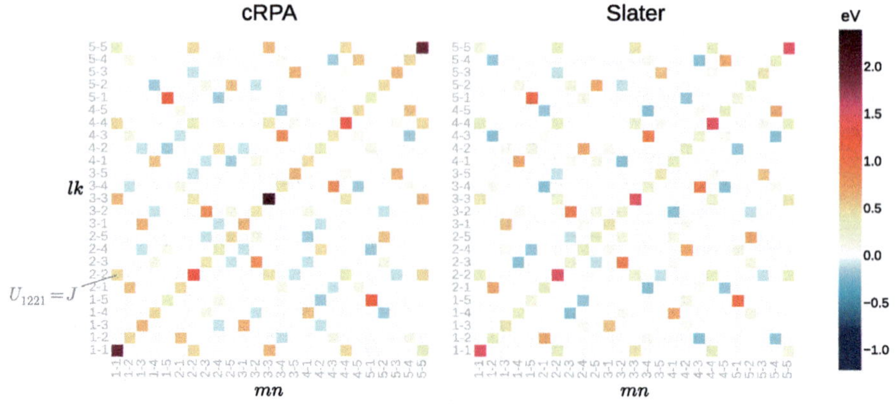

Fig. 3.9 Coulomb matrix for the d orbitals of the Co adatom obtained with the cRPA method (left) and subsequently rotationally averaged by the Slater approximation (right). The order of the orbitals is given by U_{mlkn}. The outlined element U_{1221} corresponds to the term $U_{1221}c^{+}_{1\uparrow}c^{+}_{2\downarrow}c_{2\downarrow}c_{1\uparrow}$, the index notation 1–5 refers to the orbital ordering (d_{xy}, d_{yz}, $d_{3z^2-r^2}$, d_{xz}, $d_{x^2-y^2}$)

We can now discuss the concept of Hund's impurities [31] for the case of a 3d metal adatom on different substrates (Fig. 3.8). An isolated 3d atom corresponds to integer electronic configurations with occupation of different orbitals in accordance with the first Hund's rule. The different 3d states first are filled with spin-up electrons, and finally with spin-down electrons (Fig. 3.8). The reason for such configurations with maximal total spin is related with Hund's rule exchange energy J, which prevents a spin-flip process to a non-magnetic local configuration. For the case of a 3d transition metal adatom on a metallic substrate, electrons hybridize with the bath of conduction electrons with DOS $\rho_{\text{substrate}}$. This hybridization leads to fluctuations of the charge on the adatom. The strength of hybridization V_{dk} between the 3d adatom and itinerant k bands [32], and degree of valence fluctuations [22], will define whether the electronic state of the adatom can be described by atomic multiplets, itinerant bands, or both with specific correlation effects. In the case of weak hybridization $V_{\text{dk}} \approx 0$, the adatom electronic structure can be analyzed in terms of atomic multiplets with crystal field splittings Δ_{CF} and spin-orbit coupling ξls with integer valence occupations. For small hybridization the adatom still has the integer valency, but the Coulomb correlations lead to the formation of a Kondo singlet.

For large hybridization and the case of a single magnetic orbital, the adatom spin moment would be simply quenched by the conduction electrons. Nevertheless, for the case of multi-orbital 3d adatoms with relatively strong hybridization but smaller effects of J, profound charge fluctuations can coexist with large local magnetic moments, which are strongly coupled to the substrate. This situation may be referred to the Hund's impurity regime [4, 31]. It is characterized by a complex interplay of charge fluctuations, crystal field splitting, spin-orbit coupling, and electron correlations. The investigation of this regime was a challenge for the newly developed theoretical methods which we have explained [6]. Moreover, the experimental

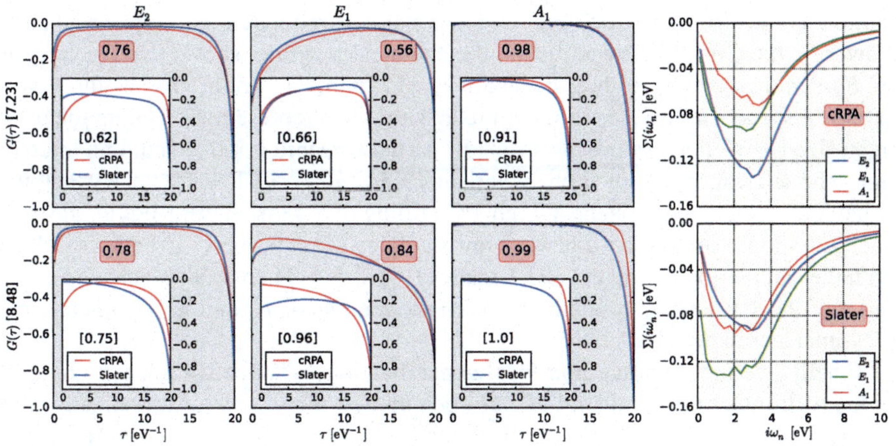

Fig. 3.10 Comparison between the representation-resolved imaginary-time Green's functions $G(\tau)$ for five d orbitals of the Co impurity calculated with the cRPA and the Slater Coulomb matrix. Total impurity occupations are given in squared brackets, orbital ones in round boxes. The upper row contains the full QMC Green's functions at lower filling together with the atomic solution in the insets (orbital occupations in brackets), and the lower row contains the corresponding Green's functions at higher filling. The filling can be adjusted by the double-counting chemical potential. The calculations were performed at $\beta = 20\,\text{eV}^{-1}$, with the Matsubara self-energy $\Sigma(i\omega_n)$ (last column) for the different symmetry representations calculated with the cRPA (upper) and the Slater Coulomb matrix (lower), both at higher co filling

realization of a Hund's impurity, and, more importantly, the full control over all the relevant parameters, i.e. magnetic anisotropy, hybridization, temperature and magnetic field, had remained incomplete so far.

An important theoretical problem for correlated adatoms on substrates is related to the realistic representation for the Coulomb interaction vertex. The effective Coulomb interaction matrix screened by conduction electrons can be calculated using the so-called cRPA method [33]. All 625 elements of the cRPA Coulomb matrix U_{mlkn} constructed in this way are shown in Fig. 3.9 for the case of the 3d orbitals of cobalt adatoms on the graphene surface. If we compare the full U-matrix with the atomic-like Slater parametrisation based on the $U - J$ average interactions, one can clearly see the lower symmetry of the Coulomb vertex obtained from cRPA calculations.

Moreover, we can still find the numerically exact solution of the multi-orbital quantum impurity problem with anisotropic hybridization functions and the full Coulomb U-matrix using the CT-QMC scheme [6]. The imaginary-time of multi-orbital Green's functions obtained by the CT-QMC impurity solver are shown in Fig. 3.10. Let us compare results obtained with the cRPA and the Slater Coulomb matrix. Rotationally averaging the Coulomb matrix by the Slater approximation slightly reduces and redistributes the overall weight of the interaction strength (Fig. 3.9). The most pronounced differences occur for the higher filling considered, $n_{\text{tot}} = 8.48$, especially in the A_1 representation. The hybridization in A_1 is small, thus the effect entirely comes from the Coulomb interaction and its reduction in the spherical case. Lowering the Co filling by adjusting the chemical potential shows that the orbitals of E_1 symmetry change their occupation, and its main weight crosses the Fermi level. This is a consequence of the orbitals within this representation being the most hybridized as well as having the strongest partial screening of the Coulomb interaction. The self-energies show a characteristic Hund's impurity low-frequency metallic behaviour. As Co on graphene at higher Co filling of 8.48 is a Fermi-liquid, the self-energies should tend to zero at very low energies. This property is better resolved with the calculations using the cRPA matrix (Fig. 3.10). There is also a change of the order of the self-energy strengths between the orbitals of E_1 and E_2 symmetry, and they intersect in the cRPA case.

Finally, we mention that non-local generalizations of effective impurity models in the path-integral formalism [34–41] open up new directions for investigations of magnetic correlations and Kondo fluctuations [42–44]. The effects of long-range interactions are very important in graphene-based systems [45–53] and can be compared with different experimental data [54–58].

Acknowledgements We would like to thank Alexey Rubtsov, Mikhail Katsnelson, Tim Wehling, Frank Lechermann, Vladimir Mazurenko, Livio Chioncel, Hartmut Hafermann, Sergey Iskakov, Evgeny Gorelov, Alexander Rudenko, Yaroslav Kvashnin, Alexander Shick, Jindřich Kolorenč, Philipp Werner and Olle Eriksson for the intense and fruitful cooperation over the years. Financial support of this work by the Deutsche Forschungsgemeinschaft through the Sonderforschungsbereich 668 (project A3) is gratefully acknowledged.

References

1. N.W. Ashcroft, N.D. Mermin *Solid-State Physics* (Holt, Rinehart and Winston, University of California 1976), pp. 1–826
2. V.I. Anisimov, F. Aryasetiawan, A.I. Lichtenstein, J. Phys. Condens. Matter **9**, 767 (1997)
3. A. Damascelli, Z. Hussain, Z.-X. Shen, Rev. Mod. Phys. **75**, 473 (2003)
4. L. de Medici, J. Mravlje, A. Georges, Phys. Rev. Lett. **107**, 256401 (2011)
5. G. Kotliar, S.Y. Savrasov, K. Haule, V.S. Oudovenko, O. Parcollet, C.A. Marianetti, Rev. Mod. Phys. **78**, 865 (2006)
6. E. Gull, A.J. Millis, A.I. Lichtenstein, A.N. Rubtsov, M. Troyer, P. Werner, Rev. Mod. Phys. **83**, 349 (2011)
7. A. Hausoel, M. Karolak, E. Şaşıoğlu, A. Lichtenstein, K. Held, A. Katanin, A. Toschi, G. Sangiovanni, Nat. Commun. **8**, 16062 (2017)
8. R. Cardias, A. Szilva, A. Bergman, I. Di Marco, M.I. Katsnelson, A.I. Lichtenstein, L. Nordström, A.B. Klautau, O. Eriksson, Y.O. Kvashnin, Sci. Rep. **7**, 4058 (2017)
9. M.I. Katsnelson, VYu. Irkhin, L. Chioncel, A.I. Lichtenstein, R.A. de Groot, Rev. Mod. Phys. **80**, 315 (2008)
10. L. Chioncel, E. Arrigoni, M.I. Katsnelson, A.I. Lichtenstein, Phys. Rev. B **79**, 235123 (2009)
11. L. Chioncel, Y. Sakuraba, E. Arrigoni, M.I. Katsnelson, M. Oogane, Y. Ando, T. Miyazaki, E. Burzo, A.I. Lichtenstein, Phys. Rev. Lett. **100**, 086402 (2008)
12. A. Melnikov, I. Razdolski, T.O. Wehling, E.T. Papaioannou, V. Roddatis, P. Fumagalli, O. Aktsipetrov, A.I. Lichtenstein, U. Bovensiepen, Phys. Rev. Lett. **107**, 076601 (2011)
13. E.Y. Vedmedenko, N. Mikuszeit, T. Stapelfeldt, R. Wieser, M. Potthoff, A. Lichtenstein, R. Wiesendanger, Eur. Phys. J. B **80**, 331 (2011)

14. P.E. Blöchl, Phys. Rev. B **50**, 17953 (1994)
15. F. Lechermann, A. Georges, A. Poteryaev, S. Biermann, M. Posternak, A. Yamasaki, O.K. Andersen, Phys. Rev. B **74**, 125120 (2005)
16. B. Amadon, F. Lechermann, A. Georges, F. Jollet, T.O. Wehling, A.I. Lichtenstein, Phys. Rev. B **77**, 205112 (2008)
17. M. Karolak, T.O. Wehling, F. Lechermann, A.I. Lichtenstein, J. Phys. Condens. Matter **23**, 085601 (2011)
18. E. Gorelov, T.O. Wehling, A.N. Rubtsov, M.I. Katsnelson, A.I. Lichtenstein, Phys. Rev. B **80**, 155132 (2009)
19. B. Surer, M. Troyer, Ph. Werner, A.M. Läuchli, T.O. Wehling, A. Wilhelm, A.I. Lichtenstein, Phys. Rev. B **85**, 085114 (2012)
20. A.C. Hewson, J. Phys. Soc. Jpn. **74**, 8 (2005)
21. N. Knorr, M.A. Schneider, L. Diekhöner, P. Wahl, K. Kern, Phys. Rev. Lett. **88**, 096804 (2002)
22. S. Gardonio, M. Karolak, T.O. Wehling, L. Petaccia, S. Lizzit, A. Goldoni, A.I. Lichtenstein, C. Carbone, Phys. Rev. Lett. **110**, 186404 (2013)
23. V.V. Mazurenko, S.N. Iskakov, A.N. Rudenko, I.V. Kashin, O.M. Sotnikov, M.V. Valentyuk, A.I. Lichtenstein, Phys. Rev. B **88**, 085112 (2013)
24. A.N. Rudenko, V.V. Mazurenko, V.I. Anisimov, A.I. Lichtenstein, Phys. Rev. B **79**, 144418 (2009)
25. A.B. Shick, F. Máca, A.I. Lichtenstein, Phys. Rev. B **79**, 172409 (2009)
26. J. Honolka, A.A. Khajetoorians, V. Sessi, T.O. Wehling, S. Stepanow, J.-L. Mi, B.B. Iversen, T. Schlenk, J. Wiebe, N. Brookes, A.I. Lichtenstein, Ph. Hofmann, K. Kern, R. Wiesendanger, Phys. Rev. Lett. **108**, 256811 (2012)
27. M. Izquierdo, M. Karolak, C. Trabant, K. Holldack, A. Föhlisch, K. Kummer, D. Prabhakaran, A.T. Boothroyd, M. Spiwek, A. Belozerov, A. Poteryaev, A. Lichtenstein, S.L. Molodtsov, Phys. Rev. B **90**, 235128 (2014)
28. M. Karolak, M. Izquierdo, S.L. Molodtsov, A.I. Lichtenstein, Phys. Rev. Lett. **115**, 046401 (2015)
29. S. Gardonio, T.O. Wehling, L. Petaccia, S. Lizzit, P. Vilmercati, A. Goldoni, M. Karolak, A.I. Lichtenstein, C. Carbone, Phys. Rev. Lett. **107**, 026801 (2011)
30. E. Gorelov, J. Kolorenč, T. Wehling, H. Hafermann, A.B. Shick, A.N. Rubtsov, A. Landa, A.K. McMahan, V.I. Anisimov, M.I. Katsnelson, A.I. Lichtenstein, Phys. Rev. B **82**, 085117 (2010)
31. A.A. Khajetoorians, M. Valentyuk, M. Steinbrecher, T. Schlenk, A. Shick, J. Kolorenč, A.I. Lichtenstein, T.O. Wehling, R. Wiesendanger, J. Wiebe, Nat. Nanotech. **10**, 958 (2015)
32. C. Carbone, M. Veronese, P. Moras, S. Gardonio, C. Grazioli, P.H. Zhou, O. Rader, A. Varykhalov, P. Gambardella, S. Lebegue, O. Eriksson, M.I. Katsnelson, A.I. Lichtenstein, Phys. Rev. Lett. **104**, 117601 (2010)
33. F. Aryasetiawan, M. Imada, A. Georges, G. Kotliar, S. Biermann, A.I. Lichtenstein, Phys. Rev. B **70**, 195104 (2004)
34. A.N. Rubtsov, M.I. Katsnelson, A.I. Lichtenstein, Phys. Rev. B **77**, 033101 (2008)
35. S. Brener, H. Hafermann, A.N. Rubtsov, M.I. Katsnelson, A.I. Lichtenstein, Phys. Rev. B **77**, 195105 (2008)
36. H. Hafermann, M.I. Katsnelson, A.I. Lichtenstein, Europhys. Lett. **85**, 37006 (2009)
37. H. Hafermann, S. Brener, A.N. Rubtsov, M.I. Katsnelson, A.I. Lichtenstein, JETP Lett. **86**, 677 (2007)
38. M. Harland, M.I. Katsnelson, A.I. Lichtenstein, Phys. Rev. B **94**, 125133 (2016)
39. A.N. Rubtsov, M.I. Katsnelson, A.I. Lichtenstein, Ann. Phys. **327**, 1320 (2012)
40. H. Hafermann, G. Li, A.N. Rubtsov, M.I. Katsnelson, A.I. Lichtenstein, H. Monien, Phys. Rev. Lett. **102**, 206401 (2009)
41. H. Hafermann, Ch. Jung, S. Brener, M.I. Katsnelson, A.N. Rubtsov, A.I. Lichtenstein, Europhys. Lett. **85**, 27007 (2009)
42. N. Néel, J. Kröger, R. Berndt, T. Wehling, A. Lichtenstein, M.I. Katsnelson, Phys. Rev. Lett. **101**, 266803 (2008)

43. N. Néel, R. Berndt, J. Kröger, T.O. Wehling, A.I. Lichtenstein, M.I. Katsnelson, Phys. Rev. Lett. **107**, 106804 (2011)
44. M. Karolak, D. Jacob, A.I. Lichtenstein, Phys. Rev. Lett. **107**, 146604 (2011)
45. T.O. Wehling, S. Yuan, A.I. Lichtenstein, A.K. Geim, M.I. Katsnelson, Phys. Rev. Lett. **105**, 056802 (2010)
46. T.O. Wehling, E. Şaşıoğlu, C. Friedrich, A.I. Lichtenstein, M.I. Katsnelson, S. Blügel, Phys. Rev. Lett. **106**, 236805 (2011)
47. T.O. Wehling, A.I. Lichtenstein, M.I. Katsnelson, Phys. Rev. B **84**, 235110 (2011)
48. B. Sachs, T.O. Wehling, M.I. Katsnelson, A.I. Lichtenstein, Phys. Rev. B **84**, 195414 (2011)
49. T.O. Wehling, M.I. Katsnelson, A.I. Lichtenstein, Chem. Phys. Lett. **476**, 125 (2009)
50. T.O. Wehling, I. Grigorenko, A.I. Lichtenstein, A.V. Balatsky, Phys. Rev. Lett. **101**, 216803 (2008)
51. T.O. Wehling, A.V. Balatsky, A.M. Tsvelik, M.I. Katsnelson, A.I. Lichtenstein, Europhys. Lett. **84**, 17003 (2008)
52. T.O. Wehling, A.V. Balatsky, M.I. Katsnelson, A.I. Lichtenstein, K. Scharnberg, R. Wiesendanger, Phys. Rev. B **75**, 125425 (2007)
53. T.O. Wehling, A.I. Lichtenstein, M.I. Katsnelson, Appl. Phys. Lett. **93**, 202110 (2008)
54. T.O. Wehling, K.S. Novoselov, S.V. Morozov, E.E. Vdovin, M.I. Katsnelson, A.K. Geim, A.I. Lichtenstein, Nano Lett. **8**, 173 (2008)
55. S.J. Altenburg, J. Kröger, T.O. Wehling, B. Sachs, A.I. Lichtenstein, R. Berndt. Phys. Rev. Lett. **108**, 206805 (2012)
56. L.V. Dzemiantsova, M. Karolak, F. Lofink, A. Kubetzka, B. Sachs, K. von Bergmann, S. Hankemeier, T.O. Wehling, R. Frömter, H.P. Oepen, A.I. Lichtenstein, R. Wiesendanger, Phys. Rev. B **84**, 205431 (2011)
57. M. Gyamfi, T. Eelbo, M. Waśniowska, T.O. Wehling, S. Forti, U. Starke, A.I. Lichtenstein, M.I. Katsnelson, R. Wiesendanger, Phys. Rev. B **85**, 161406 (2012)
58. T. Eelbo, M. Waśniowska, P. Thakur, M. Gyamfi, B. Sachs, T.O. Wehling, S. Forti, U. Starke, C. Tieg, A.I. Lichtenstein, R. Wiesendanger, Phys. Rev. Lett. **110**, 136804 (2013)

Chapter 4
Local Physical Properties of Magnetic Molecules

Alexander Schwarz

Abstract Advanced atomic force microscopy based techniques were developed to investigate local properties of individual well-separated adsorbed molecules, which can be applied to all kinds of supporting substrates independent of their conductivity. First, we find that due to the Smoluchowski effect a localized electrostatic dipole moment is present at the end of metallic tips. Since its positive pole points towards the surface, we are able to identify the chemical species in atomically resolved images on polar surfaces. We employed such tips to determine the exact adsorption geometry of single molecules on ionic bulk insulators. Moreover, we were able to detect magnitude and direction of the electrostatic dipole moment of adsorbed molecules. Secondly, if the tip is magnetic, we are even able to probe the short-range electron-mediated magnetic exchange interaction between the foremost tip apex atom and the sample atom directly below and thus established magnetic exchange force microscopy as a novel method to study magnetic sample systems with atomic resolution. By applying this new kind of magnetically sensitive force microscopy to a paramagnetic organo-metallic complex adsorbed on an antiferromagnetic bulk insulator, we find indications for a superexchange-mediated coupling between molecule and substrate.

4.1 High-Resolution Atomic Force Microscopy

Atomic force microscopy (AFM) is sensitive to all kinds of long- and short-ranged electromagnetic interactions between a sample surface and a sharp tip that is attached to a flexible cantilever. These are, e.g., van der Waals interactions, magnetostatic interactions, electrostatic interactions, magnetic exchange interactions, chemical interactions, Pauli repulsion, elastic interactions, etc. Various modes of operation exist to sense these interactions by measuring the mechanical response of the cantilever, e.g., by detecting the static deflection, or, for an oscillating cantilever, the change of its

A. Schwarz (✉)
Department of Physics, University of Hamburg, Jungiusstr. 11,
20355 Hamburg, Germany
e-mail: aschwarz@physnet.uni-amburg.de

R. Wiesendanger (ed.), *Atomic- and Nanoscale Magnetism*, NanoScience and Technology, https://doi.org/10.1007/978-3-319-99558-8_4

amplitude, frequency or phase [1]. As a near field technique, tip-sample distance and effective size of the probing tip limit the lateral resolution.[1] Thus, it is easy to imagine that with an atomically sharp tip, atomic resolution can be obtained. However, a tip apex only stays atomically sharp, as long as strong repulsive forces do not rearrange the foremost atoms close to the sample surface. Therefore, operation at small tip-sample separations while still being in the non-contact or weak repulsive regime is mandatory.[2] Since for a given spring constant k the minimal detectable force is fundamentally limited by the temperature T (it scales with $\sqrt{T}$ [2]), performing experiments at low temperatures provides higher force sensitivity. Moreover, imaging conditions are generally more stable than at room temperature, because all thermally activated processes, e.g., spontaneous rearrangements of atoms at the tip apex or thermal drift of the relative position between tip and sample, are inhibited.

The following experiments were performed with home-built microscopes, operated in dedicated ultra-high vacuum (UHV) compatible cryostat systems that are equipped with superconducting magnets for field-dependent studies [3–5]. In addition to the lower thermal noise of the cantilever, all molecules (as well as single adatoms) become immobile at low temperatures and hence do not aggregate—a prerequisite to probe local properties of individual well-separated molecules and atoms.

Figure 4.1 shows (a) a cryostat with ^{3}He insert, (b) the microscope body, (c) atomically resolved NaCl(001), and (d) a Mn monolayer pseudomorphically grown on W(110) with single well-separated Co adatoms and one CO molecule. The image in (c) was recorded at a temperature of 548 mK. Note that the ^{3}He stage alone achieves a minimal temperature of about 300 mK. The higher temperature during measurements stems from the additional thermal load due to light absorption of the interferometric detection system inside the cryostat. To prepare single well-separated adatoms on Mn/W(110) as imaged in Fig. 4.1d, the substrate was loaded into the microscope and precooled to about 5 K with the insert located inside the cryostat. Thereafter, the insert was moved downwards into the chamber below the cryostat. From an evaporator unit attached to the cryostat chamber a small amount of Co atoms was thermally evaporated onto the substrate. During this procedure, the substrate temperature rises, but stays below 30 K, which is cold enough to prevent aggregation. The CO molecules originate from the residual gas. Their peculiar donut-shape is explained in Sect. 4.2.

For high resolution imaging in the non-contact regime, cf. Fig. 4.1c, d, the frequency modulation technique is employed [2]. In this mode of operation, the cantilever always oscillates self-excited at a constant amplitude A_0 with its resonance frequency f. This frequency deviates from the unperturbed eigenfrequency f_0 of the cantilever in the presence of a tip-sample interaction F_{ts} by $\Delta f = f - f_0$. For negative frequency shifts $\Delta f < 0$ the total tip-sample interaction is attractive. If Δf

---

[1]The effective size of a tip is always related but not necessarily identical to its geometric size. For example, the lateral resolution of magnetic force microscopy (MFM) is related to the spreading of the magnetic field that emanates from the tip, which is much larger than its geometric size. However, sharp tips generally produce a more localized field.

[2]True atomic resolution in the non-contact mode or in the weakly repulsive regime allows to detect point defects. AFM in contact mode only resolves the periodicity of a surface, because many atoms are in contact with the surface.

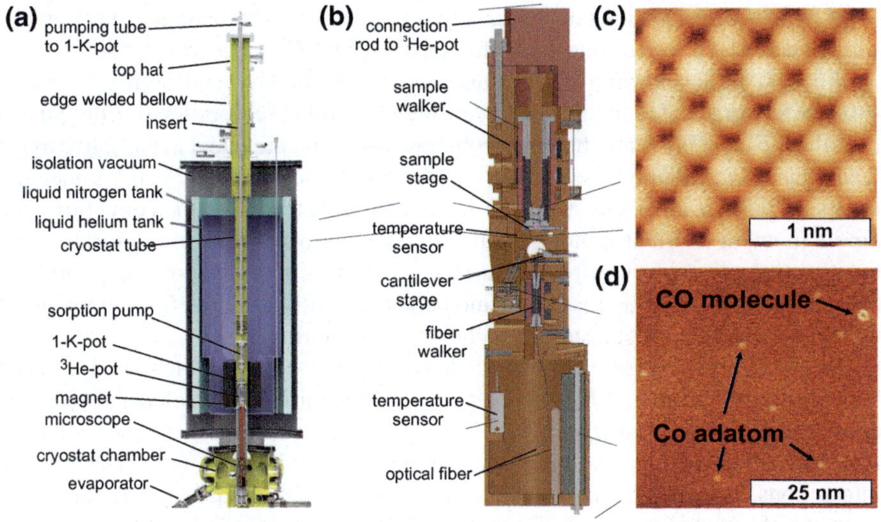

Fig. 4.1 **a** Cryostat with a movable ^{3}He evaporation insert. **b** Microscopy body that is attached to the ^{3}He pot. In-situ tip and sample exchange is possible by lowering the insert into the chamber located below the cryostat. **c** Atomically resolved NaCl(001) recorded at 548 mK. **d** Co adatoms and one CO molecule on a Mn monolayer on W(110)

is kept constant (constant Δf imaging), the surface topography $z(x, y)$ is obtained. Meanwhile, the tip apex stays in the so-called non-contact regime, implying that even in the lower turnaround point of each oscillation cycle the tip does not touch the sample surface. On the atomic or molecular scale the z-corrugation reflects the magnitude of the short-range forces between the foremost tip apex atoms and the surface atoms underneath. Its magnitude is distance- as well as tip-dependent. If the z-feedback is not active during atomic resolution imaging (constant height imaging) the variation $\Delta f(x, y)$ is recorded instead. In this imaging mode, atomic resolution is also possible in the weak repulsive regime, where the foremost tip apex atom slightly contacts the sample surface at the lower turnaround point of each oscillating cycle but the tip apex remains structurally stable.[3] The distance dependence of the tip-sample interaction can be obtained quantitatively in the so-called spectroscopy mode, by recording a $\Delta f(z)$-curve at a predefined (x, y)-position. With our typical amplitudes of $A \approx 1$ nm we usually operate in the so-called large amplitude regime, i.e., A is larger than the characteristic decay length of the tip-sample interaction, which is on the order of 0.1 nm for electron-mediated interactions. Therefore, we cannot simply use the small amplitude approximation $\Delta f(z) = -f_0/2k \cdot \partial F_{\text{ts}}/\partial z$ to calculate the tip-sample force gradient $\partial F_{\text{ts}}(z)/\partial z$ and obtain the tip-sample force $F_{\text{ts}}(z)$ and the tip-sample interaction energy $E_{\text{ts}}(z)$ by subsequent integration. However,

[3]The repulsive regime is not accessible in a controllable manner employing constant Δf imaging, because the working curve of the z-feedback loop, i.e., $\Delta f(z)$, is not monotonous, but follows the general form of a Lennard–Jones potential.

employing well-established numerical algorithms [6–9] experimental $\Delta f(z)$-curves can anyway be converted into $F_{ts}(z)$ force curves and $E_{ts}(z)$ energy curves.

Silicon cantilevers with eigenfrequencies $f_0 \approx 180\,\text{kHz}$ and spring constants $\approx 160\,\text{N/m}$ were coated in-situ with a few nm of titanium, chromium or iron. Such a metallic top layer prevents charging problems, which are prevalent in the native oxide layer that covers as-purchased silicon cantilevers. As described in Sect. 4.2, metallic tip apices possess an electrostatic dipole moment with a well-defined direction: The dipole moment is oriented along the tip axis, i.e., perpendicular to the surface, and the positive pole points towards the surface.[4] Iron and chromium coated tips are also magnetically sensitive and, as we demonstrate in Sect. 4.3, allow for a mapping of the magnetic structure of a surface with atomic resolution.

To characterize AFM tips and to determine the contact potential difference between tip and sample, $\Delta f(U)$-curves should be recorded. The voltage U_{CPD} at the apex of the parabola corresponds to the average contact potential difference, which is then applied between tip and sample to minimize disturbing long-range electrostatic interactions. Jumps in the curves indicate the presence of localized states at the tip apex, which are charged and discharged, depending on the applied voltage [10]. Such imperfections could stem, e.g., from tip-crashes. To further characterize AFM tips, the dissipated energy should be analyzed, which can be calculated from the excitation amplitude a_{exc} [11], that is required to keep A_0 constant. Apart from intrinsic dissipation, non-conservative tip-sample interactions cause energy dissipation [12]. If the dissipated energy varies on the atomic scale, adhesion hysteresis is the most likely cause [13] and hints towards a structurally unstable tip apex. As we could show, adhesion hysteresis can also be spin-dependent [14]. If it varies on the atomic scale, distance dependent $\Delta f(z)$ spectroscopy data recorded with such structurally unstable tips have to be analyzed very cautiously [15].

4.2 Utilizing the Smoluchowski Effect to Probe Surface Charges and Dipole Moments of Molecules with Metallic Tips

Functionalized tips are very beneficial to disentangle a specific tip-sample interaction from the plethora of all interactions that are present. For example magnetic coatings are used to sense magnetostatic forces or magnetic exchange forces [16]. Recently, non-reactive tips were prepared by functionalizing them with noble gas atoms like xenon or inert molecules like carbon monoxide. Due to their inertness, no chemical bonds are formed between tip apex and sample surface. Therefore, such tips are able to obtain intramolecular atomic resolution by only probing the short-range Pauli

[4]The situation can be more complex, if the tip apex atoms are not close-packed (usually a justified assumption for stable tip apices) or if more than one chemical species are present at the tip apex.

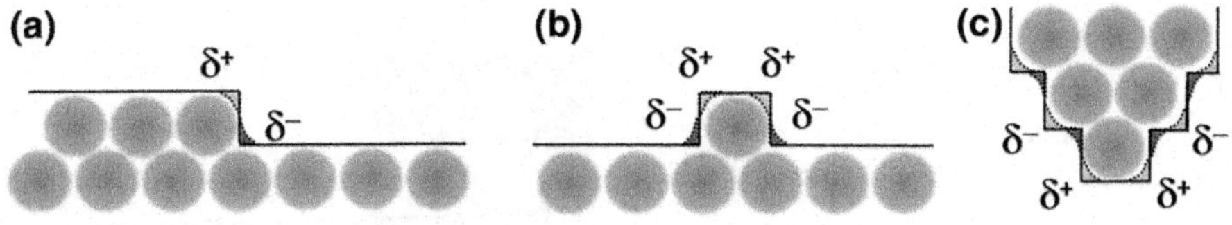

Fig. 4.2 Development of a localized electrostatic dipole moment at **a** a step edge, **b** a single adatom on a surface, and **c** a close-packed pyramidal nanotip

repulsion between tip apex and sample surface [17].[5] Here, we show that metallic tips, due to the Smoluchowski effect [18], exhibit an electrostatic dipole moment that is oriented with its positive pole towards the surface. On polar surfaces it can be utilized to identify the chemical species that exhibit an attractive and repulsive electrostatic interaction with the tip, respectively [10]. As we will show in Sect. 4.4, this feature is very useful to determine the exact adsorption geometry of a molecule on ionic surfaces [19, 20]. Moreover, it is possible to probe the dipole moment of adsorbed molecules with such tips [21, 22].

We start with explaining the Smoluchowski effect at step edges. In a simplified picture the sudden change of the potential at a step edge cannot be perfectly screened by the electrons. Their density changes more smoothly, as depicted in Fig. 4.2a. As a result, electrons accumulate on the lower terrace, while they are depleted on the upper terrace. This charge rearrangement can be represented by an electrostatic dipole moment p at the step edge, which is oriented normal to the surface and points with its positive pole away from the surface. The same phenomenon occurs for adatoms and at the end of atomically sharp tips, as sketched in Fig. 4.2b, c, respectively.

Density functional theory (DFT) based calculations support this qualitative picture. For pyramidal close-packed Cr nano-tips this is shown in Fig. 4.3a. Each layer of the pyramid contributes about 1 Debye (D) to the total electrostatic dipole moment. Its magnitude can be several Debye and thus be quite large. However, as expected, it saturates for large pyramid heights. In (b) the calculated electrostatic potential emanating from a three-layer pyramid is displayed. The plot in (c) demonstrates that the electrostatic potential calculated from DFT (red) can be very well represented by a classical dipole moment. This general behavior is very helpful to interpret qualitatively atomically resolved images on polar surfaces.

Since on close-packed chemically homogeneous metallic tips the dipole moment always points with its positive pole towards the sample, such tips can be used on ionic surfaces to identify the anion and cation sublattices, which exhibit either an attractive or a repulsive electrostatic interaction with the tip [10]. In Sect. 4.4 we use this feature to determine the adsorption geometry of Co-Salen on NaCl(001) and NiO(001). Moreover, we could show that atomically resolved data obtained

[5] Apparently, the formation of a chemical bond disturbs the mechanism that leads to intramolecular atomic resolution, because reactive tips, e.g., pure metal tips, are seemingly unable to achieve a similar resolution. If approached too close, the strong chemical interaction probably induces tip changes or sample modifications.

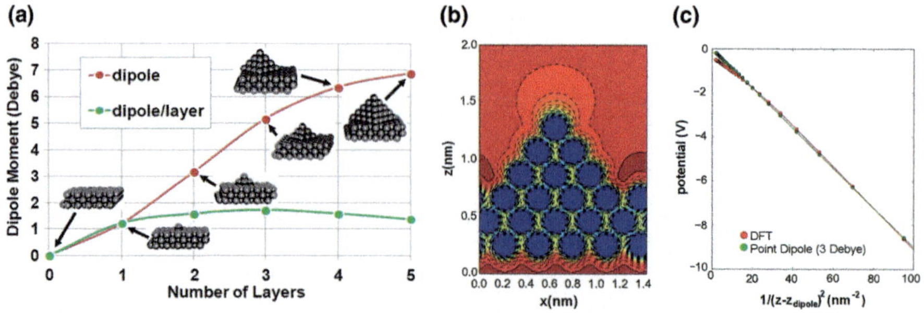

Fig. 4.3 The electrostatic dipole moment of close-packed Cr nanotips. **a** Size dependence of the magnitude of the dipole moment obtained from first principles using density functional theory. **b** Electrostatic potential emanating from the three layer pyramid displayed in **a**. Comparison of the calculated distance dependence of the electrostatic potential of the three-layer pyramid and a simulation using a simple dipole of 3 D located at the foremost tip apex atom

experimentally on ionic surfaces can be used to quantify and calibrate the dipole moment at the tip apex [21].

The procedure is shown in Fig. 4.4 for a metallic tip. In (a) an atomically resolved image of NiO(001) obtained with a metallitzed tip is show. The same tip is used to scan the larger scale image in (b) with a low coverage of CO molecules. They exhibit a characteristic donutr shape. As demonstrated in (c) the atomic corrugation can be well simulated by a dipole of 7 D. Note that the experimental curve is the average of 23 profiles taken across maxima and minima of image (a) along the [100]-direction, representing the oxygen and nickel sites, respectively. The simulated curve was obtained using the virtual AFM code (vAFM) [23], which includes the whole data acquisition process and thus exhibits noise as the experimental data does. We also consider the long-range van der Waals interaction, which we obtained from a fit to an experimental $\Delta f(z)$-curve recorded with the same tip up to 10 nm away from the surface. Using the same dipole moment of 7 D, we could also fit the line profile across the CO molecules as demonstrated in Fig. 4.4d. The characteristic line shape, indicating a locally repulsive interaction directly above the CO molecule and an attractive interaction further outside, is very well reproduced.

CO adsorbs on top of the surface oxygen atoms and possesses a dipole moment of 0.4 D that points with its positive pole towards the tip. For large tip-sample distances the attractive van der Waals interaction dominates [22]. However, at smaller separations as in (a) the interaction between the positive poles of the metallic tip and the CO molecule results in a repulsion on top of the molecule. Due to the angular dependence of the dipole–dipole interaction, i.e., $E_{\text{dip-dip}} = \frac{p_{\text{CO}} \cdot p_{\text{tip}}}{4\pi\epsilon_0} \frac{(1-3cos^2(\theta))}{r^3}$, with θ being the angle of the connecting line between two collinear dipole moments $\boldsymbol{p}_{\text{CO}}$ and $\boldsymbol{p}_{\text{tip}}$, respectively, an attractive ring is present around the center resulting in the characteristic donut shape visible in (b). This situation is displayed in the insets in Fig. 4.4d. Note that in the most general case of non-collinear and non-coplanar dipole moments their tilting angles ϕ_1 and ϕ_2 relative to the connecting line as well

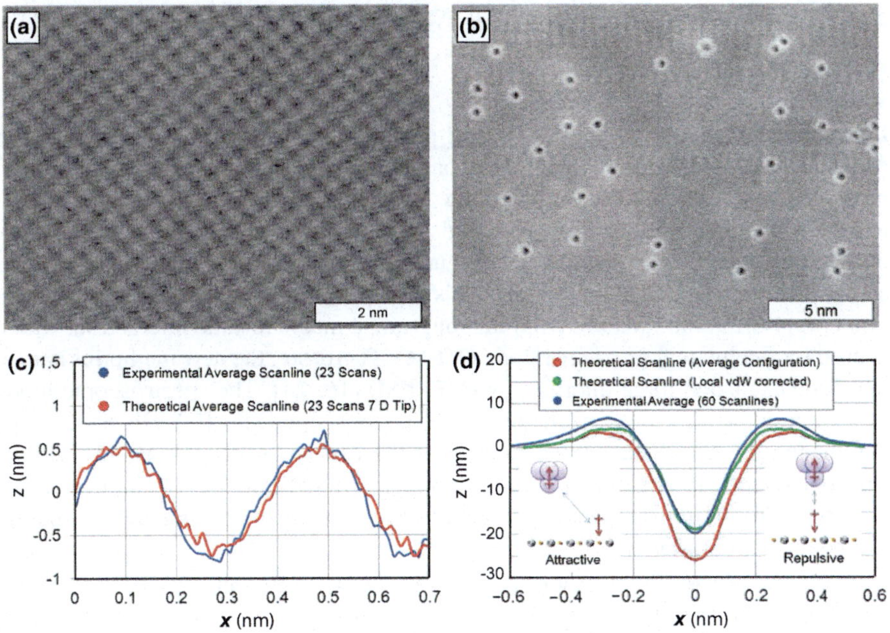

Fig. 4.4 **a** Atomically resolved image of NiO(001) obtained with a metalized tip. **b** Large scale overview image recorded with the same tip showing CO molecules with their characteristic donut-shaped appearance. **c** Averaged scan line (red) obtained from image **a** along the [100]-direction. Blue shows the simulated data using a tip dipole moment of 7 D. **d** Profile across a CO molecule with blue being the average of 60 experimental scan lines, while red is a simulated scan line using the same dipole moment of 7 D as in **b**. The green curve additionally considers local van der Waals interactions, resulting in an even better agreement

as their relative rotational angle ϕ have to be considered. The average of scan lines across 60 different CO molecules is displayed as blue curve in (d). The red curve was obtained by assuming the same dipole moment of 7 D, which was determined from the simulation in (c). We obtained an even better agreement (green curve) by taking the local van der Waals interaction between the CO molecule and the metallic tip apex into account.

Our results demonstrate that the well-defined direction of the dipole moment of an atomically sharp metallic tip is very beneficial to determine the sign of localized charges at the surface. On polar surfaces, this feature allows ion identification and on molecules the presence, direction, and even the magnitude of an electrostatic dipole moment can be determined reliably.

4.3 Magnetic Exchange Force Microscopy and Spectroscopy

In a somewhat simplified but very useful picture, achieving atomic resolution with force microscopy requires tip-sample separations, for which the orbital of the foremost tip apex atom and the orbital of the surface atom underneath overlap significantly. Only then the electron mediated short-range chemical interaction can be detected. If both, tip and sample, are magnetic, the spins of the interacting electrons also contribute to the total interaction. This short-ranged magnetic exchange interaction is very different in nature from the long-range magnetostatic interaction, which is used for magnetic force microscopy (MFM). Therefore, the new method is named magnetic exchange force microscopy (MExFM) [16, 24]. The spectroscopic mode of operation, i.e., magnetic exchange force spectroscopy (MExFS), allows to quantify magnitude and distance dependence of the exchange interaction [15]. Unlike spin-polarized scanning tunneling microscopy (SP-STM), this new technique can be applied to insulating samples as well. Moreover, chemical and magnetic structure can be readily correlated.

Figure 4.5 shows three examples, i.e., (a) the antiferromagnetic bulk insulator NiO(001) [24], (b) the itinerant antiferromagnetic Fe monolayer on W(001) [25] and (c) the nanoskyrmion lattice on the Fe monolayer on Ir(111) [26]. For all three cases, the AFM images display the atomic structure alone, while the MExFM images contain both, the magnetic information as well as the structural (and chemical) information. This is evident in the corresponding two-dimensional Fourier transformation (2D-FT), in which all periodicities are revealed that are above the noise level. In all three examples, peaks that belong to the magnetic as well as to the structural unit cell are present. Thus, it is easily possible to correlate atomic and magnetic structure with each other.

Without a magnetically sensitive tip the atomic scale pattern on NiO(001) reflects the quadratic arrangement of nickel and oxygen atoms on the surface visible in the second row of Fig. 4.5a. Since the tip was metallized, the protrusion and depressions can be identified as oxygen and nickel ions, respectively; cf. Sect. 4.2. With a magnetic tip, all protrusions, i.e., oxygen atoms, still have the same corrugation amplitude. On the other hand, the corrugation amplitude of the depressions, representing the nickel atoms at which the spins are localized, exhibit a row-wise modulation that reflects the antiferromagnetic order at the surface. The chemical corrugation amplitude in the MExFM image is about 4 pm with an additional magnetic corrugation amplitude of 1.5 pm between neighboring nickel rows. In the 2D-FT the four outer spots and the two inner spots represent the (1×1) surface unit cell and the (2×1) magnetic unit cell, respectively. Note that the MExFM image is unit-cell averaged to enhance the small magnetic contrast. The small magnetic corrugation is related to the strong localization of the spin-carrying d-electrons, which do not reach far into the vacuum. Therefore, the overlap with the spin carrying d-states of the magnetic tip is small.

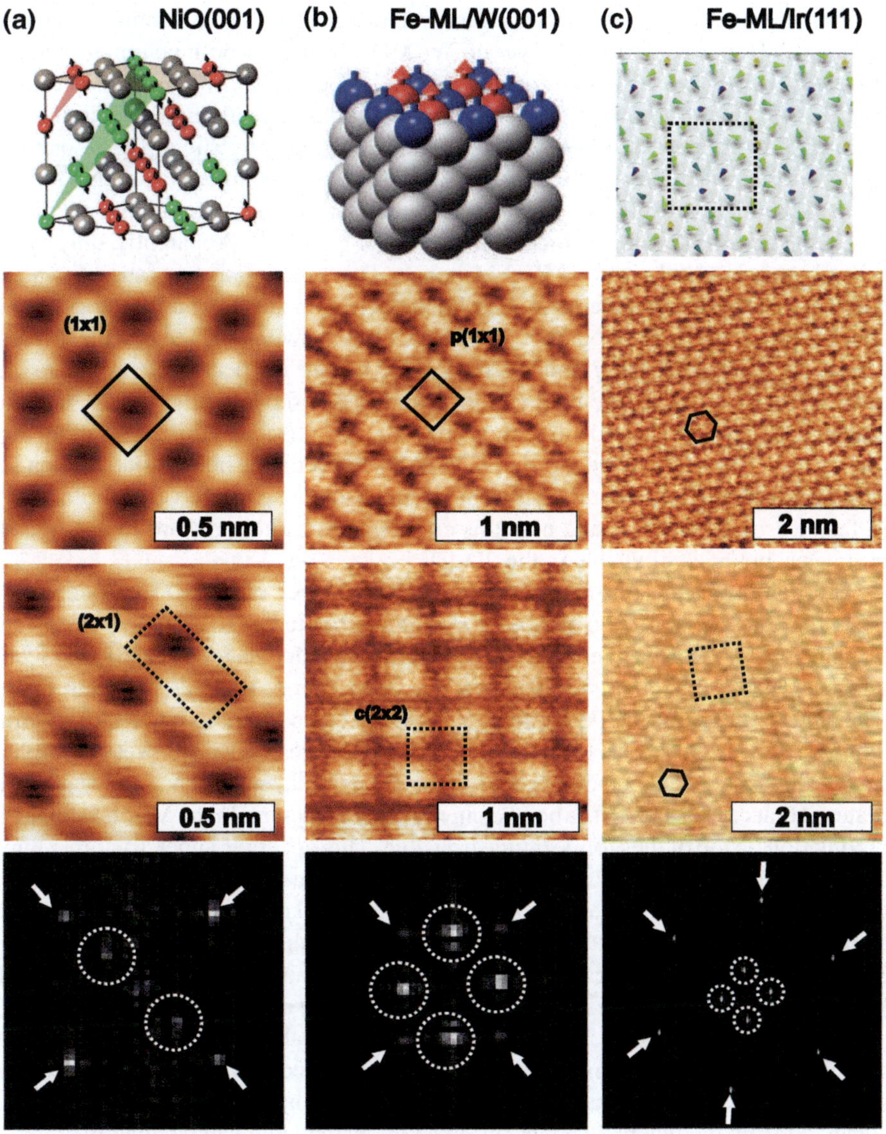

Fig. 4.5 Magnetic structure (first row), AFM data (second row), MExFM data (third row), and corresponding two-dimensional Fourier transformation (2D-FT) of the MExFM data (forth row) of **a** the row-wise antiferromagnetic order of the bulk insulator NiO(001), **b** the itinerant Fe monolayer on W(001) with its antiferromagnetic checkerboard structure, and **c** the Fe monolayer on Ir(111) with its complex non-collinear skyrmionic spin texture. The structural and magnetic unit cells are outlined in the images and the corresponding peaks representing the atomic (arrows) and magnetic (circles) periodicities are indicated in the 2D-FT

Iron is eponymous to describe ferromagnetic ordering of atomic magnetic moments.[6] However, as a monolayer on W(001) strong spin-orbit coupling with the substrates drives it into an antiferromagnetic checkerboard configuration sketched in the upper row of Fig. 4.5b [27]. While with a non-magnetic tip every iron atom is imaged as protrusion, a magnetic tip reveals the characteristic $c(2 \times 2)$ pattern. The typical corrugation amplitude of about 10 pm is much larger than on NiO(001), because the itinerant spin carrying d-states reach far into the vacuum and are therefore better accessible than the localized d-states of NiO(001). Apparently, only every second atom is visible in the MExFM image. This is surprising because chemical and magnetic exchange interaction are both electron mediated and thus appear in a similar distance regime. Note that it is impossible to switch off the chemical interaction, so it should be there. Indeed, the 2D-FT reveals the peaks that are related to the atomic $p(1 \times 1)$ surface unit cell, though they are significantly weaker than the magnetic peaks. DFT based simulations explain this phenomenum. They show that for a certain distance regime the chemical interactions on the hollow sites and on Fe atoms with parallel orientation between tip and sample spins are very similar [25]. In fact, a cross over exists and near this crossover both sites are hard to distinguish in images, but are still visible in the 2D-FT.

The non-collinear spin texture of the Fe monolayer on Ir(111) is more complex than the previous two collinear magnetic structures. It originates from the Dzyaloshinskii–Moriya interaction as well as higher order magnetic exchange mechanisms, which are, for this system, on the same energy scale as the Heisenberg exchange [29]. As for the Fe monolayer on W(001) a normal tip images every iron atom as protrusion; cf. second row in Fig. 4.5c. Interestingly, the pseudomorphic growth induces a hexagonal structure in the Fe monolayer, which is unusual for a material that crystallizes in the body centered cubic structure. In the MExFM image a modulation on top of the atomic lattice is visible. This quadratic superstructure represents a skyrmionic spin texture. The diagonal of the square lattice with a periodicity of about 1 nm is oriented along one of the close-packed directions. Hence, three energetically equivalent rotational domains exist. Peculiarly, the magnetic structure is incommensurate, which can be straightforwardly obtained by analyzing the 2D-FT [26, 29]. Hence, the spin texture is decoupled from the atomic positions [28, 29].

These three examples demonstrate the imaging capabilities of MExFM and that chemical and magnetic signals are detected simultaneously. In the following, we show how the chemical interaction (and all other non-magnetic interactions) can be separated from the magnetic contribution and how the latter can be quantified.

To quantify the distance dependence of the magnetic contribution to the total interaction for parallel and antiparallel tip-sample spins, we recorded $\Delta f(z)$ on Fe atoms with opposite spins [15]. The corresponding sites can be easily identified on images as displayed in Fig. 4.5b. Figure 4.6a shows two such curves $\Delta f_{\mathrm{min}}(z)$ (blue) and $\Delta f_{\mathrm{max}}(z)$ (red). The difference $\Delta f_{\mathrm{ex}}(z) = \Delta f_{\mathrm{max}}(z) - \Delta f_{\mathrm{min}}(z)$ (black) is displayed as well. Since all other non-magnetic long- and short-ranged contributions to the total interaction are identical, they cancel each other out. In (b) we converted

[6]Ferrum in Latin.

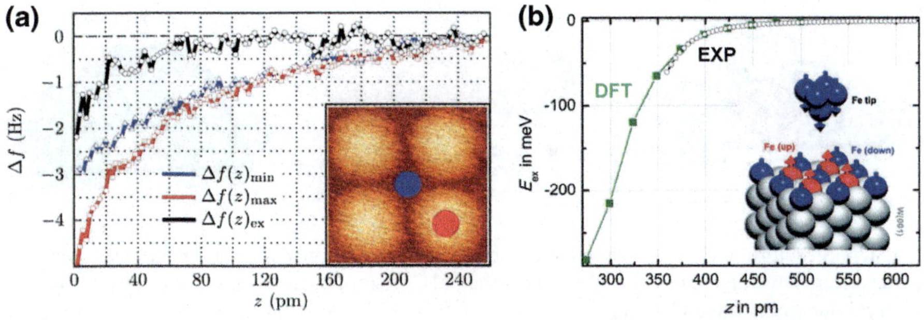

Fig. 4.6 Magnetic exchange force spectroscopy data on the Fe-monolayer on W(001). **a** Experimental $\Delta f(z)$ curves recorded on Fe atoms with opposite spin directions (blue and red, see inset). Both curves reflect the total tip-sample interaction. The difference (black) reflects the magnetic contribution $\Delta f_{\mathrm{ex}}(z)$ alone. **b** Magnetic exchange energy (black), obtained after converting the $\Delta f_{\mathrm{ex}}(z)$ curve into the magnetic exchange interaction $E_{\mathrm{ex}}(z)$, and the corresponding calculated curve (green) using density functional theory (DFT), respectively. The agreement is excellent

the experimental $\Delta f_{\mathrm{ex}}(z)$-curve into the exchange interaction energy $E_{\mathrm{ex}}(z)$ (black). The green data points represent the result of a DFT calculation using a close-packed tip pyramid made of 14 Fe atoms. The procedure to obtain the theoretical curve was analogous to the experimental approach: Between 500 and 300 pm away from the surface the tip-sample distance was reduced in steps of 25 pm. For each distance, the tip-sample system was fully relaxed and the interaction energies were calculated above Fe atoms of opposite spins. From the difference the theoretical $E_{\mathrm{ex}}(z)$ was obtained. Since the absolute distance is only known in the theory, we shifted the experimental curve to obtain the best overlap. The agreement regarding the magnitude of the interaction energy and its distance dependence is excellent. Note that we did not record experimental data for very small tip-sample separations to avoid tip changes. Knowing the distance dependence and magnitude of the magnetic exchange interaction enables us to use this interaction in a controlled manner, e.g., to switch between imaging and manipulation on a sample under investigation.

4.4 Adsorption Geometry of Co-Salen

An individual molecule in its adsorbed state will generally possess different properties compared to a free molecule it the gas phase. How these properties are modified will depend critically on how exactly the molecule is adsorbed to the surface. For example, the spin of a paramagnetic molecule could become aligned via interaction with a magnetic surface. The details of the magnetic coupling will certainly depend on the adsorption site and the adsorption geometry, cf. Sect. 4.5. Local probe techniques are ideal tools to determine the exact adsorption geometry of individual molecules with atomic resolution and atomic force microscopy is the only method that is able to do that on insulating surfaces.

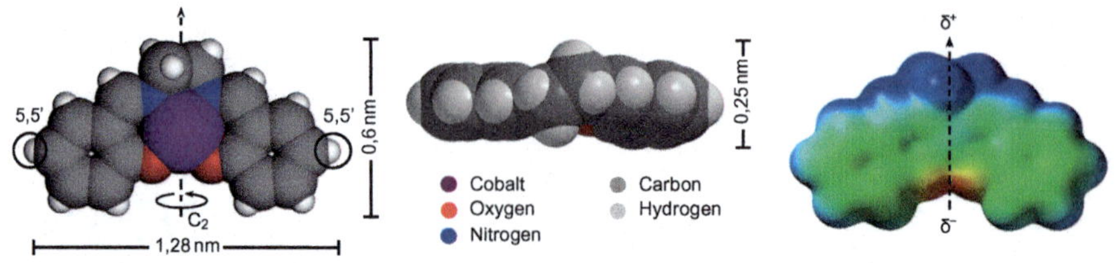

Fig. 4.7 Structure of Co-Salen in top and side view and the charge distribution with the orientation of the electrostatic dipole moment along the principal C_2 axis of the molecule. Note that the molecule is chiral due to the tilted C_2H_4 bridge

To demonstrate how the adsorption geometry of a molecule can be determined, we selected Co-Salen, a paramagnetic $S = 1/2$ organo-metallic complex. Its structure is depicted in Fig. 4.7. Two O- and two N-atoms surround the central Co atom in a planar-quadratic arrangement. Two six-membered carbon rings are attached to the side and the C_2H_4-bridge resulting in a banana-shape appearance. The molecule possesses an electrostatic dipole moment of 6D oriented along the principal C_2 axis. As substrates for our studies we selected bulk single crystals of NaCl(001), the prototypical wide band gap ionic insulator, and bulk single crystals of NiO(001), an isostructural but antiferromagnetic insulator.

To investigate the growth in the monolayer regime we deposited Co-Salen onto a substrate kept at room temperature. On NaCl(001), for example, Co-Salen exhibits a peculiar bimodal growth [30]. The island growth indicates that the molecule-molecule interaction is stronger than the molecule substrate interaction. On NiO(001) we find a layer-by-layer growth [20], which suggests a stronger molecule-substrate interaction than on NaCl(001). In the following, we concentrate on the adsorption of single well-separated molecules. To prepare such samples, we designed a dedicated miniaturized crucible that fits into the cantilever stage of our microscope [31]. In this manner we could deposit molecules onto a cold substrate ($T_{\text{substrate}} \approx 30\,\text{K}$), which immobilizes the molecules and thus prevents aggregation. Thereafter, we performed the experiments at 8 K, the base temperature of the microscope used.

Figures 4.8 and 4.9 summarize our experimental findings and the corresponding DFT calculations for both substrates. In the overview images, i.e., Figs. 4.8a and 4.9a, respectively, the banana-shaped molecules can be easily identified. To determine the orientation of their principal C_2 axis relative to the crystallographic directions of the substrates, we recorded atomically resolved images of the bare substrate and retrieved the orientation of the substrate with respect to the x- and y-scan directions. Combining this information, we find on NaCl(001) that the molecular axis is either rotated $\pm 5°$ away from the $\langle 110 \rangle$ directions (71%) or $\pm 5°$ away from the $\langle 100 \rangle$ directions (29%). DFT calculations that include van der Waals interactions reproduce the two experimentally observed orientations, cf. Fig. 4.8c, d. They, also explain their unequal distribution: For the $\langle 110 \rangle$ orientation the calculated adsorption energy (0.68 eV) is larger than for the $\langle 100 \rangle$ orientation (0.60 eV), meaning that the former is only a metastable configuration [19]. Interestingly, on NiO(001) only orientations

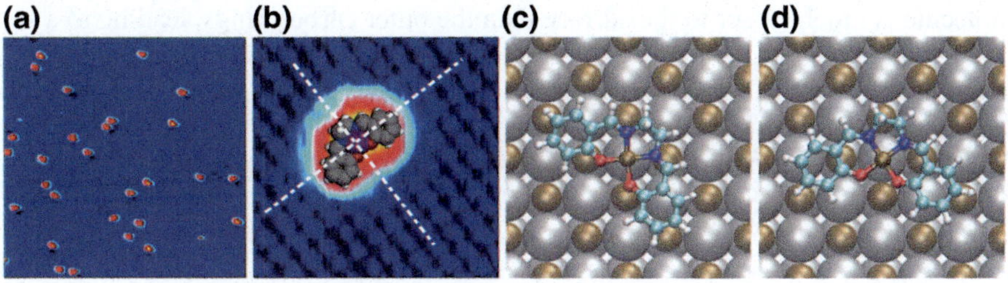

Fig. 4.8 **a** Overview image of randomly distributed Co-Salen molecules on NaCl(001). **b** Atomic resolution image of NaCl(001) together with a single molecule. **c** and **d** Calculated adsorption geometries of Co-Salen on NaCl(001). In agreement with the experimental data the principal C_2 axis is tilted away by $\pm 5°$ either from the $\langle 110 \rangle$ directions (**c**) or the $\langle 100 \rangle$ directions (**d**)

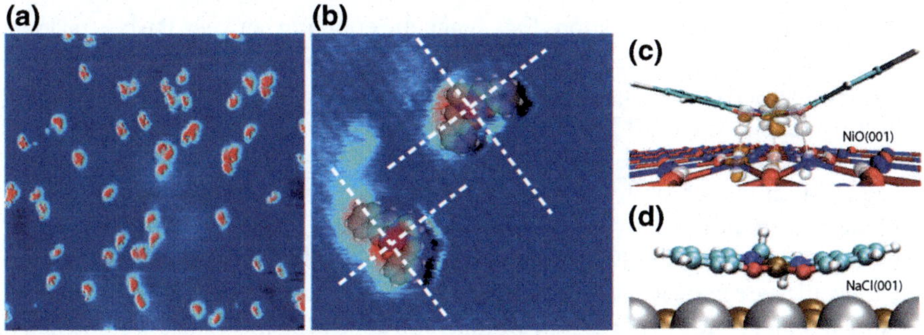

Fig. 4.9 **a** Overview image of randomly distributed Co-Salen molecules on NiO(001). **b** Atomic resolution image of NiO(001) together with a single molecule. **c** Visualization of the three covalent bonds between the central part of the molecule and the NiO(001) surface, which leads to an up-bending of the outer carbon rings. **d** For comparison: On NaCl(001) the mostly electrostatic attraction is much weaker. Thus, the molecule remains more planar

rotated about $\pm 5°$ away from the $\langle 110 \rangle$ directions are present. DFT calculations corroborate this finding and certify that no other metastable orientation exists on this substrate. The adsorption energy (1.31 eV) is much larger than on NaCl(001), which also agrees with the different growth mode observed for larger coverage [20].

To determine the adsorption site we imaged individual molecules while obtaining atomic resolution on the substrate as shown in Fig. 4.8b on NaCl(001) and in Fig. 4.9b on NiO(001). Since we used metal-coated tips the maxima in these constant Δf images can be identified with the anion sublattice, as explained in Sect. 4.2. In agreement with the DFT results on both substrates, the central Co atom adsorbs on the anion, i.e., chlorine and oxygen, respectively. However, as mentioned above the adsorption energy on NiO(001) is about twice as large as on NaCl(001). While on NaCl(001) the binding is mostly of electrostatic origin, on NiO(001) the Co-Salen forms three covalent bonds with the substrate: surface oxygen with the central Co atom and the two molecular oxygen atoms with the corresponding surface nickel atoms below, cf. Fig. 4.9c. The calculations also show that the central part of the molecule is much closer to the surface than the outer carbon rings, leading to a significant distortion of the molecules on NiO(001). For comparison, the more planar configuration on NaCl(l001) is displayed in Fig. 4.9c. The up-bending of the outer carbon rings is much less pronounced.

As we will show in the following final section, the adsorption geometry of a magnetic molecule and the type of bonding is crucial to understand a possible magnetic coupling between molecule and substrate.

4.5 Evidence for a Magnetic Coupling Between Co-Salen and NiO(001)

On NiO(001) we found indications for a magnetic coupling between the central metal atom of the Co-Salen molecule and the antiferromagnetically ordered nickel ions in the subsurface layer via superexchange across surface oxygen atoms. Figure 4.10a shows atomically resolved NiO(001) with ten Co-Salen molecules, imaged with a chromium coated tip. Note that no magnetic resolution as shown in Sect. 4.3 (a) is achieved on the substrate. In agreement with our findings discussed in Sect. 4.4 Co-Salen molecules adsorb with their central Co atom on protrusions, i.e., on oxygen, and are oriented ±5° away from ⟨110⟩-directions. In (b) line sections across all ten molecules are shown and sorted in two categories: four low appearing molecules (central height about 110 pm ± 6%) and six high appearing molecules (central height about 150 pm ± 16%). Since due to symmetry arguments all eight possible orientations of Co-Salen on NiO(001) are structurally equivalent, this hints to the presence of a possible additional magnetic signal. Such a magnetic signal would depend on the relative orientation between the magnetic moments of tip and molecule. To support our assumption that the height difference is related to a magnetic contribution to the total tip-sample interaction, we analyzed the height distribution of Co-Salen molecules adsorbed on non-magnetic NaCl(001). We find that within a typical standard deviation of 10% they all exhibit the same height level for a given tip and identical scanning parameters. In fact, there is even no significant height difference between molecules adsorbed along the ⟨110⟩- and ⟨100⟩-directions, respectively.

Since the central Co atom adsorbs on top of surface oxygen, the magnetic coupling has to be achieved via a superexchange mechanism to a subsurface nickel atom below the surface oxygen. Considering that the antiferromagnetic coupling between nickel sites in nickel oxide itself takes place via a superexchange mechanism across bridging oxygen atoms, such a presumption is certainly justified. In the following we discuss how the magnetic interaction between Co-Salen and NiO(001) could take place and how it is conceivable that a magnetic signal is detected from the molecule, but not from the substrate.

In the gas phase, Co-Salen is paramagnetic with the spin located in the Co d_{xy}-orbital, cf. Fig. 4.10c. On a magnetic substrate, the previously para-magnetic spin can couple to the ordered sample spins. Such a substrate induced alignment of the spin

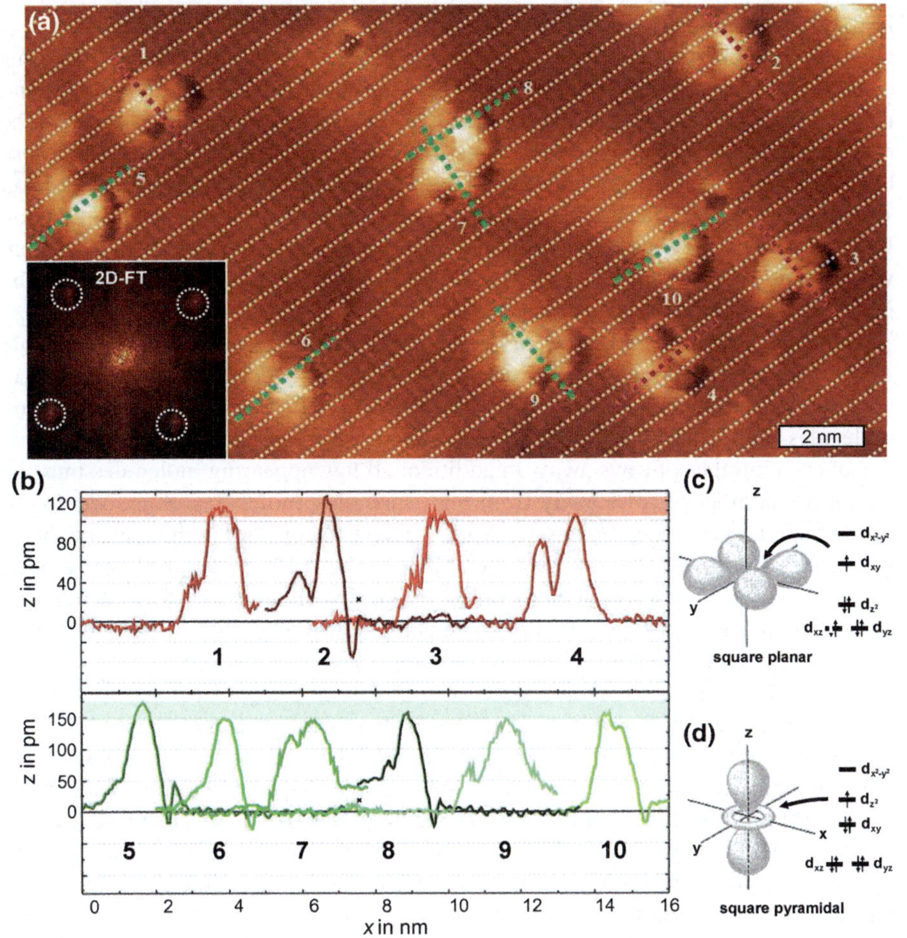

Fig. 4.10 **a** Atomically resolved NiO(001) with ten Co-Salen molecules. Dotted lines mark every second oxygen row along the [1$\bar{1}$0]-direction. The inset displays the 2D-FT showing that atomic resolution is achieved but without magnetic contrast. **b** Line sections along the C_2 axis of the ten Co-Salen molecules. They are sorted into low and large height molecules according to their maximum height in their center, i.e., at the position of the Co atom. **c** Energy level order in the square planar configuration and symmetry of the single occupied d_{xy} orbital. **d** Energy level order in the square pyramidal configuration and symmetry of the single occupied d_{z^2} orbital

of paramagnetic molecules has been observed on metallic magnetic substrates [32–34]. Note that a charge transfer, which could render the paramagnetic $S = 1/2$ to a diamagnetic $S = 0$ molecule, as e.g., observed for Co-Pthalocyanine on the itinerant metallic ferromagnetic Fe double-layer on W(110) [35], does not take place on bulk insulators like NiO(001). However, the presence of a surface changes the crystal field from square planar to square pyramidal, which changes the order of the d-levels, cf.

Fig. 4.10c, d. Now the d_{z^2}-orbital is singly occupied, which, unlike the planar d_{xy}-orbital, points into the vacuum region, i.e., in the direction of the magnetic tip. In this configuration, an orbital overlap between the spin carrying states of the tip apex and the molecule, which is required to detect a magnetic signal, is easily possible. On the other hand, as already pointed out in Sect. 4.3, nickel oxide is an insulator, because the d-states do not form a conduction band. Instead, they are localized at the nickel sites and do not reach far into the vacuum region. Very likely, a magnetic interaction with the tip requires a smaller tip-sample separation on NiO(001) than on Co-Salen, which could explain why in Fig. 4.10a a magnetic signal can be detected above the molecule, but not on the substrate.

If the occurrence of two different height levels is related to a magnetic tip-sample interaction, the positions of the molecules with identical height levels must reflect the row-wise antiferromagnetic order of NiO(001). Therefore, all high appearing molecules must either be located on the same oxygen row or on oxygen rows that are an even number of rows away. In addition, all low appearing molecules must be an uneven number of rows away from high appearing molecules. Since we do not see any magnetic contrast on the substrate, we have to consider both cases, i.e., row-wise antiferromagnetic order along the [110]- or the [1$\bar{1}$0]-direction. If we assume that the row-wise antiferromagnetic order propagates along the [1$\bar{1}$0]-direction, the before-mentioned conditions are full-filled consistently, while for the other direction the distribution of molecules with low and large heights is random.

Our analysis shows that the appearance of low and high Co-Salen molecules on structurally equivalent adsorption sites is consistent with the assumption of a magnetic coupling to the antiferromagnetic NiO(001) substrate. According to the Goodenough–Kanamori rules, a 180° superexchange type of coupling via surface oxygen should result in an antiferromagnetic alignment between subsurface Ni and the molecular Co metal center. In principal, a 90° superexchange magnetic coupling path via surface Ni and molecular oxygen to the Co atom is possible as well, because they are also connected via covalent bonds, cf. Fig. 4.9c. In this case, systematic height variations due to a 90° ferromagnetic coupling, which would depend on the spin directions of the surface Ni sites below the two molecular oxygen atoms would occur. Our statistics is not sufficient to conclude about the existence of such an effect. Furthermore, the height variations can be expected to be much smaller because a 90° ferromagnetic coupling is much weaker than for a 180° antiferromagnetic coupling.

References

1. H. Hölscher, A. Schirmeisen, H. Fuchs, in *Nanoprobes*, ed. by H. Fuchs. Nanotechnology, vol. 6 (Wiley, Weinheim, 2010), p. 49
2. T.R. Albrecht, Appl. Phys. Lett. **69**, 668 (1991)
3. W. Allers, A. Schwarz, U.D. Schwarz, R. Wiesendanger, Rev. Sci. Instrum. **69**, 221 (1998)
4. M. Liebmann, A. Schwarz, S.M. Langkat, R. Wiesendanger, Rev. Sci. Instrum. **73**, 3508 (2002)
5. H. von Allwörden, K. Ruschmeier, A. Köhler, T. Eelbo, A. Schwarz, R. Wiesendanger, Rev. Sci. Instrum. **87**, 073702 (2016)

6. U. Dürig, Appl. Phys. Lett. **75**, 1203 (1999)
7. F.J. Giessibl, Appl. Phys. Lett. **78**, 123 (2001)
8. J. Sader, S.P. Jarvis, Appl. Phys. Lett. **84**, 1801 (2004)
9. H. Hölscher, S.M. Langkat, A. Schwarz, R. Wiesendanger, Appl. Phys. Lett. **81**, 4428 (2002)
10. G. Teobaldi, K. Lämmle, T. Trevethan, M. Watkins, A. Schwarz, R. Wiesendanger, A.L. Shluger, Phys. Rev. Lett. **106**, 216102 (2011)
11. B. Anczykowski, B. Gotsmann, H. Fuchs, J.P. Cleveland, V.B. Elings, Appl. Surf. Sci. **140**, 376 (1999)
12. H. Hölscher, B. Gotsmann, W. Allers, U.D. Schwarz, H. Fuchs, R. Wiesendanger, Phys. Rev. B **64**, 075402 (2001)
13. N. Sasaki, M. Tsukada, Jpn. J. Appl. Phys. **39**, L1334 (2000)
14. E.Y. Vedmedenko, Q. Zhu, U. Kaiser, A. Schwarz, R. Wiesendanger, Phys. Rev. B **85**, 174410 (2012)
15. R. Schmidt, U. Kaiser, C. Lazo, A. Schwarz, R. Wiesendanger, S. Heinze, Phys. Rev. Lett. **106**, 257202 (2011)
16. A. Schwarz, R. Wiesendanger, Nano Today **3**, 28 (2008)
17. L. Gross, F. Mohn, N. Moll, P. Liljeroth, G. Meyer, Science **325**, 1110 (2009)
18. R. Smoluchowski, Phys. Rev. **60**, 661 (1941)
19. K. Lämmle, T. Trevethan, A. Schwarz, M. Watkins, A. Shluger, R. Wiesendanger, Nano Lett. **10**, 2965 (2010)
20. A. Schwarz, D.Z. Gao, K. Lämmle, J. Grenz, M.B. Watkins, A.L. Shluger, R. Wiesendanger, J. Phys. Chem. C **117**, 1105 (2013)
21. D.Z. Gao, J. Grenz, M.B. Watkins, F.F. Canova, A. Schwarz, R. Wiesendanger, A.L. Shluger, ACS Nano **8**, 5339 (2014)
22. A. Schwarz, A. Köhler, J. Grenz, R. Wiesendanger, Appl. Phys. Lett. **105**, 011606 (2014)
23. F. Canova, A.S. Foster, M.K. Rasmussen, K. Meinander, F. Besenbacher, J.V. Lauristen, Nanotechnology **23**, 1 (2012)
24. U. Kaiser, A. Schwarz, R. Wiesendanger, Nature **446**, 522 (2007)
25. R. Schmidt, C. Lazo, H. Hölscher, U.H. Pi, V. Caciuc, A. Schwarz, R. Wiesendanger, S. Heinze, Nano Lett. **9**, 200 (2009)
26. J. Grenz, A. Köhler, A. Schwarz, R. Wiesendanger, Phys. Rev. Lett. **119**, 047205 (2017)
27. A. Kubetzka, P. Ferriani, M. Bode, S. Heinze, G. Bihlmayer, K. von Bergmann, O. Pietzsch, S. Blügel, R. Wiesendanger, Phys. Rev. Lett. **94**, 87204 (2005)
28. K. von Bergmann, S. Heinze, M. Bode, E.Y. Vedmedenko, G. Bihlmayer, S. Blügel, R. Wiesendanger, Phys. Rev. Lett. **96**, 167203 (2006)
29. S. Heinze, K. von Bergmann, M. Menzel, J. Brede, A. Kubetzka, R. Wiesendanger, G. Bihlmayer, S. Blügel, Nat. Phys. **7**, 713 (2011)
30. S. Fremy, A. Schwarz, K. Lämmle, M. Prosenc, R. Wiesendanger, Nanotechnology **20**, 405608 (2009)
31. K. Lämmle, A. Schwarz, R. Wiesendanger, Rev. Sci. Instrum. **81**, 053902 (2010)
32. H. Wende, M. Bernien, J. Luo, C. Sorg, N. Ponpandian, J. Kurde, J. Miguel, M. Piantek, X. Xu, P. Eckhold, W. Kuch, K. Baberschke, P.M. Panchmatia, B. Sanyal, P.M. Oppeneer, O. Eriksson, Nat. Mat. **6**, 516 (2007)
33. C.F. Hermanns, K. Tarafder, M. Bernien, A. Krüger, Y.-M. Chang, P.M. Oppeneer, W. Kuch, Adv. Mat. **25**, 3473 (2013)
34. C. Nistor, C. Krull, A. Mugarza, S. Stepanow, C. Stamm, M. Soares, S. Klyatskaya, M. Ruben, P. Gambardella, Phys. Rev. B **92**, 184402 (2015)
35. J. Brede, N. Atodiresei, G. Hoffmann, S. Kuck, P. Lazic, V. Caciuc, Y. Morikawa, S. Blügel, R. Wiesendanger, Phys. Rev. Lett. **105**, 047204 (2010)

Chapter 5
Magnetic Properties of One-Dimensional Stacked Metal Complexes

Tabea Buban, Sarah Puhl, Peter Burger, Marc H. Prosenc and Jürgen Heck

Abstract Cooperative effects such as ferro- or antiferromagnetic interactions are accessible through tailor-made molecular structures of linearly arranged paramagnetic complexes. Since it is well-known that subtle changes in the molecular structure can cause distinct changes in the magnetic interaction, the inter-metal distances were varied as well as the number of stacked complexes. In addition the metal centers were changed in order to vary the numbers of interacting unpaired electrons. The final target was an investigation of the properties of stacked magnetic molecules on a substrate.

5.1 Introduction

The study of molecular magnetic materials is an important issue in view of spintronic applications. In particular the following molecular materials were investigated and showed promising characteristics [1, 2]: Prussian Blue [3–5], spin-crossover systems [6, 7], tetracyanoethylene (TCNE) salts [8–10], single-molecular magnets [11–13] and single-chain magnets [14, 15]. Contributions to the research field of coupling mechanisms between paramagnetic sandwich compounds were made for various types of complexes [16]. Examples are metallocene and oligometallocene complexes [17, 18], decorated with a different number of unpaired electrons and which are directly linked [19–21] through a saturated [22, 23] or unsaturated [24, 25] bridge, or are part of a cyclophane entity [26–29].

T. Buban · S. Puhl · P. Burger (✉) · J. Heck
Institute of Inorganic and Applied Chemistry, University of Hamburg,
Martin-Luther-King-Platz 6, 20146 Hamburg, Germany
e-mail: burger@chemie.uni-hamburg.de

J. Heck
e-mail: heck@chemie.uni-hamburg.de

M. H. Prosenc
Institute of Physical Chemistry, TU Kaiserslautern, Erwin-Schrödinger-Str. 52,
67663 Kaiserslautern, Germany
e-mail: prosenc@chemie.uni-kl.de

R. Wiesendanger (ed.), *Atomic- and Nanoscale Magnetism*, NanoScience and Technology, https://doi.org/10.1007/978-3-319-99558-8_5

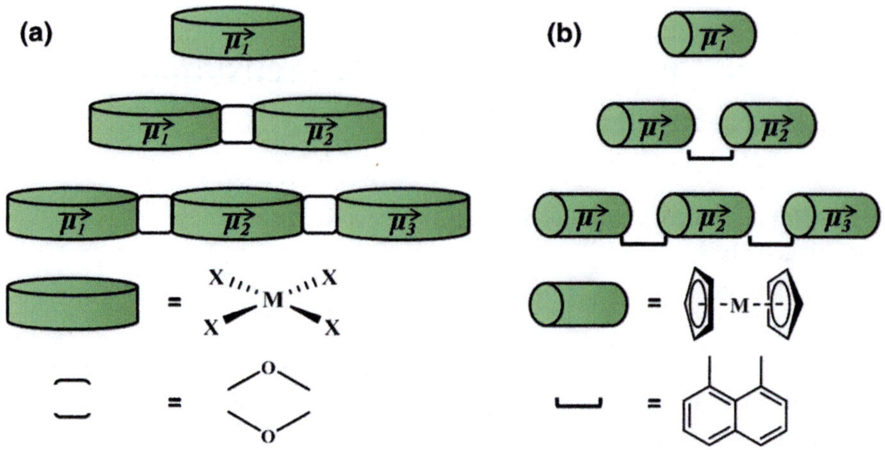

Fig. 5.1 Stacking of paramagnetic complexes **a** salen, salophen, pyridine-diimine complexes and **b** metallocenes; $\vec{\mu_i}$: magnetic moment of the subunit

Spin filters and spin-based logic devices open the door to exciting new applications and are based on magnetic rather than electric interactions. Currently, such logic devices are realized with individually arranged atoms on surfaces and require low temperatures due to their small magnetic coupling [30]. Here molecules come into play, which allow to access significantly larger magnetic interactions of the spin centers. Furthermore, they offer the opportunity to adjust the size and sign, i.e. antiferro- versus ferromagnetic coupling by tailor-made design of synthetic structures. In this regard one-dimensional stacked complexes are promising candidates [31–33]. We investigated two different types of paramagnetic 3d transition metal systems, which will be discussed in separate sections. In the first part of this chapter, paramagnetic Schiff-base complexes with salene and pyridine-diimine (PDI) ligands are discussed, in which the metal centers are placed in the plane of the ligand's π-bonds (Fig. 5.1a). A particular focus is placed on the magnetic coupling pathways of the individual molecules and in the bulk, in solution as well as in the solid state. Their surface deposition, orientation and magnetic interaction, e.g. surface-mediated Ruderman–Kittel–Kasuya–Yoshida (RKKY) coupling, will also be discussed. In the second section, di-, tri- and tetranuclear cofacially stacked sandwich complexes are reported, in which the metal centers are located perpendicular to the π-plane of the ligand (Fig. 5.1b).

5.2 Towards Molecular Spintronics

Recent advances in atom manipulation led to structures and devices suitable for storage and processing of spin information [34]. However, these structures require single atom manipulation techniques as well as very low temperatures.

If laterally linked molecular complexes were used, they need to be robust, paramagnetic and depositable on a surface. With this goal in mind, we screened suitable complexes and ligand strategies [35, 36]. While molecular cobalt salen complexes appear to be mobile on a Cu(111) surface, chlorinated derivatives arrange via self-assembly forming small six-membered aggregates to extended domains on surfaces (Fig. 5.2c).

Deposition of salen complexes on Cu yielded slightly mobile complexes for salen and methylsalen derivatives. Thus, further studies were performed using salophen complexes with a phenylendiamine bridge, which was deemed to yield more rigid complexes.

Deposition of dibromo salophen complexes on Au(111) revealed more stable complexes, which arrange in four-membered aggregates forming long bands on Au(111) (Fig. 5.3). Upon heating they initially loose bromine atoms and convert to small to long chemically bonded chains by C–C bond formation (Fig. 5.4) [37]. The lengths of the chains can be controlled by surface occupancy and temperature [38]. If in addi-

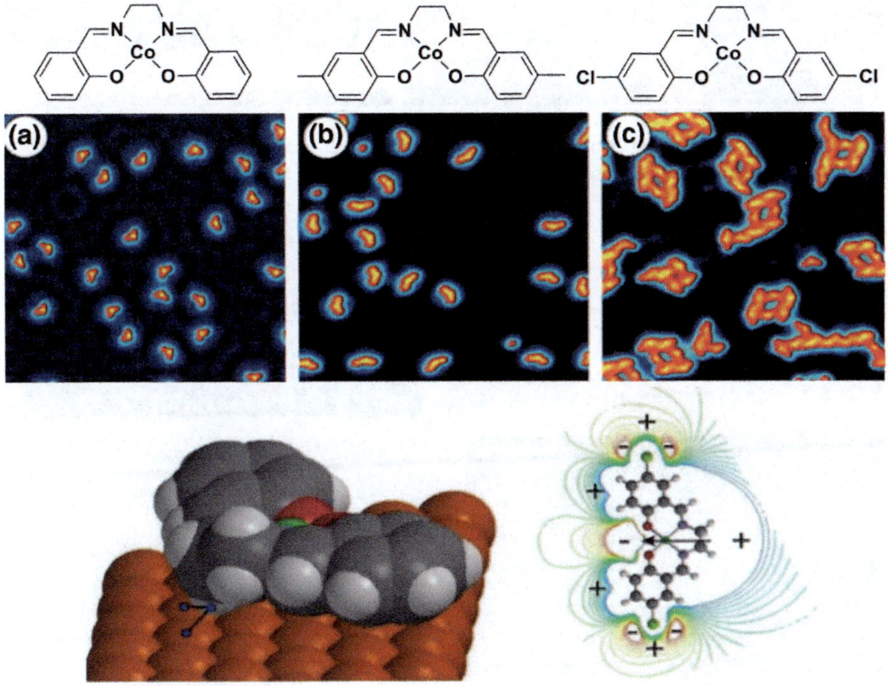

Fig. 5.2 Cobalt salen complexes on a Cu(111) surface. **a** cobalt salen, **b** 4,4′ dimethyl cobalt salen and **c** 4,4 dichloro cobalt salen complexes. For all complexes 12 orientations were found representing the six-fold symmetry of the surface together with the delta and lambda configuration. On the bottom one optimized structure of the cobalt salen complex is depicted exhibiting weak CH...Cu interactions. Calculation of the charge distribution resulted in the accumulation of negative charges on the salen oxygen atoms while the α- and β-hydrogen atoms exhibit positive charges important for self-assembly of the chloro complexes

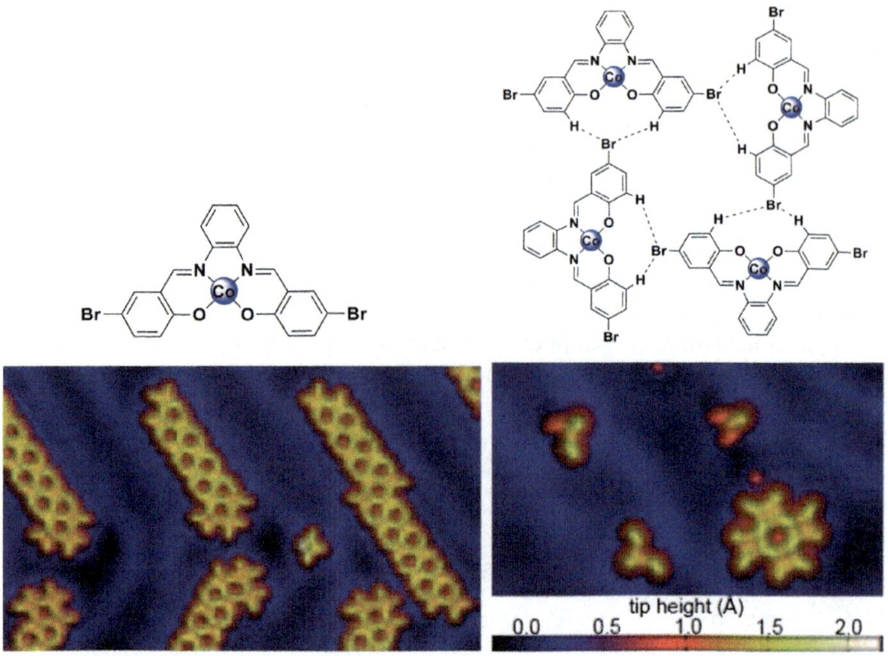

Fig. 5.3 Dibromo salophen cobalt complex (upper left), self assembled to form four-membered rings (upper right), and their observation on Au(111) surfaces by STM techniques [37] (bottom)

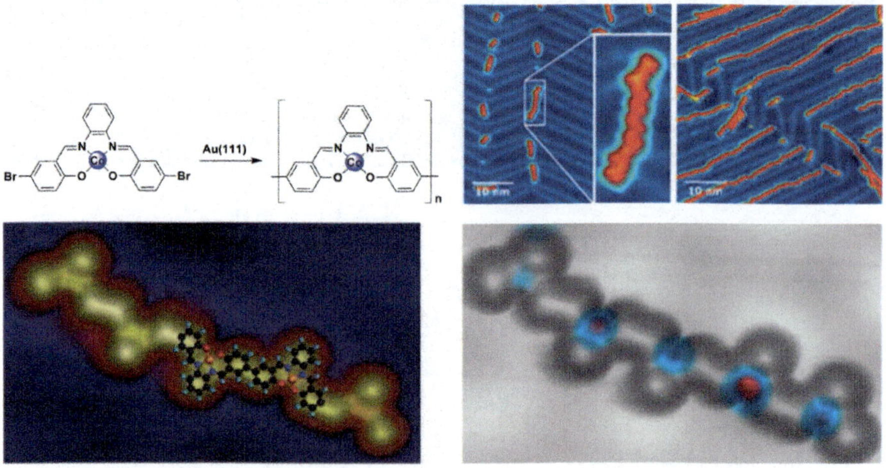

Fig. 5.4 On surface oligomerization of dibromo cobalt salophen. Small to long chains can be achieved depending on the amount of the monobromo cobalt salophen, which terminates the chains (top). On the bottom one selected chain and Kondo-measurements revealing the antiferromagnetic structure are depicted

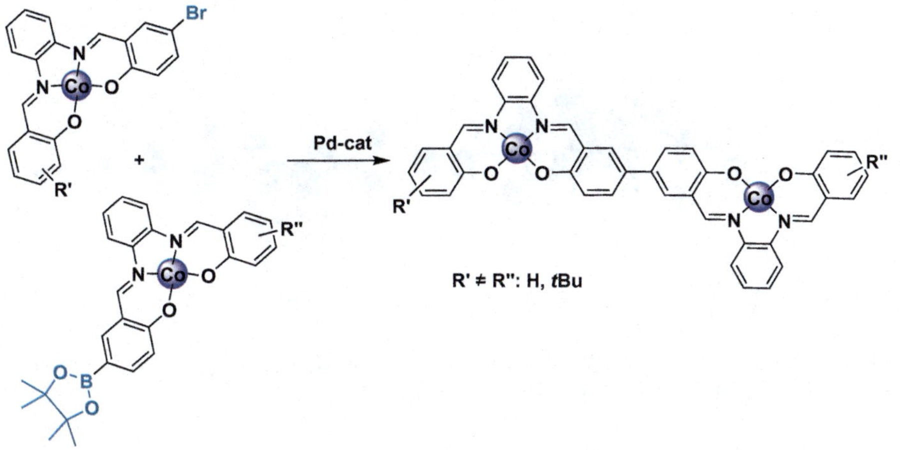

Scheme 5.1 Synthesis of disalophen complexes according to [39]

tion monobromo salophen complexes were deposited, more defined chains formed up to about 100+ Co-salophen units [38]. These cobalt complex chains consist of chemically connected paramagnetic Co(II) complexes. This raises the question about the strength and type of magnetic coupling between the Co(II) centers.

Kondo measurements of the cobalt salophen chains on surfaces revealed a dependency of the Kondo temperature on the parity of the chain length and consequently an antiferromagnetic coupling among the metal centers [37]. In order to obtain information about the magnetic coupling in the bulk material, we synthesized dicobalt- and dicopper-disalen complexes (Scheme 5.1) and investigated their magnetic properties [39, 40].

Magnetic measurements by the SQUID-methods revealed strong (Co) and weak (Cu) antiferromagnetic couplings between the metal centers [39, 40]. These results are in agreement with measurements of the magnetic properties of the chains deposited on Au(111) surfaces [37].

From X-ray crystal structure data as well as density function theory (DFT) calculations, the rotation angle around the central C–C bond linking the two salophen units is close to 0°. DFT calculations of the spin density of a dicobalt complex revealed that the two unpaired electrons are distributed in the π-electron system of the complex (Fig. 5.5). A rotation around the central C–C bond would reduce the overlap between the orbitals at the bridgehead carbon atoms and thus the coupling between the unpaired electrons at the metal spin centers.

A maximum magnitude of J was found in a coplanar arrangement, which was also found for Co-salophen complexes deposited on Au(111) [37–39, 41, 42]. From XAS-, STM-measurements and DFT calculations it became evident that the coupling between the complexes on the surface is dominated by the coupling between the the Co(II) spin centers through the ligand's π-system rather than surface mediated by RKKY interactions. The observed antiferromagnetic coupling together with the

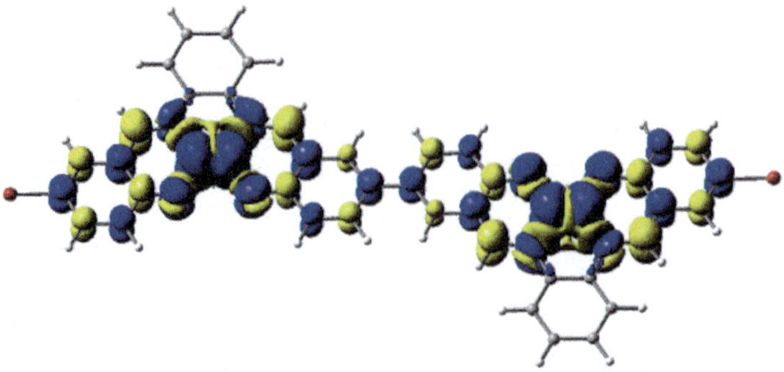

Fig. 5.5 DFT calculated spin-density of a dibromo dicobalt salophen complex. The spin density is located in π-orbitals perpendicular to the molecular plane

formation of long one-dimensional chains raises the question whether these systems could be employed in spintronic devices [41].

A simple spintronic device could be a logic gate, which transfers the information from two leads to an exit gate [41]. To build such a device, we synthesized a tri-cobalt-triplesalophen complex [41, 43] with three bromine atom substituents to be copolymerized with mono- and dibromocobaltsalophen complexes on a metal surface [38, 41]. Such a spintronic device is depicted in Fig. 5.6.

In conclusion, we were able to develop new spin devices from salophen complexes. Initial studies on salen complexes revealed the high thermal and chemical stability of cobalt complexes on Cu(111) and Au(111) surfaces. Self-assembly of halogen terminated complexes and further chemical transformation of the bromo derivative into complex chains with antiferromagnetically coupled unpaired electrons revealed the first molecular based spin-chains on metal surfaces with high Kondo temperatures depending on the chain length. Triple-cobaltsalophen complexes were used in addition to mono- and dibromosalophen cobalt complexes and resulted in structures suitable for a model of a molecular based spintronic device.

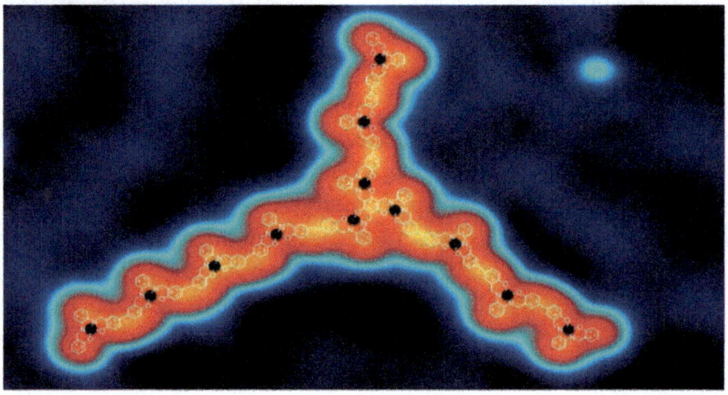

Fig. 5.6 STM-representation of the synthesized spintronic device. The central triple-cobaltsalophen complex is chemically connected to cobalt salophen complex-chains

5.3 Paramagnetic 3d-Transition-Metal Complexes with Terdentate Pyridine-Diimine Ligands

Pyridine-Diimine (PDI) ligands are classified as non-innocent" ligands whose complexes can feature interesting electronic and magnetic properties [44–48]. Spacer-connected PDI ligands provide access to dinuclear complexes with adjustable metal-metal distances (Fig. 5.7).

Due to the perpendicular alignment of the planes of the PDI systems and the bonded aromatic spacer, a magnetic coupling of the metal centers via super-exchange is likely to be prevented. DFT-calculated spin densities of the anthracenyl-bridged complex support this assumption [49]. It was therefore anticipated that the systems presented in Fig. 5.7 would allow the investigation of the dependence of the dipolar coupling on the metal-metal distances.

5.3.1 Synthesis of Novel Mono-, Di- and Trinuclear Iron(II) Complexes

The key building blocks for the desired novel ligand systems are the borylated pyridine-diimine precursors **TB1** and **TB2**. These compounds were obtained by direct regioselective borylation with an iridium catalyst and were subjected to a consecutive Suzuki–Miyaura [50, 51] cross-coupling reaction (Scheme 5.2).

Complexation to the iron(II)-compounds was successfully accomplished using a synthetic protocol of Campora et al. (Scheme 5.3) [52]. For ligands **TB3** and **TB4**

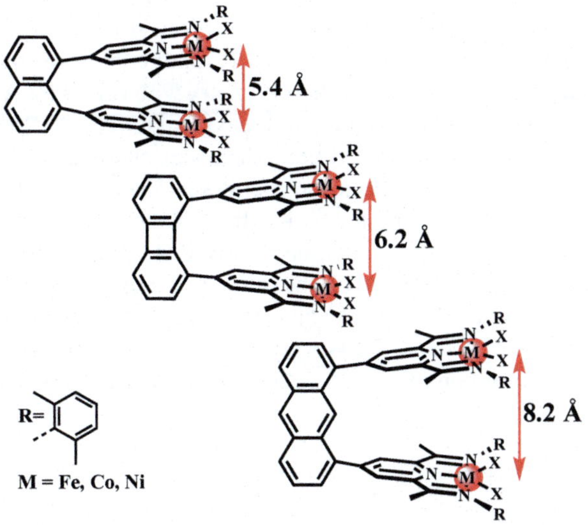

Fig. 5.7 Novel complexes with different metal-metal distances

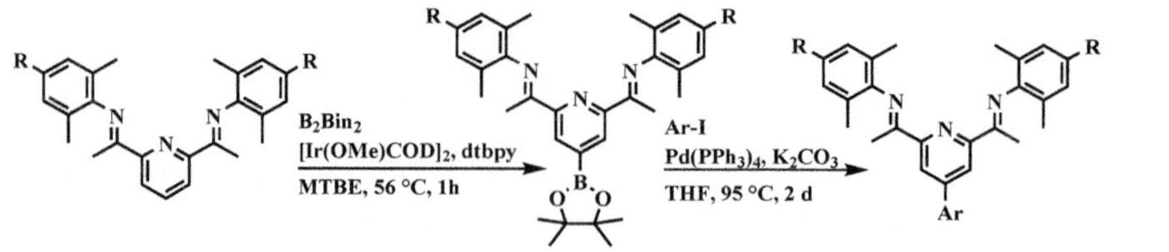

R = CH$_3$, *t*Bu

TB1: R = CH$_3$, 85 %
TB2: R = *t*Bu, 76 %

TB3: R = CH$_3$, Ar = phenyl, 59 %
TB4: R = *t*Bu, Ar = biphenyl, 65 %
TB5: R = CH$_3$, Ar = anthracenyl, 61 %

Scheme 5.2 Synthesis of ligands **TB5–TB7**

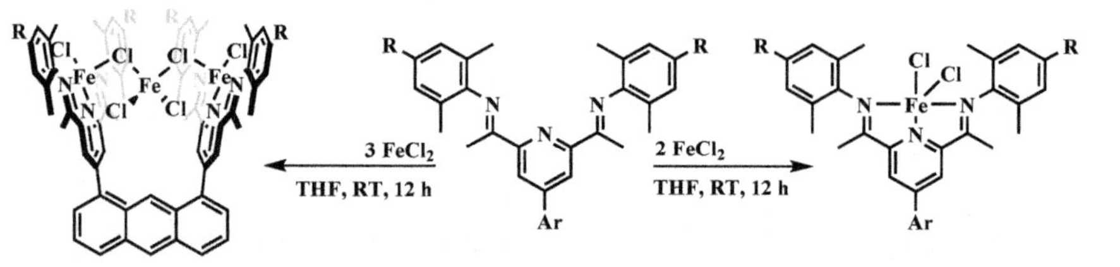

TB 8: R = CH$_3$, Ar = anthracenyl, 96 %

TB 3: R = CH$_3$, Ar = phenyl
TB 4: R = *t*Bu, Ar = biphenyl
TB 5: R = CH$_3$, Ar = anthracenyl

TB 6: R = CH$_3$, Ar = phenyl, 81 %
TB 7: R = *t*Bu, Ar = biphenyl, 89 %

Scheme 5.3 Complexation of FeCl$_2$ by ligands **TB3**, **TB4** and **TB5** to form mono-, di- and trinuclear complexes

the reaction led to the expected mono- and dinuclear complexes **TB6** and **TB7**. In the case of the ligand **TB5**, a trinuclear complex with a bridging FeCl$_2$-group was obtained.

The molecular structures of the iron(II) complexes (Fig. 5.8) could be unambiguously established through X-ray crystal structure analyses. Selected bond distances and angles are listed in Table 5.1. In all complexes the iron atoms exhibit a slightly distorted square-pyramidal geometry with values for Addison's parameter $\tau_5 \leq 0.15$. The Fe–Cl bond length is in the typical range of 2.25–2.28 Å for iron(II) complexes [53, 54]. The observed bond distances and angles of the PDI ligand compare well with structural data in the literature [55–57]. The C–N and C–C distances of the diimine groups and the exocyclic C–C bonds of the pyridine group clearly speak in favour of a neutral, i.e. innocent pyridine-diimine ligand [48, 58, 59]. The observed inter-metal distance of the dinuclear compound (Fe1–Fe2 = 7.235 Å) is larger than the DFT derived value (6.2 Å), whereas the distance in the trinuclear complex (Fe1–Fe3 = 7.804 Å) differs slightly from the calculated distance of a related dinuclear complex (8.2 Å), due to the interlinked iron(II) chlorido group.

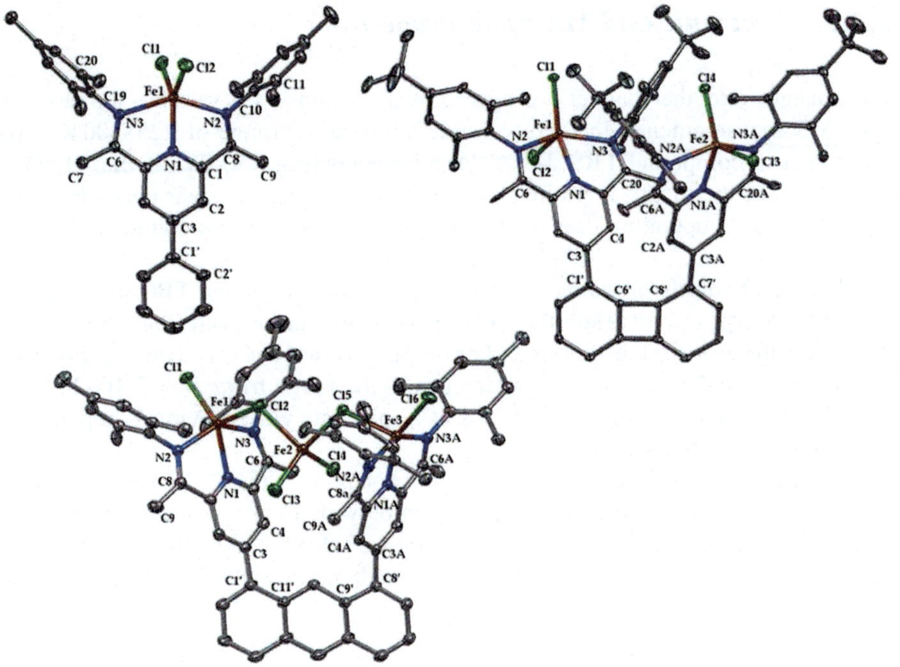

Fig. 5.8 X-ray crystal structures of **TB6–TB8**; co-crystallized solvent molecules and hydrogen atoms are partially omitted for clarity

Table 5.1 Selected distances (Å) and angles (°) for **TB6**, **TB7** and **TB8** with estimated standard deviation (esd) in parentheses

	TB6	**TB7**	**TB8**
Fe1–Cl1	2.2882(11)	2.270(2)	2.2513(1)
Fe1–Cl2	2.2876(10)	2.310(3)	2.3943(1)
Fe1–N1	2.103(3)	2.099(7)	2.110(3)
Fe1–N2	2.272(3)	2.258(7)	2.221(3)
Fe1–N3	2.254(3)	2.229(7)	2.195(3)
N3–C6	1.280(5)	1.295(11)	1.288(5)
N2–C8	1.284(4)	1.276(11)	1.279(5)
C1–C8	1.494(5)	1.498(11)	1.479(6)
C5–C6	1.491(5)	1.497(11)	1.485(6)
N2–Fe1–N3	145.40(10)	73.3(2)	144.60(1)
N1–Fe1–Cl1	125.39(9)	148.6(2)	149.45(9)
N1–Fe1–Cl2	125.80(9)	95.9(2)	101.70(9)

5.3.2 Electronic and Magnetic Properties

First insight into the magnetic properties was obtained by variable temperature ^{1}H NMR measurements. In the accessible temperature range of 225–300 K, only the trinuclear compound **TB8** shows Curie behavior (Fig. 5.9). While the ^{1}H NMR resonances of the mono- and dinuclear complexes display a linear dependence on the reciprocal temperature, the corresponding intercepts deviate significantly from zero.

This is particularly pronounced for the mononuclear complex **TB6** and might be attributed to large zero field splitting (ZFS) parameters and/or low-lying excited states mixed into the ground state. For iron PDI complexes magnetic moments ranging from $\mu_{\text{eff}} = 5.0\,\mu_B$ to $5.8\,\mu_B$ and ZFS parameters in the range from $D = -10$–$20\,\text{cm}^{-1}$ with d^6-configured iron centers with a $S = 2$ spin state were previously reported in the literature [52, 60, 61].

To establish the electronic ground (and potential excited) states of the iron complexes, Mössbauer spectroscopy was employed for the novel complex **TB6** (Fig. 5.10). The derived parameters are listed in Table 5.2. At 80 K, the spectrum displays the expected doublet as reported in the literature [48, 62] with an isomeric

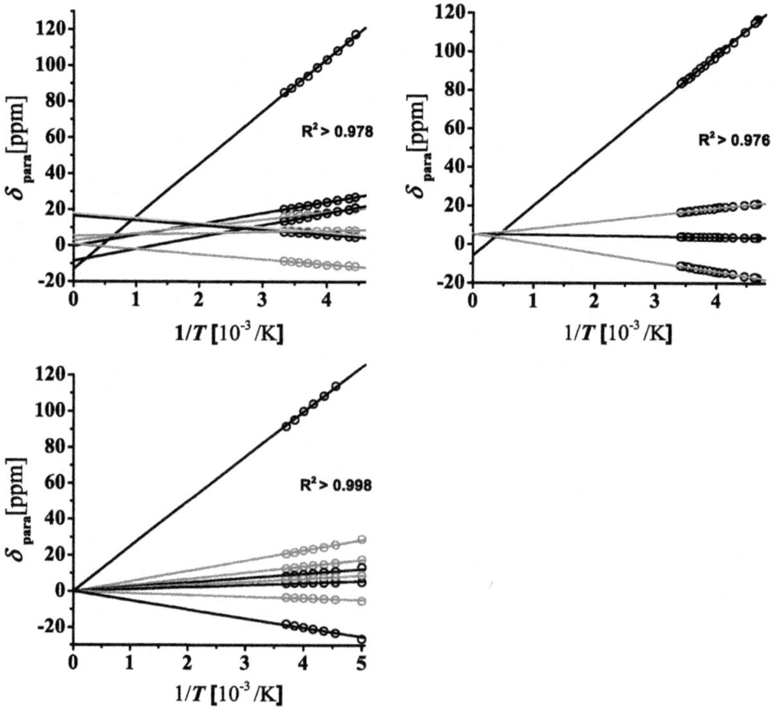

Fig. 5.9 ^{1}H NMR shifts from **TB8** (left), **TB9** (middle) and **TB10** (right)

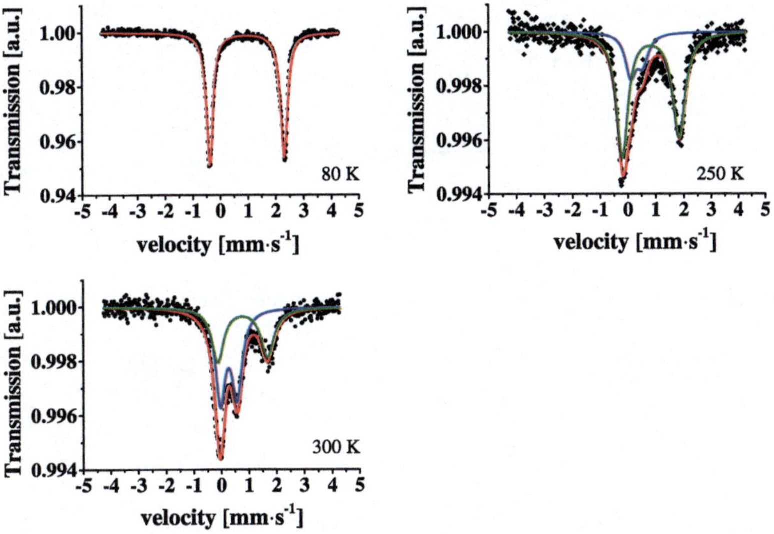

Fig. 5.10 Mössbauer spectra of complex **TB6** at 80, 250, and 300 K

Table 5.2 Parameters of the Mössbauer spectra at different temperatures

T (K)	δ_{IS} (mms^{-1})	ΔE_Q(mms^{-1})	Assignment	Population (%)
80	1.06	2.68	Fe(II)	100
250	0.92	2.03	Fe(II)	80.3
250	0.43	0.44	Fe(III)	19.7
300	0.88	1.80	Fe(II)	52.8
300	0.17	0.60	Fe(III)	47.2

shift of $\delta_{IS} = 1.06\,mms^{-1}$ and a quadrupole splitting of $\Delta E_Q = 2.68\,mms^{-1}$ typical for high-spin iron(II) center with $S = 2$ [63, 64].

With increasing temperature, a second doublet emerges, indicating two different states of the complex. The isomeric shift of $\delta_{IS} = 0.037\,mms^{-1}$ of the second doublet is typical for d^5-configured iron high-spin complexes [62], while the quadrupole splitting value of $\Delta_E Q = 2.68\,mms^{-1}$ is significantly smaller than values previously reported for real iron(III) PDI-complexes [62]. The quadrupole splitting of the aforementioned quintet decays at higher temperatures indicating a small change of the coordination geometry [65]. The observed temperature dependence of the isomeric shift is not consistent with the expected influence of a second-order Doppler effect [66] and is instead accounted for by dynamic processes between the two states.

Iron(II) PDI complexes with a reduced, i.e. non-innocent ligand are known to exhibit smaller quadrupole splittings [48]. Therefore, it is suggested that at higher temperatures an iron(III) complex bearing an anionic non-innocent ligand is populated. This indicates a case of valence tautomerism as shown in Scheme 5.4 [67].

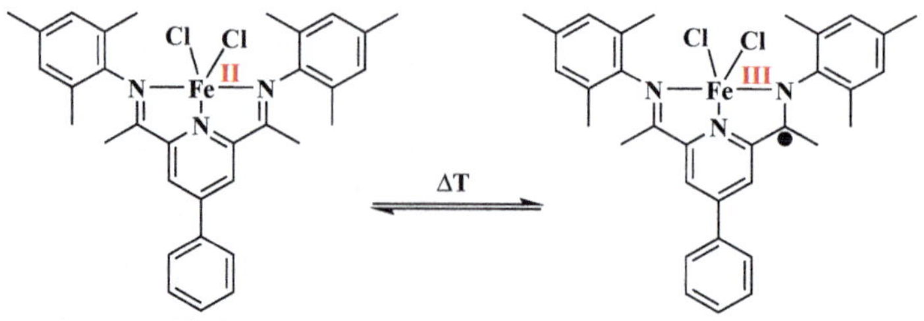

Scheme 5.4 Assumed temperature dependent valence tautomerism in complex **TB6**

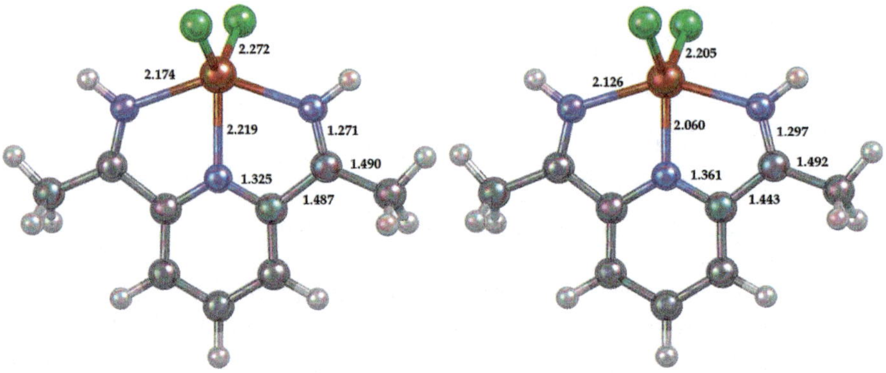

Fig. 5.11 Calculated molecular structures of Fe(II)-complex (left) and Fe(III)-complex (right) with selected bond-distances in Å

Starting from our DFT calculations, Granovsky carried out multi-reference calculations for both iron(II) and iron(III) centers with $S = 2$ and $S = 3$ spin states at the XMCQDPT2/SAS-CASSCF [13e, 11o] level [68] for the model system shown in Fig. 5.11. The energy difference between the $S = 2$ ground state and the $S = 3$ excited state is calculated to be $\Delta E = 7.3\,\text{kcal}\,\text{mol}^{-1}$. Preliminary results for calculations with larger basis sets and the full ligand system suggest that the energy difference is even smaller for the substituted real complex. Therefore, it is anticipated that the $S = 3$ excited state is populated to a significant extent above 300 K.

The calculations predicted an alteration in the bond length of the ligand, i.e. the C-N distances of the imine groups are elongated by 0.03 Å and the exocyclic C–C bonds of the pyridine group is shortened by 0.04 Å. These theoretically derived changes are smaller than for reported non-innocent" PDI ligands [58]. It was therefore not surprising that we were not able to detect changes in the bond lengths in temperature dependent single crystal structure measurements.

Figure 5.12 displays the temperature dependence of the effective magnetic moment μ_{eff} for compounds **TB6**, **TB7** and **TB8**. The results of the Curie–Weiss analysis and fitting parameters of the variable temperature (vt) magnetic susceptibility data obtained with the JulX [69] program are given in Table 5.3. For all three complexes

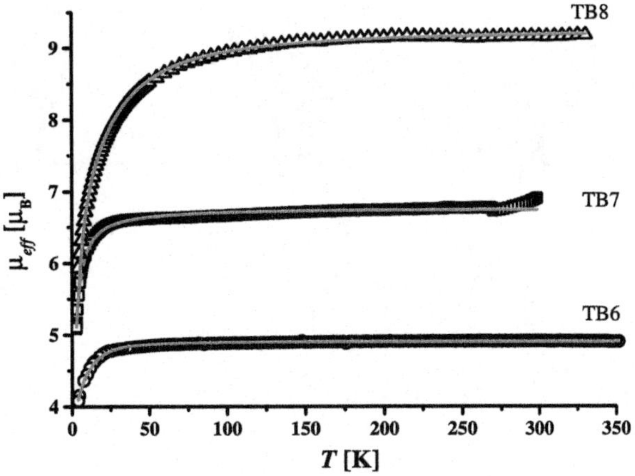

Fig. 5.12 Plot of μ_{eff} versus *T* for **TB6** (bottom), **TB7** (middle) and **TB8** (top)

Table 5.3 Parameters obtained from fitting the magnetic data of **TB6**, **TB7** and **TB8**

Compound	*S*	*g*-Value	*D* (cm^{-1})	*J* (cm^{-1})	Θ_W (K)	μ_{eff} (μ_B)
TB6	2	2.052	−8.414	–	1.400	4.90
TB7	2	1.954	−0.209	0.001	−1.949	6.77
TB8	2	2.180	−24.435	−0.390	7.98	9.17

the magnetic moment increases with higher temperatures until it reaches saturation at 70 K for **TB6**, 27 K for **TB7** and 140 K for **TB8**.

The magnetic moment of $\mu_{\text{eff}} = 4.9\,\mu_B$ for the mononuclear complex **TB6** at RT is in the range for iron(II) high-spin complexes and corresponds to the calculated spin-only value. For systems displaying valence tautomerism, a change of the magnetic moment is typically observed above 250 K [70–72]. This was, however, not observed for compound **TB6** and may be attributed to strong antiferromagnetic coupling between the iron $S = 3$ centre and the ligand radical. This could quench the additional magnetic moment as was previously reported for an iron complex by Banerjee et al. [73]. Simulations proposed a strong coupling constant of $J \approx -250\,\text{cm}^{-1}$ thus preserving the magnetic moment.

The determined ZFS parameter of $D = -8.4\,\text{cm}^{-1}$ matches the established values reported in the literature for related PDI iron complexes [62]. At room temperature a magnetic moment of $\mu_{\text{eff}} = 6.77\,\mu_B$ is observed for the dinuclear complex **TB7**. This value indicates an uncoupled system with a $S = 2$ ground state for the independent iron centres and is in full agreement with the theoretical value of $\mu_{\text{theo}} = 6.93\,\mu_B$. This is consistent with the fit of the magnetic susceptibility data, which revealed a negligible antiferromagnetic coupling between the iron centers ($J = -0.001\,\text{cm}^{-1}$). The ZFS parameter ($D = -0.209\,\text{cm}^{-1}$) is exceptionally small compared to the mononuclear compounds and could not be explained up to now. The magnetic moment of $\mu_{\text{eff}} = 9.17\,\mu_B$ for the trinuclear compound **TB8** is comparable to the spin-only value for an uncoupled system containing three

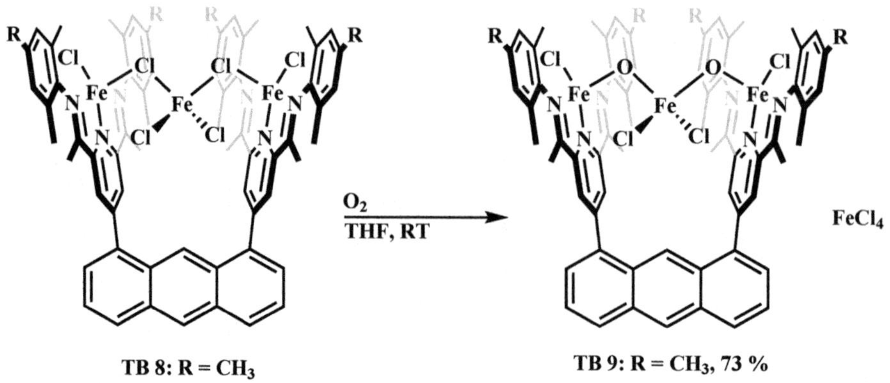

Scheme 5.5 Reaction to oxo-complex **TB9** from **TB8**

high-spin iron(II) centers ($\mu_{\text{theo}} = 8.40\,\mu_B{}^1$). For the data fit a spin state of $S_1 = S_2 = S_3 = 2$ was assumed in agreement with the Mössbauer measurements. Based on symmetry arguments, equal parameters were assigned to the outer iron atoms. The derived ZFS parameter of $D = -48.0\,\text{cm}^{-1}$ compares well with the expected value of (ideal) square-pyramidal coordination geometry ($\tau_5 = 0.08$) [77]. The fit also shows an antiferromagnetic coupling between the outer and inner metal centres by super-exchange via the μ-chlorido bridges.

5.3.3 *Molecules on Surfaces*

The MALDI mass spectra of the di- and trinuclear complexes display only peaks for the mononuclear fragments with $m/z = 599.1$ for **TB6**, $m/z = 1238.7$ for **TB7** and $m/z = 1059.45$ for **TB8**. Complex **TB8** was also examined via ESI mass spectrometry to check its accessibility for electro-spray deposition on surfaces. Rather than the expected mother iron peak, however, the recorded spectra and the isotropic pattern of the observed signal at $m/z = 1311.22$ hinted at an oxidized fragment.

The control reaction of compound **TB8** with oxygen led indeed and instantaneously to the μ-oxo-bridged iron(III) cationic complex **TB9** (Scheme 5.5), which was unambiguously confirmed by X-ray single crystal structure analysis. The spin state of $S = 5/2$ for all three iron centres could be established by Mössbauer spectroscopy.

Due to its inherent stability, the oxidized complex TB9 was deposited on a Au(111)-surface by the electro-spray-deposition method [41, 78, 79]. Figure 5.13 shows topographical images of the measurements. Based on the crystal structure an intact molecule possesses a diameter of 1 nm. Most of the observed conformations

[1]Calculated with $\mu_M^2 = \sum \mu_i^2$ [74–76].

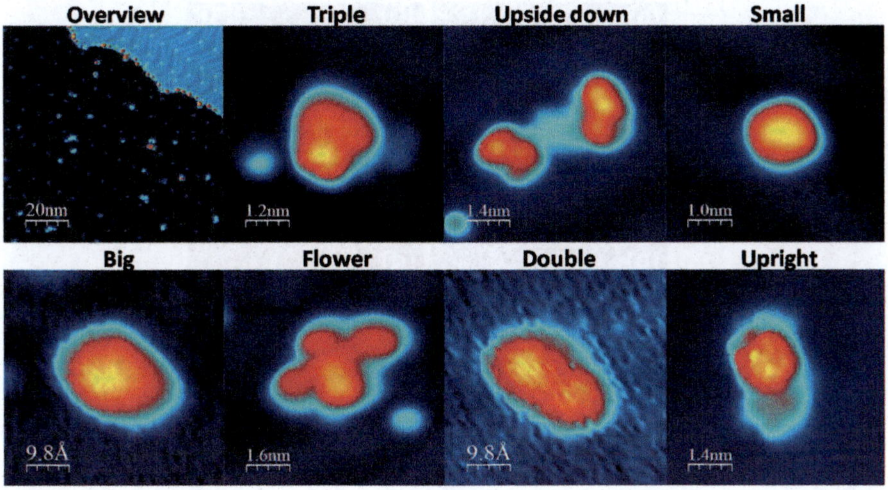

Fig. 5.13 Scanning tunnelling microscope images of the molecules deposited onto a Au(111) surface

are not robust and exist in many variants on the surface. Many of the objects agglomerate into pairs or trimers adsorbed at the elbows of the herringbone reconstruction of the Au(111) surface.

Theoretical calculations to determine the favored orientation of the molecules on the surface were performed by Hermanowicz [80] with the SIESTA DFT program package employing the PBE functional. These calculations revealed an energetic preference for the side-on orientation of 121 kcal/mol over the upside-down orientation (Fig. 5.14).

The STM measurements clearly revealed that the molecule lands in all possible rotational orientations on the surface. Similar results were obtained for a Fe_4 complex by Burgess [81], where an assignment of the DFT calculated structure to a topographical image was possible due to characteristic spin excitation energies. In our case a comparison of the measured topography images to the ones predicted by DFT calculations was not yet possible.

5.4 One-Dimensional Stacked Metallocenes

One final target was to stack paramagnetic metallocenes head-to-head and fix them in *peri*-position of a naphthalene unit. Herein, we present the synthesis, molecular structure, magnetic properties and theoretical calculations of bis- and oligo(metallocenyl)naphthalene complexes displaying different ground states.

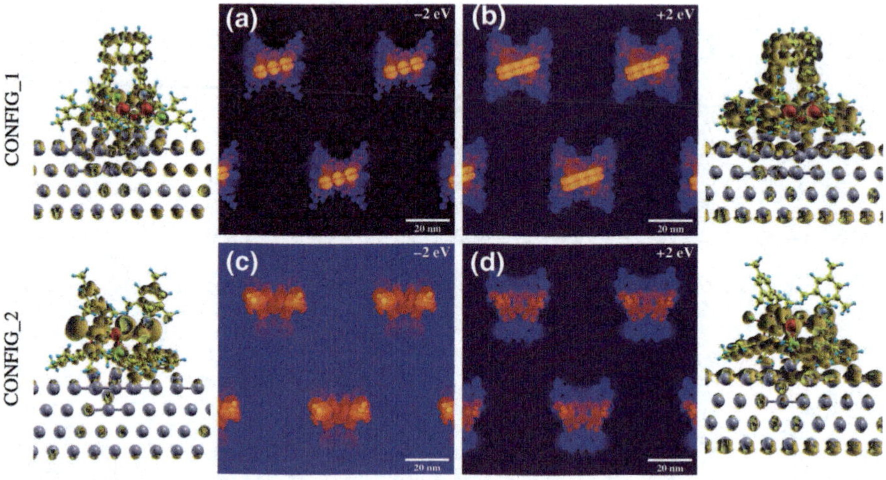

Fig. 5.14 Calculated maps as Tersoff–Hamann [82] style STM images (**a**–**d**) with corresponding 3D distributions (side views). Side on and upside orientation

5.4.1 Different Metal Centers

Depending on the metal used, the number of unpaired electrons per metallocenyl unit is predefined. Therefore the synthesized naphthalene-bridged biscobaltocenyl complex **$[Co]_2$** [83], the bisvanadocenyl complex **$[V]_2$** [84] as well as the decamethyl biscobaltocenyl **$[Co^*]_2$** [83] and bisnickelocenyl compounds **$[Ni^*]_2$** [85] (Fig. 5.15) exhibit different ground states. As a consequence, different magnetic responses should be expected by the combination of different metals.

Synthesis

The dinuclear cobaltocene complex **$[Co]_2$** was prepared in a three-step synthesis (Scheme 5.6) using 1,8-diiodonaphthalene as starting material [86]. A two-fold iodine-lithium exchange followed by a nucleophilic attack at cobaltocenium iodide led to a naphthalene-bridged dinuclear cobalt(I) complex. Hydride abstraction using a tritylium salt yielded the dinuclear cobaltocenium complex, which was readily reduced to the desired biscobaltocenyl complex using decamethylcobaltocene [83].

For the synthesis of the corresponding decamethyl biscobaltocenyl complex **$[Co^*]_2$** [83] the disodium salt of a cyclopentadienyl functionalized naphthalene

Fig. 5.15 Naphthalene-bridged bismetallocenyl complexes **$[M]_2$** (M = Co, V; R = H) and **$[M^*]_2$** (M = Co, Ni; R = Me)

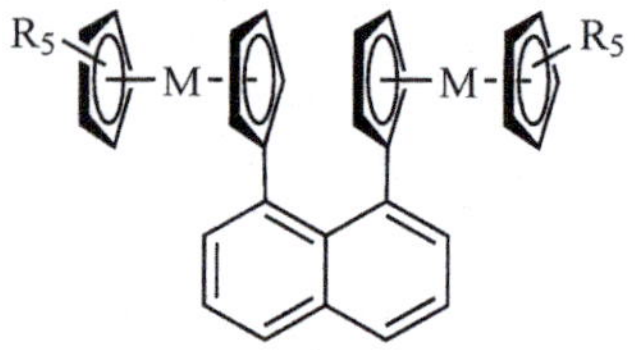

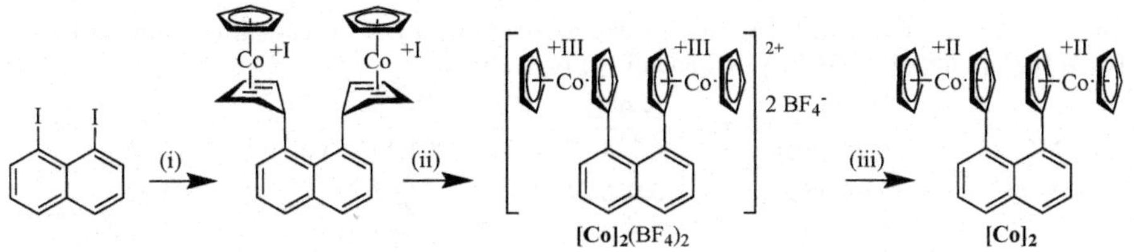

Scheme 5.6 Synthesis of the naphthalene-bridged biscobaltocenyl complex **[Co]$_2$** [83, 86]; reaction conditions: (i): (1) *n*-BuLi, Et_2O, (2) $[CoCp_2]I$, Et_2O; (ii): $Ph_3C^+BF_4^-$, dichloromethane (dcm); (iii): Cp_2^*Co, tetrahydrofurane (thf)

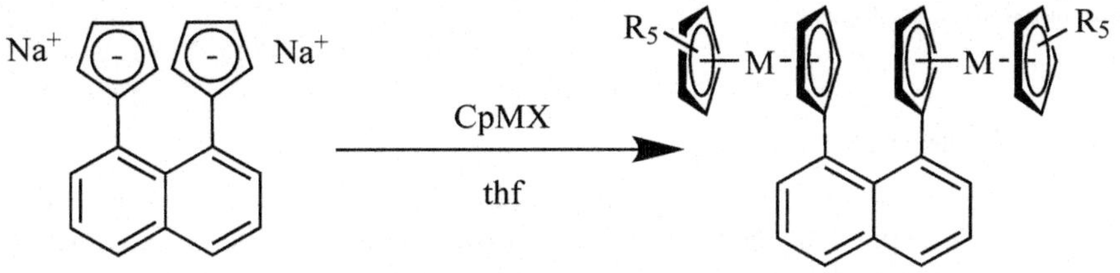

CpMX = Cp*Co(acac)	**[Co*]$_2$**:	M = Co;	R =	Me
Cp*Ni(acac)	**[Ni*]$_2$**:	Ni;		Me
$[CpV(\mu_2\text{-Cl})(thf)]_2$	**[V]$_2$**:	V;		H

Scheme 5.7 Synthesis of the naphthalene-bridged bismetallocenyl complexes **[Co*]$_2$** [83], **[Ni*]$_2$** [85] and **[V]$_2$** [84]

was used as starting material (Scheme 5.7). Reaction with a Cp*Co transfer reagent yielded the desired dinuclear compound via salt metathesis. In analogous manner, the related decamethyl nickel complex **[Ni*]$_2$** [85] and the bisvanadocenyl complex **[V]$_2$** [84] were accessible.

Molecular Structures

The molecular structures of the naphthalene-bridged bismetallocenyl complexes (Fig. 5.16) are dominated by a distortion (Table 5.4) due to the steric demand of the two metallocenyl entities in the *peri*-positions of the naphthalene linker.

The repulsion of the metallocenyl substituents is reflected in the angle between the linked cyclopentadienyl (Cp) ligands (∠Cp–Cp) and the torsional angle (Fig. 5.17). The rotational angle of the adjacent Cp ring and the corresponding six-membered subunit of the naphthalene linker (∠Cp–Ar) displays a deviation from the expected ideal head-to-head arrangement of the metallocenes. This might strongly influence the coupling pathway [83–85] that is either through space and/or through bond.

Table 5.4 Selected distances [Å] and angles [°] determined by X-ray crystal structure analysis of **[Co]₂** [83], **[Co*]₂** [83], **[Ni*]₂** [85] and **[V]₂** [84] with esd

	[Co]₂	**[Co*]₂**	**[Ni*]₂**	**[V]₂**
M1–M2	6.7392(4)	6.7244(7)	6.9705(3)	7.1212(3)
ipso–ipso	2.940(2)	3.018(4)	2.987(2)	2.974(1)
peri–peri	2.559(2)	2.563(5)	2.562(3)	2.565(1)
torsion†	29.7(1)	36.8(2)	27.6(1)	27.19(8)
∠Cp–Cp	28.46(6)	33.7(1)	31.48(7)	26.68(4)
∠Cp–Ar	40.28(5)	28.0(1)	42.40(6)	47.93(4)

†For the definition of the angle between the best-fit planes [87] of the corresponding atoms (∠) and the torsional angle see Fig. 5.17

Magnetic Properties

The magnetic behavior of the bismetallocenyl complexes in solution was studied by temperature dependent ^{1}H NMR spectroscopy. All naphthalene-bridged bismetallocenyl complexes displayed Curie behavior in the observed temperature range [83–85]. However, unusual diamagnetic chemicals shifts δ_{dia} were obtained from a linear fit of the experimental chemical shift (5.1) possibly indicating small exchange interactions.

In the solid state all compounds displayed antiferromagnetic behavior. The simulation [69] of the magnetic data revealed a strong influence of the metal center on the exchange interaction between the two spin centers (Table 5.5). While for **[Co]₂** [83] and **[Ni*]₂** [85] similar weak antiferromagnetic exchange interactions were determined (-28.1 and $-31.5\,\text{cm}^{-1}$) only a very weak coupling was found for **[Co*]₂** [83] and **[V]₂** [84]. While the decreased interaction in **[Co*]₂** compared to **[Co]₂** was attributed to geometric changes of the complexes in the solid state [83], the very weak interaction in **[V]₂** can be attributed to electronic effects [84]. In cobaltocene and

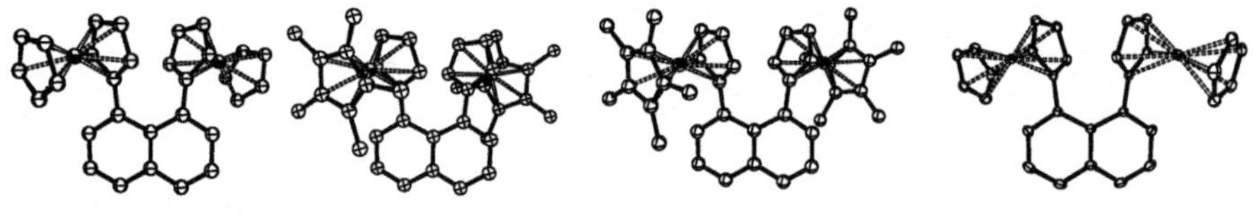

Fig. 5.16 Molecular structures of **[Co]₂** [83], **[Co*]₂** [83], **[Ni*]₂** [85] and **[V]₂** [84] (left to right); hydrogen atoms are omitted for clarity

Fig. 5.17 Schematic representation of the molecular structure of **[M]₂**

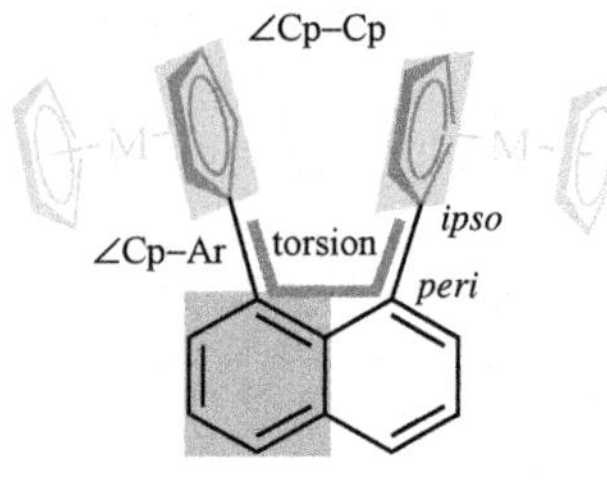

Table 5.5 Experimental and calculated DFT parameters obtained from fitting of the magnetic data of **[Co]$_2$** [83, 88], **[Co*]$_2$** [83], **[Ni*]$_2$** [85] and **[V]$_2$** [84] in the solid state using the Heisenberg model $\hat{\mathcal{H}} = -2J_{12}S_1S_2$ and coupling constants calculated with the TPSSH functional (def2-TZVP basis set)

	[Co]$_2$	**[Co*]$_2$**	**[Ni*]$_2$**	**[V]$_2$**
$S_1 = S_2$	1/2	1/2	1	3/2
J_{12} (cm^{-1}) (DFT)	−28.1	−5.9	−31.5	−2.00 (−0.5 † −0.4 ‡)
$g_1 = g_2$	1.85	2.04	1.81	1.96
Θ_W (K)	−1.6	−4.9	1.40	−3.94
T_N (K)	≈48	≈4	≈101	≈8
$D_1 = D_2$ (cm^{-1})	–	–	n/a	2.83
μ_{eff} (¯B) (300 K)	2.15	2.47	3.20	5.28 (270 K)

†Obtained from the Greens-function approach [89, 90]; ‡obtained from the broken-symmetry approach [91]; Θ_W: *Weiss* constant; T_N: *Néel* temperature; D: zero-field splitting parameter

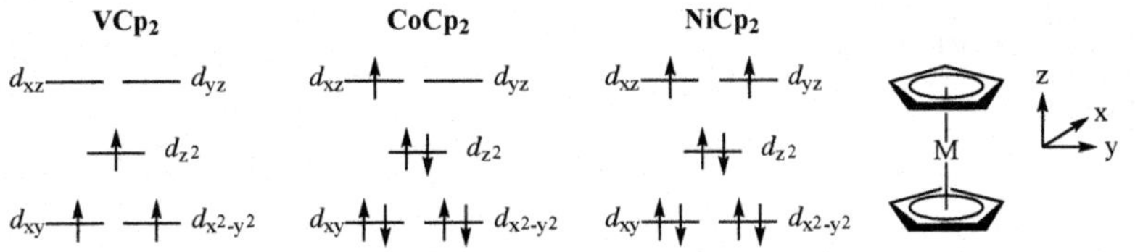

Fig. 5.18 Occupation of the d-orbitals in archetype mononuclear metallocenes

nickelocene the unpaired electrons are located in the d_{xz} and d_{yz} orbitals (Fig. 5.18), which have sufficient overlap with the orbitals of the Cp ligands allowing a distribution of the spin density throughout the ligands and the aromatic linker [83–85]. In vanadocene the unpaired electrons occupy the d_{xy}, $d_{x^2-y^2}$ and d_{z^2} orbitals. Due to the missing overlap of these orbitals with the orbitals of the Cp ligands hardly any spin transfer occurs to the ligands or the aromatic linker and an exchange interaction is strongly reduced [84].

The clear distinction between intra- and intermolecular exchange interactions of bismetallocenyl complexes is challenging considering the crystal packing in the solid state. For the decamethyl bisnickelocenyl complex **[Ni*]$_2$** the molecules form chains in the crystalline state, resulting in similar inter- and intramolecular Ni–Ni distances (Fig. 5.19) [85]. The analysis of diamagnetically diluted samples as well as DFT calculations revealed that the magnetic behavior in the solid state of the naphthalene-bridged bismetallocenyl complexes is dominated by an intramolecular exchange interaction [83–85].

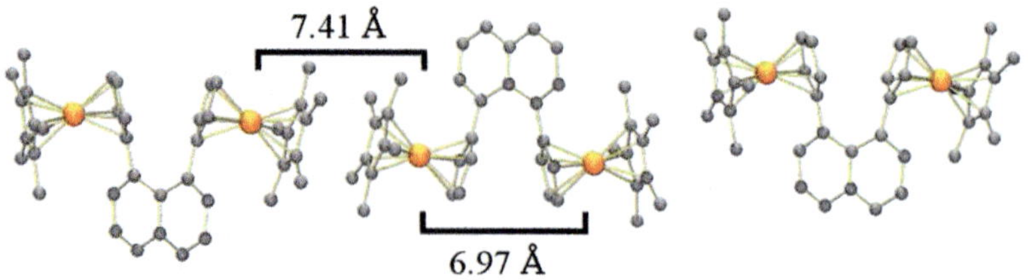

Fig. 5.19 Intra- versus intermolecular Ni–Ni distance in the crystal packing of **[Ni*]**$_2$ [85]

5.4.2 More Stacking

The increase of the number of stacked metallocenes in a spin chain might also have a strong influence on the intramolecular exchange interactions. Therefore, the naphthalene-bridged tri- and tetranuclear cobaltocenyl complexes served as model compounds, bearing three and four $S = 1/2$ spin centers, respectively.

Synthesis

In order to increase the number of naphthalene-bridged cobaltocenes stacked in one direction, the asymmetrically functionalized key compound **2** was synthesized using 1,8-diiodonaphthalene (**1**) as starting material. It was then subjected to a mono iodine-lithium exchange reaction followed by a nucleophilic attack of the in situ formed lithium organyl at cobaltocenium iodide [92] (Scheme 5.8(i)).

The remaining iodo substituent was replaced with a cyclopentadiene substituent in a cross coupling reaction with cyclopentadienyl zinc chloride in the presence of copper(I) iodide. The resulting mixture of **3** was transferred to the cationic trinuclear cobaltocenium complex **[Co]**$_3$(PF_6)$_3$ by deprotonation of the CpH substituent and addition of cobalt(II) chloride.

The oxidation of the formed cobaltocene complex was followed by a two-fold hydride abstraction. Finally, the reduction with decamethylcobaltocene yields the trinuclear air- and moisture sensitive target compound **[Co]**$_3$ [92] (Scheme 5.8(ii), (iv)).

In order to synthesize the tetranuclear cobaltocene complex, the key compound **2** was transferred into a nucleophile via an iodine-lithium exchange and was allowed to attack the biscobaltocenium complex **[Co]**$_2$(BF_4)$_2$ leading to the tetranuclear cobalt(I) complex **4**. A four-fold hydride abstraction yields the tetranuclear cobaltocenium complex **[Co]**$_4$(BF_4)$_4$, which can be reduced to the desired air- and moisture sensitive tetranuclear cobaltocene complex **[Co]**$_4$ using decamethylcobaltocene (Scheme 5.8(v)).

Molecular Structures

X-ray crystal structure determination revealed small structural variations of the stacked cobaltocenium complexes with increasing number of stacked cobaltocenium moieties **[Co]**$_2$(BF_4)$_2$[86], **[Co]**$_3$(PF_6)$_3$ and **[Co]**$_4$(BF_4)$_4$ (Fig. 5.20). The distance between cobalt atoms in close proximity varies between 6.37 Å in **[Co]**$_3$(PF_6)$_3$ and

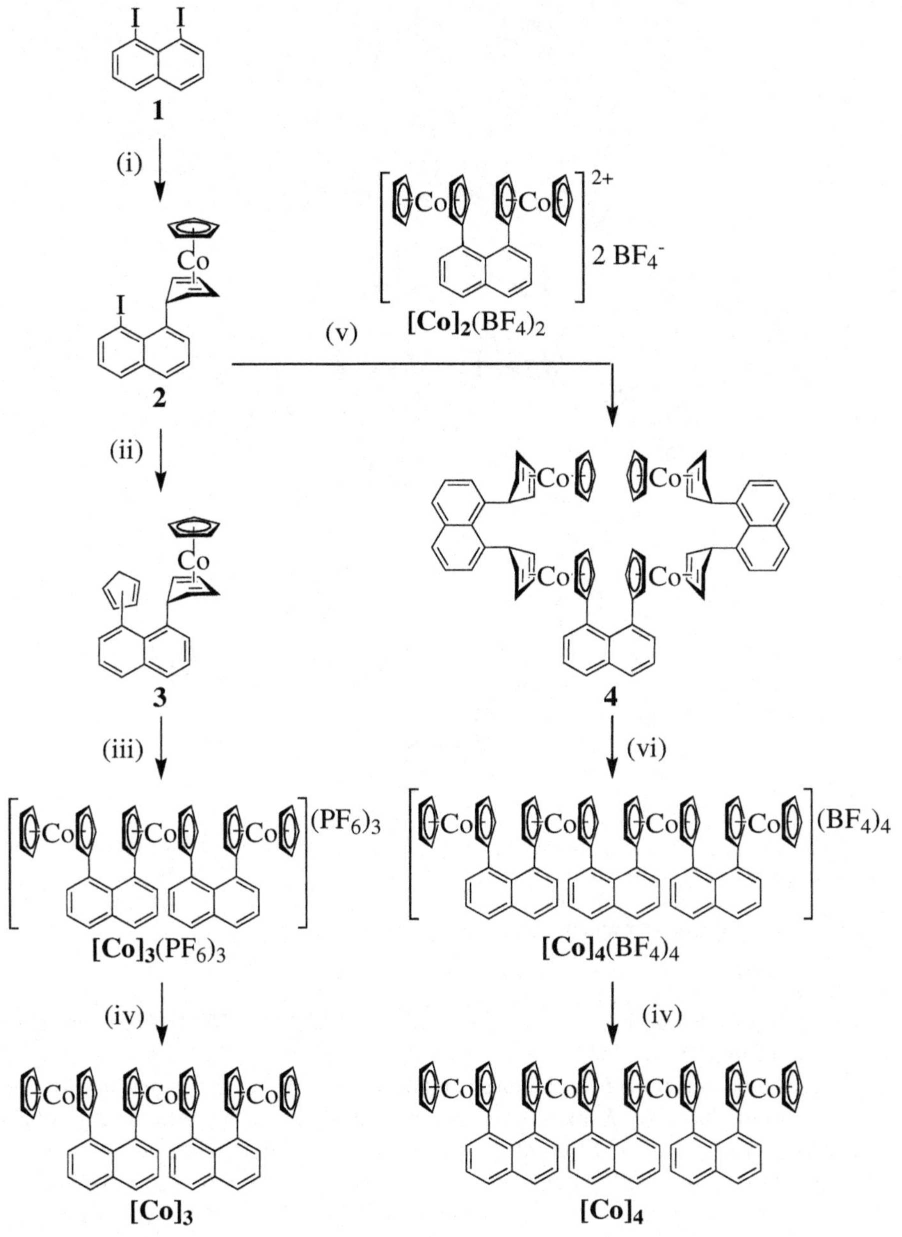

Scheme 5.8 Synthesis of the tri- and tetranuclear cobaltocenyl complexes **[Co]$_3$** [92] and **[Co]$_4$**; reaction conditions: (i): (1) *n*-BuLi, Et_2O, (2) $[CoCp_2]I$, Et_2O; (ii): CpZnCl, CuI, thf; (iii): (1) *n*-BuLi, thf, (2) $CoCl_2$, thf, (3) H_2O, NH_4PF_6, O_2, (4) $Ph_3C^+PF_6^-$, dcm; (iv): $CoCp^*_2$, thf; (v): (1) *n*-BuLi, thf, (2) **[Co]$_2$**$(BF_4)_2$; (vi): $Ph_3C^+BF_4^-$

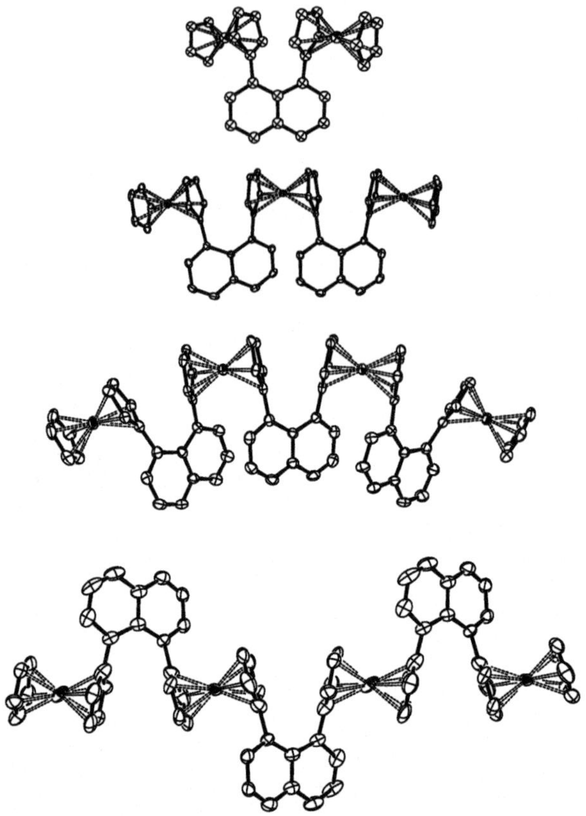

Fig. 5.20 Molecular structures of **[Co]$_2$**$(BF_4)_2$ (Co1–Co2 6.38 Å) [86], **[Co]$_3$**$(PF_6)_3$ (Co1–Co2 6.37 Å, Co1–Co3 12.72 Å), **[Co]$_4$**$(BF_4)_4$ (Co1–Co2 6.44 Å, Co1–Co3 12.66 Å, Co1–Co4 18.39 Å) and **[Co]$_4$**$(BPh_4)_4$ (Co1–Co2 6.7 Å, Co1–Co3 13.41 Å, Co1–Co4 20.11 Å); hydrogen atoms, counterions and co-crystallized solvent molecules are omitted for clarity

6.44 Å in **[Co]$_4$**$(BF_4)_4$. The Co1–Co3 distance of 12.72 Å in **[Co]$_3$**$(PF_6)_3$ drops to 12.66 Å in **[Co]$_4$**$(BF_4)_4$. The remarkable bent structure of **[Co]$_4$**$(BF_4)_4$ is attributed to the fact that the three naphthalene linkers are placed on the same side of the cobaltocenium chain. An X-ray crystal structure analysis of the tetranuclear cobaltocenium complex with a different counterion **[Co]$_4$**$(BPh_4)_4$ revealed an alternating placement of the naphthalene linkers with an increased Co1–Co2 distance of 6.70 Å (Fig. 5.20). Correspondingly, the distance of the outer cobalt atoms Co1–Co4 (20.11 Å) is increased compared to the value in **[Co]$_4$**$(BF_4)_4$ (18.39 Å).

Redox Properties

The redox properties of the di-, tri- and tetranuclear cobaltocenyl complexes were studied by means of square wave voltammetry (SWV). The measurements of **[Co]$_2$**$(BF_4)_2$, **[Co]$_3$**$(PF_6)_3$ and **[Co]$_4$**$(BF_4)_4$ revealed two, three and four separated,

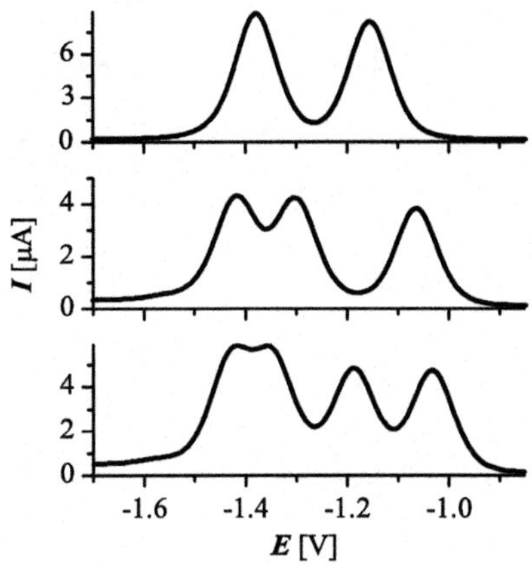

Fig. 5.21 SWV of **[Co]**$_2$(BF_4)$_2$, **[Co]**$_3$(PF_6)$_3$ and **[Co]**$_4$(BF_4)$_4$ (top to bottom); MeCN, RT, [^{n}Bu$_4$N][B(C_6F_5)$_4$] (0.1 M), versus Fc/Fc$^+$, ν = 100 mV/s

Table 5.6 SVW data of **[Co]**$_2$(BF_4)$_2$, **[Co]**$_3$(PF_6)$_3$ and **[Co]**$_4$(BF_4)$_4$

	[Co]$_2$(BF_4)$_2$	**[Co]**$_3$(PF_6)$_3$	**[Co]**$_4$(BF_4)$_4$
$E_{1/2}$ (1)	−1.160	−1.064	−1.044
$E_{1/2}$ (2)	−1.382	−1.296	−1.195
$E_{1/2}$ (3)		−1.419	−1.342
$E_{1/2}$ (4)			−1.430
$\Delta E_{1/2}$ (1/2); (K_c)	0.222 (5.65 × 10^3)	0.232 (8.35 × 10^3)	0.151 (3.57 × 10^2)
$\Delta E_{1/2}$ (2/3); (K_c)		0.123 (1.20 × 10^2)	0.171 (7.77 × 10^2)
$\Delta E_{1/2}$ (3/4); (K_c)			0.050 (7.00)
$\Delta E_{p/2}$ (1)	0.102	0.105	0.116
$\Delta E_{p/2}$ (2)	0.101	0.099	0.104
$\Delta E_{p/2}$ (3)		0.113	0.097
$\Delta E_{p/2}$ (4)			0.103
$\Delta E_{p/2}$ (Fc/Fc+)	0.104	0.102	0.110

MeCN, RT, [^{n}Bu4N][B(C_6F_5)$_4$] (0.1 M), working electrode: Pt-disk, counter electrode: Pt-rod, reference electrode: Pt-wire, versus Fc/Fc+, ν = 100 mV/s, frequency 10 Hz, step potential 5 mV, potentials in (V) ± 0.005 V, $K_c = \exp(nF\Delta E_{1/2}/(RT))$

reversible redox events (Fig. 5.21) in the typical range of the cobaltocene/cobaltocenium redox couple [93] (Table 5.6), indicating electronic communication between the cobalt centers [92].

Magnetic Behavior

The magnetic behavior of the di -, tri- and tetranuclear cobaltocene complexes in solution was studied by ^{1}H NMR spectroscopy at variable temperature (Fig. 5.22). **[Co]**$_2$ [83], **[Co]**$_3$ and **[Co]**$_4$ revealed a linear correlation between the experimental chemical shift and the reciprocal temperature in the observed temperature range according to (5.1) [94].

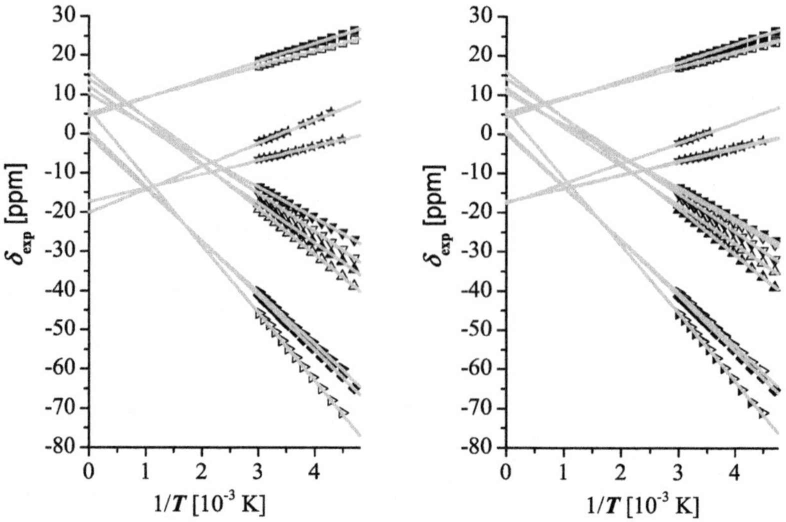

Fig. 5.22 Curie plot of the ^{1}H NMR measurements of **[Co]**$_3$ $R^2 = 0.994$–1.000 (left) and **[Co]**$_4$ $R^2 = 0.984$–0.999 (right); toluene-d$_8$, 400 MHz (213–333 K)

$$\delta_{\text{exp}} = \frac{A g_e \mu_B S(S+1)}{3\frac{\gamma_H}{2\pi} k_B T} + \delta_{\text{dia}}$$
$$A: \text{ hyperfine coupling constant,}$$
$$\gamma_H: \text{ gyromagnetic ratio of the proton} \quad (5.1)$$

The unusual intercepts, representing the diamagnetic shift δ_{dia}, might be attributed to the cobaltocene anomaly [94] or even to a small intramolecular exchange interaction between the spins, which cannot be simulated for the spin systems due to the small temperature range of measurement.

In the solid state the magnetic susceptibility of **[Co]**$_3$ and **[Co]**$_4$ was measured by a vibrating sample magnetometer (VSM) between 3 and 300 K. For both complexes the temperature dependence of the effective magnetic moment indicate an antiferromagnetic exchange interaction (Fig. 5.23).

For both complexes the temperature dependence of the effective magnetic moment indicate an antiferromagnetic exchange interaction

The magnetic data of the trinuclear complex could neither be simulated for a simple $S_1 = S_2 = S_3 = 1/2$ spin system with $g_1 = g_3$ nor treated as a Heisenberg chain, indicating a more complex intermolecular exchange interaction. In the case of the tetranuclear complex, the data can be satisfactorily fitted [69] for a $S_1 = S_2 = S_3 = S_4 = 1/2$ spin system with $g_1 = g_4$ and $g_2 = g_3$ under the assumption that there is only an antiferromagnetic exchange interaction between adjacent spin centers (Fig. 5.24).

The exchange interactions obtained from the fit indicate an increased intramolecular coupling of -58.7 and $-103.2\,\text{cm}^{-1}$ (Table 5.7) compared to the dinuclear com-

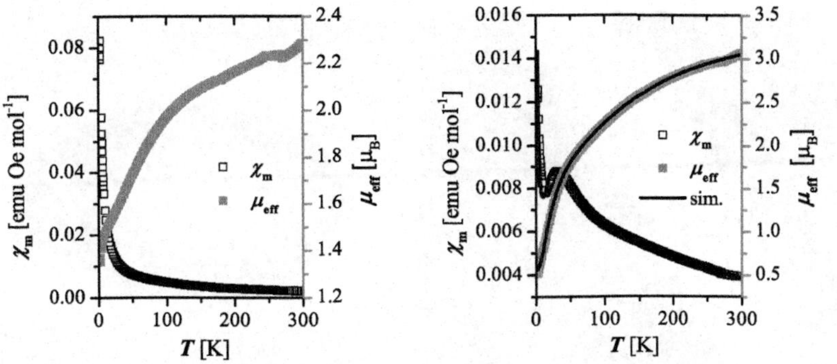

Fig. 5.23 VSM measurements of **[Co]3** (left) and **[Co]4** (right) in the solid state (field-cooled, 1 T)

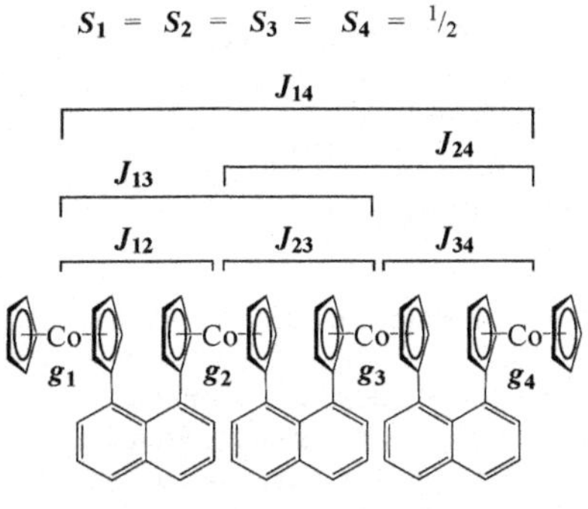

Fig. 5.24 The spin system of **[Co]4**

Table 5.7 Parameters obtained from fitting [69] of the magnetic data of **[Co]4** in the solid state

$g_1 = g_4$	$g_2 = g_3$	$J_{12} = J_{34}$ (cm^{-1})	J_{23} (cm^{-1})	Θ_W (K)	μ_{eff} (μ_B) (300 K)
2.002	2.34	−58.7	−103.2	−11.6	3.04, 3.46†

†Expected value for a non-interacting $S_1 = S_2 = S_3 = S_4 = 1/2$ spin system according to the spin-only formula [9]

pound **[Co]2** (−28.1 cm^{-1}). The antiferromagnetic exchange interaction was also confirmed by DFT calculations [88]. Since no exchange interaction for the isolated molecules in solution was observed, it is likely that the increased antiferromagnetic coupling is influenced by intermolecular interaction in the solid state.

Molecules on Surfaces

The di- and trinuclear complexes can be successfully deposited on surfaces by using the air- and moisture stable related cobaltocenium complexes [92] (Fig. 5.25). The identification of the oligonuclear complexes on the surface is, however, challenging considering that different conformers are possible. This is best illustrated for the molecular structure of the corresponding cobaltocenium complex of **[Co]4** (Fig. 5.20).

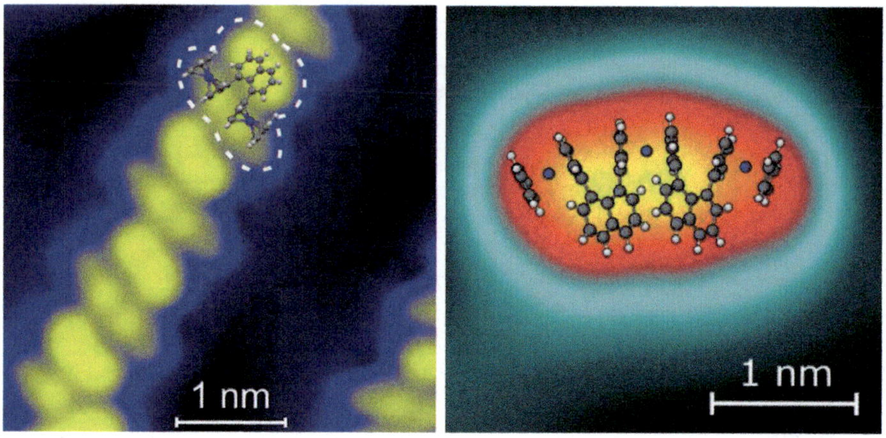

Fig. 5.25 Scanning tunnelling microscope images of **[Co]**$_2$ on a Au(111) surface (left) and **[Co]**$_3$ on a Cu(111) surface

References

1. S.J. Blundell, Lect. Notes Phys. **697**, 345 (2006)
2. O. Kahn, *Molecular Magnetism* (Wiley-VCH, Weinheim, 1993)
3. A.N. Holden, B.T. Matthias, P.W. Anderson, H.W. Lewis, Phys. Rev. **102**, 1463 (1956)
4. S. Ferlay, T. Mallah, L. Ouhahes, P. Veillet, M. Verdaguer, Nature **378**, 701 (1995)
5. S.D. Holmes, G.S. Girolami, J. Am. Chem. Soc. **121**, 5593 (1999)
6. S. Blundell, *Magnetism in Condensed Matter* (Oxford University Press, Oxford, 2001)
7. P. Gütlich, Y. Garcia, H.A. Goodwin, Chem. Soc. Rev. **29**, 419 (2000)
8. S. Chittipeddi, K.R. Cromack, J.S. Miller, A.J. Epstein, Phys. Rev. Lett. **58**, 2695 (1987)
9. J.S. Miller, A.J. Epstein, Angew. Chem. Int. Ed. Engl. **33**, 385 (1994)
10. J.S. Miller, Inorg. Chem. **39**, 4392 (2000)
11. T. Lis, Acta Crystallogr. Sect. B **36**, 2042 (1980)
12. R. Sessoli, D. Gatteschi, A. Caneschi, M.A. Novak, Nature **365**, 141 (1993)
13. R. Sessoli, H.L. Tsai, A.R. Schake, S. Wang, J.B. Vincent, K. Folting, D. Gatteschi, G. Christou, D.N. Hendrickson, J. Am. Chem. Soc. **115**, 1804 (1993)
14. A. Caneschi, D. Gatteschi, N. Lalioti, C. Sangregorio, R. Sessoli, G. Venturi, A. Vindigni, A. Rettori, M.G. Pini, M.A. Novak, Angew. Chem. **113**, 1810 (2001)
15. R. Clérac, H. Miyasaka, M. Yamashita, C. Coulon, J. Am. Chem. Soc. **124**, 12837 (2002)
16. J.S. Miller, M. Drillon, *Magnetism: Molecules to Materials* (Wiley-VCH, Weinheim, 2001)
17. S. Altmannshofer, E. Herdtweck, F.H. Köhler, R. Miller, R. Mölle, E.W. Scheidt, W. Scherer, C. Train, Chem. Eur. J. **14**, 8013 (2008)
18. E. König, V.P. Desai, B. Kanellakopulos, F.H. Köhler, Chem. Phys. **80**, 263 (1983)
19. H. Hilbig, P. Hudeczek, F.H. Köhler, X. Xie, P. Bergerat, O. Kahn, Inorg. Chem. **37**, 4246 (1998)
20. P. Hudeczek, F.H. Köhler, Organometallics **11**, 1773 (1992)
21. G.E. McManis, R.M. Nielson, M.J. Weaver, Inorg. Chem. **27**, 1827 (1988)
22. S.C. Jones, S. Barlow, D. O'Hare, Chem. Eur. J. **11**, 4473 (2005)
23. G. Laus, C.E. Strasser, M. Holzer, K. Wurst, G. Puerstinger, K.H. Ongania, M. Rauch, G. Bonn, H. Schottenberger, Organometallics **24**, 6085 (2005)
24. C. Elschenbroich, M. Wolf, O. Schiemann, K. Harms, O. Burghaus, J. Pebler, Organometallics **21**, 5810 (2002)

25. M. Rosenblum, H.M. Nugent, K.S. Jang, M.M. Labs, W. Cahalane, P. Klemarczyk, W.M. Reiff, Macromolecules **28**, 6330 (1995)
26. C. Elschenbroich, B. Kanellakopulos, F.H. Köhler, B. Metz, R. Lescouëzec, N.W. Mitzel, W. Strauß, Chem. Eur. J. **13**, 1191 (2007)
27. H. Hopf, J. Dannheim, Angew. Chem. **100**, 724 (1988)
28. R. Roers, F. Rominger, B. Nuber, R. Gleiter, Organometallics **19**, 1578 (2000)
29. C. Schaefer, G. Scholz, R. Gleiter, T. Oeser, F. Rominger, Eur. J. Inorg. Chem. 1274–1281 (2005)
30. A.A. Khajetoorians, J. Wiebe, B. Chilian, R. Wiesendanger, Science **332**, 1062 (2011)
31. R.A. Layfield, Organometallics **33**(5), 1084 (2014)
32. J.R. Friedman, M.P. Sarachik, Annu. Rev. Condens. Matter Phys. **1**(1), 109 (2010)
33. L. Bogani, W. Wernsdorfer, Nat. Mater. **7**, 179 (2008)
34. A.A. Khajetoorians, J. Wiebe, B. Chilian, R. Wiesendanger, Science (New York, N.Y.) **332**(6033), 1062 (2011)
35. S. Kuck, S.H. Chang, J.P. Klöckner, M.H. Prosenc, G. Hoffmann, R. Wiesendanger, ChemPhysChem **10**, 2008 (2009)
36. S. Fremy, A. Schwarz, K. Lämmle, M.H. Prosenc, R. Wiesendanger, Nanotechnology **20**, 405608 (2009)
37. A. DiLullo, S.H. Chang, N. Baadji, K. Clark, J.P. Klöckner, M.H. Prosenc, S. Sanvito, R. Wiesendanger, G. Hoffmann, S.W. Hla, Nano Lett. **12**, 3174 (2012)
38. E. Sierda, M. Abadia, J. Brede, M. Elsebach, B.E.C. Bugenhagen, M.H. Prosenc, C. Rogero, M. Bazarnik, R. Wiesendanger, ACS Nano **11**, 9200 (2017)
39. B.E.C. Bugenhagen, M.H. Prosenc, Dalton Trans. **45**, 7460 (2016)
40. B.E.C. Bugenhagen, M.H. Prosenc (Manuscript in preparation)
41. M. Bazarnik, B. Bugenhagen, M. Elsebach, E. Sierda, A. Frank, M.H. Prosenc, R. Wiesendanger, Nano Lett. **16**(1), 577 (2016)
42. C. Garcia-Fernandez, E. Sierda, M. Abadia, B.E.C. Bugenhagen, M.H. Prosenc, R. Wiesendanger, M. Bazarnik, J.E. Ortega, J. Brede, E. Matito, A. Arnau, J. Phys. Chem. C (submitted)
43. J. Oldengott, A. Stammler, H. Bögge, T. Glaser, Dalton Trans. **44**, 9732 (2015)
44. P.H.M. Budzelaar, Eur. J. Inorg. Chem. **2012**(3), 530 (2012)
45. S.K. Russell, A.C. Bowman, E. Lobkovsky, K. Wieghardt, P.J. Chirik, Eur. J. Inorg. Chem. **2012**(3), 535 (2012)
46. S.C.E. Stieber, C. Milsmann, J.M. Hoyt, Z.R. Turner, K.D. Finkelstein, K. Wieghardt, S. DeBeer, P.J. Chirik, Inorg. Chem. **51**(6), 3770 (2012)
47. D. Sieh, M. Schlimm, L. Andernach, F. Angersbach, S. Nückel, J. Schöffel, N. Šušnjar, P. Burger, Eur. J. Inorg. Chem. **2012**(3), 444 (2012)
48. S.C. Bart, K. Chłopek, E. Bill, M.W. Bouwkamp, E. Lobkovsky, F. Neese, K. Wieghardt, P.J. Chirik, J. Am. Chem. Soc. **128**(42), 13901 (2006)
49. T. Buban, Untersuchungen zu abstandsabhängigen magnetischen Wechselwirkungen in schwach gekoppelten Mehrkernsystemen. Dissertation, Universität Hamburg, Hamburg, 2017
50. N. Miyaura, A. Suzuki, Chem. Rev. **95**(7), 2457 (1995)
51. W. Dammann, T. Buban, C. Schiller, P. Burger, Dalton Trans. in press http://pubs.rsc.org/en/content/articlepdf/2018/DT/C8DT02347F
52. J. Cámpora, A.M. Naz, P. Palma, A. Rodrı́guez-Delgado, E. Álvarez, I. Tritto, L. Boggioni, Eur. J. Inorg. Chem. **2008**(11), 1871 (2008)
53. E.A.H. Griffiths, G.J.P. Britovsek, V.C. Gibson, I.R. Gould, Chem. Commun. **1999**(14), 1333 (1999)
54. B.L. Small, M. Brookhart, A.M.A. Bennett, J. Am. Chem. Soc. **120**(16), 4049 (1998)
55. Y. Huang, J. Chen, L. Chi, C. Wei, Z. Zhang, Z. Li, A. Li, L. Zhang, J. Appl. Polym. Sci. **112**(3), 1486 (2009)
56. R.-Q. Fan, D.-S. Zhu, L. Mu, G.-H. Li, Q. Su, J.-G. Ni, S.-H. Feng, Chem. Res. Chin. Univ. **21**(4), 496 (2005)
57. R.-Q. Fan, Y.-L. Yang, Z.-W. Lü, Chem. Res. Chin. Univ. **24**(1), 4 (2008)
58. Q. Knijnenburg, S. Gambarotta, P.H.M. Budzelaar, Dalton Trans. **2006**(46), 5442 (2006)

59. F.H. Allen, O. Kennard, D.G. Watson, L. Brammer, A.G. Orpen, R. Taylor, J. Chem. Soc. Perkin Trans. **2**(12), S1 (1987)
60. B.L. Small, M. Brookhart, J. Am. Chem. Soc. **120**(28), 7143 (1998)
61. T.M. Baker, T.L. Mako, A. Vasilopoulos, B. Li, J.A. Byers, M.L. Neidig, Organometallics **35**(21), 3692 (2016)
62. G.J.P. Britovsek., K. Clentsmith, V.C. Gibson, D.M.L. Goodgame, S.J. McTavish, Q.A. Pankhurst, Catal. Commun. **3**(5), 207 (2002)
63. P. Gütlich, Mössbauer spectroscopy and transition metal chemistry (2011)
64. Y. Yoshida, G. Langouche, *Mössbauer Spectroscopy* (Springer, Berlin, 2013)
65. P. Gütlich, Chem. unserer Zeit **4**(5), 133 (1970)
66. N.N. Greenwood, T.C. Gibb, *Mössbauer Spectroscopy* (Springer, Netherlands, Dordrecht, 1971)
67. T. Buban, P. Burger (Manuscript in preparation)
68. A.A. Granovsky, T. Buban, P. Burger, Manuscript in preparation. (FireFly version 8, http://classic.chem.msu.su/gran/firefly/index.html
69. E. Bill, julX simulation of molecular magnetic data (2008)
70. E. Evangelio, D. Ruiz-Molina, Eur. J. Inorg. Chem. **2005**(15), 2957 (2005)
71. A.S. Attia, O.-S. Junga, C.G. Pierpont, Inorg. Chim. Acta **226**(1), 91 (1994)
72. C.G. Pierpont, Coord. Chem. Rev. **216–217**, 99 (2001)
73. N. Shaikh, S. Goswami, A. Panja, X.-Y. Wang, S. Gao, R.J. Butcher, P. Banerjee, Inorg. Chem. **43**(19), 5908 (2004)
74. A. Gerli, K.S. Hagen, L.G. Marzilli, Inorg. Chem. **30**(24), 4673 (1991)
75. C. Fukuhara, K. Tsuneyoshi, N. Matsumoto, S. Kida, M. Mikuriya, M. Mori, J. Chem. Soc. Dalton Trans. **1990**(11), 3473 (1990)
76. V. Kasempimolporn, H. Okawa, S. Kida, Bull. Chem. Soc. Jpn. **52**(7), 1928 (1979)
77. I. Nemec, H. Liu, R. Herchel, Z. Xuequan, Z. Trávníček, Synth. Met. **215**, 158 (2016)
78. A. Hinaut, R. Pawlak, E. Meyer, T. Glatzel, Beilstein J. Nanotechnol. **6**, 1927 (2015)
79. O.V. Salata, Curr. Nanosci. **1**(1), 25 (2005)
80. M. Hermanowicz, Private communication (21.12.2016)
81. J.A.J. Burgess, L. Malavolti, V. Lanzilotto, M. Mannini, S. Yan, S. Ninova, F. Totti, S. Rolf-Pissarczyk, A. Cornia, R. Sessoli, S. Loth, **6**, 8216 EP (2015)
82. J. Tersoff, D.R. Hamann, Phys. Rev. B **31**(2), 805 (1985)
83. N. Pagels, O. Albrecht, D. Görlitz, A.Y. Rogachev, M.H. Prosenc, J. Heck, Chem. Eur. J. **17**, 4166 (2011)
84. S. Puhl, T. Steenbock, R. Harms, C. Herrmann, J. Heck, Dalton Trans. **46**, 15494 (2017)
85. S. Trtica, M.H. Prosenc, M. Schmidt, J. Heck, O. Albrecht, D. Görlitz, F. Reuter, E. Rentschler, Inorg. Chem. **49**, 1667 (2010)
86. N. Pagels, J. Heck, J. Organomet. Chem. **693**, 241 (2008)
87. K. Brandenburg, *Diamond*, Crystal Impact GbR, Bonn, version 3.1 edn. (1997–2005)
88. T. Steenbock, Understanding and controlling exchange spin coupling: insights from first-principles calculations. Ph.D. thesis, University of Hamburg, 2016
89. T. Steenbock, C. Herrmann, J. Comput. Chem. **39**, 81 (2018)
90. T. Steenbock, J. Tasche, A. Lichtenstein, C. Herrmann, J. Chem. Theory Comput. **11**, 5651 (2015)
91. L. Noodleman, J. Chem. Phys. **74**, 5737 (1981)
92. M. Gruber, T. Knaak, S. Puhl, F. Benner, A. Escribano, J. Heck, R. Berndt, J. Phys. Chem. C. **121**(48), 26777 (2017)
93. R.S. Stojanovic, A.M. Bond, Anal. Chem. **65**, 56 (1993)
94. I. Bertini, C. Luchinat, G. Parigi, *Solution NMR of Paramagnetic Molecules - Applications to Metallobiomolecules and Models* (Elsevier, New York, 2001)
95. F.H. Köhler, J. Organomet. Chem. **121**, C61 (1976)

Chapter 6
Designing and Understanding Building Blocks for Molecular Spintronics

Carmen Herrmann, Lynn Groß, Bodo Alexander Voigt, Suranjan Shil and Torben Steenbock

Abstract Designing and understanding spin coupling within and between molecules is important for, e.g., nanoscale spintronics, magnetic materials, catalysis, and biochemistry. We review a recently developed approach to analyzing spin coupling in terms of local pathways, which allows to evaluate how much each part of a structure contributes to coupling, and present examples of how first-principles electronic structure theory can help to understand spin coupling in molecular systems which show the potential for photo- or redoxswitching, or where the ground state is stabilized with respect to spin flips by adding unpaired spins on a bridge connecting two spin centers. Finally, we make a connection between spin coupling and conductance through molecular bridges.

6.1 Introduction

In electronics on a very small scale, heating due to electron currents flowing through thin wires has become a major problem [1]. In spintronics, information is stored, transferred and processed employing the spin rather than the charge degree of freedom. Spintronics offers, in principle, a solution to the heating challenge: By building up chains of spins which are coupled to their neighbors by (super-)exchange or by Ruderman–Kittel–Kasuya–Yosida (RKKY) interactions (when adsorbed on a metal surface), the flip of a spin on one end of the chain can be passed on along the chain, thus transferring information but not charge. This has been exploited in an experiment by Khajetoorians et al. [2], in which it was demonstrated that chains of iron atoms deposited on a copper substrate can be used to build a spin logic gate. The inputs are controlled by cobalt clusters of different size whose magnetization can be individually switched by an external magnetic field due to their different coercivities, and the output is read out by the tip of a spin-polarized scanning tunneling microscope

C. Herrmann (✉) · L. Groß · B. A. Voigt · S. Shil · T. Steenbock
Institute for Inorganic and Applied Chemistry, University of Hamburg,
Martin-Luther-King-Platz 6, 20146 Hamburg, Germany
e-mail: carmen.herrmann@chemie.uni-hamburg.de

R. Wiesendanger (ed.), *Atomic- and Nanoscale Magnetism*, NanoScience and Technology, https://doi.org/10.1007/978-3-319-99558-8_6

Fig. 6.1 Schematic representation of two spin centers (e.g., metal atoms) linked by a ligand adsorbed on a surface, with two possible competing spin–spin interactions

(STM). The atoms are coupled antiferromagnetically by RKKY interactions mediated by the conduction electrons. Controlling this interaction requires controlling the distance between the atoms on the surface, which is achieved by manipulating them with the STM tip. A simpler approach to constructing spin chains is provided by molecular self-assembly on surfaces. Molecules consisting of spin-polarized metal ions and organic ligands can be covalently linked into antiferromagnetically coupled chains and branched structures by thermally activated surface-mediated debromination [3–5]. Chain lengths of up to 81 nm could be achieved [5]. Linking metal atoms via ligands in this way introduces two competing pathways for spin coupling: one is through the surface (RKKY) and one through the bridging ligand(s) (see Fig. 6.1). To find out which of these two dominates, one usually resorts to density functional theory (DFT) calculations, comparing spin coupling for pairs of molecules on the substrate with pairs of molecules in the vacuum. In the covalently linked molecular chains discussed above, this procedure suggested that spin coupling is mediated solely through the ligands [3]. For a checkerboard structure of molecular spin centers whose ligands were *not* covalently linked to each other, the same approach suggested that only RKKY interactions are responsible for spin coupling [6]. For similarly non-linked charged tetracyano-*p*-quinonedimethane molecules adsorbed on graphene/Ru(0001), in contrast, interactions between the molecules rather than the substrate were suggested as being responsible for spin interactions [7].

Taking away the substrate will not strongly modify the electronic structure of the adsorbed molecule if molecule–substrate interactions are sufficiently weak. For cases where these interactions affect the electronic structure of the adsorbate and for the sake of efficiency, it is desirable to have a computational scheme which allows to evaluate the dominant coupling pathway more directly. Such a scheme is also helpful for disentangling the different contributions to spin coupling that may arise *within* a given molecule (e.g., because different bridging ligands are present). In Sect. 6.2.2, such a local decomposition scheme is presented [8]. It is based on a Green's function approach to evaluating spin coupling from the electronic structure of one spin state only (rather than from energy differences between two spin states) which has been established in solid-state science [9] and only occasionally been applied to molecules [10–12] (with spin densities clearly localized on the metal atoms rather than partially delocalized onto ligands). Therefore, it had to be ensured that the approach works generally well for molecules [13]. For this purpose, it was brought into a form suitable for interfacing with quantum chemical electronic structure codes [13] based on previous work by Han, Ozaki and coworkers [12]. Both is summarized in Sect. 6.2.1.

As an additional advantage, spin coupling between or within molecules can be controlled chemically to a large degree. The term "chemical control" can refer to optimizing molecular bridges and/or spin centers in terms of chemical constitution, substituents and molecular topology, and to constructing structures that can be switched by external stimuli, thus modifying spin coupling by, e.g., illumination with visible or ultraviolet light [14–19]. We discuss a candidate for such photoswitching of spin coupling in Sect. 6.3.1, and elucidate, with the help of DFT calculations, possible reasons for the unfavorable switching behavior of the complex. In Sect. 6.3.2, an alternative switching mechanism is discussed: a ferrocene unit bridging two organic radicals is oxidized, so that an additional unpaired spin on the bridge is introduced. As for the photoswitching in the example above, this does not affect the type of spin coupling (i.e., there is no change between ferro- and antiferromagnetic coupling), but it strongly changes its magnitude (i.e., the energy difference between the ferro- and antiferromagnetically coupled states). Such oxidation switching of spin coupling has also been studied experimentally and theoretically for different types of bridges [20–22]. The effect of oxidation may also be relevant, for example, when comparing isolated molecules with molecules adsorbed on surfaces, as the interaction with a metal substrate may lead to charge transfer between molecule and surface. To isolate the effect of an additional spin on the bridge from the effect of charging, a comparative study on neutral bridges with and without unpaired spin is finally presented in Sect. 6.3.3, which points to achieving delocalization of the spin density onto the bridge as a major goal for synthetic efforts towards molecules or molecular chains with large spin coupling. Introducing spin on the bridge as a means to achieve larger spin coupling has gained increase interest in recent years [23].

The logic gate described above operates at a temperature of 0.3 K. Employing molecules and optimizing their interactions may lead to devices with higher operating temperatures. At the same time, these potential technological applications are by far not the only reason why we are interested in understanding and designing spin interactions within and between molecules. Such spin interactions are important, e.g., for catalysts and biological systems, and ligands mediating them may be understood as a specific example of communication through molecular bridges, which is also relevant for, e.g., electron transfer and transport through such bridges [24–30]. Section 6.4 summarizes several examples of how first-principles electronic structure calculations can help to draw analogies between spin coupling and conductance, and to understand conductance in cases where spin plays an important role.

6.2 Local Pathways in Exchange Spin Coupling

In contrast to various properties such as charge [31, 32], spin [33–35], electric dipole moments [36–38], electron transport and transfer properties [39–46], Raman or vibrational Raman optical activity intensities [47, 48], and X-ray absorption intensities [49, 50], there is no straightforward scheme available for analyzing which parts of a molecular structure contribute to coupling between local spins. We present here such a scheme, based on a Green's function approach from solid-state physics [9].

6.2.1 Transferring a Green's Function Approach to Heisenberg Coupling Constants J from Solid State Physics to Quantum Chemistry

In quantum chemistry, the coupling between spin centers is usually evaluated from the energy difference between a ferromagnetically coupled and an antiferromagnetically coupled state (see Fig. 6.2). If one assumes a cosine dependence of the electronic energy on the angle between the two spin vectors located at the spin centers (see Fig. 6.2), one can estimate this energy difference by looking at the electronic structure of one spin state only: as illustrated in Fig. 6.2, the larger the energy difference, the larger the curvature of the energy at one of the extrema corresponding to ferro- or antiferromagnetic coupling. This was the basis for the approach developed for solid-state structures [9]. We have checked whether the energy does indeed show such a cosine behavior and found that as long as the spin does not strongly delocalize onto the bridge, with bridge atoms sharing delocalized spin from different spin centers, this is usually the case [13]. Of course, differences in molecular orbitals (MOs) and molecular structures in different spin states are neglected by such an approach. Nonetheless, it was found to give reliable spin couplings for a wide range of structures. Compared to an approach by Peralta and coworkers [51, 52], in which orbital relaxation upon spin rotation is taken into account by solving the coupled perturbed Kohn–Sham equations, the Green's function approach sacrifices some accuracy for the advantage of being a straightforward postprocessing scheme.

If magnetic anisotropy is low, spin coupling is usually well described by a Heisenberg Hamiltonian

$$\hat{H} = -2J \sum_{A>B} \hat{\mathbf{S}}_A \cdot \hat{\mathbf{S}}_B, \tag{6.1}$$

where J refers to the spin coupling constant (which is positive for ferromagnetic and negative for antiferromagnetic coupling), and $\hat{\mathbf{S}}_A$ and $\hat{\mathbf{S}}_B$ to the local spin operators for spin centers A and B. By comparing the energy change due to a small spin rotation between a Greens-function energy expression and the Heisenberg model, using the local force theorem, and introducing local projection operators onto the spin

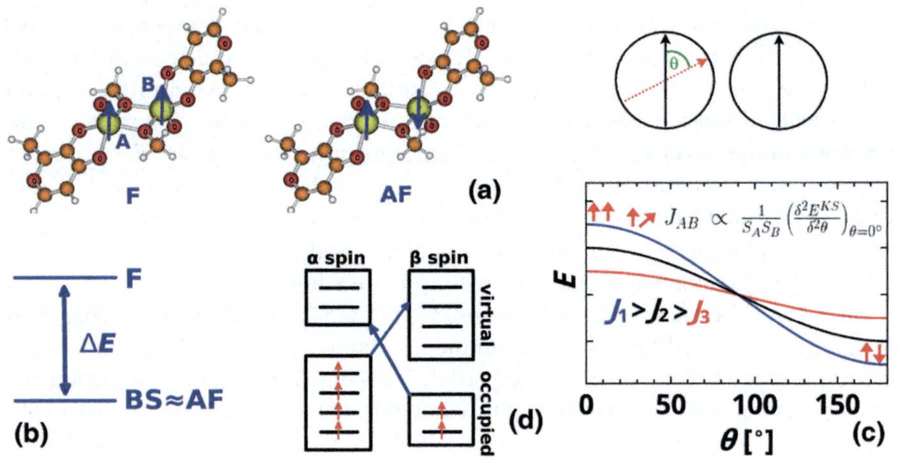

Fig. 6.2 The standard approach to evaluating spin coupling in quantum chemistry is based on evaluating energy differences between ferromagnetically (F) and antiferromagnetically (AF) coupled states (**a**), where the AF state is typically modeled by a Broken-Symmetry (BS) determinant in Kohn–Sham DFT (**b**). In the approach from solid-state physics adopted here, spin coupling is rather evaluated from the curvature of the potential energy as a function of the angle θ between two local spin vectors (**c**). The larger the energy difference, the larger the curvature. The resulting expression involves sums over pairs **d** of occupied spin-up or α orbitals and unoccupied spin-down or β orbitals, or vice versa ("spin-flip excitations"; compare (6.2)).

centers to define so-called on-site potentials, we arrive at the following equation for J (evaluated by processing the electronic structure of the ferromagnetically coupled state),

$$\begin{aligned}
J(\mathrm{F}) = &-\frac{1}{4S_A S_B} \sum_{\substack{i \in \mathrm{occ} \\ k \in \mathrm{virt}}} \sum_{\substack{\{\mu,\nu\} \in A \\ \{\mu',\nu'\} \in B}} C^{\alpha}_{\nu i} \left(F^{\alpha}_{\mu\nu} - F^{\beta}_{\mu\nu}\right) C^{\beta *}_{\mu k} \\
&C^{\alpha *}_{\mu' i} \left(F^{\alpha}_{\mu'\nu'} - F^{\beta}_{\mu'\nu'}\right) C^{\beta}_{\nu' k} \frac{1}{\epsilon^{\alpha}_{i} - \epsilon^{\beta}_{k}} \\
&-\frac{1}{4S_A S_B} \sum_{\substack{k \in \mathrm{occ} \\ i \in \mathrm{virt}}} \sum_{\substack{\{\mu,\nu\} \in A \\ \{\mu',\nu'\} \in B}} C^{\alpha}_{\nu i} \left(F^{\alpha}_{\mu\nu} - F^{\beta}_{\mu\nu}\right) C^{\beta *}_{\mu k} \\
&C^{\alpha *}_{\mu' i} \left(F^{\alpha}_{\mu'\nu'} - F^{\beta}_{\mu'\nu'}\right) C^{\beta}_{\nu' k} \frac{1}{\epsilon^{\beta}_{k} - \epsilon^{\alpha}_{i}}, \\
= &\sum_{ik} j(i,k).
\end{aligned} \tag{6.2}$$

For a detailed derivation, see [9, 13]. The sums run over pairs of occupied and unoccupied ("virtual") orbitals of opposite spin and over single-particle basis functions

μ, ν located on the spin centers A and B. $F^{\sigma}_{\mu'\nu'}$ refers to the elements of the Fock matrices for electrons of spin $\sigma \in \{\alpha, \beta\}$, where α refers to spin-up or majority spins and β refers to spin-down or minority spins, in a basis of Löwdin-orthogonalized atom-centered single-particle basis functions. In other words, this is a matrix representation of the effective single-particle operators in Kohn–Sham DFT. $C^{\sigma}_{\nu i}$ denotes the molecular orbital coefficients for a given spin σ in the same basis. S_A and S_B refer to the local spin quantum numbers on the spin centers A and B. So far, we have used ideal local spin quantum numbers rather than the local spins $\langle \hat{S}_{zA} \rangle = \frac{1}{2}(N^{\alpha}_A - N^{\beta}_A)$, where N^{σ}_A is the number of electrons with spin σ on atom A.[1]

In our case, Löwdin projectors were used [31, 53, 54], but other choices, may have their benefits as well (e.g., local partitioning schemes based on three-dimensional Cartesian space rather than on single-particle basis functions)—compare the long list of methods for analyzing partial atomic charges (population analysis). It should also be noted that there is a certain ambiguity in defining the on-site potentials (which reflect the difference in potential a spin-up electron experiences on an atom compared with a spin-down electron) in terms of local projection operators. This is discussed in more detail in [13] and shown to not strongly affect the resulting coupling constants. In all cases considered here, the two spin centers were chosen as the two atoms on which the unpaired spins are formally located (i.e., the metal atoms). In particular in strongly delocalized systems such as organic radicals, including more atoms can be advisable.

If the electronic structure of the antiferromagnetically coupled state was employed, an expression equivalent to (6.2) would be obtained, except for a sign change (J(AF) $= -J$(F)). In Kohn–Sham DFT, the wave function of the noninteracting reference system in the antiferromagnetically coupled state is usually modeled by a so-called Broken-Symmetry determinant [55], which breaks spin symmetry. There is some debate in the literature on whether this is formally correct [56–59], in particular since it is not clear how to evaluate the total spin in Kohn–Sham DFT [60, 61]. In practice, the Broken-Symmetry approach has been very successful in modeling molecular structures and energetics of antiferromagnetically coupled systems [56, 62], and whether spin projection is considered necessary or not typically has a much smaller effect on the resulting J than the choice of approximate exchange–correlation functional. We found that when applied to a Broken-Symmetry determinant in Kohn–Sham DFT, the Green's function approach ((6.2) with a sign change) does not even consistently produce qualitatively reliable coupling constants (see Fig. 6.3). Figure 6.3 shows data for two transition metal complexes which are particularly challenging, as the unpaired spin is partially delocalized from the metal atoms onto the ligands.

The lower panels of Fig. 6.3 also illustrate the challenge resulting from structural differences in the minimum-energy structures in the two spin states: depending on which molecular structure is chosen, the sign of the predicted coupling constant

[1] Ideal local spin quantum numbers would be $S_A = \frac{1}{2}$ for a spin center with formally one unpaired electron, while local spins reflect the decrease of this number that results from delocalization of unpaired spin density onto neighboring atoms such as ligands.

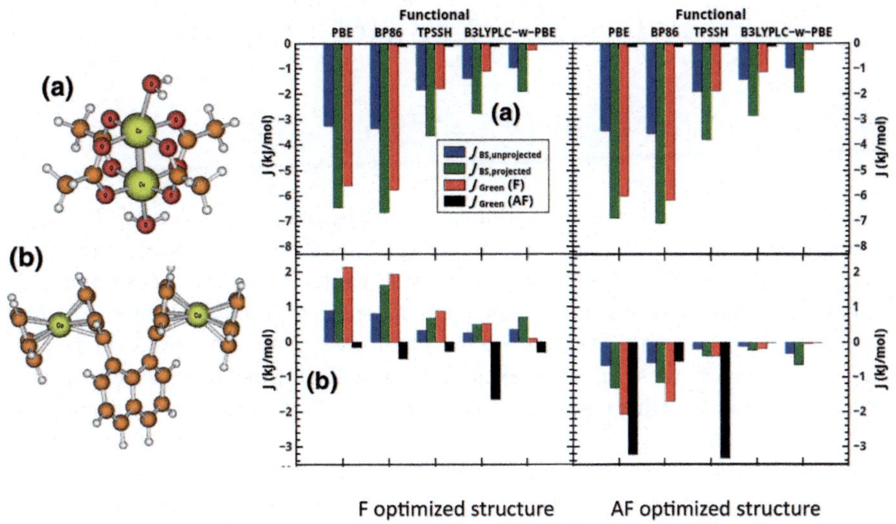

Fig. 6.3 Exchange coupling constants J for a $[Cu_2]$ (top) and a bis-cobaltocene complex (bottom), which show non-negligible delocalization of spin density from the metal atoms onto the ligands [13]. J is evaluated employing either the "traditional" energy-difference method between a ferromagnetically coupled and a Broken-Symmetry (BS) determinant with or without spin projection, or from (6.2) based on a ferromagnetically (F) or antiferromagnetically coupled (AF) electronic structure. The molecular structures were optimized either in the F state (left) or in the AF state (right). Different approximate exchange–correlation functionals were used (see x axis) in combination with a def-TZVP basis set. The experimental data are $J = -1.54$ kJ/mol for $[Cu_2]$ [63] and $J = -0.33$ kJ/mol for bis-cobaltocene [64]. In other words, for the bis-cobaltocene a prediction of the spin multiplicity of the ground state is not possible with the Green's function approach employing the chosen DFT settings. This is, however, the only case we have encountered so far in which J was this sensitive to the electronic structure

changes. This was the only case observed in our studies so far, but it would clearly be valuable for future work to establish a reliable measure or rule for when such structural differences play a role.

6.2.2 Decomposing J into Local Contributions

Equation (6.2) consists of sums over pairs of occupied spin-up (or α) and unoccupied spin-down (or β) orbitals, or vice versa (see Fig. 6.2d). These may be considered as "spin-flip excitations" $j(i, k)$,

$$j(i,k) = \frac{q}{4S_A S_B} \sum_{\substack{\{\mu,\nu\}\in A \\ \{\mu',\nu'\}\in B}} C^{\alpha}_{\nu i}\left(F^{\alpha}_{\mu\nu} - F^{\beta}_{\mu\nu}\right) C^{\beta *}_{\mu k}$$

$$C^{\alpha *}_{\mu' i}\left(F^{\alpha}_{\mu'\nu'} - F^{\beta}_{\mu'\nu'}\right) C^{\beta}_{\nu' k} \frac{1}{\epsilon^{\alpha}_{i} - \epsilon^{\beta}_{k}}, \tag{6.3}$$

where i refers to the index of an α orbital and k to the index of a β orbital. The factor q is equal to 1 if i is occupied and k unoccupied, and -1 if i is unoccupied and k occupied. For pairs of occupied or unoccupied orbitals, q is zero. These molecular orbitals are often predominantly localized on a certain part of a molecular structure, for example on a certain bridging ligand or on one or several spin centers. This suggests to employ these spin-flip excitations directly for the analysis of local contributions to spin coupling pathways. This is not straightforward, since the individual $j(i,k)$ make large contributions of opposite signs, which then barely cancel to result in the total coupling constant. This is analyzed in detail for the H_2 molecule in [8].

As an alternative, one can focus on, e.g., the occupied MOs, and sum over all spin-flip excitations from each occupied MO,

$$j^{\alpha}_{\text{MO}}(i) = \sum_{k\in\beta,\text{virt}} j(i,k) \tag{6.4}$$

$$j^{\beta}_{\text{MO}}(k) = \sum_{i\in\alpha,\text{virt}} j(i,k) \tag{6.5}$$

so that

$$J(\text{F}) = \sum_{i\in\text{occ},\alpha} j^{\alpha}_{\text{MO}}(i) + \sum_{k\in\text{occ},\beta} j^{\beta}_{\text{MO}}(k). \tag{6.6}$$

Other choices such as focusing on, e.g., all α orbitals (occupied or unoccupied) would be equally valid in principle. One argument for selecting occupied MOs of both spins would be that these are variationally optimized, while the virtual orbitals are not (except for being orthogonal to the occupied ones). These MO contributions are typically much smaller in absolute value than the (nearly canceling) spin-flip excitations they are constructed from, so that they appear to be a more reasonable choice for further analysis.

One can then proceed as follows: First, the most important contributions j^{σ}_{MO} from occupied orbitals are selected (with a cutoff chosen either as a certain percentage of J, or as an absolute value). Then, the largest spin-flip excitations contributing to those can be analyzed further. The advantage of such an orbital-based approach is that it can generate insight in terms of orbital symmetry. There is often some ambiguity in deciding which orbital resides on which part of the structure, so that this scheme is not ideally suited for an automated decomposition. Also, if one is predominantly interested in contributions from different atom-centered basis functions on the spin centers (in particular those corresponding to d orbitals on the metal atoms), an alternative scheme in which the individual terms resulting from the double sum over the basis functions located on the spin centers in (6.2) are analyzed may be more promising [65, 66].

A more straightforward scheme for analyzing the contributions of different regions of space to spin coupling is an atomic decomposition, which can be based on defining weights of each molecular orbital i of spin σ on an atom or molecular fragment A,

$$\omega_A^\sigma(i) = \frac{\sum_{\mu \in A} \left| C_{\mu i}^\sigma \right|^2}{\sum_\nu \left| C_{\nu i}^\sigma \right|^2}, \quad (6.7)$$

$$\sum_A \omega_A^\sigma(i) = 1, \quad (6.8)$$

(where $C_{\nu i}^\sigma$ again refers to MO coefficients w.r.t. a Löwdin-transformed basis) and then defining a fragment contribution to J such that the spin-flip excitations are weighted according to the average weight of the two orbitals involved on the fragment under consideration,

$$j_{\text{Frag}}(A) = \sum_{i,k} \left(\frac{\omega_A^\alpha(i) + \omega_A^\beta(k)}{2} \right) j(i,k), \quad (6.9)$$

$$J(\text{F}) = \sum_A j_{\text{Frag}}(A). \quad (6.10)$$

Again, there is some degree of arbitrariness in this approach, and one might, for example, consider a density-based weighting scheme in future work, or taking the geometric rather than the arithmetic average between the two fragment weights. From our experience so far, the approach described above works well for the purpose of qualitative analysis we have in mind (see below and [8]).

The calculation and local decomposition of exchange coupling constants J was implemented in our program package ARTAIOS [67] (see Fig. 6.4), which can post-process output from various electronic-structure codes.

6.2.3 Application to Bismetallocenes: Through-Space Versus Through-Bond Pathways

As an illustrative example, we show here the analysis of through-bond versus through-space coupling pathways for naphthalene-bridged bis-metallocenes synthesized by Heck and coworkers [68, 69] (see Fig. 6.5).

We compare the fragment decomposition scheme according to (6.7)–(6.10) with an alternative approach, in which the bridge is removed to evaluate the pure through-space contribution (see Table 6.1). As discussed in the introduction, this has the

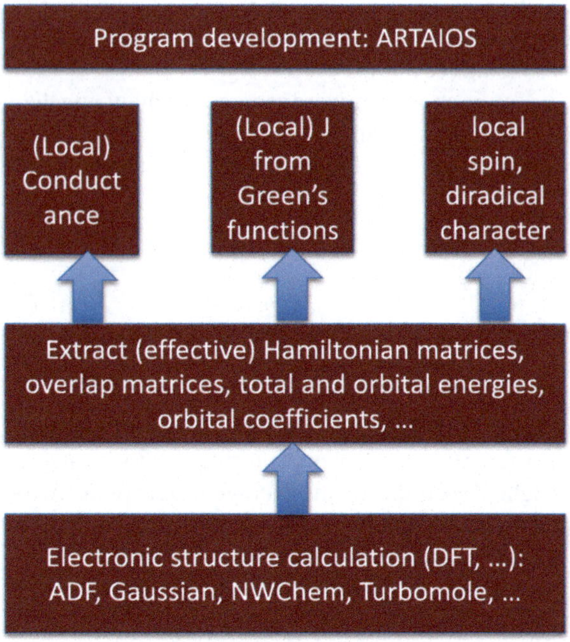

Fig. 6.4 Schematic workflow in our program package ARTAIOS [67]

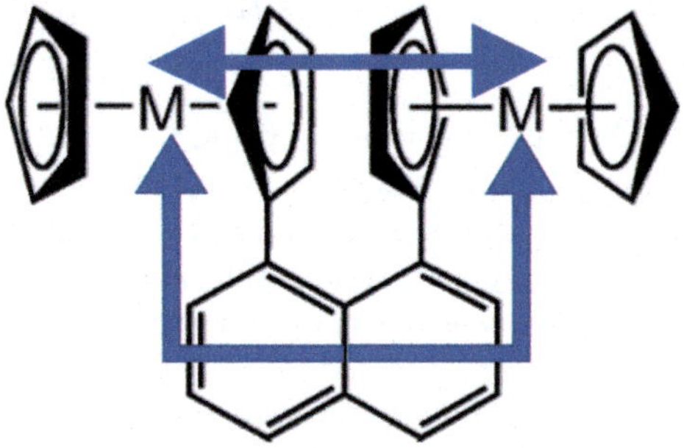

Fig. 6.5 Illustration of two spin coupling pathways in bis-metallocenes (top: through-space, bottom: through-bond)

disadvantage of modifying the electronic structure of the spin centers somewhat, so that removing the bridge may also modify the through-space interactions to some extent. Qualitatively, both approaches result in the same picture: For vanadocene, the overall coupling is weak because both pathways contribute little (which can be attributed to the small delocalization of unpaired spin from the spin centers onto the ligands [69]), while for nickelocene, the antiferromagnetic coupling is dominated by the through-space interaction (with the bridging contributing a smaller equally antiferromagnetic term).

Table 6.1 Through-space (TS) and through-bond (TB) contributions to the Heisenberg spin coupling constant J (in cm^{-1}) for naphthalene (NP)-bridged bis-cobaltocenes. The values in the second column were obtained by evaluating H for a molecule in which the bridge was taken out and replaced by two hydrogen atoms to saturate the dangling bonds (but all other nuclear coordinates were the same as in the molecule including the bridge; see Fig. 6.6). For the values in the third column, the resulting J was subtracted from J for the full molecule (including the bridge). In all calculations TPSSH / def2-TZVP was used

System	TS (no bridge)	TB (total − no bridge)	TS (6.9)	TB (6.9)
V–NP–V	−0.2	−0.2	−1.2	+0.7
Ni–NP–Ni	−24.3	−1.3	−17.3	−8.3

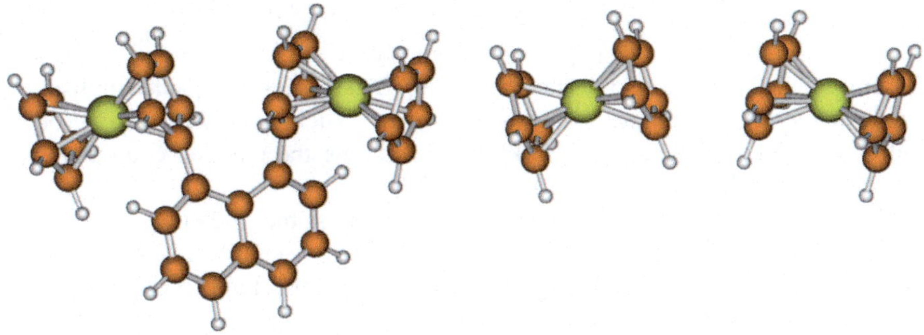

Fig. 6.6 Ball-and-stick representation of a bis-metallocene with and without a naphthalene bridge (in the structure without the bridge, the dangling bonds are saturated with hydrogen atoms)

6.3 Chemically Controlling Spin Coupling

In the ideal case, chemical control can be exerted by external stimuli, e.g., by modifying the chemical structure of a bridge between two spin centers through photo- or redoxswitching. Nonetheless, comparing different bridges that are not directly interconvertible, as presented in the last part of this section, is very helpful for establishing structure–property relationships.

6.3.1 Photoswitchable Spin Coupling: Dithienylethene-Linked Biscobaltocenes

Bismetallocenes with dithienylethene (DTE) linkers promise a combination of photo- and redox-switching, and for metallocenes with unpaired electrons, photoswitchable spin coupling. This has been demonstrated for DTE-linked organic radical spin centers, and occasionally also between metal centers [70]. Attempts at bringing a cobaltocene–DTE-cobaltocene molecule (see Fig. 6.8, top right), where each cobal-

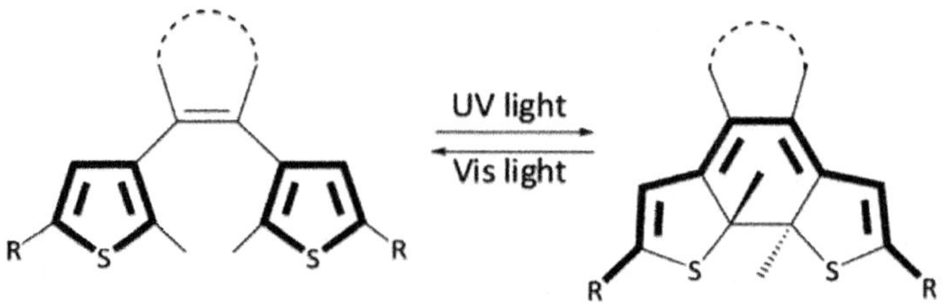

Fig. 6.7 Photoswitching a dithienylethene bridge with different substituents R. π-conjugation is indicated by the thick lines. The dotted line indicates one or two methylene units, resulting in five-membered and six-membered rings, respectively, which were both studied without showing much difference in their switching behavior

tocene carries one unpaired spin, to photoswitch between an uncoupled "open" and a coupled closed form were not successful [19]. By studying a sequence of disubstituted dithienylethene bridges and comparing their switching behavior with ground- and excited-state potential energy (PES) scans along the reaction coordinate, we could shed some light on why this is so.

For chlorine-substituted DTE, photoswitching is possible, and this is true for both a five-membered and a six-membered ring [71]. The advantage of a six-membered ring is its potential for chiral functionalization. When attaching (diamagnetic) ferrocene substituents for R in Fig. 6.7, the switching behavior was considerably poorer than for the chlorine-substituted compound. This could be attributed to an increased number of accessible excited electronic states, only one of which results in the desired photoreaction [72]. This increased number of excited-state pathways not leading to ring closure or opening is even more pronounced when moving to a DTE bridge with two attached paramagnetic cobaltocene units. This is in contrast to the case where the π systems of substituents and bridge are disconnected by a sp^3-hybridized carbon atom (see Fig. 6.8). Accordingly, while the latter system can be photoswitched, it was not possible to switch the corresponding bis-cobaltocene [19].

6.3.2 Redox-Switchable Spin Coupling: Ferrocene as Bridging Ligand

Ferrocene was mentioned above as a substituent on a photoswitchable bridge. Here, we exploit its redox properties and employ it as a bridge between two radical substituents (see Fig. 6.9). In the neutral state, ferrocene is diamagnetic, while in the oxidized state, it has one unpaired electron. This unpaired electron can interact with the unpaired electrons on the two radical substituents, so that one would expect an increase in overall spin coupling (i.e., energy difference between the electronic ground state and excited states obtained by spin flips). This is indeed the case: In the neutral form, the spins on the radical substituents are weakly ferromagnetically

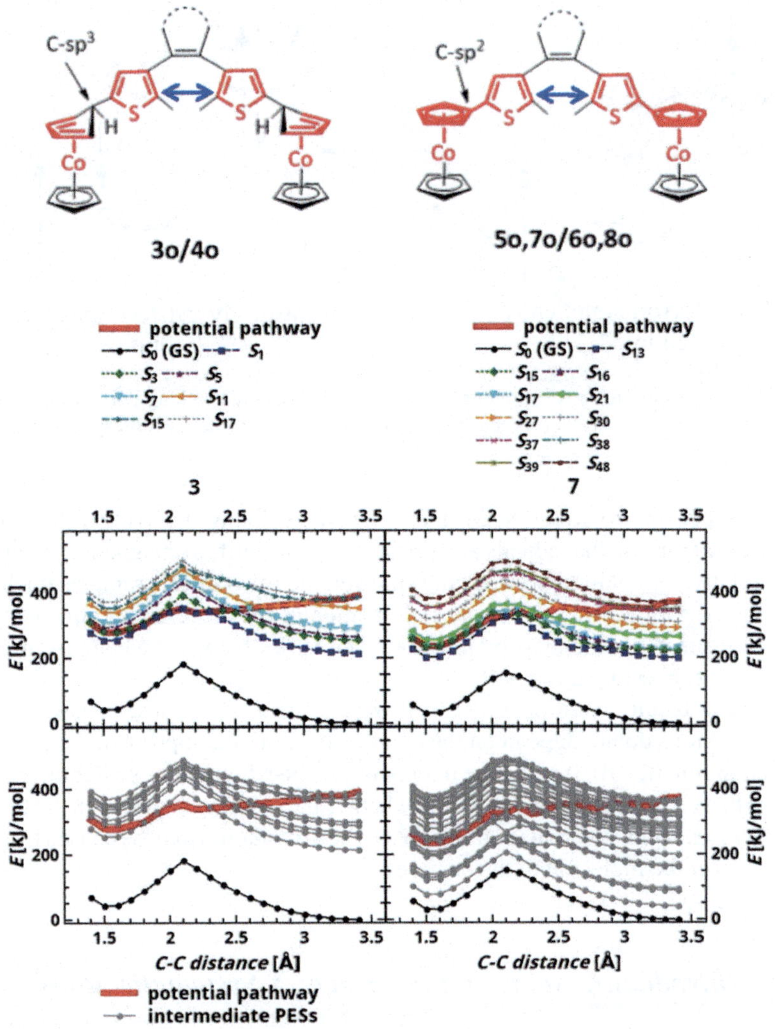

Fig. 6.8 Top: Lewis structures of DTE-linked complexes with diamagnetic Co(I) centers attached via sp^3-hybridized carbon atoms (left) and with paramagnetic cobaltocene (Co(II) centers attached via sp^2-hybridized carbon atoms, resulting in π conjugation between DTE and the cyclopentadienyl ligands (right). The dotted line indicates one or two methylene units, which were both studied without showing much difference in their switching behavior. The data shown below are for the five-membered ring resulting from one methylene unit. Middle: Total energy as a function of the distance between the reactive carbon atoms involved in ring closure / opening (indicated by the blue arrows in the top panels), for the ground state and for those excited states which were considered as potentially contributing to the photoreaction due to their relatively large transition dipole moment from the electronic ground state (see [19] for more details). Bottom: The same plots including all excited states under consideration. In the molecular structure optimizations, all nuclear coordinates were allowed to relax except for the distance between the reactive carbon atoms, which was held fixed at the values indicated on the x axis (B3LYP-D3/ def-TZVP)

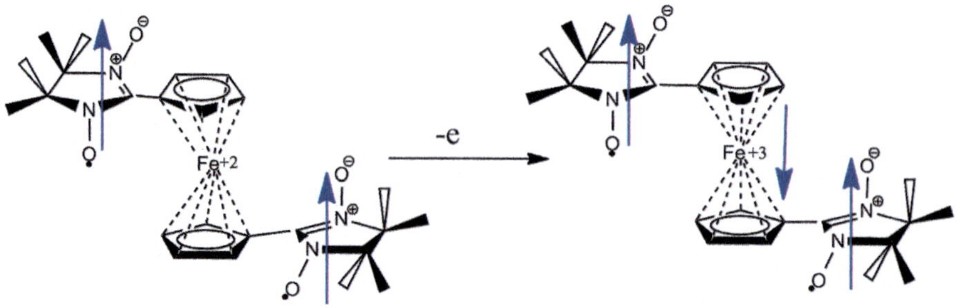

Fig. 6.9 Illustration of redox switching a bridging ferrocene unit linking two nitronyl nitroxide radicals to achieve an overall energetic stabilization of the ground state with respect to spin flips

coupled according to Kohn–Sham DFT. When oxidizing the bridge, the coupling between the spins on the radicals and on the bridge is antiferromagnetic, so that overall the relative orientation between the spins on the two radicals remains unchanged (i.e., aligned parallel). The resulting stabilization of the electronic ground state with respect to spin flips is at least by a factor of three (with some exchange–correlation functionals suggesting up to 300).

All these results were obtained from Kohn–Sham DFT (where it was verified that conclusions do not depend on the choice of a particular approximate exchange–correlation functional). It would be interesting to see whether this switching behavior can also be verified experimentally, and whether the stronger coupling in the oxidized state plays a role for adsorbates on surfaces (where oxidation may be caused by charge transfer form the molecule to the surface).

6.3.3 Introducing Spins on the Bridge: A Systematic Study

To study more systematically the effect of introducing a spin center on the bridge, we investigated a series of nitronyl nitroxide (NNO)–bridge–semiquinone (SQ) compounds, where the bridge is a *meta*-phenylene with different closed-shell and radical substituents (see Fig. 6.10) [73]. Again, introducing a spin on the bridge leads to an energetic stabilization of the ground state with respect to spin flips by a factor of three to six. There is a clear correlation between the amount of spin density that gets delocalized from the radical substituent onto the bridge and the amount of spin-state stabilization. Since in the potential synthetic target system (bottom right in Fig. 6.10), this delocalization is much smaller than in the model systems under study (the other panels in the figure), this points to controlling spin delocalization in synthetically accessible molecules as an important goal when aiming at a stabilization of coupled spin systems with respect to spin flips.

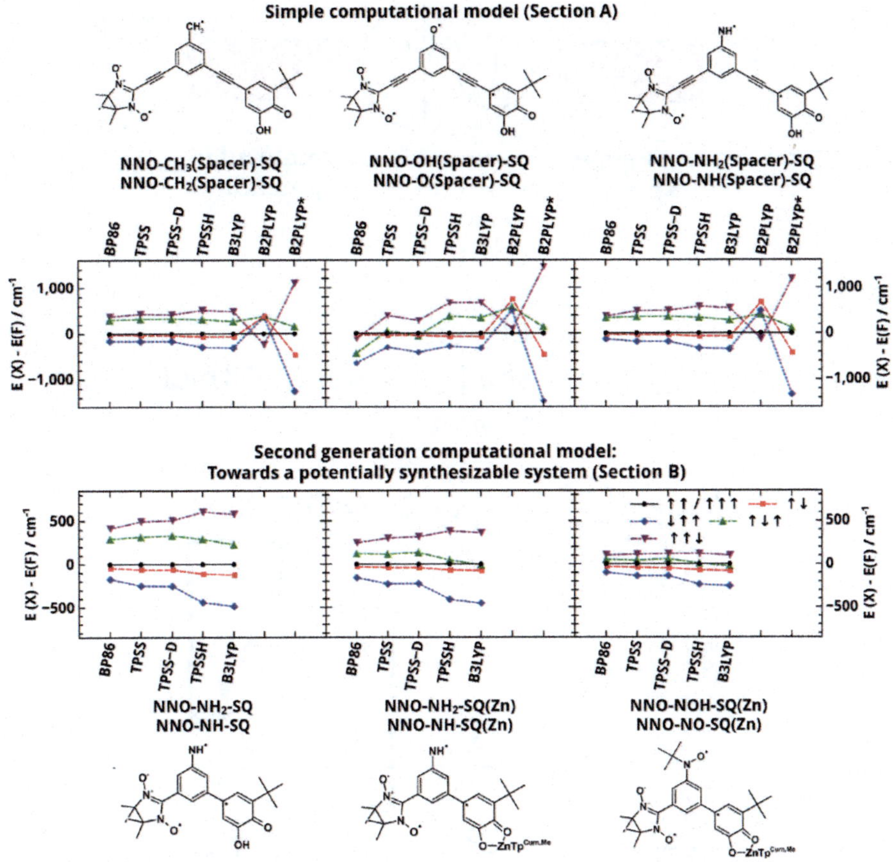

Fig. 6.10 (Top) Relative spin state energies of meta-connected ethynyl-bridged model radicals for different exchange–correlation functionals (legend: bottom right plot). (Bottom) Relative spin state energies of a potential synthetic target system. The Lewis structure shows structures with a radical substituent on the bridge, resulting in triradicals; these are compared with the spin-state energetics of analogous compounds with an added hydrogen atom changing the substituent into a closed-shell one, resulting in diradicals (compare the text below/above the Lewis structures). For both diradicals and triradicals, energies are given with respect to the ferromagnetically coupled state (↑↑ or ↑↑↑). B2PLYP* refers to the B2PLYP functional employing 100% DFT correlation rather than the original 27% admixture of MP2 correlation

6.4 From Spin Coupling to Conductance

The relation between spin coupling and electron transfer or transport has been studied for some time [24, 26–30]. Recently, it was also pointed out that there is a connection between the existence of diradicals and the occurrence of quantum interference in molecular wires [74]. We showed that comparing conductance and spin coupling from a molecular-orbital point of view results in the common trends reported before (with

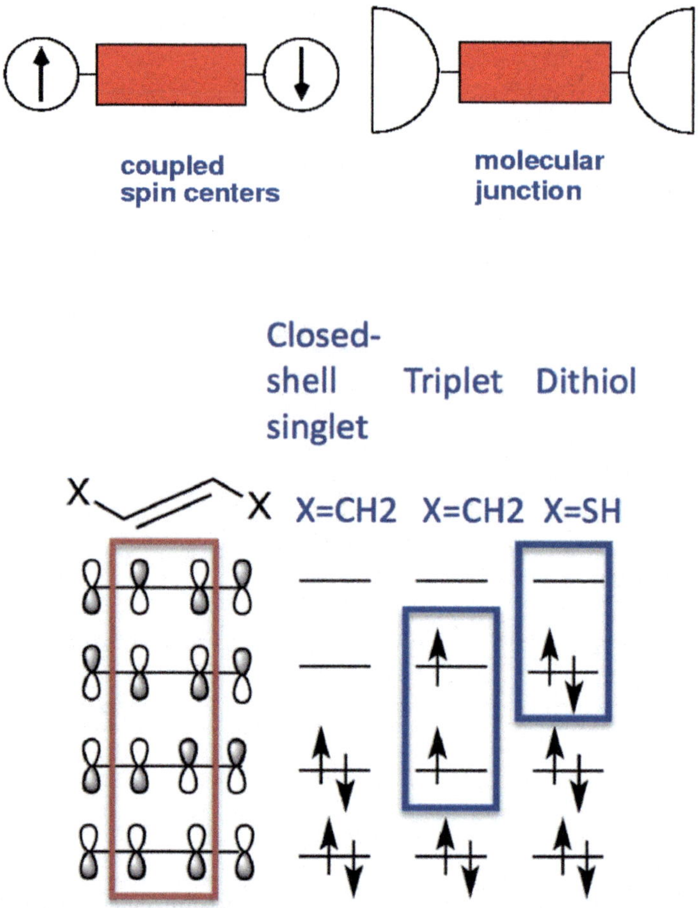

Fig. 6.11 Top: Schematic comparison of a molecular bridge (red rectangle) between two spin centers versus one between macroscopic electrodes (consisting, e.g., of gold with the bridge attached via thiolate linkers [75]). Bottom: Illustration of the frontier orbitals relevant for understanding spin coupling (middle) and molecular conductance (right). A larger energy splitting between the two singly occupied orbitals in the triplet diradical (blue box in middle molecular orbital scheme) indicates larger antiferromagnetic coupling, and a larger splitting between the two frontier orbitals in a molecular dithiolate bridge (blue box in the right-hand orbital scheme) indicates smaller conductance

large conductance corresponding to large antiferromagnetic coupling), but this is due to pairs of frontier orbitals showing opposite trends (see Fig. 6.11): Antiferromagnetic coupling gets larger as orbital energy splittings increase, while conductance gets smaller in that case (provided the electronic coupling to the electrodes remains the same). This apparent contradiction is explained by the fact that in a typical thiolate–bridge–thiolate molecular wire, there are two relevant electrons more than for a typical spin–bridge–spin system, so the relevant frontier orbitals are different [25].

We also studied radical and closed-shell adsorbates on carbon nanotubes, where they may affect conductance [76]. This effect is often called chemical gating, and it is attributed to charge transfer between adsorbates and nanotube, the effect of the adsorbate dipole moments on the nanotube electronic structure, or a combination of both. Therefore, this study required the derivation and implementation of a generalized origin-independent approach to evaluating local dipole moments [36, 37] (see Fig. 6.12).

When considering the relation between conductance and spin-dependent properties, other exciting phenomena are magnetoresistance [77] and effects resulting from spin–orbit coupling, such as the Rashba effect which in colloidal PbS nanosheets leads to a circularity dependent photo-galvanic effect [78]. For the magnetoresistance measured for a TEMPO-radical-substituted oligophenylene-ethynylene (OPE) molecular wire in a mechanically controlled break junction, electron transport does not go through the radical substituent (see Fig. 6.13), but yet the presence of the radical strongly increases magnetoresistance compared with the unsubstituted OPE molecule. This leaves several open questions for future work.

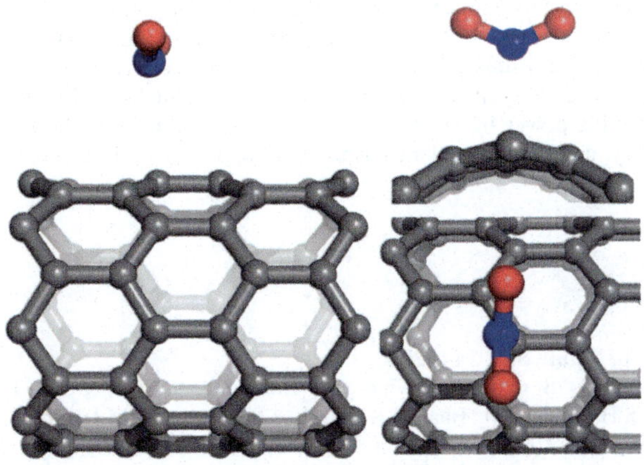

Fig. 6.12 Optimized structure of a spin-polarized NO_2 adsorbate on a (8,0) carbon nanotube (periodic boundary conditions, PBE-D2) [76]

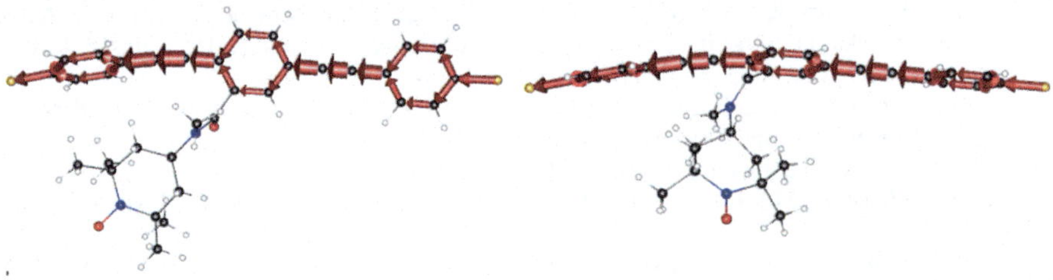

Fig. 6.13 Schematic illustration of local contributions to electron transmission in two nearly degenerate conformations of a TEMPO-substituted oligophenylene-ethynylene molecular wire [39, 77]

6.5 Conclusion

We have reviewed methodological and computational efforts towards understanding and designing spin interactions in molecular systems, which may be important for functional units in nanoscale spintronics. In particular, our focus was on analyzing local contributions to spin coupling (i.e., coupling pathways), on photoswitching and redoxswitching spin coupling, on quantifying the effect of unpaired spins located on bridging units, on the stabilization of the ground state with respect to spin flips, and on pointing out common aspects of spin coupling and conductance through molecular bridges. In the future, it would be valuable to employ the concepts and methods suggested here to specifically design (switchable) spin coupling in molecules and molecular chains, both isolated and on surfaces or in other environments that may facilitate their use for nanospintronics applications.

Acknowledgements For funding, we thank the DFG via SFB 668 (project B17). We are grateful for the support by and discussions with our collaborators within the SFB 668 and beyond, in particular Jürgen Heck, Alexander Lichtenstein, Elke Scheer, Christian Klinke, Martin L. Kirk, David A. Shultz, and their groups. Furthermore, we thank all Ph.D., master and bachelor students who have contributed to this project during the last years, in particular: Marc Philipp Bahlke, Martin Sebastian Zöllner, Joscha Nehrkorn, Conrad Stork, Aaron Bahde, Mariana Hildebrandt, Jos Tasche, Jonny Proppe, Jan Elmisz, Lea Freudenstein, Lawrence Rybakowski. We also gratefully acknowledge administrative support by Andrea Beese, Heiko Fuchs, and Beate Susemihl, and IT support and computing power by HLRN, the HPC cluster and team at the Regional Computing Center at University of Hamburg, and the chemistry IT service at University of Hamburg.

References

1. M.M. Waldrop, Nature **530**, 144 (2016)
2. A.A. Khajetoorians, B. Chilian, J. Wiebe, R. Wiesendanger, Science **332**, 1062 (2011)
3. A. DiLullo, S.H. Chang, N. Baadji, K. Clark, J.P. Klöckner, M.H. Prosenc, S. Sanvito, R. Wiesendanger, G. Hoffmann, S.W. Hla, Nano Lett. **12**, 3174 (2012)
4. M. Bazarnik, B. Bugenhagen, M. Elsebach, E. Sierda, A. Frank, M.H. Prosenc, R. Wiesendanger, Nano Lett. **16**(1), 577 (2016)
5. E. Sierda, M. Abadia, J. Brede, M. Elsebach, B. Bugenhagen, M.H. Prosenc, M. Bazarnik, R. Wiesendanger, ACS Nano. (2017)
6. J. Girovsky, J. Nowakowski, M.E. Ali2, M. Baljozovic, H.R. Rossmann, T. Nijs, E.A. Aeby, S. Nowakowska, D. Siewert, G. Srivastava, C. Wäckerlin, J. Dreiser, S. Decurtins, S.X. Liu, P.M. Oppeneer, T.A. Jung, N. Ballav, Nat. Commun. **8**, 15388 (2017)
7. M. Garnica, D. Stradi, S. Barja, F. Calleja, C. Diaz, M. Alcami, N. Martin, A.L.V. de Parga, F. Martin, R. Miranda, Nat. Phys. **9**, 368 (2013)
8. T. Steenbock, C. Herrmann, J. Comput. Chem. (2017). Accepted
9. A.I. Liechtenstein, M.I. Katsnelson, V.P. Antropov, V.A. Gubanov, J. Magn. Magn. Mater. **67**, 65 (1987)
10. D.W. Boukhvalov, V.V. Dobrovitski, M.I. Katsnelson, A.I. Lichtenstein, B.N. Harmon, P. Kögerler, Phys. Rev. B **70**, 054417 (2004)
11. D.W. Boukhvalov, A.I. Lichtenstein, V.V. Dobrovitski, M.I. Katsnelson, B.N. Harmon, V.V. Mazurenko, V.I. Anisimov, Phys. Rev. B **65**, 184435 (2002)
12. M. Han, T. Ozaki, J. Yu, Phys. Rev. B **70**, 184421 (2004)

13. T. Steenbock, J. Tasche, A. Lichtenstein, C. Herrmann, J. Chem. Theory Comput. **11**, 56515664 (2015)
14. K. Matsuda, M. Irie, J. Am. Chem. Soc. **122**, 7195 (2000)
15. K. Matsuda, M. Irie, Chem. Eur. J. **7**, 3466 (2001)
16. K. Takayama, K. Matsuda, M. Irie, Chem. Eur. J. **9**, 5605 (2003)
17. S. Nakatsuji, Chem. Soc. Rev. **33**, 348 (2004)
18. W.R.B. Ben L. Feringa, *Molecular Switches* (Wiley VCH, Hoboken, 2007)
19. A. Escribano, T. Steenbock, C. Stork, C. Herrmann, J. Heck, Chem. Phys. Chem. **18**, 596 (2017)
20. A. Ito, R. Kurata, D. Sakamaki, S. Yano, Y. Kono, Y. Nakano, K. Furukawa, T. Kato, K. Tanaka, J. Phys. Chem. A **117**, 12858 (2013)
21. M.E. Ali, V. Staemmler, F. Illas, P.M. Oppeneer, J. Chem. Theor. Comput. **9**, 5216 (2013)
22. E.D. Piazza, A. Merhi, L. Norel, S. Choua, P. Turek, S. Rigaut, Inorg. Chem. **54**, 6347 (2015)
23. S. Demir, I.R. Jeon, J.R. Long, T.D. Harris, Coord. Chem. Rev. **289**, 149 (2015)
24. C. Herrmann, J. Elmisz, Chem. Commun. **49**, 10456 (2013)
25. J. Proppe, C. Herrmann, J. Comput. Chem. **36**, 201 (2015)
26. M.L. Kirk, D.A. Shultz, E.C. Depperman, C.L. Brannen, J. Am. Chem. Soc. **129**, 1937 (2007)
27. M.L. Kirk, D.A. Shultz, D.E. Stasiw, G.F. Lewis, G. Wang, C.L. Brannen, R.D. Sommer, P.D. Boyle, J. Am. Chem. Soc. **135**(45), 17144 (2013)
28. G. Blondin, J.J. Girerd, Chem. Rev. **90**, 1359 (1990)
29. P. Bertrand, Chem. Phys. Lett. **113**(1), 104 (1985)
30. T.C. Brunold, D.R. Gamelin, E.I. Solomon, J. Am. Chem. Soc. **122**, 8511 (2000)
31. A.E. Clark, E.R. Davidson, Int. J. Quantum Chem. **93**, 384 (2003)
32. S.M. Bachrach, in *Reviews in Computational Chemistry*, vol. 5, ed. by K.B. Lipkowitz, D.B. Boyd (VCH Publishers, New York, 1994)
33. A.E. Clark, E.R. Davidson, J. Chem. Phys. **115**(16), 7382 (2001)
34. C. Herrmann, M. Reiher, B.A. Hess, J. Chem. Phys. **122**, 034102 (2005)
35. E. Ramos-Cordoba, E. Matito, I. Mayer, P. Salvador, J. Chem. Theory Comput. **8**, 1270 (2012)
36. L. Groß, C. Herrmann, J. Comput. Chem. **37**, 2324 (2016)
37. L. Groß, C. Herrmann, J. Comput. Chem. **37**, 2260 (2016)
38. A. Krawczuk, D. Prez, P. Macchi, J. Appl. Cryst. **47**, 1452 (2014)
39. G.C. Solomon, C. Herrmann, T. Hansen, V. Mujica, M.A. Ratner, Nat. Chem. **2**, 223 (2010)
40. T.H.T. Hansen, G.C. Solomon, J. Chem. Phys. **146**, 092322 (2017)
41. N. Sai, N. Bushong, R. Hatcher, M. di Ventra, Phys. Rev. B **75**, 15410 (2007)
42. M. Ernzerhof, J. Chem. Phys. **125**, 124104 (2006)
43. T.N. Todorov, J. Phys.: Condens. Matter **14**, 3049 (2002)
44. A. Pecchia, A.D. Carlo, Rep. Prog. Phys. **67**, 1497 (2004)
45. V. Pohl, L.E.M. Steinkasserer, J.C. Tremblay, arXiv:1707.07635
46. J.N. Onuchic, D.N. Beratan, J.R. Winkler, H.B. Gray, Ann. Rev. Biophys. Biomol. Struct. **21**, 349 (1992)
47. W. Hug, Chem. Phys. **264**, 53 (2001)
48. C. Herrmann, K. Ruud, M. Reiher, Chem. Phys. **343**, 200 (2008)
49. M.U. Delgado-Jaime, S. DeBeer, M. Bauer, Chem. Eur. J. **19**, 15888 (2013)
50. S.W. Oung, J. Rudolph, C.R. Jacob, Int. J. Quantum Chem. (2017)
51. J.E. Peralta, V. Barone, J. Chem. Phys. **129**, 194107 (2008)
52. J.J. Philips, J.E. Peralta, J. Chem. Phys. **138**, 174115 (2013)
53. P.O. Löwdin, J. Chem. Phys. **18**(3), 365 (1950)
54. G. Bruhn, E.R. Davidson, I. Mayer, A.E. Clark, Int. J. Quantum Chem. **106**, 2065 (2006)
55. L. Noodleman, J. Chem. Phys. **74**(10), 5737 (1981)
56. E. Ruiz, J. Cano, S. Alvarez, P. Alemany, J. Comput. Chem. **20**, 1391 (1999)
57. I. de P. R. Moreira, F. Illas, Phys. Chem. Chem. Phys. **8**, 1645 (2006)
58. J. Perdew, A. Savin, K. Burke, Phys. Rev. A **51**, 4531 (1995)
59. C. Jacob, M. Reiher, Int. J. Quantum Chem. **112**, 3661 (2012)
60. A.J. Cohen, D.J. Tozer, N.C. Handy, J. Chem. Phys. **126**, 214104 (2007)

61. J. Wang, A.D. Becke, V.H. Smith Jr., J. Chem. Phys. **102**(8), 3477 (1995)
62. C.J. Cramer, D.G. Truhlar, Phys. Chem. Chem. Phys. **11**, 10757 (2009)
63. A.P. Gingsberg, Inorg Chim. Acta Rev. **5**, 45 (1971)
64. N. Pagels, O. Albrecht, D. Görlitz, A.Y. Rogachev, M.H. Prosenc, J. Heck, Chem. Eur. J. **17**, 4166 (2011)
65. Y.O. Kvashnin, R. Cardias, A. Szilva, I. Di Marco, M. Katsnelson, A. Lichtenstein, L. Nordström, A. Klautau, O. Eriksson, Phys. Rev. Lett. **116**(21), 217202 (2016)
66. D.M. Korotin, V. Mazurenko, V. Anisimov, S. Streltsov, Phys. Rev. B **91**(22), 224405 (2015)
67. M. Deffner, L. Groß, T. Steenbock, B.A. Voigt, M.S. Zöllner, G.C. Solomon, C. Herrmann (2008–2017)
68. S. Trtica, M.H. Prosenc, M. Schmidt, J. Heck, O. Albrecht, D. Grlitz, F. Reuter, E. Rentschler, Inorg. Chem. **49**, 1667 (2010)
69. S. Puhl, T. Steenbock, R. Harms, C. Herrmann, J. Heck, Dalton Trans. (2017). (Under revision)
70. M. Irie, T. Fukaminato, K. Matsuda, S. Kobatake, Chem. Rev. **114**, 12174 (2014)
71. T. Steenbock, A. Escribano, J. Heck, C. Herrmann, Chem. Phys. Chem. **16**, 1491 (2015)
72. A. Escribano, T. Steenbock, C. Herrmann, J. Heck, Chem. Phys. Chem. **17**, 1881 (2016)
73. T. Steenbock, D.A. Shultz, M.L. Kirk, C. Herrmann, J. Phys. Chem. A **121**, 216 (2017)
74. Y. Tsuji, R. Hoffmann, M. Strange, G.C. Solomon, Proc. Natl. Acad. Sci. **113**(4), E413 (2016)
75. J.C. Cuevas, E. Scheer, *Molecular Electronics: An Introduction to Theory and Experiment*. World Scientific Series in Nanotechnology and Nanoscience, vol. 1 (World Scientific, Singapore, 2010)
76. L. Groß, M.P. Bahlke, T. Steenbock, C. Klinke, C. Herrmann, J. Comput. Chem. **38**, 861 (2017)
77. R. Hayakawa, M.A. Karimi, J. Wolf, T. Huhn, M.S. Zöllner, C. Herrmann, E. Scheer, Nano Lett. **16**, 4960 (2016)
78. M.M.R. Moayed, T. Bielewicz, M.S. Zöllner, C. Herrmann, C. Klinke, Nat. Commun. **8**, 15721 (2017)

Chapter 7
Magnetic Properties of Small, Deposited 3d Transition Metal and Alloy Clusters

Michael Martins, Ivan Baev, Fridtjof Kielgast, Torben Beeck, Leif Glaser, Kai Chen and Wilfried Wurth

Abstract Clusters are structures in the nano- or sub-nanometer regime ranging from a few atoms up to several thousand atoms per cluster. Supported metal clusters and adatoms are interesting systems for magnetic studies as their magnetic properties can strongly depend on the size, composition and the substrate due to quantum size effects. This offers rich possibilities to tailor systems to specific applications by choosing the proper size and composition of the cluster. In this article the magnetic properties of small 3d metal and alloy clusters in the few atom limit measured by X-ray magnetic circular dichroism and how their spin and orbital moments depend on size and composition are discussed. Special emphasis is put on the non-collinear magnetic coupling in the clusters resulting in complex spin structures and the influence of oxidation on the magnetic properties of the clusters.

7.1 Introduction

When thinking about magnetism usually ferromagnetic coupling is considered, a phenomenon known for a long time. Since its discovery it plays an important role in various areas of technology, e.g., navigation or nowadays to store information. In a ferromagnetically coupled material the magnetic moments of the individual atoms within a domain are all oriented parallel to each other. However, there is also antiferromagnetism, with antiparallel orientation of the magnetic moments and more complicated magnetic structures, for example spin spirals [1, 2]. The coupling of a ferromagnet and an antiferromagnet can result in new effects such as the giant magneto-resistance observed in Cr-Fe multilayers [3–5]. Nevertheless, even more complex magnetic ordering exists, which is known from domain walls, showing a non-collinear coupling of the magnetic moments. Furthermore, in recent years another very complex type of non-collinear magnetic ordering has been observed, namely nano-scale magnetic skyrmions. These structures are topologically stabilized and may lead to new approaches for storing digital information [6]. As these

M. Martins (✉) · I. Baev · F. Kielgast · T. Beeck · L. Glaser · K. Chen · W. Wurth
Institut Für Experimentalphysik, Universität Hamburg, Hamburg, Germany
e-mail: michael.martins@uni-hamburg.de

R. Wiesendanger (ed.), *Atomic- and Nanoscale Magnetism*, NanoScience and Technology, https://doi.org/10.1007/978-3-319-99558-8_7

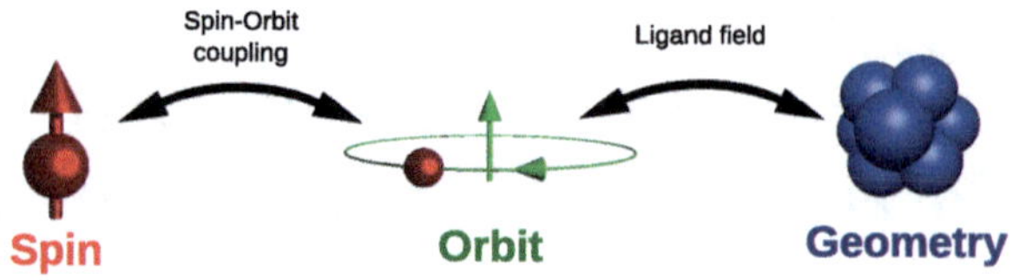

Fig. 7.1 Principle of the spin-lattice coupling in clusters and solids. Creative Common Attribution 3.0 License in J. Phys. C 28, 503002 (2016)

skyrmions can have very small sizes in the nm or even sub-nm regime, they may have potential as ultra-high density storage devices.

Nowadays, storing huge amounts of digital information in magnetic storage devices is still performed using small ferromagnetic domains. By decreasing the size of the magnetic particles storing the information enormous progress has been made in the last decades in increasing the available magnetic storage density. However, by decreasing the size of the particles further they will become superparamagnetic, i.e., the magnetic moments of the individual atoms in the particle are still ferromagnetically coupled, but the resulting total magnetic moment of the particle can rotate freely. The energy barrier which has to be overcome to rotate the magnetisation from the easy axis to another direction, e.g., switching from zero to one, is given by the anisotropy energy.

If information is stored in ferromagnetic particles the natural size limit is a single atom. Several studies have been performed on the magnetic properties of adatoms on different surfaces [7–15] and large anisotropy energies per atom have been found for some adatoms. However, a stable magnetisation of adatoms might have been observed recently for a single Holmium atom at cryogenic temperatures [8].

Magnetic particles with large magnetic moments and a sufficient anisotropy energy to store information at room temperature might be tailored by selecting a special number of atoms in the particle, e.g., using clusters with only a few atoms or magnetic molecules. As the physical and chemical properties of small clusters are strongly affected by quantum size effects these properties can strongly vary with the size even by removing or adding only a single atom ("each atom counts") [16], which has been shown for chemical [16, 17] as well as magnetic properties [18, 19] and opens a way to tailor the magnetic anisotropy. For this, the microscopic origin of the magnetic anisotropy in small clusters needs to be understood. In Fig. 7.1 the mechanism how the spin moment is coupled to the geometry via the spin-orbit coupling and the ligand field is sketched. The total magnetic moment of a cluster is mostly given by the spin moment of the atoms. As the spin has no direct coupling to the geometry of the cluster, the coupling of the magnetic moments must be mediated via the magnetic orbital moment of the atoms. The coupling to the spin is caused by the spin-orbit interaction. The orbital moment is coupled to the geometry of the cluster via the ligand or crystal field. As the ligand field is in general asymmetric, this asymmetry is transferred via the orbital moment to the magnetic spin moment, resulting in the magnetic asymmetry energy [20, 21]. Thus, the asymmetry might be enhanced by an increased spin-orbit coupling, a larger magnetic orbital moment or by a stronger ligand field. This mechanism has been demonstrated by

Gambardella et al. [7] for small Co adatoms and non size selected Co clusters on a Pt(111) surface. They found a clear correlation between the magnetic anisotropy energy per atom and the orbital magnetic moments of the Co atoms within the cluster.

Hence, experiments are mandatory which can measure both the spin and orbital magnetic moments of the supported clusters. Furthermore, experiments on such clusters have to be performed on mass selected clusters, as the magnetic properties can depend strongly on the exact number and geometry of atoms. Such experiments can be realised by X-ray spectroscopic methods like X-ray absorption (XAS) and especially X-ray magnetic circular dichroism (XMCD) spectroscopy [22]. Therefore, in this chapter the magnetic properties of mass selected, supported 3d transition metal and alloy clusters in the size range from the adatom up to a dozen atoms per cluster studied by X-ray magnetic circular dichroism will be discussed. A special emphasis is put on the non-collinear ordering in the small cluster and how non-collinear ordering is emerging or changing in small systems in the few atom limit. Furthermore, hybridization effects due to alloying of the clusters to tailor the orbital moments and the magnetic anisotropy will be discussed.

Small clusters are also interesting from a theoretical point of view, as they are large enough to show complex physical properties and small enough to use highly sophisticated theoretical methods. For a detailed understanding of their magnetic properties including the interaction within particles as well as with the substrate the systems should be studied also theoretically.

This chapter is structured as following. We will first discuss the experimental methods to produce and deposit mass selected clusters and will give a brief introduction on how to measure their magnetic properties using X-ray spectroscopy. As examples the magnetic properties of small Cr and Co clusters will be discussed. The influence of alloying Co clusters with 4d and 5d metals will be presented in the following section. Finally the influence of oxygen on the magnetic properties of Co and Co alloy clusters will be discussed.

7.2 Experiments

7.2.1 Cluster Sample Preparation

To perform X-ray spectroscopy on mass selected, supported metal clusters three steps are mandatory. The clusters have to be produced and subsequently mass selected, hence, the clusters should be produced as ions for an easy mass selection. Finally, the clusters have to be deposited on a substrate. In Fig. 7.2 the setup used in our experiment is sketched, consisting of cluster production, mass selection and deposition as well as spectroscopic investigation.

Within our setup [23], Xe ions with up to 30 keV kinetic energy are used to sputter a target with the cluster material put on high voltage. Atomic and cluster ions produced due to the sputtering process are accelerated typically to 500 eV kinetic energy and

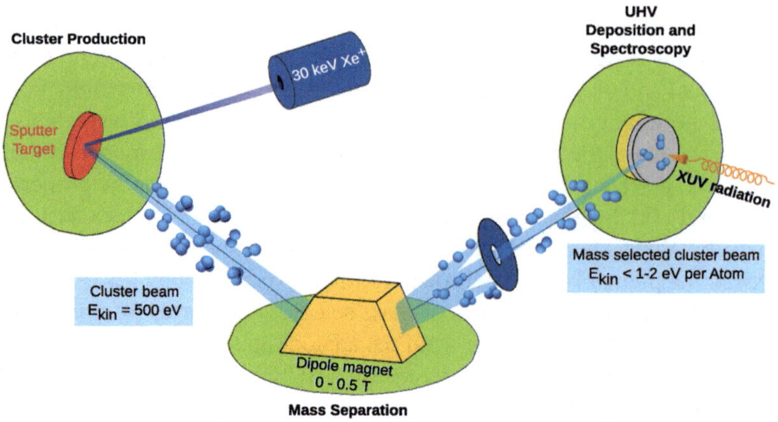

Fig. 7.2 Experimental setup for X-ray spectroscopy of deposited, mass selected clusters. Creative Common Attribution 3.0 License in J. Phys. C 28, 503002 (2016)

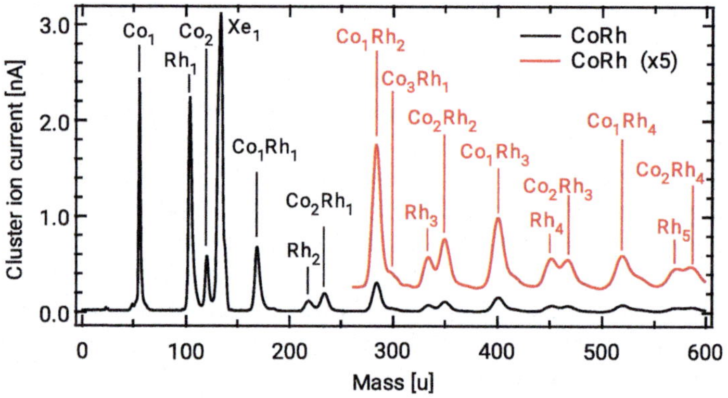

Fig. 7.3 Mass separated ion yield from a CoRh target produced by the ICARUS cluster source [23]. Creative Common Attribution 3.0 License in New J. Phys. 18, 113007 (2016)

are collimated by an electrostatic lens system. As the source is operated at a Xe background pressure around $1 \cdot 10^{-7}$ mbar with a base pressure of $\cong 1 \cdot 10^{-8}$ mbar or better the contamination of the often highly reactive small cluster ions with oxygen can be largely avoided. A typical example for the clusters which can be produced by the source is depicted in Fig. 7.3 for a CoRh alloy target. With increasing cluster size the yield is strongly decreasing. However, special size effects for some specific cluster sizes can increase the yield.

The mass selection of the clusters is achieved using a dipole magnet field. In particular for bimetallic alloy clusters a good mass resolution is required to select a specific size and composition. Here also the natural isotope distribution of the elements has to be taken into account, as this can strongly reduce the achievable

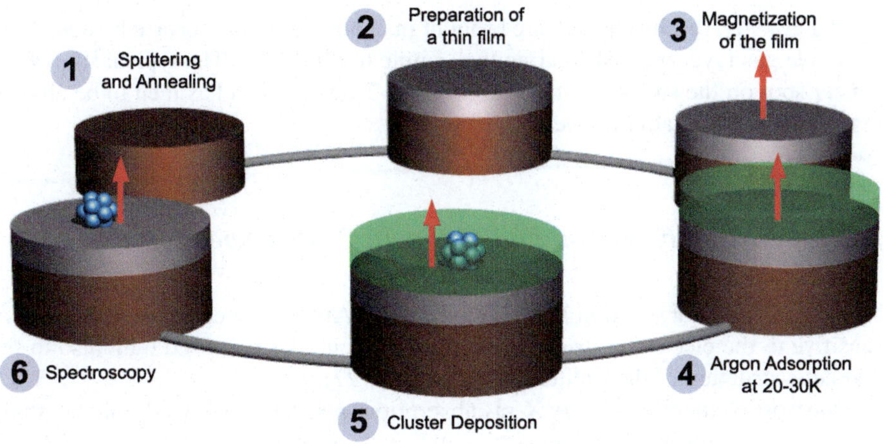

Fig. 7.4 Preparation of samples with mass selected clusters using the soft landing scheme described in the text. Creative Common Attribution 3.0 License in J. Phys. C 28, 503002 (2016)

mass resolution. From the measured ion current the density of the clusters on the surface can directly be calculated, if the size of the cluster spot is known.

To investigate the size dependency of the cluster properties, the coalescence or the interaction of the mass selected supported clusters have to be avoided. This can be realised by depositing only a small amount of clusters in the order of a few percent of a monolayer on the surface. However, to achieve a sufficient count rate in X-ray spectroscopy, a reasonable number of particles is required. The typical coverage given by the atom density in our experiments is on the order of 3% of a monolayer.

To deposit the metal clusters on a surface without fragmentation or changing the surface a soft landing scheme is used. Within this soft landing scheme the mass selected cluster ions are decelerated to a kinetic energy around 1 eV per atom. By using a rare gas buffer layer on the substrate of typically 5–10 monolayers the remaining kinetic energy of the clusters can be efficiently transferred to this buffer layer which is partially desorbed. Finally the rare gas layers are desorbed by flash heating the sample and the deposited clusters are in contact with the surface. Using molecular dynamics simulations Cheng and Landman [24, 25] have shown, that by this procedure, metal clusters can be landed on a surface without destruction.

The scheme to prepare the cluster samples in situ is depicted in Fig. 7.4. A Cu(100) single crystal is cleaned and prepared by sputter and anneal cycles (1). The clusters are deposited on thin magnetic films to align the cluster moments, prepared in the second step (2). In a third step (3) the films are out-of-plane magnetised to remanence. The clusters are deposited in the soft landing scheme in a thin noble gas layer of typical 5–10 monolayers (ML) (4) and the clusters are landing in the noble gas matrix with a kinetic energy less than 1 eV per atom (5). With typical retarded cluster currents between 10 pA up to 1 nA for the different cluster sizes the deposition process takes 5–60 min for a cluster coverage around 3% of a monolayer.

In a last step (6) the clusters are brought in contact with the surface by desorbing the noble gas layer by flash heating the sample to 80–100 K. The typical size of the cluster spot on the surface is in the order of 1–2 mm^2, which has then to be aligned to the synchrotron radiation beam.

7.2.2 X-Ray Absorption and Magnetic X-Ray Spectroscopy

Near edge X-ray absorption spectroscopy (NEXAFS) has been shown to be very sensitive to the electronic structure and by using circular polarised light also to the magnetic structure of the excited atoms [22, 26, 27].

Due to its element specificity X-ray absorption spectroscopy is well suited to study the electronic and magnetic properties of small, deposited clusters. The localised core electrons are excited into the unoccupied valence states and due to the dipole transition selection rules are excited to first order valence states with a specific angular moment. In Fig. 7.5b this excitation is sketched for the 3d metals, where the spin-orbit split $2p_{3/2}$ and $2p_{1/2}$ core electrons are excited into exchange split unoccupied 3d states.

To study dilute systems, such as submonolayer systems, a high intensity, stable and tunable X-ray source is required which can be produced by using undulators at third generation storage ring facilities [28]. Especially APPLE II [29] like undulator sources have been proven to be excellent sources for circular polarised synchrotron radiation over a wide energy range covering in particular L_{23}-edges of the 3d transition metals. In Fig. 7.5a the principle measurement scheme used in our experiments is depicted. The sample with the deposited clusters is magnetised out-of-plane and X-ray spectroscopy is carried out at normal incidence. The X-ray absorption is measured by recording the total electron yield via the sample current.

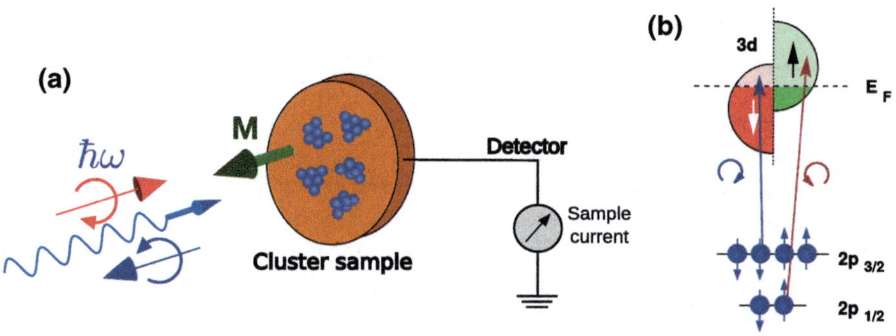

Fig. 7.5 Principle of X-ray and magnetic X-ray absorption spectroscopy. Exciting the spin-orbit split 2p electrons of a 3d metal atom of a magnetised sample with left and right polarised light (**a**) will create spin-polarised electrons, which probe the exchange coupling split empty *d* states (**b**). Creative Common Attribution 3.0 License in J. Phys. C 28, 503002 (2016)

Magnetic X-ray spectroscopy using circular polarised synchrotron radiation (XMCD) is especially useful to measure the spin and orbital magnetic moments which can be estimated using the sum rules [30, 31] given in (7.1) and (7.2) for $2p \rightarrow 3d$ transitions.

$$\mu_{\mathrm{S}}^{\mathrm{eff}} = -2\mu_B \cdot n_h \frac{A - 2B}{C} + T_z \tag{7.1}$$

$$\mu_\ell = -\frac{4}{3}\mu_B \cdot n_h \frac{A + B}{C} \tag{7.2}$$

with

$$C = C_{L3} + C_{L2} \tag{7.3}$$

A and B are the integrated dichroism signals at the L_3 and L_2 edge marked in Fig. 7.6, respectively. The normalizing factor C is the integrated intensity C_{L3} and C_{L2} over the L_2 and L_3 white lines, respectively. n_{h} is the number of d holes. For bulk materials this number is usually quite well known, however, it is not known for small clusters, where n_{h} might also change with the size of the cluster. Hence, all magnetic moments given are divided by n_{h}. The effective spin moment $\mu_{\mathrm{S}}^{\mathrm{eff}}$ includes the term T_z which is a magnetic dipole term and is a measure of the asphericity of the spin magnetisation [22]. In a system with cubic symmetry this term is in general negligible, but this does not hold for clusters and adatoms deposited on surfaces [7, 22, 32, 33] and the given spin moments $\mu_{\mathrm{S}}^{\mathrm{eff}}$ always include this usually unknown contribution.

The absolute spin and orbital magnetic moments can be estimated for the later 3d elements (Fe, Co, Ni) with an error around 10% [34]. For the other 3d elements, the error for the absolute value can be 50% or above, as the 2p spin-orbit splitting is decreasing and the L_3 and L_2 white lines start to overlap [35–37]. However, relative changes between different systems, e.g., different cluster sizes, of the magnetic moments can still be obtained.

In X-ray absorption spectra also the direct, non-resonant excitation into s-like states is included. This does not contribute to the magnetic properties, shows no resonant feature and can be described by a step function. In Fig. 7.6 X-ray absorption spectra of Fe_7 clusters deposited on a magnetised Ni/Cu(100) substrate are shown as an example. The L_3 and L_2 white lines at 708 and 720 eV due to the excitation of $2p_{3/2}$ and $2p_{1/2}$ electrons into unoccupied states are situated on a slowly varying background from the Ni/Cu substrate.

7.3 3d Metal Cluster

In the following section the magnetic properties of small chromium and cobalt clusters deposited on thin magnetised films will be discussed. Cr and Co have been

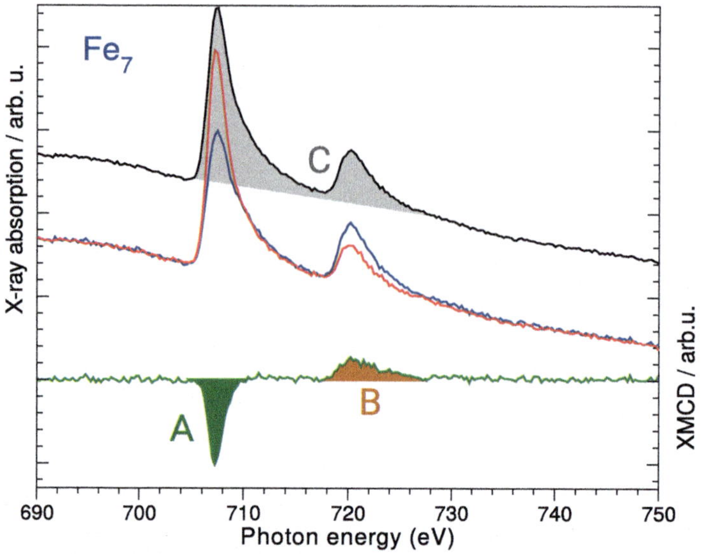

Fig. 7.6 X-ray absorption and magnetic circular dichroism (XMCD) spectrum of Fe_7 clusters deposited on an out-of-plane magnetised Ni/Cu(100) substrate. The red and blues curves are the X-ray absorption spectra recorded with the two light helicities. The upper (black) trace is the average absorption spectrum corresponding to the absorption of unpolarised or linear polarised light. The low (green) trace is the difference spectrum of both helicities, e.g., the XMCD spectrum. The shaded areas A, B and C are the areas used in the sum rules in (7.1) and (7.2). Creative Common Attribution 3.0 License in J. Phys. C 28, 503002 (2016)

chosen as examples for materials showing an antiferromagnetic and ferromagnetic coupling in the bulk, respectively.

7.3.1 Chromium Clusters

The clusters discussed in this chapter are in general too small for complex spin structures like skrymions, however, already a dimer on a magnetised surface might show a complex non-collinear magnetic coupling, if the magnetic coupling relative to the surface and within the dimer is antiferromagnetic. In Fig. 7.7 the possible magnetic structure for such adatoms and dimers deposited on a magnetised surface are depicted. For the dimer a ferromagnetic, an antiferromagnetic as well as a non-collinear coupling might be realised depending on the coupling strength between the adatoms J_{aa} and an adatom and the surface J_{as}.

Such ultra small non-collinear spin structures can be realised with chromium clusters, as chromium is a prototype system for an antiferromagnetic coupling in the bulk.

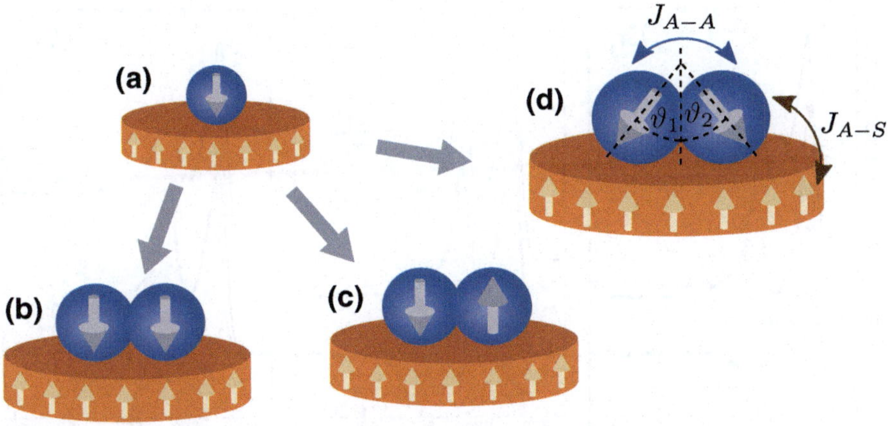

Fig. 7.7 Schematic illustration of the possible non-collinear coupling of a dimer on a magnetised surface. **a** antiferromagnetic coupling to the substrate of the adatom, **b** ferromagnetic coupling of the dimer, **c** antiferromagnetic coupling of the dimer, **d** non-collinear coupling of the dimer with angles $\vartheta_{1,2}$ and exchange coupling constants $J_{\mathrm{A-A}}$ and $J_{\mathrm{A-S}}$

In Fig. 7.8 the XMCD spectra of Cr adatoms and Cr_n clusters for $n = 3, 4$ deposited on a magnetised three monolayer Fe film, deposited on Cu(100) are depicted [38]. A strong XMCD signal is found, which shows a double peak structure at the L_3 edge. As the positive part is much stronger in total this XMCD signal corresponds to the expected antiferromagnetic coupling. This double peak structure is identical to the structure found for polarised Cr atoms in the gas phase by Prümper et al. [39]. In the gas phase the two peaks with opposite sign can be assigned to two different spin configurations of the 3d electrons, e.g., a ^7P and a ^5P multiplet [40].

The spin and orbital magnetic moments in the size range from the Cr adatom up to 13 Cr atoms per cluster deposited on a magnetised Fe film are depicted in Fig. 7.9. The orbital moment is close to zero and the spin moment is strongly decreasing from 0.4 μ_B for the Cr adatom down to 0.10–0.15 μ_B for Cr_n with $n \geq 8$. This strong decrease is a result of the increasing antiferromagnetic ordering of the Cr cluster with increasing size. The almost vanishing orbital moment for chromium can be attributed to the d^5 high spin configuration, which will have a total angular momentum $L = 0$ according to the Pauli principle.

From Fig. 7.7 for a Cr dimer on a magnetic surface already a non-collinear spin structure is expected, which would result in a strongly reduced magnetic moment. However, the magnetic moment is only slightly reduced compared to the Cr adatom suggesting a ferromagnetic coupling of the Cr atoms in the dimer coupled antiferromagnetically to the substrate shown in Fig. 7.7b.

The exchange coupling J between the Cr adatoms $J_{\mathrm{A-A}}$ and the Cr atoms and the Fe substrate $J_{\mathrm{A-S}}$ has been calculated in [38] using the SPR-KKR (spin-polarised relativistic Korringa–Kohn–Rostoker) method. The classical spin Hamiltonian is given by

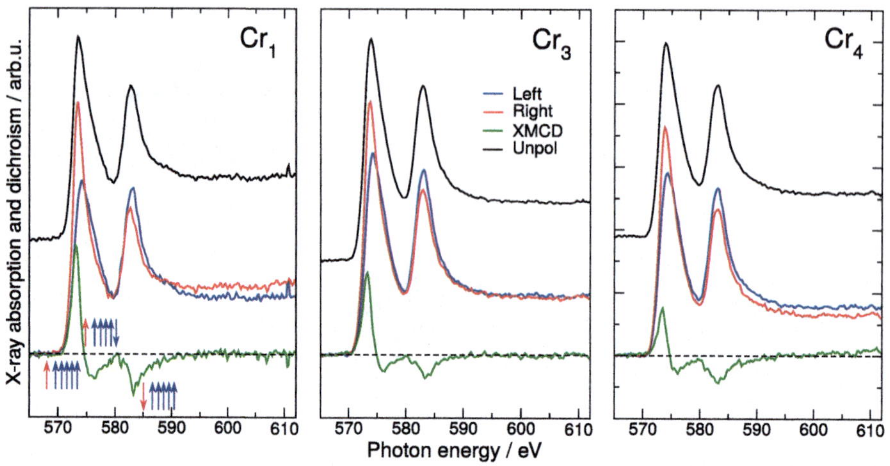

Fig. 7.8 X-ray absorption and XMCD spectra for different $Cr_{1,3,4}$ clusters deposited on a Fe/Cu(100) surface. The arrows in the Cr_1 graph depict the spin of the 2p core hole (red) and the 3d electrons (blue). The dashed line is the zero line for the XMCD as well as the left and right polarised XAS spectra. The unpolarised spectrum is shifted up for convenience. Creative Common Attribution 3.0 License in J. Phys. C 28, 503002 (2016)

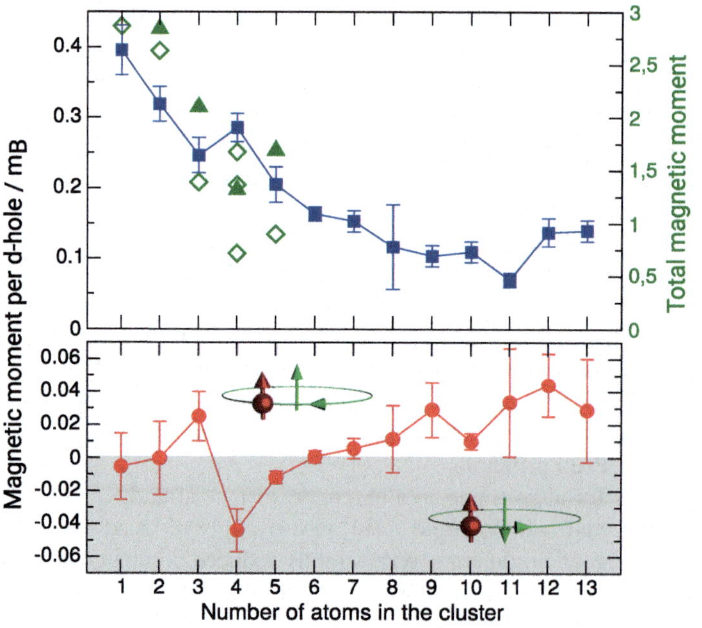

Fig. 7.9 Experimental magnetic spin μ_s (■) and orbital μ_ℓ (●) moments of chromium adatoms and small clusters in the mass range from n = 1 − 13 on a Fe(100) fcc surface. Theoretical total magnetic moments from [38] (◇) and [41] (▲). Note the different y-scales on the graphs. Creative Common Attribution 3.0 License in J. Phys. C 28, 503002 (2016)

$$\mathcal{H} = -\frac{1}{2}\sum_{i \neq j} J_{ij} e_i e_j, \tag{7.4}$$

where $e_{i,j}$ is a unit vector defining the direction of the magnetic moment and i and j indicate the Cr cluster atoms and their first Fe neighbors. Taking into account only first-neighbor interactions and neglecting the rotation of Fe moments the Hamiltonian for the Cr dimer can be described by

$$\mathcal{H} = -J_{A-A}\cos(\vartheta_1 + \vartheta_2) - 4J_{A-S}(\cos\vartheta_1 + \cos\vartheta_2), \tag{7.5}$$

with $\vartheta_{1,2}$ the angle of the Cr magnetic moments relative to the magnetisation of the Fe substrate (see Fig. 7.7). The angle defining the non-collinear solution can be obtained by minimizing (7.5) as

$$\cos(\vartheta_1) = \cos(\vartheta_2) = -2J_{A-S}/J_{A-A}. \tag{7.6}$$

If $2|J_{\mathrm{A-S}}| > |J_{\mathrm{A-A}}|$, the angle is not defined and the non-collinear solution does not exist. For Cr_2 on the Fe substrate this is realised, as $2|J_{\mathrm{Cr-Fe}}| = 2 \times 80.8\,\mathrm{meV} > |J_{\mathrm{Cr-Cr}}| = 77.6\,\mathrm{meV}$.

Hence, the exchange coupling of the individual Cr atoms to the Fe substrate is much stronger in comparison to the Cr–Cr coupling. This results in a ferromagnetic coupling of the two Cr spins which are then coupled antiferromagnetically to the Fe surface as depicted in Fig. 7.7d.

In Fig. 7.10 the calculated spin orientations for a Cr trimer and two possible geometries of Cr tetramers are depicted. A non-collinear coupling is found for all clusters. However, for the structure of Cr_4 the non-collinear coupling is less pronounced. Starting from Cr_5 a transition to an antiferromagnetically coupled Cr cluster is continuing decreasing the non-collinear contributions.

In Fig. 7.9 the results for the experimental and calculated magnetic moments of Cr_n clusters on a 3 ML Fe/Cu(100) are compared. In the calculations a Fe(100) fcc surface has been used, as the 3 ML thick Fe film is growing pseudomorphically on a fcc Cu(100) surface. Only if a non-collinear coupling due to a spin frustration within the clusters is taken into account in the calculations the principle behaviour of the experimental data can be described. For the direct comparison of the experimental and theoretical total magnetic moment per atom the number of d-holes n_{d} is required in (7.1). Furthermore, for Cr the XMCD sum rules might be wrong by 50% for the absolute values [35–37]. Hence, in Fig. 7.9 the experimental spin moments and the theoretical total moments are compared on a relative scale.

The calculations of Robles and Nordström [41] in Fig. 7.9 are using a tight-binding model for the s, p and d valence electrons in a mean-field approximation. They have calculated the magnetic coupling and the total moments from the Cr dimer up to Cr_9 clusters taking into account a collinear and a non-collinear coupling of the individual magnetic moments. The calculated total moment of the clusters is always antiferromagnetically coupled to the fcc Fe(100) surface and the non-collinear

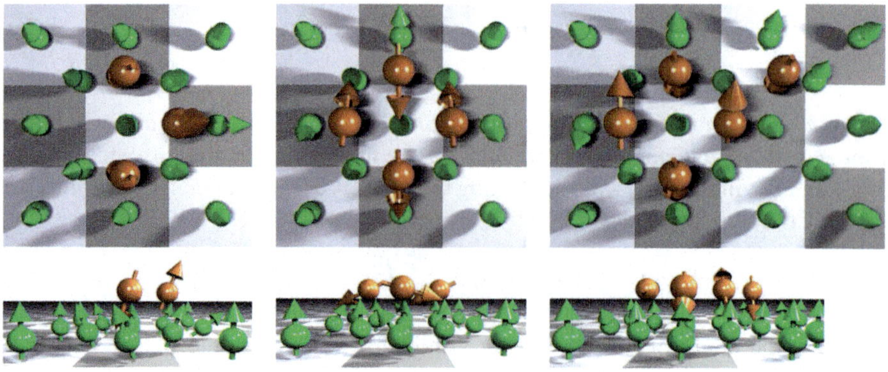

Fig. 7.10 Non-collinear coupling in small chromium clusters. Left: Cr trimer, middle: Compact Cr tetramer, right: Cr pentamer (from [38])

coupling results in a strong reduction of the magnetic moments with increasing cluster size similar to the experimental data and the SPR-KKR calculations. Robles and Nordström have also calculated the magnetic properties of Cr clusters on a bcc Fe(100) surface. The total moment of the clusters is also antiferromagnetically coupled to bcc Fe surfaces, however, with much larger magnetic moments.

Hence, the surface has a tremendous effect on the magnetic properties of the clusters, and further experiments of Cr adatoms, dimers and trimers deposited on a magnetised Ni/Cu(100) film have been performed. Only very small XMCD signals are found for all samples [42, 43], indicating an antiferromagnetic ordering relative to the magnetisation of the Ni substrate. Assuming as a first approximation a similar exchange coupling $J_{\mathrm{Cr-Cr}}$ for the Cr atoms in the dimer and a much smaller coupling of the Cr atoms to the Ni surface $J_{\mathrm{Cr-Ni}}$ compared to $J_{\mathrm{Cr-Fe}}$ above, (7.6) will give $\cos(\vartheta_{\mathrm{k}}) \approx 0$ and the magnetic moments in the Cr dimer on the Ni surface will be oriented almost perpendicular to the Ni magnetisation resulting in an almost vanishing dichroism signal measured perpendicular to the surface.

The magnetic structure of small Cr clusters on a magnetised Ni surface has been studied theoretically by Lounis et. al [44] using the SPR-KKR method. For the Cr dimer they find a large $J_{\mathrm{Cr-Cr}} \cong -200\,\mathrm{meV}$ exchange coupling compared to $J_{\mathrm{Cr-Cr}} \cong -77\,\mathrm{meV}$ for Cr on the Fe surface and a much weaker $J_{\mathrm{Cr-Ni}} \cong -5\,\mathrm{meV}$ exchange coupling to the surface. From that they calculate an antiferromagnetic ordering of the Cr spins with the collinear configuration as the ground state and the non-collinear state as a local minimum. Also for the Cr trimer on Ni/Cu(100) a collinear spin configuration parallel to the Ni magnetisation is predicted, which should show an XMCD signal in the experimental geometry. Here also a non-collinear solution perpendicular to the Ni magnetisation similar to the Cr dimer would explain the experimental result [43].

Using the orbital sum rule (7.2), the orbital moments μ_ℓ of the Cr clusters deposited on the Fe/Cu(100) substrate have been evaluated and are shown in the lower part of Fig. 7.9. Due to the half filled d shell the ground state of free Cr atoms is $^7S_3$

and thus the orbital moment $\mu_\ell = 0$. As expected, for the deposited Cr clusters very small orbital moments μ_ℓ close to zero are observed. The sign of the orbital moment represents the orientation relative to the spin moments, indicating a trend to a parallel coupling of the orbital and spin moments for larger clusters. This is in contrast to the data of Scherz et al. [36, 37], who found a very small orbital moment $\mu_{\ell,\text{film}} = -0.011\,\mu_B$ coupled anti-parallel to the spin moment for ultra-thin chromium films.

An interesting behavior is found for the Cr_3 and Cr_4 clusters. Cr_3 has a small orbital moment $\mu_{\ell,3} = 0.025(15)\,\mu_B$ parallel to the spin moment μ_S, whereas Cr_4 shows an orbital moment $\mu_{\ell,4} = -0.044(13)\,\mu_B$ with an anti-parallel coupling relative to the spin. This finding is already evident from the XMCD spectra of these clusters depicted in Fig. 7.8. Cr_4 shows a much weaker positive XMCD signal at the L_3 edge compared to Cr_1 and Cr_3, whereas the negative part at the L_3 edge and the XMCD signal at the L_2 edge is less affected. This change in the relative orientation of the spin and orbital moment with the cluster size can be explained by a small change of the number of d-holes n_d [43]. According to Hund's rule for a more than half filled shell the ground state of an atom has the maximum possible angular momentum J corresponding to a parallel coupling of the orbital and spin moment. However, for a less than half filled shell the situation changes and an anti-parallel orientation of orbital and spin moment is favored. The Cr atom with $3d^5$ has a half filled shell and the number of d-holes n_d of the Cr atoms in the clusters should be close to 5. However, a small deviation from 5 will result either in a parallel or anti-parallel coupling of the spin and orbital moments. Hence, within this simple model an increase of the number of d-holes would favor the observed anti-parallel coupling of orbital and spin moment in the Cr_4 cluster [43].

7.3.2 Cobalt Clusters

In Fig. 7.11 the magnetic spin and orbital moments for Co clusters deposited on a Ni/Cu(100) [45] and a Pt(111) [7] surface in the few atom limit are depicted. Size selected Co_n cluster with $n = 1 - 3$ have been studied on the remanently magnetised Ni/Cu(100) surface, whereas on the Pt(111) surface size averaged $Co_{\langle s \rangle}$ in a strong magnetic field has been investigated.

Spin and orbital magnetic moments are found to be larger on the Pt(111) surface compared to the Ni/Cu(100) surface. In particular for the Co adatom on Pt(111) a giant orbital moment is found, which is decreasing with increasing size $\langle s \rangle$. On the Ni/Cu(100) surface the smaller orbital moments show a non-monotonic behavior. They are decreasing from the adatom to the dimer, however, for the Co trimer the largest orbital moment is found. Hence, Co clusters show larger magnetic moments on the more weakly coupling Pt(111) surface. This might be attributed to the effect of hybridisation of the 3d orbitals with the substrate, which is in this case smaller with the filled Pt 5d band.

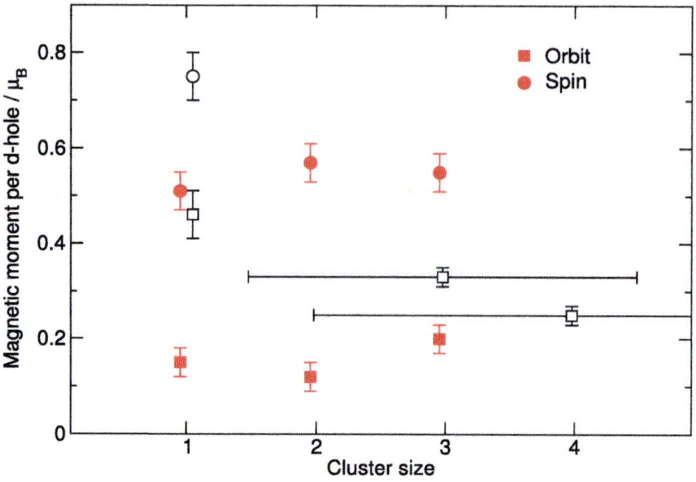

Fig. 7.11 Spin (circles ○) and orbital (squares □) magnetic moments of small Co clusters on two different surfaces; filled symbols: mass selected Co_n on Ni/Cu(100), open symbols: size averaged $Co_{\langle s \rangle}$ on Pt(111) taken from [7] with the number of d-holes $n_d = 2.40$. Creative Common Attribution 3.0 License in J. Phys. C 28, 503002 (2016)

Co adatoms and dimers on a Ni surface show a similar size dependency as the corresponding Fe adatoms and dimers with a small increase of the spin moments [18, 19, 46]. The orbital moment is slightly decreasing from Co_1 to Co_2 and is almost equal for Fe_1 and Fe_2. Adding a further Co atom to dimer increases the orbital moment strongly by 50% while the spin moment for the Co_2 and Co_3 are rather similar. In contrast to this, adding an Fe atom to the Fe_2 the orbital and spin moment drops by 50 and 25%, respectively.

7.4 Alloy Clusters

Already for clusters with only one element a strong size dependency of the magnetic moments is found. However, to enhance or tailor the orbital and spin magnetic moments or tuning the magnetic anisotropy not only clusters consisting of one element should be taken into account. 3d metal atoms have a large spin moment, however, the spin-orbit interaction is rather small. In contrast to that the corresponding 4d and 5d elements as Rh or Pt have a larger spin-orbit interaction, but in the bulk are non-ferromagnetic. However, they can be polarised and small Rh clusters show a superparamagnetic behaviour in the gas phase [47]. Also for small Ru clusters on the Fe/Ni(100) substrate a magnetic ordering has been measured [48], whereas for Ru and Rh impurities on Ag(100) and Pt(997) no magnetic moments could be found [49].

3d/4d or 3d/5d metal alloys, as e.g. Co-Cr-Pt, FePt, CoPt are often planned or already used in modern magnetic storage devices [50–52]. These alloys are interesting for applications as the anisotropy energy might be enhanced by combining the large magnetic moments of the 3d metals and the large spin-orbit interaction of the 4d/5d metals. Therefore, the magnetic spin and orbital moments have been studied on size-selected alloy clusters depending on the exact size and composition.

For all deposited clusters the exact structure is in general not known and could not be easily measured, even if the clusters are produced by atom manipulation [53]. This becomes even more critical for alloy clusters, as the number of possible geometric configurations is further increased [43]. As the typical spot size of soft X-rays is in the order of some 10 μm only an average of these different structures will be measured. For the case of Cr_4 it has already been discussed above that different geometric configurations can show different magnetic properties.

7.4.1 Co Alloy Clusters

For Co clusters three different alloy systems, CoPt [45], CoPd [54] and CoRh [55], have been studied. In Fig. 7.12 the spin and orbital magnetic moments as well as the ratio of these quantities are depicted for some Co_nM_m alloy clusters together with Co_n $(n = 1 - 3)$ clusters deposited on an out-of-plane magnetised Ni/Cu(100) substrate as described above. For both alloy clusters an increase of the orbital moment with increasing number m of 4d/5d atoms is found. For Co_2Pt also the spin moment is increasing, whereas for CoPd the spin moment is decreasing, which results in a strongly enhanced orbital to spin ratio for both systems.

Comparing the results for the Co and CoPt clusters to corresponding Co [56–58] and CoPt [59] nanoparticles as well as CoPt thin films [60, 61] slightly smaller spin moments within the small cluster are found. However, the orbital moments are strongly enhanced. Hence, the aim to increase the orbital moments anticipated by alloying 3d and 4d/5d can be achieved. Essential for this effect is the hybridisation of the 3d and the 4d or 5d electrons in an alloy cluster.

One should already note here that also the chemical reactivity of the Co clusters is strongly enhanced by alloying with the 4d and 5d metals atoms inhibiting experiments on other Co alloy clusters. This will be discussed further in Sect. 7.5.

7.4.2 FePt

To study the effect of the hybridisation of the 3d and 5d elements in more detail a study on Fe_nPt_m clusters has been performed by Chen et al. [46]. An important quantity to understand the magnetic coupling is the spin-orbit (SO) interaction of the 3d electrons as sketched in Fig. 7.1. The SO should be increased to potentially increase the orbital moments and, in turn, maybe the anisotropy energies per atom.

As has been shown by Thole and van der Laan [62, 63], the 3d spin-orbit interaction can be obtained from X-ray absorption spectroscopy by evaluating the branching ratio

$$B_r = \frac{C_{L3}}{C_{L3} + C_{L2}} \tag{7.7}$$

of the L_3 and L_2 white line intensities C_{Lx} defined in (7.3) and depicted in Fig. 7.6. The branching ratio B_r depicted in Fig. 7.13 is increasing with increasing Pt content, which is attributed to an increased 3d spin-orbit interaction of the Fe 3d electrons by hybridisation with the Pt 5d electrons.

In Fig. 7.14 the XMCD spectra, normalised to the same intensity at the L_2 edge, of Fe_2Pt_m clusters are shown. The L_3 XMCD lines have a similar height, however, with increasing number of Pt atoms in the cluster the L_3 XMCD line is getting narrower. This corresponds to a change of the unoccupied density of states of the clusters and is a result of the change in hybridisation. The spin and orbital moments calculated by the sum rules from (7.1) are shown in the left panel of Fig. 7.13. The orbital moment (□) of Fe_nPt_m is decreasing with increasing number m of Pt atoms and the spin moment (○) has a maximum for $m = 1$ and is decreasing, if a second Pt atom is added. From this one can conclude, that there is an optimal number of Pt atoms in the Fe cluster to create the maximum spin moment. This is in contrast to FePt bulk, thin films and nanoparticles [64–66], where the effective spin moment is increasing with the Pt content up to about 50 at% Pt and is rather constant for larger Pt content [64].

From Fig. 7.13 an increase of the Fe 3d spin-orbit interaction with increasing Pt content has been concluded by Chen et al. [46]. However, this increase for the Fe_nPt_m

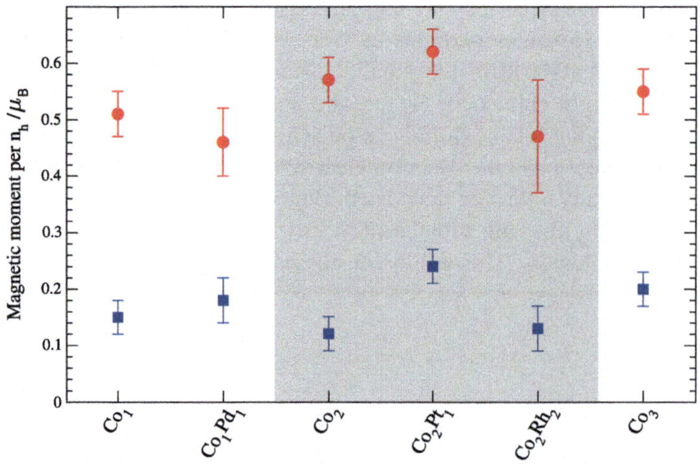

Fig. 7.12 Measured spin (•) and orbital (■) magnetic moments of Co_nM_m ($n = 1 - 3$) 3d metal alloy clusters depending on the size and composition of the cluster. Note, that the Co_2Rh_2 cluster has a ≈10% contribution from the oxidised cluster

clusters does result in a decrease of the orbital moment and an increase of the spin moment.

7.5 Magnetism and Chemical Reactivity

A critical point in the preparation of cluster samples is their possibly high chemical reactivity. In this section the chemical reactivity in terms of oxidation and its effect on the magnetic properties of the clusters will be discussed.

Using the setup described in Sect. 7.2.1 pure metallic alloy clusters with various compositions and sizes can be created. However, when the clusters are deposited in the noble gas matrix they might change their chemical state. The advantage of X-ray absorption spectroscopy in studying deposited clusters is its capability to get information about the chemical state of a 3d metal due to the varying multiplet structure.

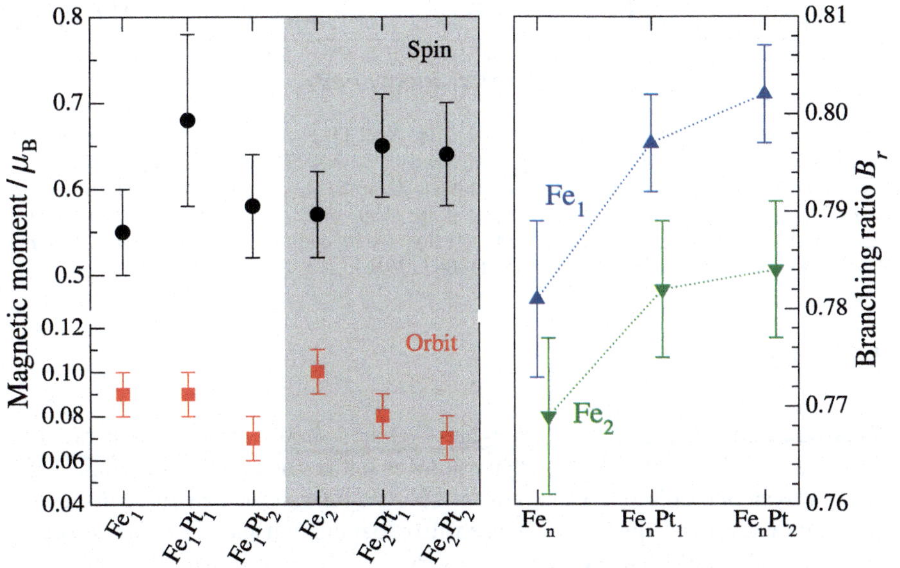

Fig. 7.13 (Left panel) Spin (•) and orbital (■) moments of Fe_nPt_m clusters deposited on a magnetised Ni/Cu(100) substrate; (right panel) Branching ratio B_r of the intensity of the L_3 and L_2 white lines for Fe adatoms, dimers and Fe_nPt_m clusters

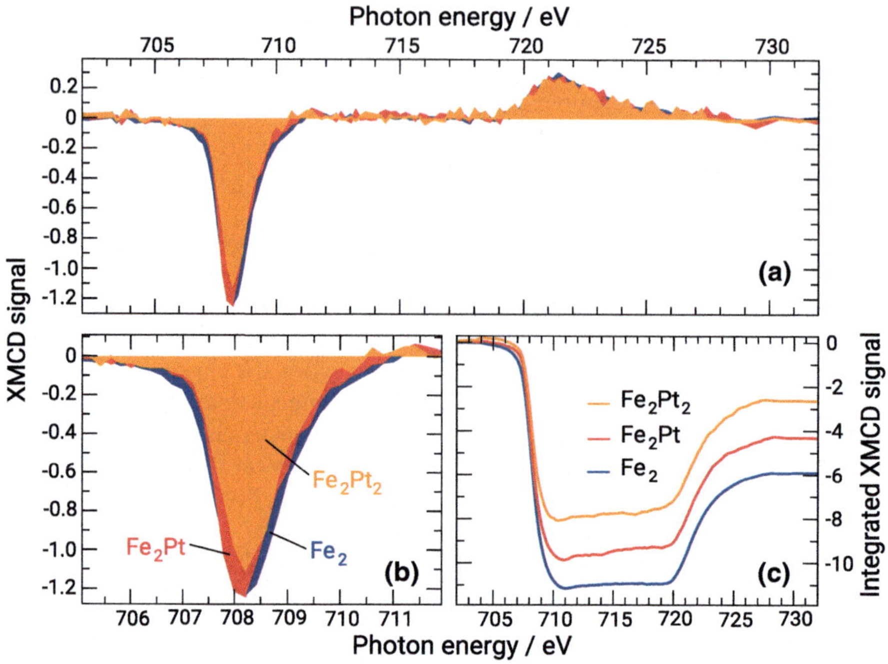

Fig. 7.14 Normalised (**a**) and integrated (**c**) L-edge XMCD spectra of Fe_2Pt_n clusters deposited on a magnetised Ni/Cu(100) surface. In **b** the normalised L_3 XMCD signal is enlarged. The XMCD spectra have been normalised to the integrated intensity of the L_2-edge. For Fe_2 and Fe_2Pt_1 the L_3 XMCD signal has the same height; however, the line shape of Fe_2Pt_1 is narrower, resulting in the observed decreasing integrated L_3 XMCD intensity in **c** by adding Pt to Fe_2. Creative Common Attribution 3.0 License in New J. Phys. 18, 113007 (2016)

7.5.1 CoO

As discussed in Sect. 7.3.2 Co clusters and Co metal show a ferromagnetic ordering. In contrast oxides such as CoO or NiO show an antiferromagnetic ordering. Hence, as a first system oxidised Co adatoms on a Ni/Cu(100) substrate will be discussed.

The pure Co metal clusters are not sensitive to oxidation and the X-ray absorption spectra of deposited clusters are usually varying only slightly with the cluster size. Usually the L_3 white line is shifting in energy but the shape is unchanged as has been shown, for the example, for Cr clusters in the size range from the adatom up to 13 atoms per cluster [67].

CoO was prepared by depositing Co cations in a noble gas matrix with a small amount of oxygen as described in [54]. A change in the oxidation state of a 3d metal can be monitored by the NEXAFS spectra at the L_3 edge, as depicted in Fig. 7.15. CoO shows a multiplet structure whereas non-oxidised Co adatoms and clusters show a metallic-like resonance. With this knowledge the oxidation state of the deposited clusters can be estimated [54]. In the right panel of Fig. 7.15 the X-ray absorption and

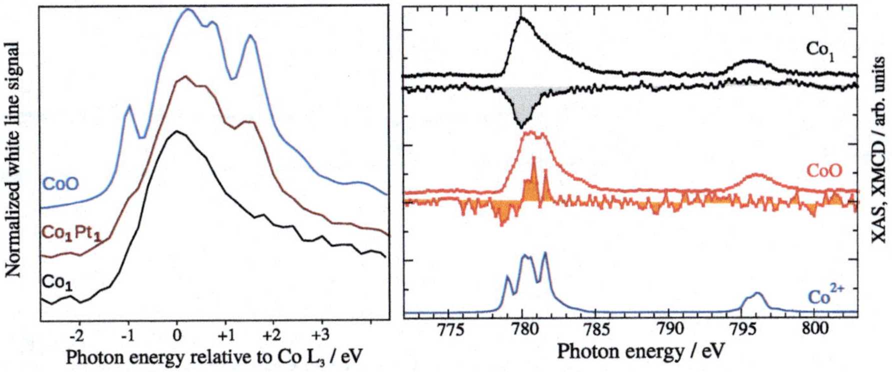

Fig. 7.15 (Left) L_3 XAS spectra of Co_1, Co_1Pt_1-O and bulk CoO; (right) XMCD and XAS spectra of Co_1 and CoO on Ni/Cu(100). The lower curve is the result of a multiplet calculation for Co^{2+}

XMCD spectra of CoO and the result of a multiplet calculation are depicted. The XAS spectrum clearly shows a multiplet structure similar to that of CoO. Compared to Co adatoms the XMCD signal is strongly suppressed and shows an antiferromagnetic ordering of the Co magnetic moment relative to the Ni substrate.

7.5.2 *CoPd Dimers*

In Fig. 7.16 the X-ray absorption and XMCD spectra of CoPd dimers on a magnetised Ni/Cu(100) film for different amounts of oxidation are depicted. From this data the XMCD spectrum of a pure oxidised CoPd-O cluster can be calculated as described in Chen et al. [54]. Similar as for CoO and for CoPdO an antiferromagnetic coupling is found, however with a magnetic spin moment $\mu_s^{\text{eff}} = -0.29(9)\,\mu_B$. For example, the spin moment in CoPd-O is only 30% compared to CoPd, but its orientation relative to the Ni magnetisation has changed.

7.5.3 *CoRh Oxidised Clusters*

The rather large number of stable Pd isotopes results in broad mass distributions of the different Co_nPd_m clusters which complicates the mass selection of clusters with a specific size and composition. Hence, Co_nRh_m clusters have been studied, as Co and Rh have only one natural stable isotope and a large number of specific compositions can be selected from the mass spectrum depicted in Fig. 7.3. Similar to the other Co_n 4d/5d alloy clusters also Co_nRh_m shows a strongly enhanced chemical reactivity.

For Co_1Rh_m ($m = 0, 1, 2$) an increase of the chemical reactivity with the number of Rh atoms in the cluster is found, as the amount of oxidised clusters on the surface

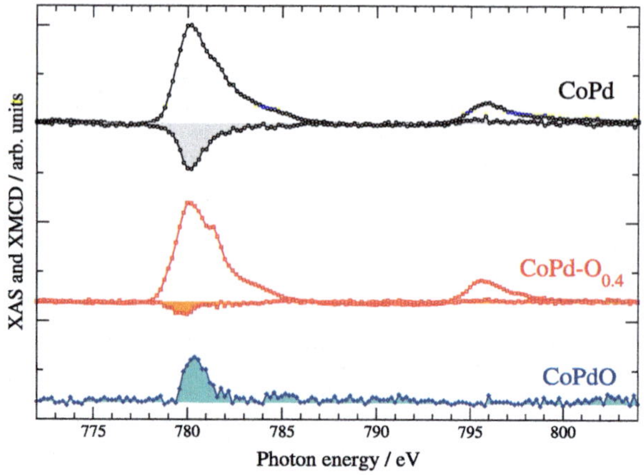

Fig. 7.16 XMCD and XAS spectrum of CoPd-O

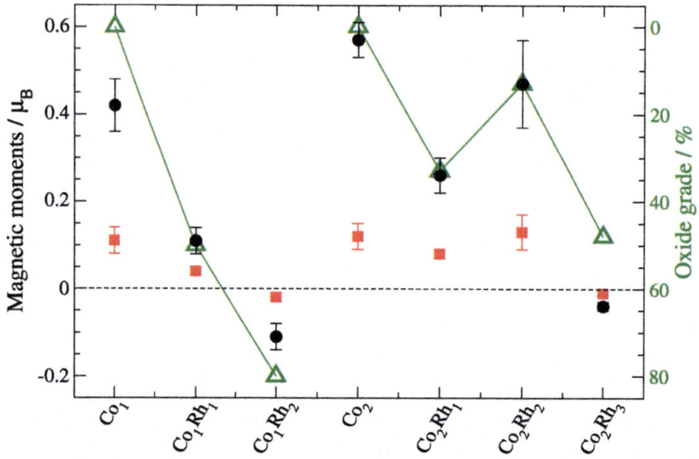

Fig. 7.17 Magnetic moments (left axis, spin •, orbit ■) and degree of oxidation (right axis, ▲) for Co_nRh_m clusters

is increasing up to 80% for Co_1Rh_2. In parallel the magnetic moment is decreasing almost linearly as depicted in Fig. 7.17 and for Co_1Rh_2−O similar to CoPd−O an antiferromagnetic coupling is observed [55].

The situation is more complicated for Co_2Rh_m. In general an increased chemical reactivity is found for these alloy clusters, however, in a non-monotonic way. Adding a Rh atom to the Co dimer results in an oxidation rate of 30% and a decreased magnetic moment, whereas a second Rh atom reduces oxidation to 13% and a third Rh atom increases it again up to 48%. The increased oxidation rate is correlated to a reduction of the magnetic moment of the Co atoms and a transition from a ferromagnetic to an antiferromagnetic alignment of the Co magnetic moments.

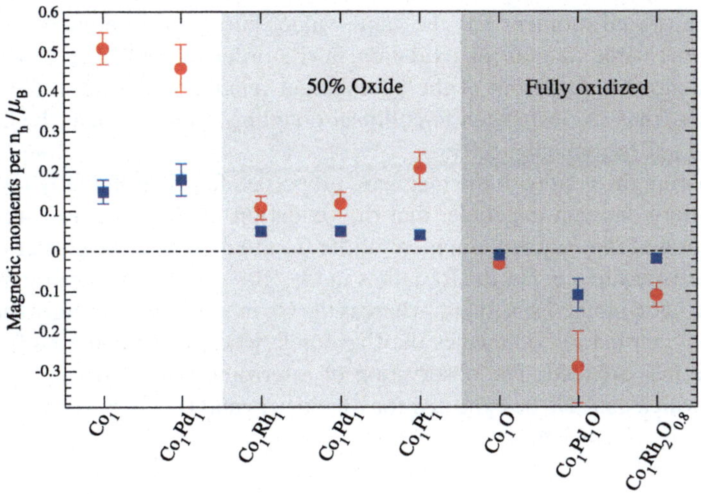

Fig. 7.18 Co spin (circle •) and orbital (square ■) magnetic moments of Co_1M_m adatoms and alloy clusters depending on the oxidation. Note, that the oxidation grade of Co_1Rh_2 is only 80%

Table 7.1 Magnetic moment of $Co_nX_1{-}O$ clusters on Ni/Cu(100)

Cluster	Oxide	μ_s	μ_ℓ	μ_ℓ/μ_s
Co_1Rh_1	50	0.11	0.04	0.32
Co_1Pd_1	40	0.12	0.05	0.37
Co_1Pt_1	50	0.21	0.04	0.17
Co_2Rh_1	33	0.26	0.08	0.30
Co_2Pt_1	≅50	0.24	0.09	0.38

For Rh atoms XMCD spectra can be recorded at the Rh $M_{2,3}$ edges similar to the 3d metal $L_{2,3}$ edges. However, as the Auger rates of the Rh 3p core holes are much larger due to the possible super-Coster-Kronig decay and the 3p-4d dipole matrix element is smaller, the measurement of the corresponding XAS and XMCD spectra of the diluted cluster samples is challenging [48]. Hence, only for the Rh rich cluster samples Co_1Rh_2 and Co_2Rh_3 XMCD spectra have been recorded [55].

Co_1Rh_2 clusters do not show any dichroism, whereas for Co_2Rh_3 an XMCD signal is found. From the negative sign of the XMCD at the Rh M_3 edge it can be concluded, that the Rh magnetic moment is ferromagnetically oriented relative to the Ni substrate and antiferromagnetically to the Co adatoms. This may result in a complex non-collinear coupling within the cluster which would explain the small magnetic moments found for the Co atoms (Fig. 7.18).

In Table 7.1 the magnetic moments of some oxidised Co clusters are summarised. Comparing the spin moments to the unoxidised alloy clusters or the Co_n clusters only a reduction of the spin moment is found. $Co_1X_1{-}O$ clusters for Rh and Pd show very similar values for the spin and orbital moments of $\mu_s \cong 0.1$ and $\mu_\ell \cong 0.04$.

For Pt the orbital moment has the same value, but the spin moment is twice as large for the same amount of oxidation in the order of 50%. For Co_1Pd_1-O an antiferromagnetic alignment of the Co moment relative to the Ni surface is found and for Co_2Rh_3-O clusters a non-collinear coupling is proposed, as the Co and Rh is coupled antiferromagnetically.

Comparing the results of the magnetic properties for the differently oxidised Co alloy clusters one can conclude, that the oxidation of the Co cluster results in a reorientation of the magnetic moment of the Co atoms within the cluster relative to the magnetised surface. For the Rh atoms in Co_2Rh_3-O a ferromagnetic orientation of the magnetic moments is found, whereas the Co moments are strongly suppressed. Hence, oxygen mainly influences the Co atoms, whereas Rh and possibly also Pd atoms are less affected. The observation of a ferromagnetic ordering of small Ru clusters, which can also be expected for Rh clusters and is observed for Co_2Rh_3-O, supports this conclusion.

7.6 Summary

To summarise, the magnetic properties of adatoms and small mass selected 3d metal and alloy clusters have been studied using soft X-rays. By using the unique possibility of magnetic circular dichroism to measure the orbital and spin magnetic moments detailed information on the magnetic properties have been obtained on the very small particles in the few atom limit.

The magnetic moments show a strong dependency on the size and composition of the deposited clusters, which proves that size selection is important and each atom counts.

A non-collinear coupling is found for small Cr_n clusters with the help of calculations based on the SPR-KKR method. The magnetic coupling depends strongly on the size and the chosen magnetised surface which can be understood by the spin frustration of the magnetic moments and the variation of the exchange coupling constants J.

For Co clusters with up to three atoms per cluster a ferromagnetic coupling relative to the magnetised surface and also within the cluster is found, which is the case if the cluster is alloyed with a 4d or 5d metal. The alloying has a strong effect on the magnetic moments of the Co atoms, however, there is also a very strong increase of the chemical reactivity of the alloy cluster against oxidation.

The oxidation results in general in a strong decrease of spin and orbital moments of the Co atoms within the cluster and for CoO a switch to an antiferromagnetic coupling of the Co atom relative to the surface is found with very small magnetic moments. This antiferromagnetic coupling is also found for oxidised Co_1Pd_1 and Co_1Rh_2 clusters with much larger magnetic moments compared to CoO.

For some CoRh clusters the magnetic properties could be measured for both, the Co and Rh atoms. Rh shows a much weaker X-ray magnetic circular dichroism effect, nevertheless, again a strong size and composition dependency is found. In particular, for Co_2Rh_3 an antiferromagnetic orientation of the Rh magnetic moments relative to the surface and small Co magnetic moments are found, which is a hint to a non-collinear coupling also within these alloy clusters.

The studied 3d metal and alloy clusters with only a few atoms do not show any remanent magnetisation, but some clusters show a complex non-collinear magnetic coupling. This non-collinear coupling can be tailored by choosing the right size, composition and substrate material. Hence, these clusters can serve maybe as the smallest model systems for large complex spin structures which are considered for magnetic storage devices.

Acknowledgements This work was supported by the Deutsche Forschungsgemeinschaft via the SFB668 project A7. We thank the BESSY II beamline staff, namely H. Pfau and F.-X. Talon, for their logistic assistance. Finally, we have to thank Graham Appleby for carefully reading the manuscript.

References

1. E. Weschke, H. Ott, E. Schierle, C. Schussler-Langeheine, D.V. Vyalikh, G. Kaindl, V. Leiner, M. Ay, T. Schmitte, H. Zabel, P.J. Jensen, Phys. Rev. Lett. **93**(15), 157204 (2004)
2. P.J. Hsu, A. Finco, L. Schmidt, A. Kubetzka, K. von Bergmann, R. Wiesendanger, Phys. Rev. Lett. **116**, 017201 (2016)
3. G. Binasch, P. Grünberg, F. Saurenbach, W. Zinn, Phys. Rev. B **39**, 48284830 (1989)
4. P. Grünberg, R. Schreiber, Y. Pang, M.B. Brodsky, H. Sowers, Phys. Rev. Lett. **57**(19), 2442 (1986)
5. R.E. Camley, J. Barnas, Phys. Rev. Lett. **63**, 664 (1989)
6. N. Romming, C. Hanneken, M. Menzel, J.E. Bickel, B. Wolter, K. von Bergmann, A. Kubetzka, R. Wiesendanger, Science **341**(6146), 636639 (2013)
7. P. Gambardella, S. Rusponi, M. Veronese, S.S. Dhesi, S. Rusponi, M. Veronese, S.S. Dhesi, P.H. Dederichs, K. Kern, C. Carbone, H. Brunel, Science **300**, 1130 (2003)
8. M. Toshio, S. Tobias, M. Tobias, B. Christopher, B. Timofey, S. Alexander, K. Christian, A. Stephan, M. Michael, H. Martin, G. Matthias, O. Sergey, H. Wolfram, M. Ingrid, S. Gerd, E. Arthur, W. Wulf, Nature **503**(7475), 242246 (2013)
9. H. Brune, P. Gambardella, Surf. Sci. **603**(1012), 18121830 (2009)
10. I.G. Rau, S. Baumann, S. Rusponi, F. Donati, S. Stepanow, L. Gragnaniello, J. Dreiser, C. Piamonteze, F. Nolting, S. Gangopadhyay, O.R. Albertini, R.M. Macfarlane, C.P. Lutz, B.A. Jones, P. Gambardella, A.J. Heinrich, H. Brune, Science **344**(6187), 988992 (2014)
11. S. Baumann, F. Donati, S. Stepanow, S. Rusponi, W. Paul, S. Gangopadhyay, I.G. Rau, G.E. Pacchioni, L. Gragnaniello, M. Pivetta, J. Dreiser, C. Piamonteze, C.P. Lutz, R.M. Macfarlane, B.A. Jones, P. Gambardella, A.J. Heinrich, H. Brune, Phys. Rev. Lett. **115**, 237202 (2015)
12. F. Donati, A. Singha, S. Stepanow, C. Wäckerlin, J. Dreiser, P. Gambardella, S. Rusponi, H. Brune, Phys. Rev. Lett. **113**, 237201 (2014). https://doi.org/10.1103/PhysRevLett.113.237201
13. T. Eelbo, M. Waniowska, P. Thakur, M. Gyamfi, B. Sachs, T.O. Wehling, S. Forti, U. Starke, C. Tieg, A.I. Lichtenstein, R. Wiesendanger, Phys. Rev. Lett. **110**, 136804 (2013). https://doi.org/10.1103/PhysRevLett.110.136804
14. M. Etzkorn, C.F. Hirjibehedin, A. Lehnert, S. Ouazi, S. Rusponi, S. Stepanow, P. Gambardella, C. Tieg, P. Thakur, A.I. Lichtenstein, A.B. Shick, S. Loth, A.J. Heinrich, H. Brune, Phys. Rev. B **92**, 184406 (2015). https://doi.org/10.1103/PhysRevB.92.184406
15. A. Lehnert, S. Rusponi, M. Etzkorn, S. Ouazi, P. Thakur, H. Brune, Phys. Rev. B **81**, 104430 (2010). https://doi.org/10.1103/PhysRevB.81.104430
16. U. Heiz, A. Sanchez, S. Abbet, W.D. Schneider, J. Am. Chem. Soc. **121**, 32143217 (1999)

17. A. Sanchez, S. Abbet, U. Heiz, W.D. Schneider, H. Häkkinen, R.N. Barnett, U. Landman, J. Phys. Chem. A **103**, 9573 (1999)
18. J. Lau, A. Föhlisch, M. Martins, R. Nietubyč M. Reif, W. Wurth. New J. Phys. **4**, 98 (2002)
19. J. Lau, A. Föhlisch, R. Nietubyč, M. Reif, W. Wurth. Phys. Rev. Lett. **89**(5), 057201 (2002)
20. P. Bruno, Phys. Rev. B **39**, 865868 (1989)
21. P. Bruno, J.P. Renard, Appl. Phys. A **49**(5), 499506 (1989)
22. J. Stöhr, J. Elec. Spec. Relat. Phenom. **75**, 253272 (1995)
23. J. Lau, A. Achleitner, H.U. Ehrke, U. Langenbuch, M. Reif, W. Wurth, Rev. Sci. Instr. **76**, 063902 (2005)
24. H.P. Cheng, U. Landman, J. Phys. C **98**, 35273537 (1994)
25. H. Cheng, U. Landman, Science **260**, 1304 (1993)
26. J. Stöhr, J. Magn. Magn. Mat. **200**, 470497 (1999)
27. J. Stöhr, H.C. Siegmann, *Solid-State Sciences*, vol. 5 (Springer, Berlin, 2006)
28. D.H. Bilderback, P. Elleaume, E. Weckert, J. Phys. B **38**(9), S773 (2005)
29. S. Sasaki, K. Miyata, T. Takada, Jpn. J. Appl. Phys. **31**, L1794 (1992)
30. B.T. Thole, P. Carra, F. Sette, G. van der Laan, Phys. Rev. Lett. **68**(12), 1943 (1992)
31. C.T. Chen, Y.U. Idzerda, H.J. Lin, N.V. Smith, G. Meigs, E. Chaban, G.H. Ho, E. Pellegrin, F. Sette, Phys. Rev. Lett. **75**(1), 152 (1995)
32. R. Wu, A.J. Freeman, Phys. Rev. Lett. **73**(14), 1994 (1994)
33. O. ipr, J. Minr, H. Ebert, EPL (Europhys. Lett.) **87**(6), 67007 (2009)
34. C. Piamonteze, P. Miedema, F.M.F. de Groot, Phys. Rev. B **80**, 184410 (2009)
35. A. Scherz, H. Wende, K. Baberschke, J. Minar, D. Benea, H. Ebert, Phys. Rev. B **66**, 184401 (2002)
36. A. Scherz, H. Wende, K. Baberschke, Appl. Phys. A **78**, 843846 (2004)
37. A. Scherz, H. Wende, C. Sorg, K. Baberschke, J. Minar, D. Benea, H. Ebert, Physica Scripta **T115**, 586588 (2005)
38. S. Lounis, M. Reif, P. Mavropoulos, L. Glaser, P. Dederichs, M. Martins, S. Blügel, W. Wurth, Eur. Phys. Lett. **81**, 47004 (2008)
39. G. Prümper, S. Kröger, R. Müller, M. Martins, J. Viefhaus, P. Zimmermann, U. Becker, Phys. Rev. A **68**, 032710 (2003)
40. M. Martins, K. Godehusen, T. Richter, P. Wernet, P. Zimmermann, J. Phys. B **39**(5), R79R125 (2006)
41. R. Robles, L. Nordstrom, Phys. Rev. B **74**(9), 094403 (2006)
42. W. Wurth, M. Martins, *in Atomic Clusters from Gas Phase to Deposited*, *The Chemical Physics of Solid Surfaces*, vol. 12, ed. by D. Woodruff (Elsevier, Amsterdam, 2007), p. 471
43. M. Martins, W. Wurth, J. Phys. Cond. Mat. **28**, 503002 (2016)
44. S. Lounis, P. Mavropoulos, P.H. Dederichs, S. Blügel, Phys. Rev. B (Condens. Matter Mater. Phys.) **72**(22), 224437 (2005)
45. L. Glaser, K. Chen, S. Fiedler, M. Wellhöfer, W. Wurth, M. Martins, Phys. Rev. B **86**, 075435 (2012)
46. K. Chen, S. Fiedler, I. Baev, T. Beeck, W. Wurth, M. Martins, New J. Phys. **14**(12), 123005 (2012)
47. A.J. Cox, J.G. Louderback, L.A. Bloomfield, Phys. Rev. Lett. **71**(6), 923 (1993)
48. J. Minr, S. Bornemann, S. Mankovsky, H. Ebert, M. Martins, M. Reif, L. Glaser, W. Wurth, Phys. Status Solidi (b) **247**(5), 11801186 (2010)
49. J. Honolka, K. Kuhnke, L. Vitali, A. Enders, K. Kern, S. Gardonio, C. Carbone, S.R. Krishnakumar, P. Bencok, S. Stepanow, P. Gambardella, Phys. Rev. B (Condens. Matter Mater. Phys.) **76**(14), 144412 (2007)
50. N.A. Frey, S. Sun, *Inorganic Nanoparticles: Synthesis, Application, and Perspective* (2010), p. 3368
51. I. McFadyen, E. Fullerton, M. Carey, Mrs Bull. **31**(05), 379383 (2006)
52. G.W. Qin, Y.P. Ren, N. Xiao, B. Yang, L. Zuo, K. Oikawa, Int. Mater. Rev. **54**(3), 157179 (2009)

53. A.A. Khajetoorians, B. Baxevanis, C. Hübner, T. Schlenk, S. Krause, T.O. Wehling, S. Lounis, A. Lichtenstein, D. Pfannkuche, J. Wiebe, R. Wiesendanger, Science **339**(6115), 5559 (2013)
54. K. Chen, T. Beeck, S. Fiedler, I. Baev, W. Wurth, M. Martins, Phys. Rev. B **93**, 144421 (2016)
55. T. Beeck, I. Baev, K. Chen, S. Palutke, W. Wurth, M. Martins, New J. Phys. **18**(11), 113007 (2016)
56. A. Kleibert, J. Passig, K.H. Meiwes-Broer, M. Getzlaff, J. Bansmann, J. Appl. Phys. **101**(11), 114318 (2007)
57. J. Bansmann, A. Kleibert, F. Bulut, M. Getzlaff, P. Imperia, C. Boeglin, K.H. Meiwes-Broer, Eur. Phys. J. D **45**(3), 521528 (2007)
58. J. Bansmann, M. Getzlaff, A. Kleibert, F. Bulut, R. Gebhardt, K. Meiwes-Broer, Appl. Phys. A **82**(1), 7379 (2006)
59. F. Tournus, A. Tamion, N. Blanc, A. Hannour, L. Bardotti, B. Prvel, P. Ohresser, E. Bonet, T. Epicier, V. Dupuis, Phys. Rev. B **77**(14), 144411 (2008)
60. W. Grange, I. Galanakis, M. Alouani, M. Maret, J.P. Kappler, A. Rogalev, Phys. Rev. B **62**(2), 11571166 (2000)
61. W. Grange, M. Maret, J.P. Kappler, J. Vogel, A. Fontaine, F. Petroff, G. Krill, A. Rogalev, J. Goulon, M. Finazzi, N.B. Brookes, Phys. Rev. B **58**(10), 62986304 (1998)
62. B.T. Thole, G. van der Laan, Phys. Rev. B **38**(5), 3158 (1988)
63. G. van der Laan, B.T. Thole, Phys. Rev. Lett. **60**(19), 19771980 (1988)
64. C. Antoniak, M. Spasova, A. Trunova, K. Fauth, F. Wilhelm, A. Rogalev, J. Minr, H. Ebert, M. Farle, H. Wende, J. Phys.: Condens. Matter **21**(33), 336002 (2009)
65. C. Antoniak, J. Lindner, M. Spasova, D. Sudfeld, M. Acet, M. Farle, K. Fauth, U. Wiedwald, H.G. Boyen, P. Ziemann, F. Wilhelm, A. Rogalev, S. Sun, Phys. Rev. Lett. **97**(11), 117201 (2006)
66. A. Kleibert, K.H. Meiwes-Broer, J. Bansmann, Phys. Rev. B **79**(12), 125423 (2009)
67. M. Reif, L. Glaser, M. Martins, W. Wurth, Phys. Rev. B **72**, 155405 (2005)

Chapter 8
Non-collinear Magnetism Studied with Spin-Polarized Scanning Tunneling Microscopy

Kirsten von Bergmann, André Kubetzka, Oswald Pietzsch and Roland Wiesendanger

Abstract Non-collinear magnetic states in nanostructures and ultra thin films have moved into the focus of research upon the experimental discovery that the interface-induced Dzyaloshinskii–Moriya interaction (DMI) can play a crucial role for the magnetic ground state. In particular, DMI-induced magnetic skyrmions, which are particle-like knots in the magnetization of two-dimensional systems, have attracted significant attention due to their potential use in future spintronic devices. Since then, research has focused both on tailoring thin-films and multilayers hosting magnetic skyrmions, and investigating specific processes such as controlled lateral movement, detection, as well as writing and deleting of single magnetic skyrmions. This chapter reviews the fundamental interactions and mechanisms for the formation of non-collinear spin textures and then introduces how scanning tunneling microscopy (STM) can be exploited to investigate such magnetic states. Next, examples of (one-dimensional) spin spirals will be discussed before the emergence of two-dimensional non-collinear spin textures is studied and characterized in detail. Finally, different mechanisms for the controlled writing and deleting of magnetic skyrmions with the STM tip are explored.

8.1 Introduction

Non-collinear magnetism [1–3] is at the heart of current spintronics research [3, 4]. Particle-like magnetic skyrmions are non-collinear entities which are topologically distinct from the surrounding ferromagnet. Non-collinear magnetic textures interact efficiently with spin(-polarized) currents and it has been shown that skyrmions can be moved by lateral currents either via spin-polarized currents within the material [5] or via spin-orbit torques from a lateral current in an adjacent metallic film [6]. This opens the opportunity to encode information in the form of skyrmions and to move a train

K. von Bergmann · A. Kubetzka · O. Pietzsch · R. Wiesendanger (✉)
Department of Physics, University of Hamburg, Hamburg, Germany
e-mail: wiesendanger@physnet.uni-hamburg.de

K. von Bergmann
e-mail: kbergman@physnet.uni-hamburg.de

R. Wiesendanger (ed.), *Atomic- and Nanoscale Magnetism*, NanoScience and Technology, https://doi.org/10.1007/978-3-319-99558-8_8

of such "bits" within a magnetic track to a stationary read and write head. In such a racetrack device, mechanically moving parts are avoided and the third dimension can be used by undulating tracks to increase storage density [7]. Other concepts, such as reservoir computing, make direct use of one more material dimension and the particle character of skyrmions. In any case, key aspects of skyrmion-based spintronic devices are skyrmion mobility, i.e. material design with minimized skyrmion pinning, as well as reliable creation, annihilation and detection mechanisms, to mention just the most obvious requirements. This chapter exemplifies how fundamental research at the nanometer scale can contribute to advance the rapidly evolving field of skyrmion-based spintronics.

8.2 Magnetic Interactions

Non-collinear spin textures arise due to the competition of different magnetic interactions. Often the leading energy term is the pair-wise Heisenberg exchange

$$E_{\mathrm{H}} = -\sum_{ij} J_{ij} \mathbf{S}_i \cdot \mathbf{S}_j \tag{8.1}$$

between adjacent atomic spins $\mathbf{S}$, which prefers ferromagnetic (FM) or antiferromagnetic (AFM) alignment depending on the sign of the interaction constant J. However, also longer range Heisenberg interactions, i.e. between more distant neighbors ij, can play a role and can lead to non-collinear magnetic order, such as spin spirals. In a spin spiral the magnetic moments rotate stepwise along the propagation direction and a 360° rotation is completed after a characteristic wavelength λ. A spin spiral due to Heisenberg exchange frustration can arise, e.g. when the nearest neighbor coupling is FM and the next-nearest neighbor coupling is AFM, or when the nearest neighbor coupling is AFM and the next-nearest neighbor coupling is also AFM.

The Dzyaloshinskii–Moriya interaction (DMI) arises due to spin-orbit interaction and is described by

$$E_{\mathrm{DM}} = -\sum_{ij} \mathbf{D}_{ij} (\mathbf{S}_i \times \mathbf{S}_j), \tag{8.2}$$

where the direction of $\mathbf{D}$ determines which orthogonal alignment between adjacent spins is preferred, i.e. the type and the direction of relative spin alignment. Also a competition between (nearest neighbor) Heisenberg exchange and DMI can lead to spin spiral states, however, in contrast to spin spirals due to frustration of exchange interactions, such spin textures exhibit a unique rotational sense [1, 2, 8, 9]. DMI can arise in bulk systems with broken inversion symmetry such as chiral crystal structures (e.g. B20) or it can be induced by a symmetry breaking due to the presence of interfaces. Interface-DMI systems typically exhibit cycloidal rotation because of the symmetry selection rules [2, 4, 10].

A uniaxial anisotropy energy is described by

$$E_{\text{ani}} = \sum_i K_i (S_i^z)^2 \tag{8.3}$$

and a sizable K can lead to a distortion of a given spin spiral. In such an inhomogeneous spin spiral the angle between nearest neighbors varies and depends on the local quantization axis. For large K the system forms magnetic domains that are separated by domain walls. When DMI plays a role the domain walls exhibit a unique rotational sense and often they are called chiral domain walls [11–14], although when cycloidal, they are identical with their mirror image.

When the non-collinearity extends to two dimensions, the spin texture can acquire a topological winding number

$$Q = \frac{1}{4\pi} \int \mathbf{n} \cdot \left(\frac{\partial \mathbf{n}}{\partial x} \times \frac{\partial \mathbf{n}}{\partial y} \right) \mathrm{d}x\mathrm{d}y, \tag{8.4}$$

where $\mathbf{n}$ is the normalized magnetic vector field and x, y are the spatial coordinates. One prominent example is a magnetic skyrmion, which is characterized by a quasi-continuous spin rotation of 180° with unique rotational sense from its center to the ferromagnetic surrounding, see sketches in Fig. 8.1(left). Different mechanisms can lead to the formation of skyrmion lattices [2, 3, 15], i.e. they can arise from higher-order interactions in zero magnetic field [16, 17] or they can be induced from a spin spiral phase by an external magnetic field [18–20]; however, the DMI is always necessary to impose the unique rotational sense.

8.3 Spin-Polarized Scanning Tunneling Microscopy

Scanning tunneling microscopy (STM) exploits the distance and bias voltage dependence of the tunnel current between two electrodes [25]. When one or both of the electrodes are magnetic several magnetoresistive effects contribute to the transport, see Fig. 8.1(top). The tunnel magnetoresistance (TMR) effect occurs when both the sample and the tip of the STM are magnetic and leads to a cosine-dependence of the tunnel current for the relative magnetization directions of the two magnetic electrodes [26, 27]. This spin-polarized (SP) tunnel current I_{SP} at lateral position r and bias voltage U can be expressed by

$$I_{\text{SP}}(\boldsymbol{r}, U) = I_0[1 + P_{\text{s}} \cdot P_{\text{t}} \cdot \cos(\boldsymbol{M}_{\text{s}}, \boldsymbol{M}_{\text{t}})], \tag{8.5}$$

where I_0 is the spin-averaged contribution to the current, and P and M are the spin-polarization and the magnetization of the sample (s) and the tip (t). When SP-STM is operated in constant-current mode, i.e. a feedback loop adjusts the tip-sample distance to keep the tunnel current at a given setpoint, the corrugation of a homogeneous flat

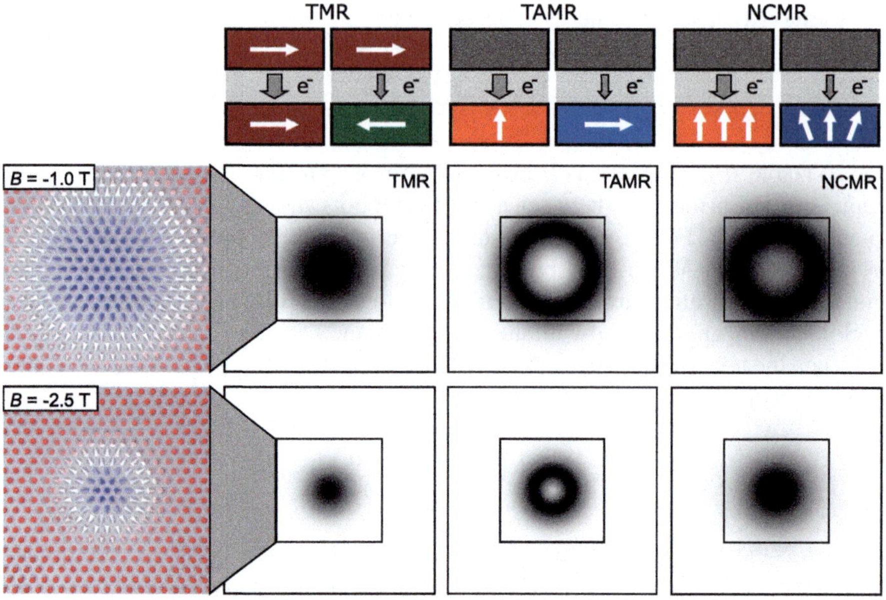

Fig. 8.1 Magnetoresistance (MR) effects of tunnel junctions. Top: sketches for a planar tunnel junction exhibiting the tunnel MR effect (TMR) between two magnetic electrodes, the (tunnel) anisotropic MR (TAMR), and the non-collinear MR (NCMR); in the latter two cases the MR occurs between a non-magnetic and a magnetic electrode. Below (left) are sketches of atomic magnetic moments of individual magnetic skyrmions in Pd/Fe on Ir(111) at the indicated magnetic field values. Next to them are simulated STM images taking the respective MR effects for the two different skyrmion sizes into account separately [21–24]. (Figure adapted from [23])

magnetic film can be interpreted as the spin-polarized contribution to the tunnel current. This kind of measurement mode is useful for studying atomic scale spin textures [1, 28, 29].

When the tip is non-magnetic, the tunnel current can still be sensitive to the local spin configuration of a magnetic sample. Due to the tunnel anisotropic magnetoresistance (TAMR) originating from spin-orbit coupling the local electronic structure of out-of-plane and in-plane magnetized sample areas of the same material is different, which is reflected in a different (spin-averaged) STM signal [22, 30]. The non-collinear magnetoresistance (NCMR) effect is due to spin mixing between canted magnetic moments compared to parallel magnetic moments and leads to a modification of the electronic states depending on the degree of non-collinearity [23, 24, 31].

Figure 8.1 shows how these different magnetoresistive effects influence the appearance of magnetic skyrmions at two different magnetic field values in the system of Pd/Fe/Ir(111). Whereas the TMR shows maximum contrast between center and surrounding of the skyrmion when $\mathbf{M}_t$ is normal to the magnetic film, the TAMR has the same signal in the center and the ferromagnetic background with a ring at the position of the in-plane spins. The appearance of the NCMR is more complex and when modeled as proportional to the local mean angle between adjacent magnetic moments, there is a transition from ring- to dot-like shape with increasing magnetic field, reflecting the position where the spin canting is maximal [23, 24]. In an STM measurement the different effects are bias-dependent and may occur simultaneously. A disentanglement is not always straightforward: whereas the TMR can be "switched off" by using a non-spin-polarized tip, the remaining TAMR and NCMR can give rise to similar images, see Fig. 8.1 [23, 24].

8.4 Spin Spirals with Unique Rotational Sense

8.4.1 A Manganese Monolayer on W(110) and W(001)

Mn monolayers on W(110) have been investigated at the atomic scale with SP-STM as well as *ab-initio* calculations and local antiferromagnetic order was found [32]. Measurements on a larger scale are shown in Fig. 8.2 [1]: while the vertical stripes in the spin-resolved images (left) are clear indications for the antiferromagnetic order, they nearly vanish periodically about every 5 nm. Measurements with differently magnetized tips at the same sample position reveal the origin of this effect: while in (a) the Fe coated W-tip is sensitive to the in-plane components of the sample

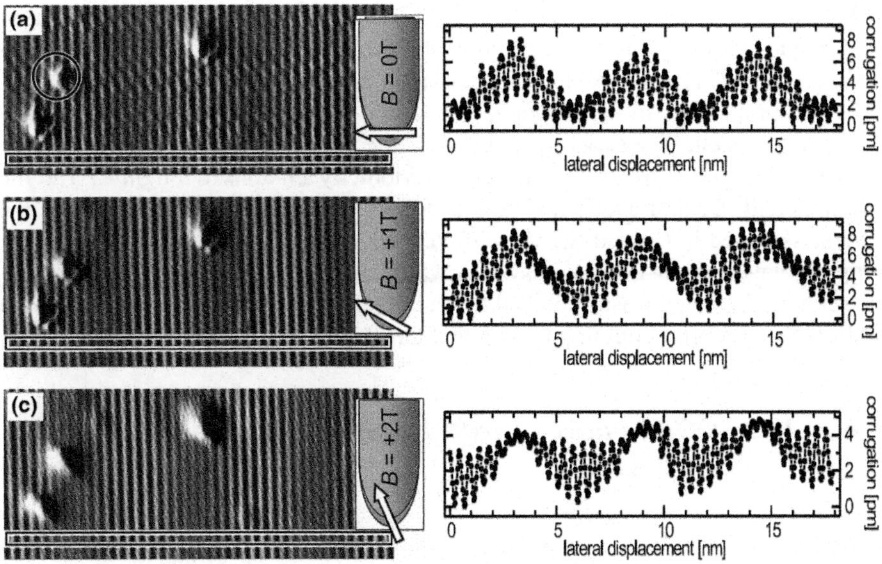

Fig. 8.2 Field-dependent SP-STM measurements. Magnetically sensitive constant-current images of the Mn monolayer on W(110) (left panels) and corresponding line sections (right panels) taken with a ferromagnetic Fe-coated tip at external fields of 0 T (**a**), 1 T (**b**) and 2 T (**c**). As sketched in the insets, the external field rotates the tip magnetization from in-plane (**a**) to out-of-plane (**c**), shifting the position of maximum spin contrast ($I = 2\,\text{nA}$, $U = 30\,\text{mV}$). (Figure adapted from [1])

magnetization, in (c) the tip is magnetized perpendicular to the surface and is therefore sensitive to the out-of-plane components. In (b) the tip magnetization has an intermediate angle. The apparent shift of the magnetic superstructure to the left with increasing external field directly shows that the Mn monolayer on W(110) is only locally close to an antiferromagnet, but the spins in fact rotate by about 173° between atomic rows, giving rise to a spin spiral along the $[1\bar{1}0]$ direction.

Two possible mechanisms for the formation of a spin spiral have been discussed in Sect. 8.2: first, a competition of different Heisenberg exchange contributions (e.g. nearest vs next-nearest neighbor coupling), or a twist between adjacent magnetic moments due to the Dzyaloshinskii–Moriya interaction (DMI), which is a result of spin-orbit interaction in an environment lacking inversion symmetry. While a purely exchange driven spin spiral is energetically degenerate with respect to the rotational sense, the DMI can lift this degeneracy, favoring one rotational sense over the other. The SP-STM signal shape of six independent Mn islands measured with the same spin-polarized tip with canted magnetization (as in Fig. 8.2b) shows the same asymmetry, i.e. position of the maximum of the magnetic contrast amplitude relative to the position of electronic contrast maximum. This directly shows that all of these investigated islands exhibit the same rotational sense imposed by the DMI as confirmed by density functional theory calculations [1].

To study the influence of the symmetry of the atomic lattice on spin spiral states the pseudomorphic Mn monolayer on W(001), which has a four-fold symmetry, was investigated. Again a spin spiral is observed in spin-resolved measurements, see Fig. 8.3, which has atoms with magnetization components in the surface plane (a) and normal to the surface (b) [8]. On a larger sized image (c) one can see a labyrinth pattern due to spin spirals propagating along the two equivalent <110> directions of the surface. This gives rise to four spots in the Fourier transform of this image (d). Note that in addition to the magnetic signal also atomic resolution is obtained for this sample system. The experimental line profile indicated by the white line in (a) is shown in (e) as open circles. The magnetic periodicity amounts to roughly 5.5 atomic distances and Fig. 8.3f shows a simulated SP-STM image of the spin spiral ground state as sketched in (g) and (h). *Ab-initio* calculations show that without spin-orbit coupling, i.e. without DMI, this system would already exhibit a spin spiral due to Heisenberg exchange frustration and that a unique rotational sense is favored by the DMI when spin-orbit coupling is switched on [8].

8.4.2 Fe and Co Chains on Ir(001): Magnetism in One Dimension

The magnetic ground state of biatomic Fe chains on the reconstructed Ir(001) surface is a spin spiral, where adjacent magnetic moments have an angle of about 120°, see sketch in Fig. 8.4a [9]. Whereas this magnetic state fluctuates due to thermal excitations at $T = 8\,\mathrm{K}$ (see Fig. 8.4b left), it can be stabilized by direct exchange coupling to a ferromagnetic Co chain (see Fig. 8.4b center). Combined SP-STM and DFT studies have demonstrated that the magnetocrystalline anisotropy axis of one

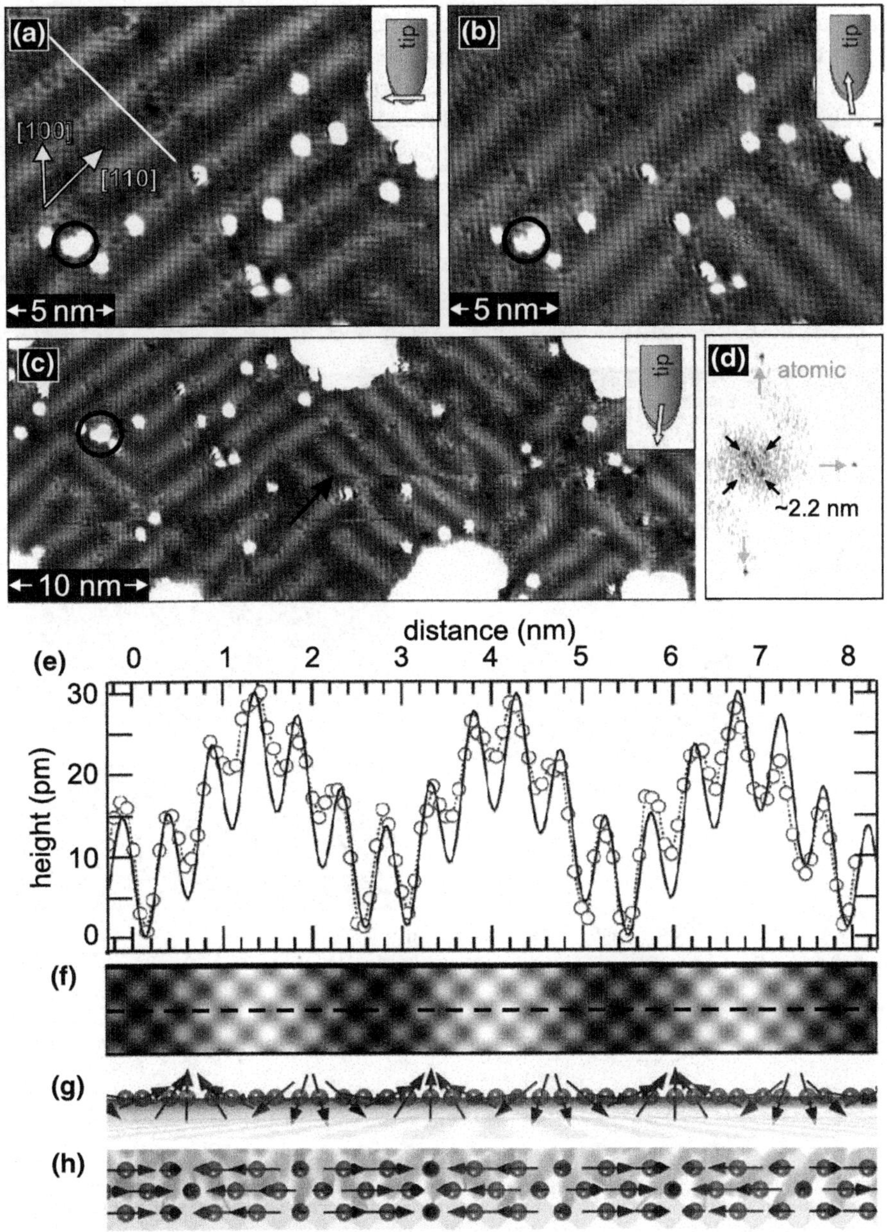

Fig. 8.3 Spin-resolved STM measurements of 1 ML Mn/W(001). Constant-current image with magnetic tip sensitive to **a** the in-plane and **b**, **c** the out-of-plane component of the sample magnetization (the black circle acts as position marker); **d** Fourier transform of a dI/dU map (not shown) of a sample area with both rotational domains; **e** experimental (circles) and simulated (solid line) profile along the lines indicated in **a** and **f**, respectively; **f** simulated SP-STM image, **g** side, and **h** top view of the corresponding model of the spin spiral. The following values of bias voltage U and current I were used: **a** and **b** $U = -0.1\,\text{V}$, $I = 1\,\text{nA}$; **c** $U = -0.1\,\text{V}$, $I = 0.1\,\text{nA}$. (Figure adapted from [8])

structural type of these biatomic ferromagnetic Co chains is canted by about 30° with respect to the surface normal, compare sketch in Fig. 8.4c [34]. This unusual behavior originates from the asymmetric adsorption sites of the two strands of the Co chain in combination with the spin-orbit coupling of the Ir substrate, as revealed by a detailed analysis of the DFT calculations [34].

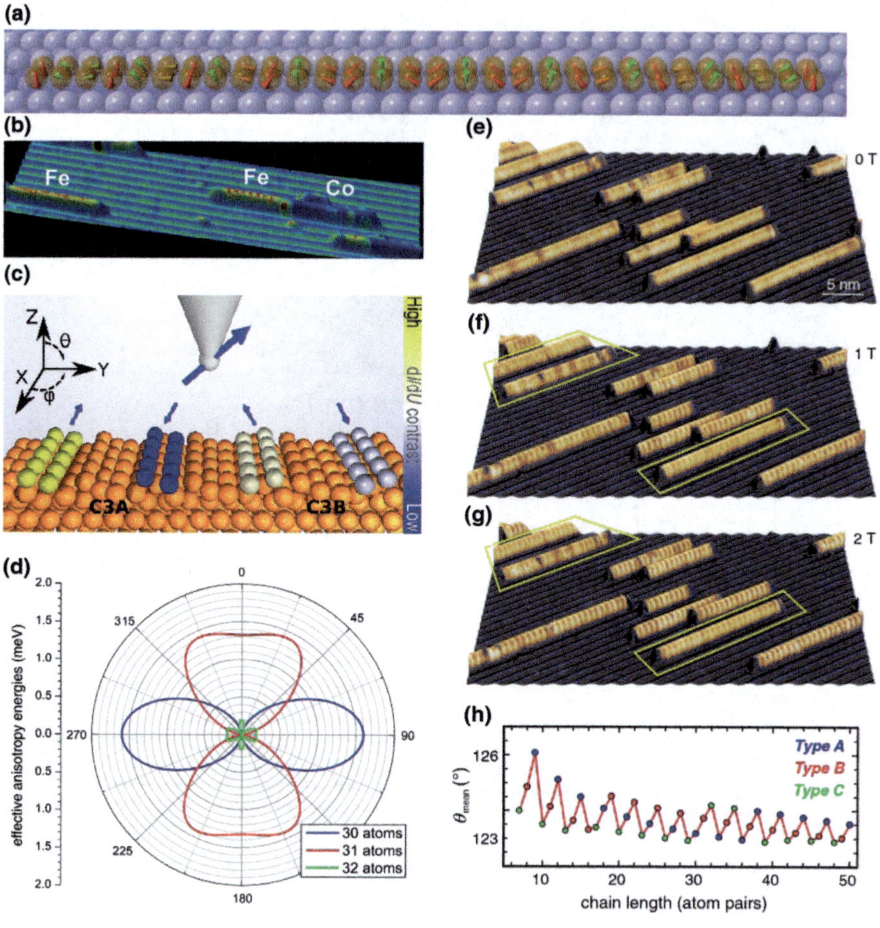

Fig. 8.4 Magnetism in biatomic Fe and Co chains on the reconstructed Ir(001) surface. **a** Sketch of the spin spiral state of an Fe chain. **b** The magnetic state of Fe chains fluctuates at a temperature of about 8 K (left) but can be stabilized by an adjacent ferromagnetic Co chain (center). **c** Sketch of the four degenerate magnetization directions of ferromagnetic biatomic Co chains with canted magnetocrystalline anisotropy. **d** Effective anisotropy energies shown for three exemplary biatomic Fe chains as obtained from micromagnetic simulations. **e, f, g** Sample area with biatomic Fe chains of different lengths at three different values of the external magnetic field as indicated. **h** Mean angle between adjacent atom pairs as a function of chain length, the three chain types categorized by their effective anisotropy are indicated. (Figure adapted from [9, 33, 34])

The effective magnetocrystalline anisotropy of Fe chains on the other hand was found to depend on the exact length of the chain [33], as demonstrated for three examples in Fig. 8.4d: micromagnetic simulations using the magnetic interaction parameters as determined by DFT show that 30-atom long chains possess an effective out-of-plane anisotropy, in contrast to the effective in-plane anisotropy of 31-atom long chains; chains with 32 atom pairs along their axis have a negligible magnetocrystalline anisotropy. This parity effect arises from the 120° spin spiral interacting with the finite chain lengths. Also in spin-resolved STM measurements a variation in the behavior is observed for different chain lengths [33]: whereas in the absence of an external magnetic field no magnetic signal is detected due to rapid thermal fluctuations of the spin spiral state, see Fig. 8.4e, with increasing magnetic field more and more chains exhibit the typical magnetic period of three atomic distances (Fig. 8.4f and g at 1 T and 2 T, respectively). The simulations predict an oscillatory behavior of the mean angle between neighboring magnetic moments as a function of chain length, Fig. 8.4h, resulting from a compromise between the chain type, i.e. characterized by the effective magnetocrystalline anisotropy, and a preferential nearest neighbor moment angle.

8.5 Nanoskyrmion Lattices in Fe on Ir(111)

A hexagonally ordered atomic Fe layer can be stabilized on the surface of an Ir(111) substrate due to pseudomorphic growth. Using a magnetic tip, which is sensitive to the out-of-plane magnetization component of the sample, a roughly square magnetic superstructure with a period of 1 nm is found for the fcc-stacked Fe monolayer

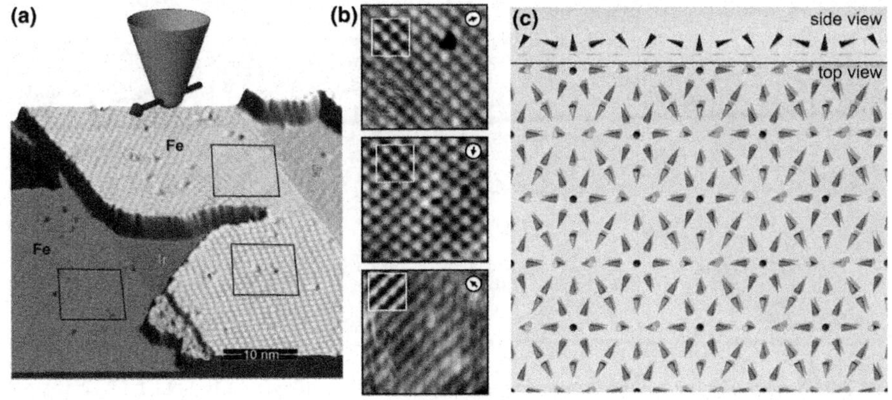

Fig. 8.5 Monolayer fcc Fe on Ir(111). **a**, **b** SP-STM measurements of all possible rotational magnetic domains imaged with an in-plane sensitive tip. **c** Sketch of the derived magnetic structure in the commensurate approximation: the nanoskyrmion lattice is nearly square with a periodicity of about 1 nm. (Figure adapted from [16])

[29, 35]. Due to the roughly square symmetry of the magnetic state on the hexagonal atom arrangement, three rotational magnetic domains are possible. SP-STM measurements with an in-plane magnetized tip on all three possible rotational domains of the two-dimensional magnetic state, Fig. 8.5a,b, reveal a square nanoskyrmion lattice as the ground state in zero magnetic field, see sketch in Fig. 8.5c. Density functional theory calculations show that the length scale of the magnetic order is governed by the interplay of magnetic exchange interactions and the DMI. In addition, the DMI imposes a unique rotational sense on the spin texture. The question arises why here, in contrast to the uniaxial spirals of the Mn monolayers on W surfaces (Sect. 8.4.1), a two-dimensional magnetic state is realized. To answer this question simulations within an extended Heisenberg model were performed and it has been shown, that the sizable higher-order four-spin interaction is responsible for the coupling of spin spiral states to form this exotic two-dimensional spin texture [16].

The fcc stacking can be grown either as stripes at the step edges at elevated deposition temperature or, when deposited at room temperature, in addition as free-standing triangular islands pointing in a specific direction, see Fig. 8.6a [29, 35, 36]. The three rotational magnetic domains are denoted by A, B, and C and a correlation between rotational domain and close-packed Fe step edges is observed, such that the diagonal of a magnetic unit cell couples to a given Fe-to-vacuum interface. In triangular islands this leads to the coexistence of the three possible rotational magnetic domains and frustration at positions where they merge, see magnified view of the lower central triangular Fe monolayer island in Fig. 8.6a [36]. When the nanoskyrmion

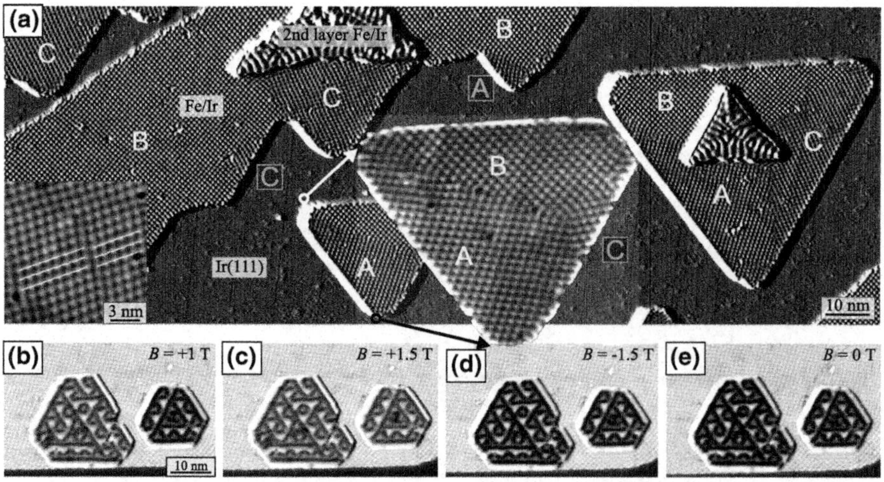

Fig. 8.6 Rotation of the nanoskyrmion lattice in the fcc-Fe monolayer on Ir(111). **a** Overview SP-STM current image of fcc Fe monolayer stripes and islands. The three possible rotational magnetic domains are labeled as A, B, C. The coupling of the skyrmion lattice to the edges of triangular islands leads to frustration and multi-domain states. **b–e** Ni islands on the fcc Fe monolayer at different external magnetic fields as indicated; near the ferromagnetic Ni/Fe bilayer patches the nanoskyrmion lattice is rotated away from its preferred state. (Figure adapted from [36])

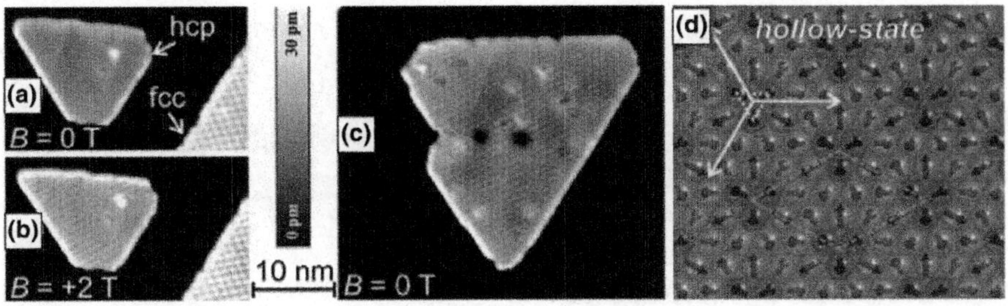

Fig. 8.7 The nanoskyrmion lattice in the hcp Fe monolayer on Ir(111). **a**, **b** hcp Fe monolayer island without and with applied magnetic field, demonstrating the susceptibility of the hexagonal nanoskyrmion lattice ground state to thermal fluctuations and external magnetic fields. **c** Larger hcp Fe island, which is magnetically stable also in zero magnetic field. **d** Sketch of the hexagonal magnetic nanoskyrmion lattice with 12 atoms in the unit cell. (Figure adapted from [17])

lattice is in direct vicinity of a ferromagnet, such as a Ni/Fe bilayer island, see Fig. 8.6b–e [37], the coupling of the diagonal of the magnetic unit cell to a close-packed row can be destroyed in favor of a coupling of the edge of the magnetic unit cell to the ferromagnet [36].

The vast majority of Fe islands, however, grow in the hcp stacking. Typically triangular islands are observed, see Fig. 8.7a and b [17]. Whereas thermal fluctuations at zero magnetic field can lead to vanishing magnetic contrast in small islands (a), with SP-STM a hexagonal superstructure is observed in external magnetic field (b) or for larger islands also at zero magnetic field (c). This hexagonal magnetic superstructure is interpreted as a commensurate hexagonal nanoskyrmion lattice, see sketch with 12 atoms per magnetic unit cell in Fig. 8.7d. Because of this symmetry there is only one rotational domain, which, in contrast to the square nanoskyrmion lattice in fcc-Fe, exhibits a remaining net magnetic moment and can thus be aligned in an external magnetic field [17]. The Fe atom stacking alone thus has a major effect on the symmetry and field dependence of the observed spin texture.

8.6 Magnetic Skyrmions in Pd/Fe on Ir(111)

8.6.1 Pd/Fe/Ir(111): Magnetic Phases

It has been shown that the choice of the substrate plays a major role for the magnetic properties of adsorbed atomic layers: completely different magnetic ground states have been found in the past for e.g. ultrathin Fe layers on different substrates, where the element, the symmetry, or the lattice constant is varied. To only slightly tune magnetic properties an alternative and more practical way is to cover a specific magnetic system with a layer of non-magnetic material.

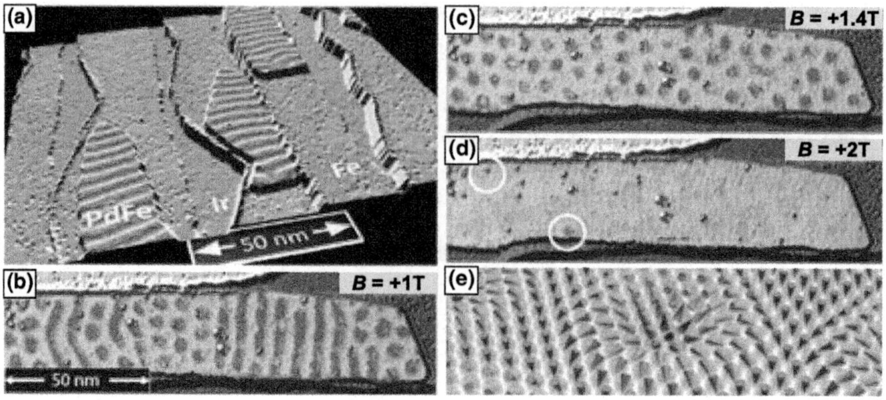

Fig. 8.8 SP-STM measurements of an Pd/Fe bilayer on Ir(111) in external magnetic fields at $T = 8\,\text{K}$. **a** $B = 0\,\text{T}$: spin spiral state, **b** $B = 1\,\text{T}$: coexistence of spin spiral and skyrmions, **c** $B = 1.4\,\text{T}$: hexagonal skyrmion lattice, **d** $B = 2\,\text{T}$: ferromagnetic phase. **e** Sketch of a magnetic skyrmion. (Figure adapted from [20])

When Pd is deposited onto the nanoskyrmion lattice of the Fe-ML on Ir(111) this leads to a change of the magnetic ground state: the Pd/Fe bilayer exhibits a spin spiral ground state with an approximately seven times larger period than the two-dimensional ground state of the uncovered Fe/Ir(111), see Fig. 8.8a. Upon application of an external magnetic field other magnetic phases can be observed: at intermediate fields a transition to a hexagonal lattice of magnetic skyrmions is found (b, c), and higher magnetic fields lead to a nearly saturated ferromagnetic state (d). This is the first example of an interface-DMI driven magnetic field induced skyrmion lattice [20], displaying the different magnetic phases which are similarly observed in typical bulk-DMI materials [18, 19].

8.6.2 Isolated Skyrmions: Material Parameters and Switching

Single magnetic skyrmions in the Pd/Fe bilayer on Ir(111) can be imaged with SP-STM [20, 38]: for out-of-plane sensitive magnetic tips, see sketch in Fig. 8.9a, they appear rotationally symmetric as in the measurement shown in (b), whereas a two-lobe structure is observed when the tip is sensitive to the in-plane magnetization of the sample as in (c), (d). The fact that all skyrmions in a Pd/Fe film imaged with the same in-plane tip show the same appearance demonstrates their unique rotational sense, which is imposed by the DMI. When the external magnetic field is inverted the rotational sense is preserved: Fig. 8.9c and d show that the magnetic contrast inverts when this measurement is performed with a Cr tip that is insensitive to applied magnetic fields. The spin structure across a skyrmion was found to nicely follow the spin

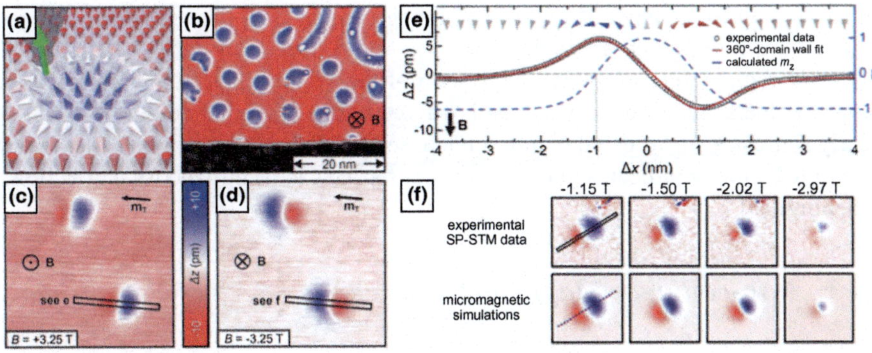

Fig. 8.9 Magnetic skyrmions in the Pd/Fe bilayer on Ir(111): magnetic field-dependence of size and shape. **a** Sketch of a magnetic skyrmion, colorized according to the observed SP-STM signal with out-of-plane magnetized magnetic tip. **b** SP-STM measurement with out-of-plane magnetic tip. **c**, **d** Two magnetic skyrmions in opposite external magnetic out-of-plane fields imaged with the same in-plane magnetized Cr tip. **e** Experimental line profile along the black rectangle in **d** together with a fit to a 360° domain wall and the extracted out-of-plane magnetization component m_z. **f** Magnetic field-dependent SP-STM measurements (top) and simulations (bottom) of a magnetic skyrmion in Pd/Fe/Ir(111). (Figure adapted from [38])

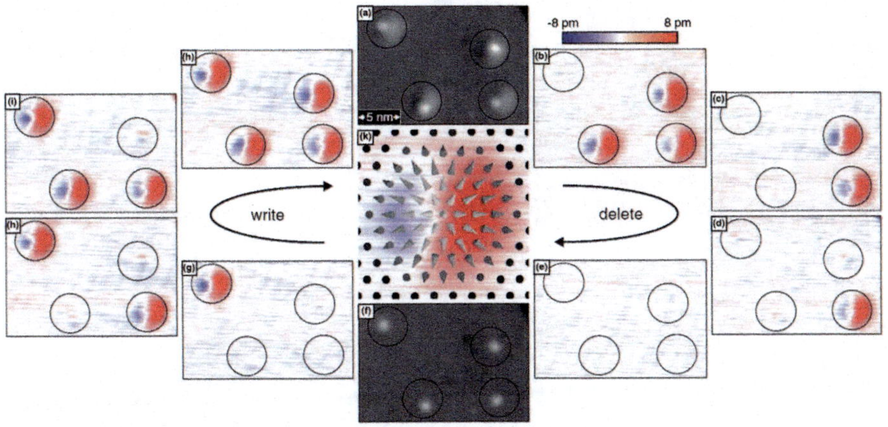

Fig. 8.10 SP-STM measurements of a group of artificially created isolated skyrmions in the fcc Pd/Fe bilayer on Ir(111) at $T = 4\,\text{K}$, $B = 3.25\,\text{T}$. Constant-current images (gray-scale) with **a** and without **f** skyrmions pinned at the four defects. The difference images (color-scale, with respect to **f**) show how the skyrmions can be deleted and written independently. The central image depicts the magnetization direction within a skyrmion. (Figure adapted from [20])

structure of two overlapping 180° domain walls [38]. A fit of such profiles to field-dependent measurements of an individual skyrmion, see Fig. 8.9e, and comparison with the energy functional for magnetic skyrmions, yields the material parameters for the Pd/Fe bilayer on Ir(111) such as exchange stiffness, DMI, and magnetocrystalline anisotropy. Micromagnetic simulations using these experimentally derived values as input parameters show very good agreement with the real-space images, compare Fig. 8.9f top and bottom.

At low temperature ($T < 8$ K) the system of the Pd/Fe bilayer on Ir(111) can be trapped in metastable states. Then, isolated skyrmions pinned at defects [39] can be locally addressed and created or annihilated, see Fig. 8.10. A systematic study of the switching behavior indicates that spin transfer torque can be utilized for a directional switching [20].

8.6.3 Non-collinear Magnetoresistance

Figure 8.11a displays a dI/dU map taken at $U = +0.7$V with a non-magnetic tip where the magnetic skyrmions in Pd/Fe/Ir(111) show a lower signal. To investigate

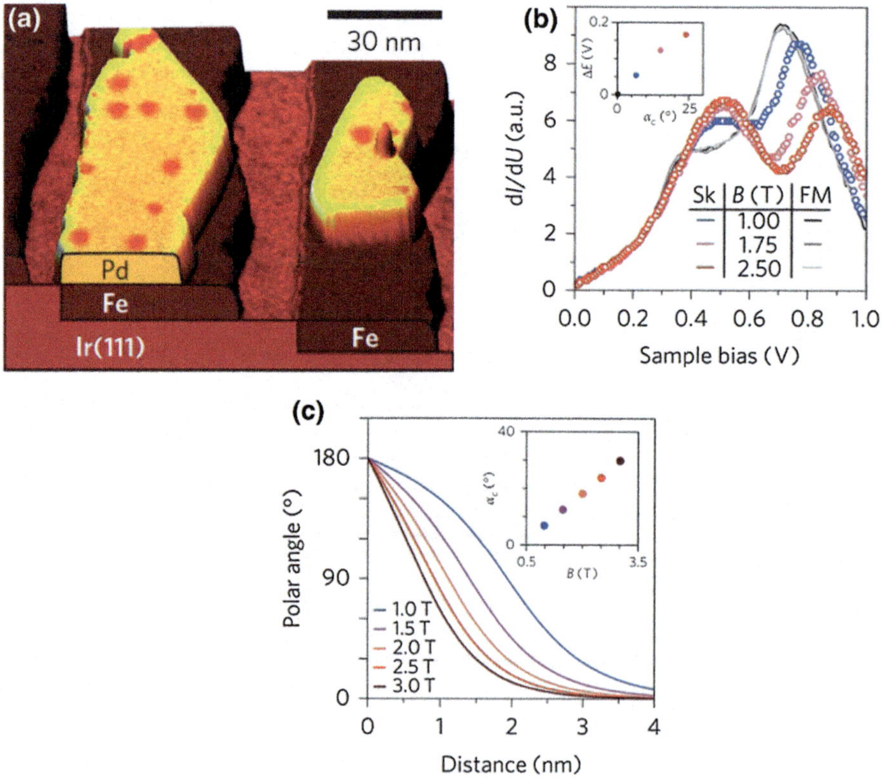

Fig. 8.11 Magnetic skyrmions in the Pd/Fe bilayer on Ir(111) measured with a non-magnetic tip. **a** Perspective view of an STM constant-current image, colorized with the dI/dU map ($U = +0.7$ V, $I = 1$ nA, $B = 1.8$ T, $T = 8$ K) **b** Spectra of differential conductance taken in the center of a skyrmion and outside at different magnetic fields. **c** Experimentally determined magnetic field-dependent size and shape of a magnetic skyrmion in Pd/Fe/Ir(111). (Figure adapted from [23])

the origin of this magnetoresistance effect, occurring with just one magnetic electrode, namely the sample, spatially resolved scanning tunneling spectroscopy was performed, see Fig. 8.11b. A significant difference between the electronic states in the center of a skyrmion and the ferromagnetic surrounding is found [23]: the position of the peak at roughly $+0.7$ V in the differential tunneling conductance dI/dU shifts towards higher energies as the non-collinearity is increased from the ferromagnetic state to a large skyrmion at 1 T and further when the skyrmion is compressed as the magnetic field is increased. By comparison of the local tunnel spectra [23] and the experimentally determined magnetic field-dependent size and shape of individual skyrmions [38], Fig. 8.11b and c, it can be shown that the non-collinearity has a significant contribution to the total magnetoresistance (see also sketches in Fig. 8.1). This non-collinear magnetoresistance (NCMR) originates from the mixing of spin-up and spin-down states due to non-collinearity [23] and the peak intensity is correlated with the local mean angle between neighboring spins [24]. Skyrmions are thus electronically distinct from their ferromagnetic surrounding, allowing new all-electrical skyrmion detection schemes.

8.7 (SP-)STM of Higher Layers of Fe on Ir(111)

8.7.1 Influence of Strain Relief and Temperature

While the Fe monolayer grows pseudomorphically on Ir(111) and hosts two-dimensional magnetic states with a period of 1 nm, the addition of more Fe layers leads to strain relief and dislocation line formation and the magnetic states become one-dimensional. The magnetic state of the Fe double layer is a spin spiral with a period of about 1.5 nm, which exhibits a zigzag wavefront and propagates along the dislocation lines in the three symmetry-equivalent $\langle 1\bar{1}2 \rangle$ directions, see Fe-DL labeled area at the top of Fig. 8.12a; the origin of the zigzag shape of the cycloidal spin spiral wavefront is the underlying atomic structure [42]. The triple layer of Fe on Ir(111) also shows dislocation lines and a spin spiral, however, two different reconstructions are found and the magnetic period increases to 4–9 nm, see Fig. 8.12a [43]. In accordance with their atomic structure models one of the reconstruction types has a straight wavefront, which is canted with respect to the dislocation lines, whereas the other one exhibits a zigzag wavefront [40]. The fact that the spin spiral period decreases with increasing line spacing, see Fig. 8.12b, is attributed to a change of the effective exchange stiffness with varying degrees of strain relief within the Fe-film [40].

The fourth layer of Fe on Ir does not exhibit a periodic magnetic state but appears to be ferromagnetic, with a locally changing quantization axis on the order of some tens of nanometers, see Fig. 8.12c. A strong temperature dependence of the magnetic

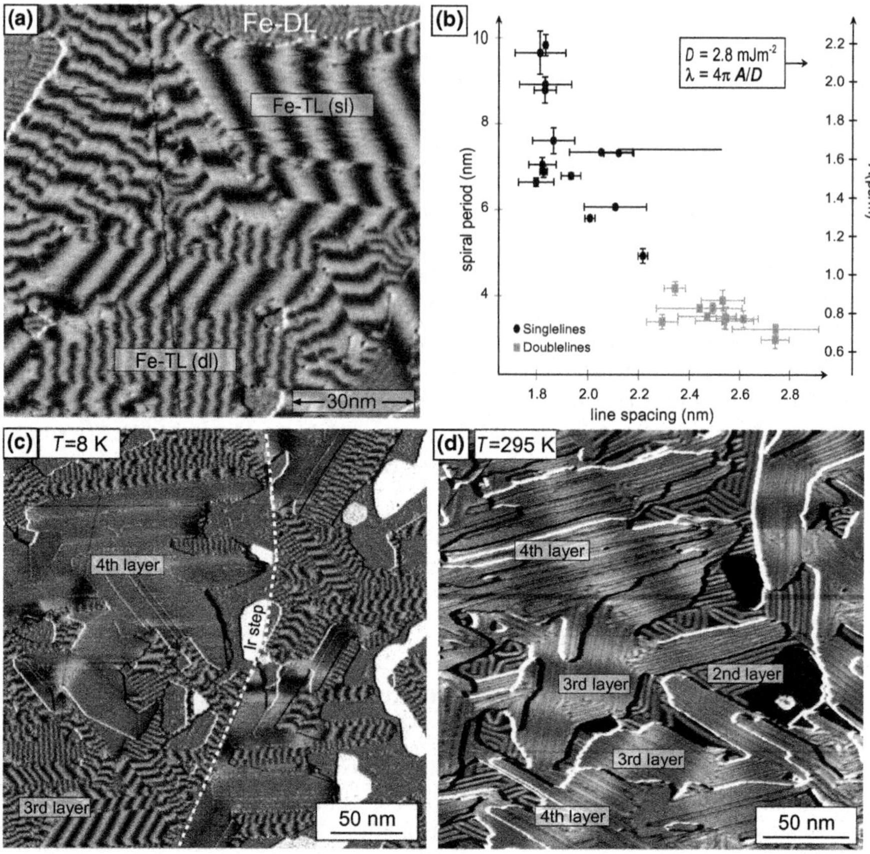

Fig. 8.12 The magnetic states of higher Fe layers on Ir(111). **a** The SP-STM measurement (dI/dU map) shows all three rotational domains of the canted and the zigzag spin spiral in the single line (sl) and double line (dl) reconstruction in the Fe triple layer (TL). **b** The spin spiral period decreases with increasing line spacing of the reconstruction, which is attributed to a concurrent change in effective exchange stiffness A. **c** SP-STM measurement (dI/dU map) of a sample with locally 2, 3 and 4 layers of Fe on Ir(111) measured at $T = 4$ K. **d** Similar sample as in **c** but measured at room temperature. (Figure adapted from [40, 41])

states of the triple layer and the fourth layer of Fe is observed, as evident from the comparison of the low temperature measurement in Fig. 8.12c and the room temperature data in (d): the magnetic period of the triple layer spin spiral has increased by nearly an order of magnitude and the fourth layer now shows roughly the same magnetic state as the triple layer [41]. To explain this increase of the magnetic period with temperature layer-dependent magnetic interaction parameters are necessary [41].

8.7.2 Influence of Magnetic and Electric Field

The spin spiral in the Fe-TL shown in Fig. 8.13a is modified in an external out-of-plane magnetic field and breaks up into small bean-shaped magnetic objects, see Fig. 8.13b. Here, the typical hexagonal skyrmion lattice is suppressed and instead the dislocation lines function as skyrmion tracks. An evaluation of measurements with a tip sensitive to the in-plane component of the sample magnetization, see Fig. 8.13c–e, demonstrates that these magnetic objects are indeed topologically distinct skyrmions; a model of the spin state is shown in Fig. 8.13f and SP-STM simulations thereof are in good agreement with the experimental data, compare Fig. 8.13c–e top and bottom [43].

Writing and deleting of these magnetic skyrmions was realized with local voltage pulses between tip and sample, see Fig. 8.13g. Measurements with a non-spin-polarized W-tip directly ruled out spin-transfer torque as dominant switching mechanism. An investigation of the threshold voltages for creation and annihilation at different tip-sample distances demonstrated the leading role of the electric field for the switching between these topologically distinct states, Fig. 8.13h. Thus, simultaneously to a global control of the energy levels of skyrmion and ferromagnet in an

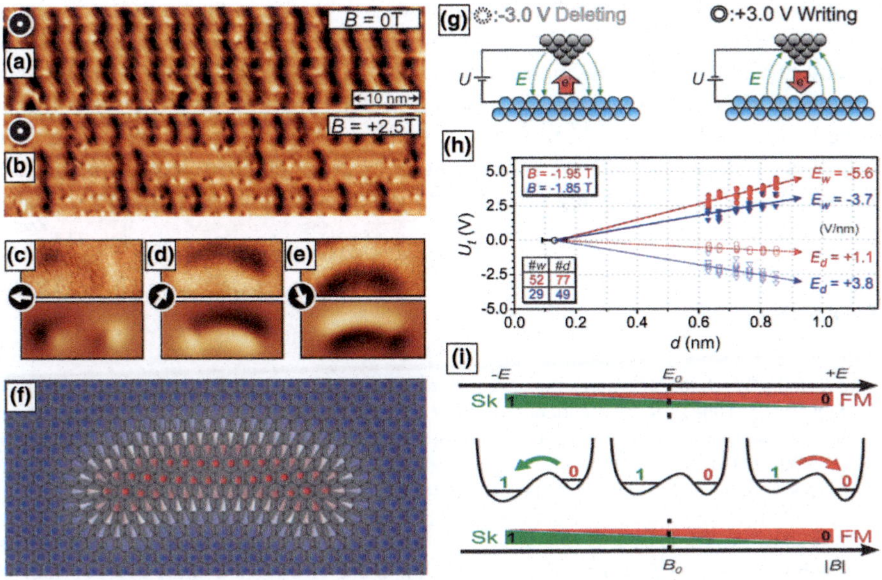

Fig. 8.13 Magnetic skyrmions in the triple-layer Fe on Ir(111) and switching with electric fields. **a** The zero magnetic field spin spiral ground state. **b** At 2.5 T single magnetic objects are formed. **c, d, e** Three rotated magnetic skyrmions were imaged with the same in-plane magnetic tip (top). The bottom row displays SP-STM simulations of the spin structure sketched in **f**. **g** Electric field in an STM setup. **h** Critical electric fields derived from distance-dependent measurements of writing and deleting of individual skyrmions with a W-tip. **i** Sketch of the modification of the double-well potential of skyrmion and ferromagnet by global magnetic or local electric fields. (Figure adapted from [43])

external magnetic field, a local tuning of the relative energy levels is possible via the local *electric* field of an STM tip, Fig. 8.13i [43]. For this mechanism, it is essential that the two states are inequivalent; an electric field cannot lift the degeneracy of, e.g. a mere up and down magnetization. Different mechanisms are possible for this magneto-electric coupling, e.g. a change of the electronic states in the near surface region accompanied by a change of inter-layer distances. As electric fields decay fast in metallic films and both the DM interaction and the crystalline anisotropy stem from the buried Fe-Ir interface, a likely scenario is that the electric field is effectively changing the local exchange interaction within the film.

8.8 Conclusion

Magnetic field-dependent SP-STM is an effective and versatile method to reveal and manipulate complex spin textures down to the atomic scale. The discovery of interface-DMI was the first step toward a novel research field, which now explores chiral domain walls, skyrmionic bubble domains and skyrmions for spintronic applications. Interface-DMI allows to tailor skyrmion properties by, e.g. designing multilayer structures of magnetic 3d layers and heavy 5d elements with large spin-orbit interaction. Atomic-scale SP-STM measurements of skyrmionic systems, e.g. the field-dependent size and shape of magnetic skyrmions, allows to test present theories and can yield the relevant material parameters, like the DMI value. The discovery of the NCMR effect shows that skyrmions are (quasi-)particles with distinct electric properties, allowing a novel all-electrical mechanism for skyrmion detection. The controlled writing and deleting of skyrmions by local *electric* fields establishes a new magneto-electric coupling scheme with potential for future spintronic devices.

References

1. M. Bode, M. Heide, K. von Bergmann, P. Ferriani, S. Heinze, G. Bihlmayer, A. Kubetzka, O. Pietzsch, S. Blügel, R. Wiesendanger, Nature **447**, 190 (2007)
2. K. von Bergmann, A. Kubetzka, O. Pietzsch, R. Wiesendanger, J. Phys. Condens. Matter **26**, 394002 (2014)
3. R. Wiesendanger, Nature Rev. Mater. **1**, 16044 (2016)
4. A. Fert, V. Cros, J. Sampaio, Nature Nanotechnol. **8**, 152 (2013)
5. F. Jonietz, S. Mühlbauer, C. Pfleiderer, A. Neubauer, W. Münzer, A. Bauer, T. Adams, R. Georgii, P. Böni, R.A. Duine, K. Everschor, M. Garst, A. Rosch, Science **330**(6011), 1648 (2010)
6. W. Jiang, P. Upadhyaya, W. Zhang, G. Yu, M.B. Jungfleisch, F.Y. Fradin, J.E. Pearson, Y. Tserkovnyak, K.L. Wang, O. Heinonen, S.G.E. te Velthuis, A. Hoffmann, Science **349**(6245), 283 (2015)
7. S.S.P. Parkin, M. Hayashi, L. Thomas, Science **320**(5873), 190 (2008)
8. P. Ferriani, K. von Bergmann, E.Y. Vedmedenko, S. Heinze, M. Bode, M. Heide, G. Bihlmayer, S. Blügel, R. Wiesendanger, Phys. Rev. Lett. **101**(2), 027201 (2008)

9. M. Menzel, Y. Mokrousov, R. Wieser, J.E. Bickel, E. Vedmedenko, S. Blügel, S. Heinze, K. von Bergmann, A. Kubetzka, R. Wiesendanger, Phys. Rev. Lett. **108**, 197204 (2012)
10. T. Moriya, Phys. Rev. **120**(1), 91 (1960)
11. M. Heide, G. Bihlmayer, S. Blügel, Phys. Rev. B **78**, 140403 (2008)
12. S. Meckler, N. Mikuszeit, A. Preßler, E.Y. Vedmedenko, O. Pietzsch, R. Wiesendanger, Phys. Rev. Lett. **103**(15), 157201 (2009)
13. S. Emori, U. Bauer, S.M. Ahn, E. Martinez, G. Beach, Nat. Mater. **12**, 611 (2013)
14. K.S. Ryu, L. Thomas, S.H. Yang, S. Parkin, Nat. Nanotechnol. **8**, 527 (2013)
15. N. Nagaosa, Y. Tokura, Nat. Nanotechnol. **8**, 899 (2013)
16. S. Heinze, K. von Bergmann, M. Menzel, J. Brede, A. Kubetzka, R. Wiesendanger, G. Bihlmayer, S. Blügel, Nat. Phys. **7**, 713 (2011)
17. K. von Bergmann, M. Menzel, A. Kubetzka, R. Wiesendanger, Nano Lett. **15**(5), 3280 (2015)
18. S. Mühlbauer, B. Binz, F. Jonietz, C. Pfleiderer, A. Rosch, A. Neubauer, R. Georgii, P. Böni, Science **323**(5916), 915 (2009)
19. X.Z. Yu, Y. Onose, N. Kanazawa, J.H. Park, J.H. Han, Y. Matsui, N. Nagaosa, Y. Tokura, Nature **465**, 901 (2010)
20. N. Romming, C. Hanneken, M. Menzel, J.E. Bickel, B. Wolter, K. von Bergmann, A. Kubetzka, R. Wiesendanger, Science **341**(6146), 636 (2013)
21. S. Heinze, Appl. Phys. A **85**(4), 407 (2006)
22. K. von Bergmann, M. Menzel, D. Serrate, Y. Yoshida, S. Schröder, P. Ferriani, A. Kubetzka, R. Wiesendanger, S. Heinze, Phys. Rev. B **86**, 134422 (2012)
23. C. Hanneken, F. Otte, A. Kubetzka, B. Dupé, N. Romming, K. von Bergmann, R. Wiesendanger, S. Heinze, Nat. Nanotechnol. **10**(12), 1039 (2015)
24. A. Kubetzka, C. Hanneken, R. Wiesendanger, K. von Bergmann, Phys. Rev. B **95**, 104433 (2017)
25. R. Wiesendanger, H.J. Güntherodt, G. Güntherodt, R.J. Gambino, R. Ruf, Phys. Rev. Lett. **65**(2), 247 (1990)
26. O. Pietzsch, R. Wiesendanger, in *Atomic-Scale Magnetism Studied by Spin-Polarized Scanning Tunneling Microscopy*, Fundamentals of Picoscience (CRC Press, Boca Raton, 2013)
27. R. Wiesendanger, Rev. Mod. Phys. **81**(4), 1495 (2009)
28. D. Wortmann, S. Heinze, P. Kurz, G. Bihlmayer, S. Blügel, Phys. Rev. Lett. **86**(18), 4132 (2001)
29. K. von Bergmann, S. Heinze, M. Bode, E.Y. Vedmedenko, G. Bihlmayer, S. Blügel, R. Wiesendanger, Phys. Rev. Lett. **96**(16), 167203 (2006)
30. M. Bode, S. Heinze, A. Kubetzka, O. Pietzsch, X. Nie, G. Bihlmayer, S. Blügel, R. Wiesendanger, Phys. Rev. Lett. **89**(23), 237205 (2002)
31. D.M. Crum, M. Bouhassoune, J. Bouaziz, B. Schweflinghaus, S. Blügel, S. Lounis, Nat. Commun. **6**, 8541 (2015)
32. S. Heinze, M. Bode, A. Kubetzka, O. Pietzsch, X. Nie, S. Blügel, R. Wiesendanger, Science **288**(5472), 1805 (2000)
33. M. Menzel, A. Kubetzka, K. von Bergmann, R. Wiesendanger, Phys. Rev. Lett. **112**, 047204 (2014)
34. B. Dupé, J.E. Bickel, Y. Mokrousov, F. Otte, K. von Bergmann, A. Kubetzka, S. Heinze, R. Wiesendanger, N. J. Phys. **17**(2), 023014 (2015)
35. K. von Bergmann, S. Heinze, M. Bode, G. Bihlmayer, S. Blügel, R. Wiesendanger, New J. Phys. **9**(10), 396 (2007)
36. J. Hagemeister, D. Iaia, E.Y. Vedmedenko, K. von Bergmann, A. Kubetzka, R. Wiesendanger, Phys. Rev. Lett. **117**, 207202 (2016)
37. D. Iaia, A. Kubetzka, K. von Bergmann, R. Wiesendanger, Phys. Rev. B **93**, 134409 (2016)
38. N. Romming, A. Kubetzka, C. Hanneken, K. von Bergmann, R. Wiesendanger, Phys. Rev. Lett. **114**, 177203 (2015)
39. C. Hanneken, A. Kubetzka, K. von Bergmann, R. Wiesendanger, New J. Phys. **18**(5), 055009 (2016)
40. A. Finco, P.J. Hsu, A. Kubetzka, K. von Bergmann, R. Wiesendanger, Phys. Rev. B **94**, 214402 (2016)

41. A. Finco, L. Rózsa, P.J. Hsu, A. Kubetzka, E. Vedmedenko, K. von Bergmann, R. Wiesendanger, Phys. Rev. Lett. **119**, 037202 (2017)
42. P.J. Hsu, A. Finco, L. Schmidt, A. Kubetzka, K. von Bergmann, R. Wiesendanger, Phys. Rev. Lett. **116**, 017201 (2016)
43. P.J. Hsu, A. Kubetzka, A. Finco, N. Romming, K. von Bergmann, R. Wiesendanger, Nat. Nanotechnol. **12**, 123 (2017)

Chapter 9
Theory of Magnetic Ordering at the Nanoscale

Elena Vedmedenko

Abstract Complex spin structures feature a periodic variation of the local alignment of non-collinear magnetic moments. The number of systems showing non-trivial magnetic structuring steadily increases. Particularly, several uniaxial as well as two-dimensionally modulated magnetic states were found with unique rotational sense due to the spin-orbit based Dzyaloshinskii-Moriya-interaction. Another source for the modulated magnetic configurations is given by the frustration and competing higher-order exchange interactions arising due to the itinerant nature of electrons. One of the most exciting recent developments with respect to these non-trivial configurations is the emergence of magnetic quasiparticles, which are characterized by their enhanced temporal and thermal stability. In this review several aspects of the static and dynamic properties of the quasiparticles and other interfacial non-collinear magnetic structures will be addressed.

9.1 Stability of Magnetic Quasiparticles

Exotic phases of condensed matter recently became a central topic of scientific activity [1]. Non-trivial phases encompass a large variety of phenomena in different areas of physics from very specific band structures to noncollinear magnetic states. Often these phases correspond to non-trivial metastable magnetic configurations. Excited magnetic configurations, which possess properties of quasiparticles, play a particularly important role among the complex magnetic structures. To the class of quasiparticle states belong three- or two-dimensional magnetic skyrmions, one-dimensional chiral magnetic domain walls, one-dimensional topological solitons or helices and zero-dimensional magnetic excitations known as "monopoles", which can also be bound by one-dimensional Dirac strings. Once created, these magnetic

E. Vedmedenko (✉)
Department of Physics, University of Hamburg, Hamburg, Germany
e-mail: vedmeden@physnet.uni-hamburg.de

R. Wiesendanger (ed.), *Atomic- and Nanoscale Magnetism*, NanoScience and Technology, https://doi.org/10.1007/978-3-319-99558-8_9

objects can only be erased with effort from a surface. This makes them valuable for the application in future data storage devices but also poses fundamental questions on the microscopic reasons for the stability. During the past decade we addressed several aspects of the stability of non-trivial magnetic states by means of analytical and numerical analysis. The work on this exciting topic has permitted us to find answers to such important questions like: Which metastable topological configurations can be achieved in ultrathin magnets or arrays of structured films? What are the lifetimes of those metastable states? How can the complex magnetic states be manipulated? This review addresses several important aspects of the phase diagrams, statics and dynamics of magnetic quasiparticles of different dimensions. Due to these investigations we were able to show that energy can be stored in metastable states of nontrivial topological matter. The review starts with a general description of magnetic structuring at the nanoscale due to exotic, higher-order magnetic interactions. Later, the emphasis will be put on two-dimensional quasiparticles, that are interfacial skyrmions, continued by the discussion of the stability and manipulation of one-dimensional quasiparticles, that are helices and domain walls, and finished by the description of curious properties of zero-dimensional excitations in magnetic arrays. The last section will be devoted to the manipulation of zero-dimensional magnetic objects with the tip of an atomic force microscope.

9.2 Higher-Order Complex Magnetic Interactions

Nanostructured $3d$ and $4d$ metallic films on different substrates often belong to the class of itinerant magnetic systems with delocalized electrons. The thermodynamic properties and magnetic ordering of itinerant systems are very important for any study of their macroscopic behaviour. It is often not possible to study the microscopic magnetic structuring of these systems by standard density-functional theory (DFT) methods because of the extremely large unit cell of the magnetic ground states. To be able to describe those nontrivial states we have used a multiscale approach, where the classical exchange Hamiltonian has been extended by higher-order terms to account for the itinerant nature of magnetic interactions. We have studied phase diagrams as well as specific experimental situations using parameters of the higher-order terms extracted from DFT calculations. The terms beyond the nearest neighboring Heisenberg exchange interaction can be derived by means of the fourth order perturbation expansion of the Hubbard model [2] when the spin-orbit interaction is not included [3]. In addition to these longer-range exchange terms, further higher order contributions are represented by the biquadratic and the four-spin exchange interaction resulting from the hopping of electrons over several neighboring atoms. An important role is also played by the antisymmetric part of the exchange tensor; that is, by the Dzyaloshinskii-Moriya (DM) interaction. Although the mathematical form of those terms was known, a general study of the magnetic structuring in the itinerant magnets remained obscure until recently. The magnetic structures due to these non-trivial interactions, however, appear to be very interesting. An example of such a magnetic configuration based on the four-particle exchange interactions and DM interactions will be described in the following.

Model

A classical atomistic spin model described by the following Hamiltonian has been considered:

$$\begin{aligned}\mathcal{H} = &\sum_{i<j} J_{ij}\mathbf{S}_i \cdot \mathbf{S}_j + J_{bi}\sum_{\langle ij\rangle}(\mathbf{S}_i \cdot \mathbf{S}_j)^2 \\ &+ D_z\sum_i (S_{zi})^2 - \omega\sum_{i<j}\frac{3(\mathbf{S}_i\cdot\mathbf{e}_{ij})(\mathbf{e}_{ij}\cdot\mathbf{S}_j) - \mathbf{S}_i\cdot\mathbf{S}_j}{r_{ij}^3} \\ &+ J_{4-\mathrm{spin}}\sum_{\langle ijkl\rangle}\left[(\mathbf{S}_i\cdot\mathbf{S}_j)(\mathbf{S}_k\cdot\mathbf{S}_l) + (\mathbf{S}_i\cdot\mathbf{S}_k)(\mathbf{S}_j\cdot\mathbf{S}_l)\right. \\ &\left.- (\mathbf{S}_i\cdot\mathbf{S}_l)(\mathbf{S}_j\cdot\mathbf{S}_k)\right] + \sum_{i<j}\mathbf{D}_{ij}(\mathbf{S}_i\times\mathbf{S}_j) \end{aligned} \qquad (9.1)$$

with $\mathbf{S}_i = \boldsymbol{\mu}_i/\mu_s$ the three-dimensional magnetic moments of unit length on a triangular lattice with open boundary conditions and μ_s the strength of the magnetic moment in Bohr magneton μ_B.

The summation in the exchange interactions goes up to 3rd nearest neighbors. $J_{ij} < 0$ means a ferromagnetic and $J_{ij} > 0$ antiferromagnetic exchange coupling. The biquadratic exchange between two nearest neighboring spins is given by the second term, while the third sum describes a crystalline anisotropy favoring the vertical z orientation of magnetization for negative D_z, and an easy-plane anisotropy if the parameter D_z is positive. The fourth sum is the long-range dipolar interaction with the coupling constant $\omega = \mu_s^2\mu_0/4\pi a^3$, the interatomic distances r_{ij} in units of the lattice constant a, and the unit vectors $\mathbf{e}_{ij}$ in the direction of $\mathbf{r}_{ij}$. To calculate the long-range manner of the dipolar interaction we use a Fast Fourier Transformation technique for decreasing the required CPU time [5]. The last two terms of the Hamilton operator represent the many-spin interaction and the DM coupling $\mathbf{D}_{ij}$ respectively. In the case of the nearest neighbor four-spin coupling, the four sites i, j, k and l involved form a minimal parallelogram, where each side is a line connecting two nearest neighbors [4].

Phase diagrams and the ground states of a system with pure biquadratic exchange interaction and the 4-spin interaction are shown in Figs. 9.1 and 9.2. The ground state of a system with pure DM interaction corresponds to a spin spiral with angle of $\pi/2$ between nearest-neighboring spins. As one can see from the presented data all higher order interactions lead to complex, non-collinear magnetic states. However, only the DM interaction breaks the rotational symmetry. Therefore, a combination of the DM interaction with ferro- or antiferromagnetic exchange interaction and other

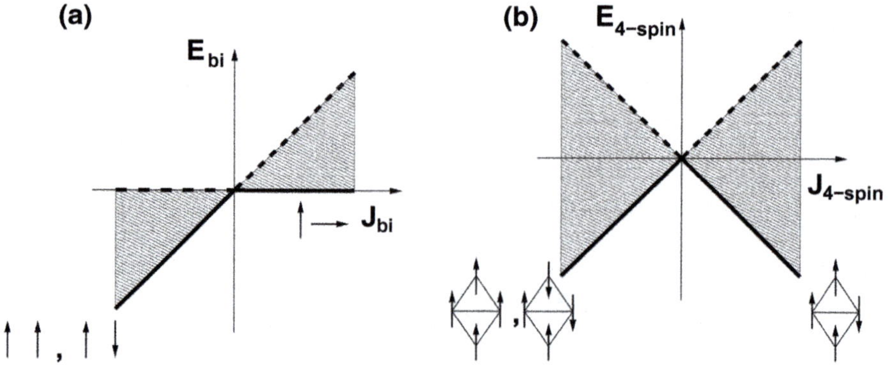

Fig. 9.1 Energy of a spin system coupled by a pure biquadratic exchange interaction (**a**) and a four-spin exchange interaction (**b**) as a function of the strength of the corresponding energy constants [4]. The solid lines correspond to the ground state energies, the dashed lines show the highest possible energy states, while insets exemplify low energy states

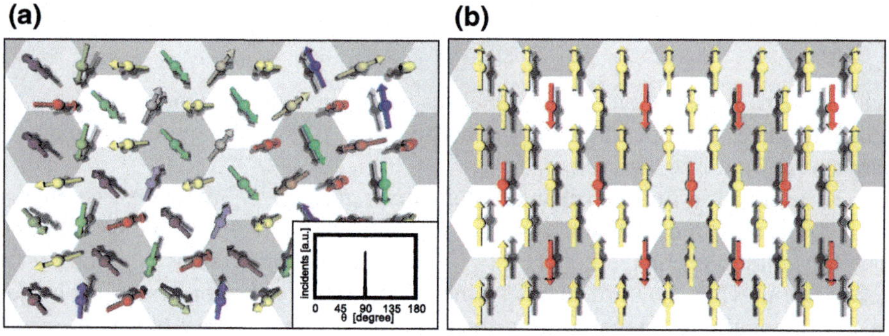

Fig. 9.2 Top view of the ground states of a system with (**a**) pure biquadratic exchange interaction ($J_{bi} > 0$) and (**b**) positive four spin exchange coupling $\left(J_{4-\text{spin}} > 0\right)$. The inset shows a frequency distribution of angles between nearest-neighboring spins. The color scheme denotes the spatial orientation of the sublattices

higher-order terms lead to the formation of complex magnetic structures with unique rotational sense as we have shown in several publications [6–9] (Fig. 9.3).

9.3 Two-Dimensional Quasiparticles: Interfacial Skyrmions

One of the most prominent complex magnetic structures with unique rotational sense are skyrmions [10–12]. There are several different mechanisms for the formation of skyrmionic phases at interfaces. They can be classified on the basis of the involved

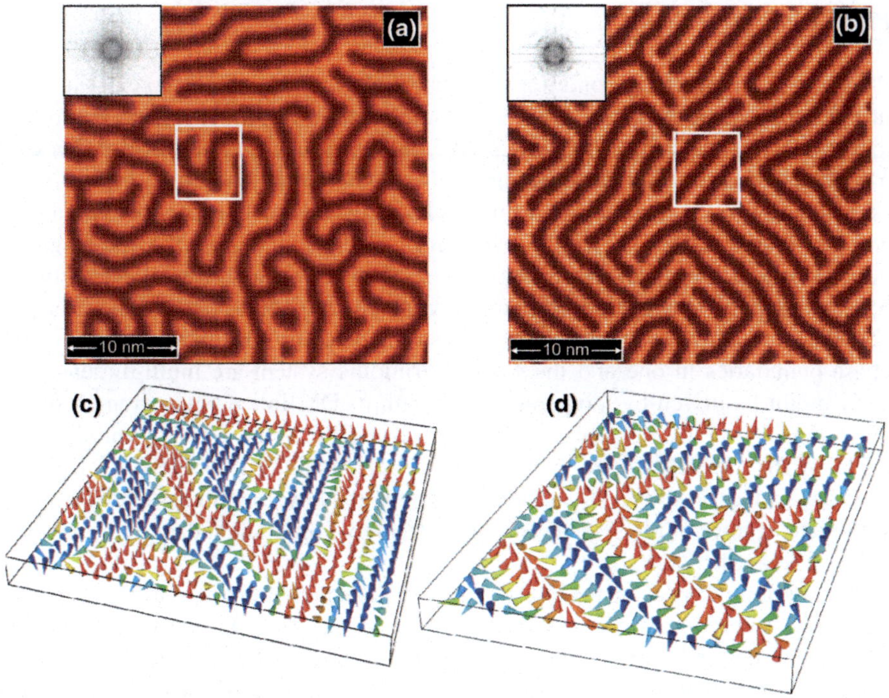

Fig. 9.3 An SP-STM image derived theoretically from the MC calculations for a Mn monolayer on W(001) at 25 K (**a**) and 13 K (**b**). The inset shows the calculated Fourier transform. **c–d** Three-dimensional representation of the area indicated in (**a–b**). The spins are shown by cones colored accordingly to their vertical magnetization, ranging from red (up) to dark blue (down)

sets of interactions [13]. The first and simplest Skyrmion Class (SC-I) results from a competition between the exchange, the DM, and the Zeeman interactions only [10]

$$H = -\sum_{i<j} J_{ij}(\mathbf{r})\mathbf{S}_i\mathbf{S}_j - \sum_{i<j} \mathbf{D}_{ij}(\mathbf{r}) \cdot (\mathbf{S}_i \times \mathbf{S}_j) - \mu_s \sum_i [\mathbf{B} \cdot \mathbf{S}_i] + O \qquad (9.2)$$

Skyrmions of this class appear in different kinds of interfacial systems like sputtered [14, 15] or epitaxial [16, 17] thin magnetic films. The class SC-II contains systems involving higher-order interactions described by (9.1) [12]. Magnetic skyrmions of the second class can appear spontaneously. They do not need any magnetic field for their formation. The size of SC-II skyrmions is very small: several atomic distances only [18]. The third class of skyrmionic interfacial systems SC-III is presented by the systems without DM interactions, but with competing exchange interactions of different range [19], while the last class SC-IV concerns the so-called skyrmionic bubbles, which appear in systems with strong dipolar coupling. In combination with the DM interaction the bubble domains acquire unique chirality and become more stable [20]. In the following, we will mainly concentrate on skyrmion classes I and II.

Phase Diagrams

Phase diagrams showing conditions under which multiple phases co/exist in equilibrium are indispensable for a systematic description of any physical system. For topological magnets they are particularly important because this exotic phase of matter appears in a rather narrow range of thermodynamic parameters. Typically, the skyrmion lattice (SkX) occupies a tiny pocket between much larger spin spiral (SS) and ferromagnetic (FM) phases, i.e. for very specific temperature (T) and magnetic field (B) values. While all phase diagrams of skyrmionic matter distinguish between SS, SkX and FM states, there is a large diversity in the coordinates of those phases [21]. There are three main reasons for this strong diversity. One of the reasons is that phase boundaries in one and the same skyrmionic system are multi-dimensional. They might be driven by temperature, strength of DM interaction, strength of applied magnetic field, pressure etc. There are also differences of the thermodynamic behavior for different lattice symmetries, thereby making the diversity of phases even larger. Hence, to investigate the stability of those non-collinear configurations systematic studies of phase diagrams are extremely important and are the subject of ongoing scientific work.

For skyrmionic systems the B-D diagram (field versus strength of the DM interaction) plays a central role, because it permits to characterize the skyrmion formation in different materials. Each material class has a characteristic value of the D/J ratio. In [17] we have calculated a B-D diagram for ultrathin films with interfacial DM interaction. This diagram is shown in Fig. 9.4. Its originality lies in the order parameters used: Additionally to the commonly considered winding number Q, we have defined a density order parameter ρ [17] by the ratio between a surface area occupied of skyrmions obtained in simulations for given parameters and the area occupied by an ideal, close packed lattice of skyrmions of identical radius R. For the phase diagrams, large samples (10^6 atomic sites) of rectangular shape, with a D/J gradient along the x-axis, have been equilibrated at various magnetic fields B. The advantage of the method of gradients is the possibility of a direct representation of the magnetic microstates in dependence of the interaction strength.

Due to the novel order parameter and the gradient method we were able not only to distinguish between the skymionic, ferromagnetic, and the spiral phases, but also define the transition from the phase of the isolated skyrmions to the skyrmion lattice phase. Particularly, a drastic change in the behavior of the skyrmion radius as a function of the D/J ratio was observed. Interestingly, R increases with the strength of the DM interaction if the D/J ratio is less than 1.05; that is, in the single skyrmion regime, while it decreases with the DM interaction for higher D/J; that is, for the skyrmion lattices.

Minimal Size and Shape of Skyrmions

The detailed investigations of the phase diagrams [17] showed that skyrmions with a radius less than a certain critical value R_c do not exist in discrete atomistic systems.

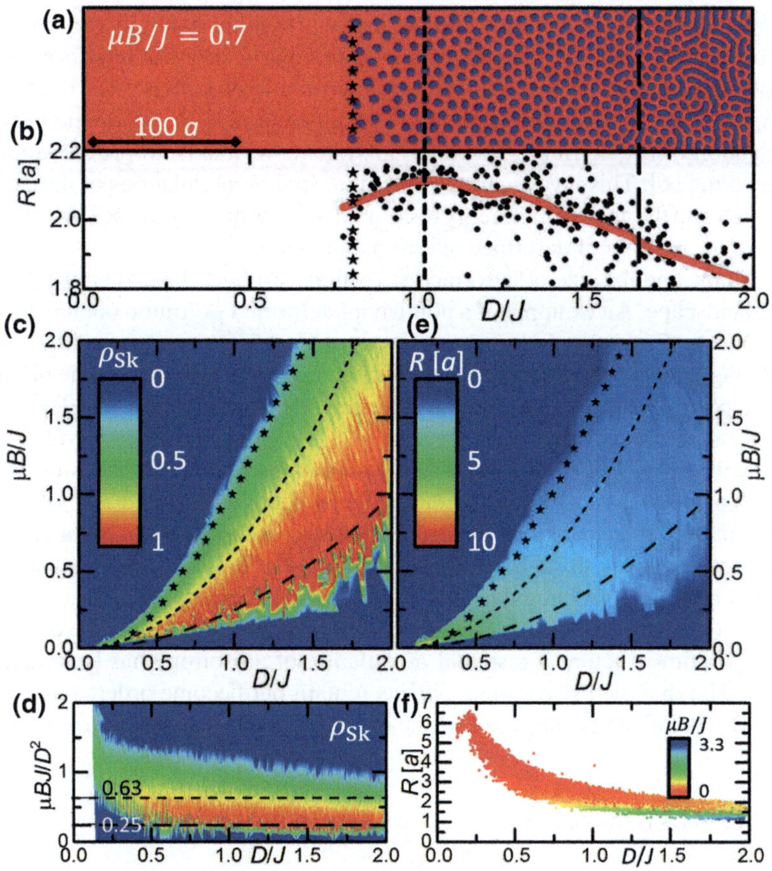

Fig. 9.4 **a** Magnetic ground states derived from extended MC calculations at $\mu B/J = 0.7$ and $k_B T/J = 8.5 \cdot 10^{-3}\,J$. There is a gradient of the strength of DM interaction along the x-axis; **b** Skyrmion radii as a function of the gradient described above. Points correspond to the numerical data, solid lines give the averaged R values; **c**, **e** $B - D$ phase diagrams using the density ρ and the radius of the skyrmions as an order parameter; **d**, **e** Parameter $\mu BJ/D^2$ and the skyrmion radii as a function of the position in phase space

At this specific skyrmion size the system overcomes the separating energy barrier and inevitably relaxes into the ferromagnetic state. With increasing strength of DM interaction R_c decreases and eventually reaches its ultimate limit. For interfacial magnetic systems like 1 ML Pd/1 ML Fe/Ir(111) the minimal skyrmion radii are very small and lie in the region of (0.7–1) lattice constants; that is, a minimal skyrmion consists of four (if the skyrmion center is between the atomic sites) or seven spins (if the skyrmion center coincides with one of the atomic sites) only. These data are in very good agreement with a recent experimental study [22].

Until now spherically symmetric skyrmions have mainly been addressed as only in this case the skyrmion radius unambiguously defines its geometrical properties.

However, various experimentally feasible material systems naturally exhibit spatially anisotropic behaviour. This phenomenon is particularly strong at interfaces [23]. An example relevant for interfacial anisotropic skyrmionic systems is given by the double and triple atomic layers of Fe on the Ir(111) substrate [24]. A systematic theoretical investigation of the skyrmion formation in systems with anisotropic environment is presented in [16]. This investigation shows that spatial modulations of the exchange interaction and the anisotropy energy in combination with an isotropic DM interaction lead to the formation of deformed skyrmionic objects.

The shape and the size of deformed skyrmions strongly depend on the particular energy landscape. An example of a non-trivial deformed skyrmion obtained with the help of Monte-Carlo simulations is analyzed in Fig. 9.5. In this case a spatial modulation of magnetocrystalline anisotropy between in-plane and out-of-plane orientation has been considered to account for the skyrmion deformation (see Fig. 9.5a). Such a modulation might occur, for instance, due to stress induced surface reconstructions. Additionally, it is known that the exchange interaction parameters J_{ij} might be modulated as a function of the orientation of the respective bond and also of the bond position in the lattice. Figure 9.5b–c show the skyrmionic structures for two different magnetic fields with such a spatial modulation of the exchange parameters, but without any anisotropy modulation. One observes ordered bent non-collinear spin states with a non-vanishing topological charge. Figure 9.5d–e show the corresponding equilibrium structure if a spatial modulation of anisotropy has been taken into account. The distorted skyrmionic objects remain but become ordered along linear tracks. The detailed spin structure of the deformed skyrmions is shown in Fig. 9.5f.

Lifetimes of Skyrmions

In many cases isolated magnetic skyrmions correspond to metastable states, which can be deleted or created by fields or currents. This metastability permits the use of topologically distinct skyrmionic and ferromagnetic states as information bits. The critical parameter for any bit of information is its stability. The stability of any state can be quantified by measuring its lifetime. The lifetimes of metastable states in turn depend on temperature, external magnetic field, and other intrinsic or extrinsic parameters. At zero temperature a skyrmion might possess an infinite lifetime. At higher temperatures the thermal energy has to be compared with the height of the energy barrier between the two states. Therefore, the interesting question is how the lifetimes of the skyrmionic and ferromagnetic states depend on the field and temperature. We addressed this important issue in [15]. In these investigations the lifetimes of the skyrmionic states have been studied by means of Monte-Carlo simulations. It has been found that the skyrmion lifetime follows an Arrhenius-like law. This conclusion is in good agreement with other investigations of this subject [25]. The success of our approach was two-fold. First, we were able to determine the energy barriers between the skyrmionic and the ferromagnetic states. By that means we were able to quantify the attempt frequencies and the lifetimes of skyrmionic (Sk) and ferromagnetic (FM) states. The ratio of the corresponding attempt frequencies

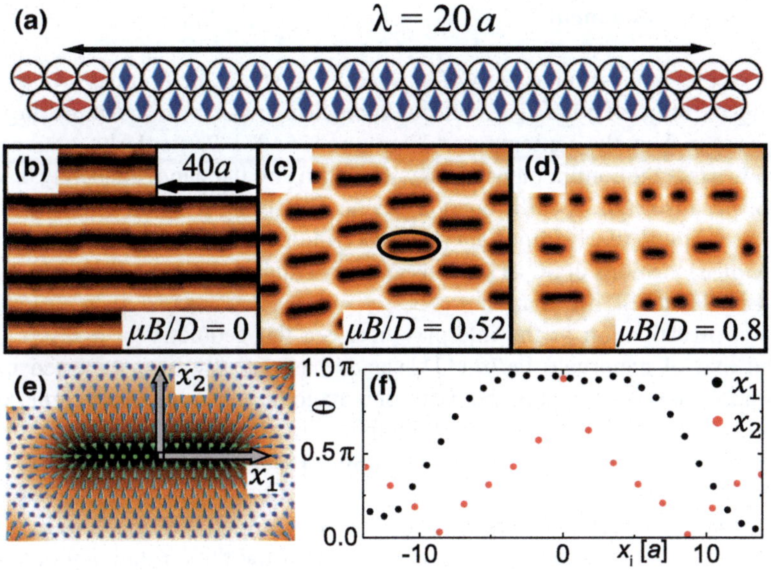

Fig. 9.5 **a** Schematic representation of the atomic lattice indicating the spatial changes of the anisotropy axis orientation. **b**, **c** Maps of the vertical component of the magnetization of equilibrium skyrmionic states for different strengths of magnetic field and spatial modulation of the exchange interaction corresponding to 3 ML Fe on Ir(111). **d**, **e** The same as in (**b**, **c**) with an additional modulation of the anisotropy according to the scheme outlined in panel (**a**). **f** Spin structure and local density of the topological charge for a deformed magnetic skyrmion

obtained from the Arrhenius fit is on the order of inverse time units (Monte Carlo steps) in the complete range of studied fields and temperatures.

Secondly, we were able to demonstrate that this large difference results from the higher entropy of the skyrmionic state. This leads us to the conclusion that the simple Arrhenius behavior would not explain the high skyrmion stability. To understand why the increased entropy leads to higher attempt frequencies one has to consider the Eyring equation, as a more general form of the Arrhenius law:

$$\tau = \tau_0 \cdot e^{\Delta S/k_B} \cdot e^{\Delta E/k_B T} = \tau_{\text{eff}} \cdot e^{\Delta E/k_B T} \tag{9.3}$$

Thirdly, by mapping our numerical results to experimental data we were able to quantify the exchange and DM parameters (7 and 2.2 meV respectively) and lifetimes for Pd/Fe/Ir(111) biatomic layers. Additionally, we were able to identify critical fields B_c, for which the skyrmionic and ferromagnetic states have the same stability.

Skyrmion's Confinement

In view of the application aspects of skyrmionic systems, theoretical investigations exploring the effect of boundaries and confinement become more and more important, as can be seen from the contemporary literature [26, 27]. Particularly, it has recently been shown that the confinement of a skyrmion in a circular nanoisland may lead to its isolation because of the specific boundary conditions induced by the DM interactions. This aspect is particularly important for the skyrmion lattices in systems of type SC-II, because the skyrmions in SC-II systems are very small and can significantly increase the density of stored information.

A first investigation of the interplay between the geometry of a skyrmion lattice and that of nanoscale Fe/Ir(111) islands has recently been reported in [18]. In this publication, interaction between one diagonal of the magnetic unit cell and the edges along the principal crystallographic directions of Fe nanostructures has been observed experimentally by means of Spin-Polarized Scanning Tunneling Microscopy and theoretically by means of Monte-Carlo simulations [18]. The details of the theoretically calculated micromagnetic structure are presented in Fig. 9.6. A clear trend of close-packed edges favouring one of the three rotational domains of the skyrmionic lattice can be seen in Fig. 9.6a, c. However, in an island of triangular shape it is impossible to orient the diagonal of a square nanoskyrmion lattice along all three edges of the island simultaneously. The mismatch of the symmetries of the skyrmionic lattice and the shape of the island leads to frustration and triple-domain states as visualized in Fig. 9.6a. On the other hand, the formation of domain walls (Fig. 9.6a) is accompanied by an energy increase with respect to a monodomain state (Fig. 9.6b).

For the identification of the energy at the rim of the sample and within the domain walls, and for the definition of the different energy contributions to the total energy of the system, spatially resolved energy maps of the triangular islands have been analyzed in Fig. 9.6d, where the average energy cost per atom with respect to the corresponding value in the interior of a very large sample in the nth atomic row being parallel to a favorable or an unfavorable edge is plotted. In total, the monodomain state has lower internal energy than the triple domain state. However, despite the lower energy of the single-domain state, multi-domain configurations show up in experiments and numerical simulations. This interesting result emerges because of the superposition of the effects of entropy and a pinning of the chiral domain walls.

9.4 One-Dimensional Quasiparticles

In addition to ultrathin magnetic films discussed in the previous section a variety of one-dimensional magnetic chains with open ends or closed chains on a variety of substrates can be experimentally addressed [28–31]. Atomic spin ensembles, magnetic and molecular nanoarrays as well as other metamaterials belong to this class of systems. Some experimental data are partially still not understood. Particularly, the

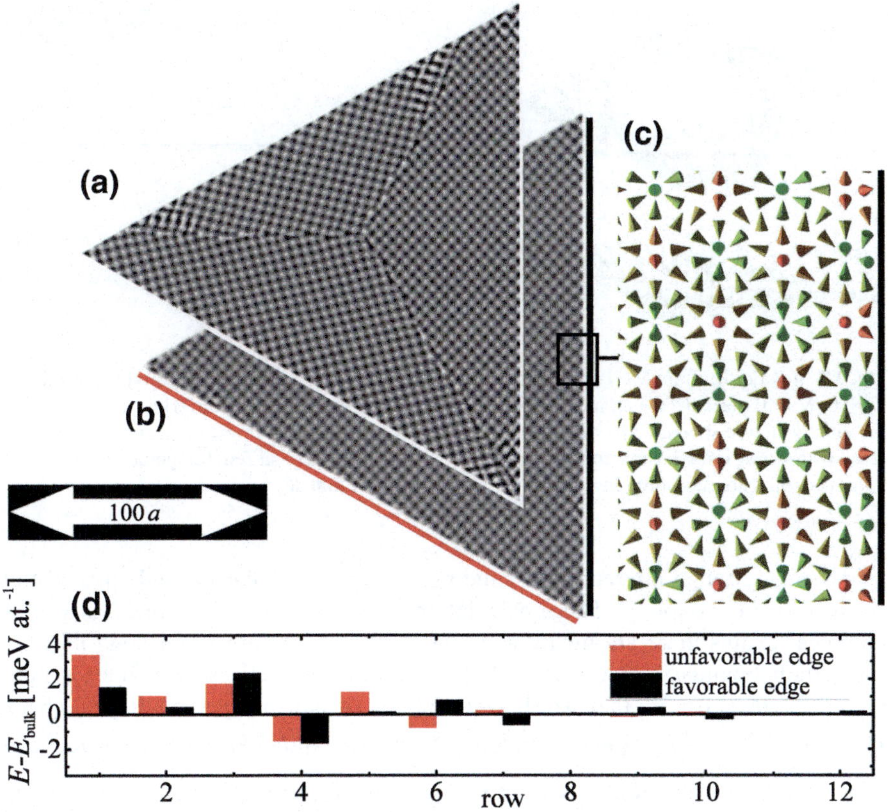

Fig. 9.6 **a** An excerpt of the spin structure of the nanoskyrmion lattice at T = 1K as obtained by MC calculations. **b**, **c** Islands with triangular boundary shape and open boundaries exhibiting multi-domain and single domain nanoskyrmion lattice states. **d** Mean energy cost in units of the bulk value per atom which belongs to the n-th atomic row parallel to the boundary. The energetically (un)favorable border is marked in (red) black in (**c**)

hysteresis curve measured on open and closed atomic magnetic chains show much higher disorder than predicted by the Ising model. Hence, the question arises whether topological magnetism can be achieved in closed linear structures with periodic or non-periodic boundaries and whether it can be responsible for the deviations of experimental results from the prediction by standard models [32]. Another important question is what is the role of the free energy for the formation of topological states?

Stability of One-Dimensional Magnetic Helices and Solitons

As reported in [30, 31, 33, 34] we have performed analytical and numerical analysis of the physical properties of one-dimensional magnetic structures that are cou-

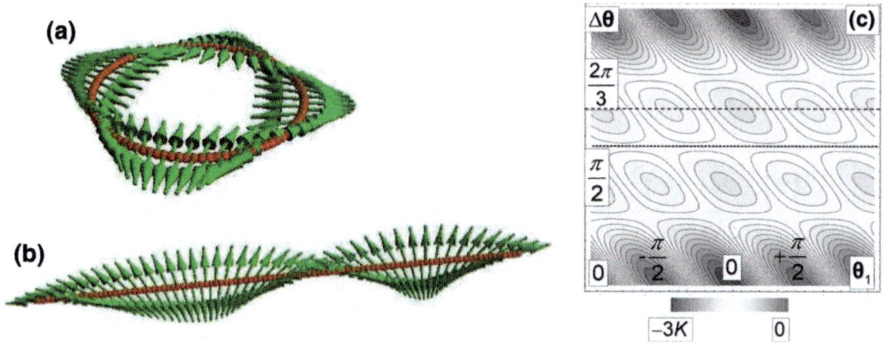

Fig. 9.7 **a**, **b** Equilibrium MC configuration at T = 0.2 K of a ring made of N = 100 spins coupled by $D = 0$, $K = 0$ meV, $J = -40$ meV: **a** AFM configuration for a closed chain (ring structure) and **b** the same structure shown with open ends for clarity. **c** Local energy minima of an antiferromagnetic chain with strong uniaxial anisotropy perpendicular to the chain axis at zero temperature. The energy minima depths are measured in units of the anisotropy constant K

pled via long-range interactions including exchange, dipolar or Ruderman-Kittel-Kasuya-Yosida coupling. These calculations have shown that in linear as well as nonlinear chains there are metastable magnetic configurations with long lifetimes for certain boundary conditions. They correspond to helices with integer number of π-twists, or, in other words, with integer or half-integer topological charge $Q = \frac{1}{4\pi}\int_S \left(\frac{\mathrm{dm}}{\mathrm{d}x} \times \frac{\mathrm{dm}}{\mathrm{dy}}\right) \cdot \mathbf{m}\mathrm{d}x\mathrm{d}y$. In some cases, for example for antiferromagnetic interactions or dipolar coupling with easy-plane anisotropy topologically stable double-helices are observed. Two examples of such double-helices for closed and open chains are given in Fig. 9.7a, b.

Similarly to skyrmions and domain walls the double-helices are topologically non-trivial objects ($Q \neq 0$). In continuous matter, once created they would be stable for infinite times. In discrete systems their lifetimes are finite. The life-times of these metastable magnetic configurations are defined via the energy barriers between the two subsequent states with different Q numbers. The internal energy of a short chain consisting of four classical Heisenberg spins coupled via an antiferromagnetic exchange interaction $J < 0$ and subject to a strong uniaxial anisotropy $K >> |J|$ is plotted in Fig. 9.7c. The angle $\Delta\theta$ denotes the angle between neighboring spins, while the angle θ_1 gives the polar orientation of the first spin in a chain. The straight lines show the band of saddle points at $\Delta\theta = \pi/2$, and the band of local energy minima at $\Delta\theta = 2\pi/3$. The entire number of local energy minima has direct proportionality with the number of sites in a chain. Interestingly, the minima correspond to the angles $\Delta\theta = 2\pi m/(N-1)$ with m integer and N the number of magnetic moments.

In contrast to two-dimensional skyrmions topological helices are one-dimensional objects. In [33] we proposed to use the metastable helix configurations with integer number of revolutions for an energy storing element that uses spins only. In order to keep the energy over time one has to twist one of the chain ends until it reaches a local energy minimum and fix the ends by local fields or other means. At a later time the magnet may be released to deliver the energy on demand. It is sufficient to let the end spins free; the introduced revolutions can be gained back. The longer a chain is, the larger number of twists can be stored. These stable chiral configurations can also be used to transfer the information. For that purpose, a knot, created at one end of the chain has to be transfered to the other end and then read out. The proposed concept can be scaled from the atomic scale to the scale of nanoscopic metamaterial or even to macroscopic objects. It might be applicable to the large diversity of systems like, e.g., magnetic multilayers, nanoarrays, colloidal films, Bose-Einstein condensates or atomic ensembles.

9.5 Zero-Dimensional Magnetic Objects

Zero-dimensional magnetic objects include quasiparticles like emerging magnetic monopoles or just single atomic spins or individual molecules on surfaces. In the following several examples of this class of magnetic structures will be reviewed.

Energy Storage in Emerging Magnetic Monopoles

Zero-dimensional quasiparticles known as "emerging magnetic monopoles" appear in spin-ices, which can be routinely created in nanomagnetic arrays [35–37]. The study of two-dimensional dipolar spin ices (2D-DSI), artificial counterparts of three-dimensional spin ices in magnetic arrays on lattices of different symmetry including quasicrystals [38, 39], is a very active and innovative field of science. One of the most popular research topics on 2D-DSI concerns the metastable defects, also known as emerging magnetic monopoles? arising at the ends of a line of reversed magnetic dipoles, the so-called Dirac string. The ultimate goal of investigations on 2D-DSI is the creation of the magnetic analog of spintronic devices utilizing these defects.

From the point of view of physics, the majority of investigations on 2D-DSI is concentrated on the behavior of unbound magnetic monopoles with vanishing tension of Dirac strings. In our recent investigation [40], the complete phase space defined by the Dirac string's tension has been investigated. In this investigation, we were able to show that in the regime of bound monopoles (BM) the effective degrees of freedom are Dirac strings rather than the monopoles. Particularly, in contrast to the common believe, the BMs do not obey the Coulomb law. For example, in contrast to standard electric Coulomb charges, the magnetic charges of opposite sign can be attracted or repulsed. The behaviour of the BM is defined by the specific quantity: the tension-to-mass ratio of the Dirac string. This ratio is found to be a fundamental quantity in strongly coupled 2D-DSI. It can be shown that the tension-to-mass ratio is defined by the fine-structure constant and lattice specific parameters of the magnetic array under consideration. Hence, it can be scaled by the geometry of a nanoarray. The path-time dependence of the kinetics of emerging magnetic monopoles at the ends of stretched and then released Dirac strings in 2D-DSI has been made and verified in computer simulations and with the help of a macroscopic experimental model. It has been shown that this string/monopole kinetics can be used to achieve avalanche-like currents of confined monopoles. This is important, because until now only field driven currents have been reported and because spontaneous currents can be used to store energy in Dirac strings. The duration of the avalanche monopole motion can be increased by geometrical means, e.g. increasing the length of a sample. Similarly to the example of topological helices, this effect can be used to store energy in Dirac strings.

Manipulation of the Zero-Dimensional Magnetic Objects

While the two- and one-dimensional magnetic objects can be successfully manipulated by external magnetic fields or currents, one has to find new ways of manipulation of zero-dimensional objects like emerging monopoles or even single magnetic spins. One of the possible ways of manipulation of these microscopic objects is the application of spin-sensitive local probe methods. Recent magnetic exchange force microscopy (MExFM) experiments demonstrated that apart from SP-STM measurements, also an atomic force microscopy (AFM) based setup can be utilized for spin mapping with atomic resolution. The MExFM brings several advantages with respect to scanning tunneling microscopy and spectroscopy. Until now the MExFM-based experiments are utilized for magnetic imaging only. Recently, however, we have developed a realistic theoretical concept for the manipulation of individual spins or other zero-dimensional magnetic objects with the tip in non-contact MExFM experiments [41–43].

In the quest for smaller magnetic data storage devices, single magnetic atoms, magnetic molecules or magnetic quasiparticles are vividly discussed in the literature as candidates for information bits. Experimentally, one needs to fix these objects onto a substrate, to stabilize the magnetic moment against fluctuations of different kind, and to read or write magnetic states. Here, we review theoretical concepts for the writing of magnetic states being based on a combination of first-principles calculations and spin dynamics simulations. The main idea hereby is to investigate the feasibility of spin switching using the tip of a MExFM.

A simple model system consisting of a MExFM probe made of a few Fe atoms and magnetic vanadium-, niobium- or tantalum-benzene complexes has been investigated in [41, 42]. This class of half-sandwiched transition-metal-benzene molecules was investigated previously and can be routinely synthesized [44]. In [42] the influence of the substrate on the molecule was neglected and the magnetic exchange coupling between the probe and the molecule has been investigated. The forces calculated within density functional theory for the Fe probe / V-benzene system with a ferro- (FM) and antiferromagnetic (AFM) ordering between Fe and V magnetic moments are displayed in Fig. 9.8a. A remarkable difference between the forces and interaction energies obtained for the two configurations in the range of 3.2–3.5 Å is seen in this plot. At a distance of 3 Å, however, the two force-distance functions become almost

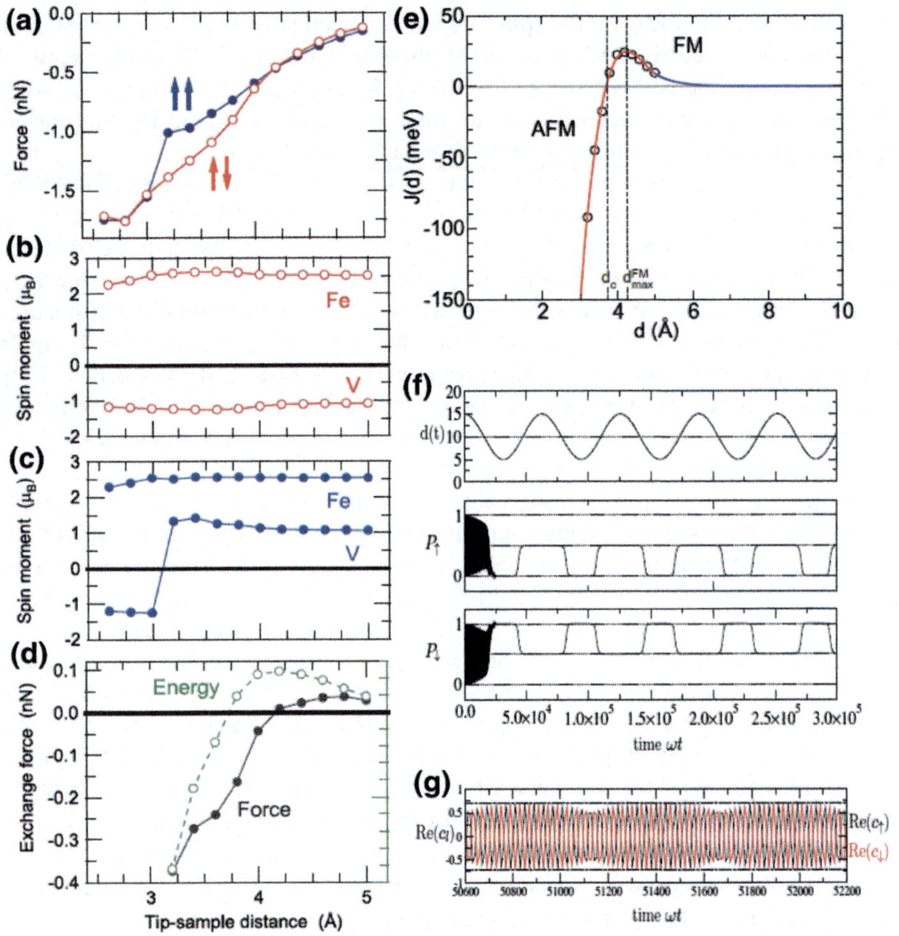

Fig. 9.8 Vertical component of the force between a vanadium-benzene molecule and a Fe probe (**a**) and the effective magnetic moments (**b**) as a function of tip-samle distance for two initial states (ferromagnetic and antiferromagnetic). In the simulations the probe-molecule separation has been relaxed numerically. When the tip is approaching, the orientation of the magnetic moment of the V molecules switches. Therefore, in (**c**) the direction of the magnetic moment of V atoms switches when approached by the tip. **d** Force and energy differences obtained for AFM and FM configuration at $d > 3.8$ Å, where an FM configuration corresponds to the ground-state, while at $d < 3.8$ Å an AFM one is energetically more stable. **e** Distance dependence of the exchange parameter J coupling tip and sample. The critical separation d_c denotes the distance at which J changes its sign and becomes AFM or FM

the same. This unexpected vanishing of the attractive force, being the difference between the forces for the two alignments, can be clarified if one analyzes magnetic moments at the MExFM tip and molecular atoms. If the spins of the Fe and V molecules are in the AFM state (Fig. 9.8b), this state remains stable when the tip is approaching. If, however, the spins are ferromagnetically aligned, a sudden jump from the FM state, Fig. 9.8c, to an AFM alignment appears. This sudden jump of magnetization emerges because of the strong direct exchange interaction between the Fe probe and the molecule making the mutual FM order unstable. For that reason, the more stable AFM solution is found in the DFT calculations.

According to the first-principles calculations of [42] the switching of the magnetic molecular state is possible by the adiabatic change of the distance between the MExFM tip and a molecule (see Fig. 9.8e), because the exchange forces between these two entities are large enough to overcome the barrier corresponding to the magnetic anisotropy. The query whether a switching event can be observed in a dynamic MExFM experiment requires investigations by atomistic spin dynamics. To apply the dynamical methods to the TM-benzene molecule disturbed by a magnetic tip the quantum mechanical Heisenberg model [42] has been adapted to the parameters from the DFT calculations and the time dependent Schrödinger equation has been solved numerically.

Depending on the tip polarization and the strength of the anisotropies six possible scenarios of the dynamical behaviour have been identified. They are defined by the combination of quantum tunneling and relaxation, and depend on the geometry of the tip-sample potential. In Fig. 9.8f–g the third scenario is exemplified. This scenario appears if D_s^x and / or D_s^y are nonzero and the minimal tip-sample distance doesn't exceed a critical distance shown in Fig. 9.8e $d_{\min} > d_c$. In this specific case both states $|\uparrow\rangle$ and $|\downarrow\rangle$ have the same energy. The wave-like oscillation of $\langle \hat{S}_s^z \rangle$ between $+1$ and -1 (black area) corresponds to the quantum-mechanical tunneling between $|\uparrow\rangle$ and $|\downarrow\rangle$ states at the crossing points of field dependent energy levels. With decreasing tip-sample distance d the degeneracy between $|\uparrow\rangle$ and $|\downarrow\rangle$ lifts up and the state $|\psi\rangle$ transforms to $|\psi\rangle = a\,|\uparrow\rangle + b\,|\downarrow\rangle$ with arbitrary a and b coefficients. After this, the system relaxes towards the equilibrium and the initial state $\langle \hat{S}_s^z \rangle$ transforms into -1. At $J(d) = 0$, $\langle \hat{S}_s^z \rangle$ vanishes, because $|\psi\rangle = \frac{1}{\sqrt{2}}\,(|\uparrow\rangle + |\downarrow\rangle)$. Hence, the leading process is the interlay between the relaxation and periodic quantum tunneling between the two stable configurations. Hereby, the $|0\rangle$ state does not play any important role as this superposition possesses higher internal energy.

Hence, the MExFM technique can be utilized to switch the magnetization of single magnetic molecules. In a classical regime; i.e., when the tip-sample distance is larger than a specific critical separation d_c, a reproducible magnetization reversal between AFM and FM alignment of spins in the MExFM probe and the molecule can be realized. In a quantum-mechanical regime; i.e., when the tip-sample distance is smaller than d_c, the magnetization switching of individual molecules under the action of the MExFM probe becomes more complex [45, 46]. Particularly, wave-like oscillations of the magnetization due to quantum-mechanical magnetization tunneling can appear.

Acknowledgements I would like to thank Mathias Schult, Nikolai Mikuszeit, Robert Wieser, Qing Chui Zhu, Yangui Zhang, Michael Karolak, David Altwein, Ansgar Siemens and Julian Hagemeister for the intense and fruitful cooperation over the years. Financial support of this work by the Deutsche Forschungsgemeinschaft through the SFB 668 (project A11) is gratefully acknowledged.

References

1. E.Y. Vedmedenko, *Competing Interactions and Pattern Formation in Nanoworld* (Wiley-VCH, Weinheim, 2007)
2. A.H. MacDonald, S.M. Girvin, D. Yoshioka, Phys. Rev. B **37**(16), 9753 (1988)
3. D. Wortmann, P. Kurz, S. Heinze, K. Hirai, G. Bihlmayer, S. Blügel, J. Magn. Magn. Mater. **240**, 57 (2001)
4. R. Wieser, E.Y. Vedmedenko, R. Wiesendanger, Phys. Rev. B **77**, 064410 (2008)
5. D. Hinzke, U. Nowak, J. Magn. Magn. Mater. **221**, 365 (2000)
6. E.Y. Vedmedenko, L. Udvardi, P. Wienberger, R. Wiesendanger, Phys. Rev. B **75**, 104431 (2007)
7. P. Ferriani, K. von Bergmann, E.Y. Vedmedenko, S. Heinze, M. Bode, M. Heide, G. Bihlmayer, S. Blügel, R. Wiesendanger, Phys. Rev. Lett. **101**, 027201 (2008)
8. S. Meckler, N. Mikuszeit, E.Y. Vedmedenko, O. Pietzsch, R. Wiesendanger, Phys. Rev. Lett. **103**, 157201 (2009)
9. K. von Bergmann, S. Heinze, M. Bode, E.Y. Vedmedenko, G. Bihlmayer, S. Blügel, R. Wiesendanger, Phys. Rev. Lett. **96**, 167203 (2006)
10. A. Bogdanov, A. Hubert, J. Magn. Magn. Mater. **138**, 255 (1994)
11. M. Ezawa, Phys. Rev. B **83**, 100408 (2011)
12. S. Heinze, K. von Bergmann, M. Menzel, J. Brede, A. Kubetzka, R. Wiesendanger, G. Bihlmayer, S. Blügel, Nat. Phys. **7**, 713 (2011)
13. E.Y. Vedmedenko, R. Wiesendanger, *Handbook on Spin Transport and Magnetism*, 2nd edn. (CRC Press Tayor and Francis, FL, 2017). Chap. Magnetic Skyrmions on Discrete Lattices
14. C. Moreau-Luchaire, C. Moutafis, N. Reyren, J. Sampaio, C.A.F. Vaz, N.V. Horne, K. Bouzehouane, K. Garcia, C. Deranlot, P. Warnicke, P. Wohlhüter, J.M. George, M. Weigand, J. Raabe, V. Cros, A. Fert, Nat. Nanotechnol. **11**, 444 (2016)
15. J. Hagemeister, N. Romming, K. von Bergmann, E.Y. Vedmedenko, R. Wiesendanger, Nat. Commun. **6**, 9455 (2015)
16. J. Hagemeister, E.Y. Vedmedenko, R. Wiesendanger, Phys. Rev. B **94**, 104434 (2016)
17. A. Siemens, Y. Zhang, J. Hagemeister, E. Vedmedenko, R. Wiesendanger, New J. Phys. **18**, 045021 (2016)
18. J. Hagemeister, D. Iaia, E. Vedmedenko, K. von Bergmann, A. Kubetzka, R. Wiesendanger, New J. Phys. **18**, 045021 (2016)
19. A.O. Leonov, M. Mostovoy, Nat. Commun. **6**, 8275 (2015)
20. N.S. Kiselev, A. Bogdanov, R. Schäfer, U.K. Rößler, J. Phys. D **44**, 392001 (2011)
21. X.Z. Yu, N. Kanazawa, Y. Onose, K. Kimoto, W.Z. Zhang, S. Ishiwata, Y. Matsui, Y. Tokura, Nat. Mater. **10**, 106 (2011)
22. N. Romming, A. Kubetzka, C. Hanneken, K. von Bergmann, R. Wiesendanger, Phys. Rev. Lett. **114**, 177203 (2015)
23. K. Shibata, J. Iwasaki, N. Kanazawa, S. Aizawa, T. Tanigaki, M. Shirai, T. Nakajima, M. Kubota, M. Kawasaki, H.P. et al., Nat. Nano **10**, 589 (2015)
24. P. Hsu, A. Kubetzka, A. Finko, N. Romming, K. von Bergmann, R. Wiesendanger, Nat. Nano. **12**, 123 (2017)
25. L. Rozsa, K. Palotas, A. Deak, E. Simon, R. Yanes, L. Udvardi, L. Szunyogh, U. Nowak, Phys. Rev. B **95**, 094423 (2017)
26. S. Rohart, A. Thiaville, Phys. Rev. B **88**, 184422 (2013)
27. R. Keesman, A.O. Leonov, P. van Dieten, S. Buhrandt, G.T. Barkema, L. Fritz, R.A. Duine, Phys. Rev. B **92**, 134405 (2015)
28. A. Stupakiewicz, E. Vedmedenko, A. Fleurence, T. Maroutian, P. Beauvillain, A. Maziewski, R. Wiesendanger, Phys. Rev. Lett. **103**, 137202 (2009)
29. A.A. Khajetoorians, J. Wiebe, B. Chilian, R. Wiesendanger, Science **332**, 1062 (2011)
30. M. Menzel, Y. Mokrousov, R. Wieser, J.E. Bickel, E.Y. Vedmedenko, S. Blügel, S. Heinze, K. von Bergmann, A. Kubetzka, R. Wiesendanger, Phys. Rev. Lett. **108**, 197204 (2012)

31. L. Zhou, J. Wiebe, S. Lounis, E.Y. Vedmedenko, F. Meier, S. Blügel, P.H. Dederichs, R. Wiesendanger, Nat. Phys. **4**, 035603 (2010)
32. P. Weinberger, E.Y. Vedmedenko, R. Wieser, R. Wiesendanger, Philso. Magn. **91**, 2248 (2011)
33. E.Y. Vedmedenko, D. Altwein, Phys. Rev. Lett. **112**, 017206 (2014)
34. S. Castillo-Sepúlveda, R.A. Escobar, D. Altbir, M. Krizanac, E.Y. Vedmedenko, Phys. Rev. B **96**, 024426 (2017)
35. A. Remhof, A. Schumann, A. Westphalen, H. Zabel, N. Mikuszeit, E.Y. Vedmedenko, T. Last, U. Kunze, Phys. Rev. B **77**, 134409 (2008)
36. A. Westphalen, A. Schumann, A. Remhof, H. Zabel, M. Karolak, B. Baxevanis, E.Y. Vedmedenko, T. Last, U. Kunze, T. Eimüller, Phys. Rev. B **77**, 174407 (2008)
37. A. Schumann, P. Szary, E.Y. Vedmedenko, H. Zabel, New J. Phys. **14**, 035015 (2012)
38. E.Y. Vedmedenko, *Reference Module in Materials Science and Materials Engineering*, 2nd edn. (Elsevier, Oxford, 2016). Chap. Quasicrystals: Magnetism
39. A. Jagannathan, B. Motz, E.Y. Vedmedenko, Philso. Magn. **91**, 2765 (2011)
40. E.Y. Vedmedenko, Phys. Rev. Lett. **116**, 077202 (2016)
41. E.Y. Vedmedenko, Q. Zhu, U. Kaiser, A. Schwarz, R. Wiesendanger, Phys. Rev. B **85**, 174410 (2012)
42. R. Wieser, V. Cacius, C. Lazo, H. Hölscher, E.Y. Vedmedenko, R. Wiesendanger, New J. Phys. **15**, 013011 (2013)
43. K. Them, T. Stapelfeldt, E.Y. Vedmedenko, R. Wiesendanger, New J. Phys. **15**, 013009 (2013)
44. Y. Mokrousov, N. Atodiresei, G. Bihlmayer, S. Heinze, S. Blügel, Nanotechnology **18**, 495402 (2012)
45. M. Krizanac, E.Y. Vedmedenko, R. Wiesendanger, New J. Phys. **18**, 033029 (2016)
46. M. Krizanac, E.Y. Vedmedenko, R. Wiesendanger, New J. Phys. **19**, 013032 (2017)

Chapter 10
Magnetism of Nanostructures on Metallic Substrates

Michael Potthoff, Maximilian W. Aulbach, Matthias Balzer, Mirek Hänsel, Matthias Peschke, Andrej Schwabe and Irakli Titvinidze

Abstract Various novel effects in nanostructures of magnetic atoms exchange-coupled to a metallic system of conduction electrons are reviewed. To this end we discuss analytical results and numerical data obtained by the density-matrix renormalization group and the real-space dynamical mean-field theory for the multi-impurity Kondo model. This model hosts complex physics resulting from the competition or cooperation of different mechanisms, such as the Kondo effect, the RKKY indirect magnetic exchange, finite-size gaps and symmetry-induced degeneracies in the electronic structure of finite metallic systems, quantum confinement in the strong-coupling limit, and geometrical frustration.

10.1 Introduction

Collective phenomena of a macroscopic physical system, resulting from strong interactions of its microscopic constituents, represent a central theme in theoretical physics, and collective magnetism is one of the most appealing examples in this context. There are uncountably many concrete realizations of spontaneous magnetic order, of related fundamental theoretical problems, of concrete realizations in different materials classes, of computational techniques, and of fascinating applications in modern technology [1, 2]. From a fundamental theoretical perspective, collective magnetic order requires at least three important ingredients: (i) The formation of local magnetic moments must be understood from the strong (Coulomb) interactions of electrons in a solid. (ii) Those moments must communicate to each other, i.e., the magnetic coupling mechanisms have to be identified to explain magnetic order on the level of an effective low-energy theory, and (iii) the magnetic coupling has

M. Potthoff (✉) · M. W. Aulbach · M. Balzer · M. Hänsel · M. Peschke · A. Schwabe
I. Titvinidze
I. Institute of Theoretical Physics, University of Hamburg, Hamburg, Germany
e-mail: michael.potthoff@physik.uni-hamburg.de

R. Wiesendanger (ed.), *Atomic- and Nanoscale Magnetism*, NanoScience and Technology, https://doi.org/10.1007/978-3-319-99558-8_10

to compete successfully with the omnipresent thermal and quantum fluctuations of the system, such that a ground state or a thermal state is realized that has a lower symmetry than the spin-rotation invariant Hamiltonian.

This highly general and challenging agenda for theory is somewhat concretized and also simplified for the case of nanostructured systems. Typically, correlated many-body problems are most complex in two spatial dimensions while one-dimensional systems or systems with a finite or even a few number of degrees of freedom are amenable to an exact solution or at least to a reliable numerical treatment, thanks to the progress that could be made in the recent decades or even years on computational techniques [3–7].

Here, we review some progress in our understanding of nanosystems which are composed of "magnetic" (3d-transition-metal) atoms on a nonmagnetic metallic surface – a topic that is driven by the progress achieved on the experimental side with the development of modern spin-resolved scanning-tunnelling-microscopy (STM) techniques and the novel abilities to manipulate magnetic systems and to detect their magnetic properties on the atomic scale [8–11]. The nano character of the theoretical systems studied here can be due to the finiteness of one or both of these two subsystems, i.e., either due to the small number of magnetic atoms, ranging, for instance, from one to several tens of atoms, and possibly due to the large but finite nonmagnetic host which the magnetic atoms are coupled to. In the context of surface physics, the realization of a finite metallic host requires that this is electronically isolated, e.g., by an insulating spacer from the supporting surface. Other realizations are conceivable as well, including individual grains [12, 13], single metallocene molecules [14], coupled quantum dots [15] or carbon nanotube pieces [16, 17]. Simulations of the systems studied will also be possible using ultra cold atoms trapped in optical lattices [18, 19].

The theoretical studies comprise various many-body model systems, including the famous Hubbard [20–24], the Ising and the Heisenberg model [25], but here we will concentrate on the multi-impurity Kondo model:

$$H_{\text{Kondo}} = -t \sum_{\langle i,j\rangle,\sigma} c_{i\sigma}^{\dagger} c_{j\sigma} + J \sum_{m=1}^{M} \boldsymbol{s}_{i_m} \boldsymbol{S}_m \, . \tag{10.1}$$

The Hamiltonian describes M spins $\boldsymbol{S}_m$, with spin-quantum number $1/2$, which are coupled locally via an antiferromagnetic exchange $J > 0$ to the local spins $\boldsymbol{s}_i$ of a system of N itinerant and non-interacting conduction electrons. The conduction electrons hop with amplitude $t \equiv 1$ between non-degenerate orbitals on neighboring sites of a lattice with L sites. Furthermore, $c_{i\sigma}$ annihilates an electron at site $i = 1, \ldots, L$ with spin projection $\sigma = \uparrow, \downarrow$, and the local conduction-electron spin $\boldsymbol{s}_i = \frac{1}{2} \sum_{\sigma\sigma'} c_{i\sigma}^{\dagger} \boldsymbol{\sigma}_{\sigma\sigma'} c_{i\sigma'}$ at i is given in terms of the vector of Pauli matrices $\boldsymbol{\sigma}$. The sites at which the impurity spins couple to the electron system are denoted by i_m, where $m = 1, \ldots, M$. Unless stated differently, we will study the system at half-filling $N = L$.

This model, in a nutshell, describes the physics of magnetic atoms (with local spins $\boldsymbol{S}_m$) coupled to the metallic substrate. It also represents an effective low-energy theory [26, 27] of the more fundamental Anderson model H_{And}, where the Kondo interaction part $\propto J$ is replaced by $\sum_{m\sigma} \varepsilon_m f^\dagger_{m\sigma} f_{m\sigma} + U \sum_m n^{(f)}_{m\uparrow} n^{(f)}_{m\downarrow} + V \sum_m f^\dagger_{m\sigma} c_{i_m\sigma} + \text{H.c.}$. Here, V is the hybridization strength between the adatom and the metal surface, ε_m the one-particle energy of the adatom orbital and U the local Hubbard-type Coulomb repulsion. The Kondo model with Kondo coupling $J = 8V^2/U$ is found for half-filled impurity orbitals in the $U \to \infty$ limit.

Among the various aspects of the intricate physics covered by the model, we will discuss (i) modern numerical techniques addressing the indirect magnetic exchange (Sect. 10.2), (ii) the competition of indirect exchange with the Kondo screening of the moments in a quantum box and modifications of the famous Doniach diagram [28] (Sect. 10.3), (iii) a theory of Kondo underscreening and overscreening in nanosystems (Sect. 10.4), (iv) a novel type of indirect magnetic exchange in the strong-coupling limit (Sect. 10.5), and (v) the competition between magnetic coupling, Kondo screening and geometrical frustration (Sect. 10.6), before the conclusions are given (Sect. 10.7).

10.2 Indirect Magnetic Exchange

In most cases, we choose the underlying lattice of the multi-impurity Kondo or Anderson model as one-dimensional ($D = 1$). This is sufficient to work out the main qualitative physics for the cases studied here and is convenient, since the one-dimensional model is amenable to a numerically exact solution by means of the density-matrix renormalization-group (DMRG) technique [29, 30]. On the other hand, due to the famous area law, DMRG is restricted to the $D = 1$ case or, for higher D, to a few impurities only. Therefore, from the very beginning it is clear that, e.g., for realistic applications and addressing concrete materials, the DMRG approach must be replaced by a more versatile method eventually.

Unfortunately, even the actually already oversimplified model, (10.1), is by far too complicated in $D = 3$ or $D = 2$ dimensions and can be rigorously treated only in the unphysical infinite-dimensional limit where the dynamical mean-field theory (DMFT) [4] applies. A formal generalization of the DMFT to systems of arbitrary dimension and geometry, and to nanosystems in particular, namely the real-space DMFT (R-DMFT) [31], is well-known and can be used to study magnetic properties. R-DMFT is a comprehensive, thermodynamically consistent and nonperturbative theory. However, it provides mean-field-type approximations only, and essentially this also holds true for the various cluster and other extensions of DMFT [5–7]. As a rule of thumb, the lower the dimension, the less applicable is DMFT, such that one would expect unacceptable results in $D = 1$ where a comparison with the DMRG is possible. Right for the case of magnetic atoms on metallic surfaces, however, this is not necessarily the case as the DMFT also treats the single-impurity case ($M = 1$)

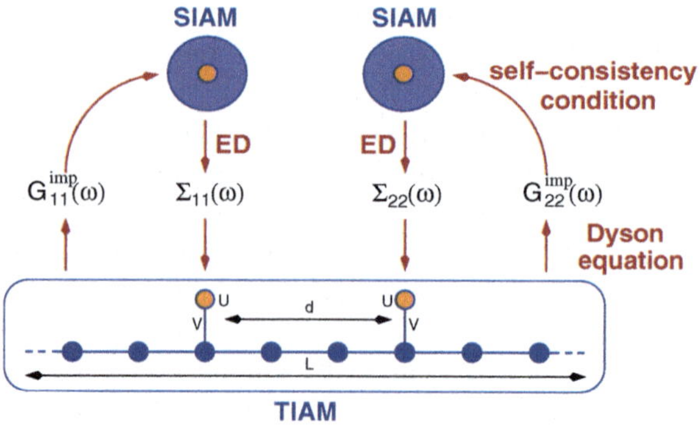

Fig. 10.1 (adapted from [33]). Real-space dynamical mean-field theory applied to a two-impurity Anderson model

exactly. The interesting case is therefore the two-impurity ($M = 2$) Anderson model (TIAM), and the distance d between the two impurities is the interesting control parameter as this bridges between the essentially single-impurity case for $d \to \infty$ and the "dense" case for neighboring impurities $d = 1$. Comparison of approximate R-DMFT with exact DMRG results as a function of d, and also as a function of the effective Kondo coupling $J = 8V^2/U$, provides us with an estimate of the reliability of the mean-field approach which will serve as a guide for systems with $M>2$ magnetic atoms in higher dimensions as well.

Figure 10.1 illustrates how to apply the R-DMFT to the TIAM (box). The TIAM is self-consistently mapped onto two single-impurity Anderson models (SIAM) which are independently solved by means of exact diagonalization (ED) [32] to get the local self-energies $\Sigma_{mm}(\omega)$ with $m = 1, 2$. These are used in the TIAM Dyson equation, the solution of which gives the local impurity Green's functions $G^{\text{imp}}_{mm}(\omega)$ which then define via the R-DMFT self-consistency conditions [31] the parameters of the impurity models. The procedure is iterated until self-consistency is achieved.

Figure 10.2 shows results for the static local ($\alpha = \beta$) and nonlocal ($\alpha \neq \beta$) adatom-adatom susceptibilities defined as

$$\chi_{\alpha\beta} = -\int_0^\infty d\tau \langle S^f_{\alpha,z}(\tau) S^f_{\beta,z}(0)\rangle \ , \tag{10.2}$$

with $\alpha = 1, 2$, which provide information on the indirect magnetic coupling. Here, $S^f_{\alpha,z} = \frac{1}{2}(n^f_{\alpha,\uparrow} - n^f_{\alpha,\downarrow})$ are magnetic moments on the adatom site α. We find almost perfect agreement with the exact DMRG data for large distances (left panel). Only for $d \lesssim 5$ do we see some significant deviations. For $d = 1$ (right panel) R-DMFT provides reasonable results for strong effective Kondo coupling $J = 8V^2/U$; it breaks down, however, for weak $V^2/U \lesssim 0.2$ and fails to maintain a Fermi-liquid ground state: Here, the screening of the magnetic moments is too weak to compensate for the ordering tendencies induced by a comparatively strong inter-adatom RKKY interaction [35–37]. We observe an artificial spontaneous symmetry breaking induced by the mean-field approximation, i.e., the adatoms' nonlocal SU(2) invariant singlet

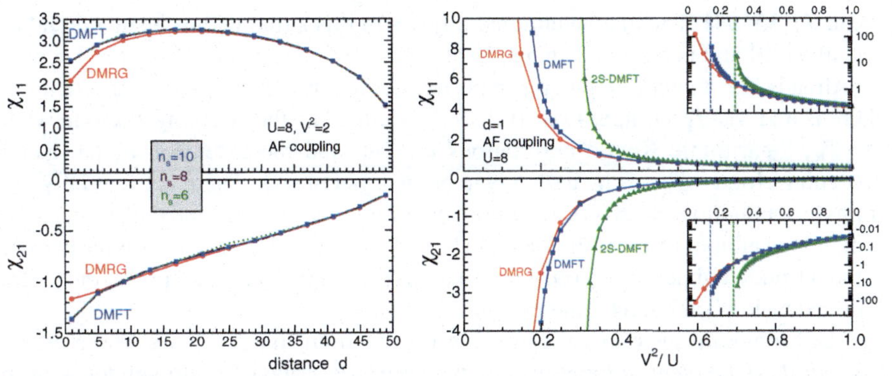

Fig. 10.2 (adapted from [33]). *Left*: χ_{11} and χ_{12}, see (10.2), as functions of the distance d (left) as obtained by R-DMFT and compared to DMRG for $L = 50$. R-DMFT computation with different number of orbitals n_s in the effective SIAM. *Right*: χ_{11} and χ_{12} as functions of the effective Kondo coupling V^2/U at $d = 1$ and $L = 50$. Results of R-DMFT and a simplified two-site DMFT (2S-DMFT) [34] compared to DMRG. $t = 1$ fixes the energy scale. *Inset*: The same on a log scale

state $(|\uparrow\downarrow\rangle - |\downarrow\uparrow\rangle)/\sqrt{2}$ for $J \to 0$ is replaced in mean-field theory by an incoherent mixture of degenerate ordered states $|\uparrow\downarrow\rangle$ and $|\downarrow\uparrow\rangle$. This qualitative failure is indicated by divergencies of χ_{11} and χ_{21}.

For complex magnetic nanostructures with several magnetic adatoms in different chain or cluster geometries on semi-infinite three-dimensional metallic surfaces, a mean-field approach is inevitable. As in ab-initio studies, the accessible system size strongly depends on the remaining, e.g., lateral spatial symmetries, and the computational effort scales nearly linearly with the number of inequivalent correlated sites only. The TIAM represents a model that is rather unfavorable to a single-site R-DMFT approach. Even for this case, however, R-DMFT can in fact almost quantitatively predict the effects of indirect magnetic exchange in competition with the Kondo screening and with geometrical effects—as long as the approximation predicts a Fermi-liquid ground state.

10.3 The Kondo-Versus-RKKY Quantum Box

A study of the competition between Kondo screening and RKKY coupling cannot be performed in a mean-field picture. What is missing here is the *feedback* of the nonlocal magnetic correlations on the one-particle quantities. An numerically exact treatment of this problem is interesting since one may expect that the generic phase diagram proposed by Doniach already in 1977 [28] must be revised for the case of a nanostructure. Here the finite-size gap Δ represents a third energy scale competing with the Kondo scale $T_K \sim e^{-1/J}$ [38] and with the RKKY scale $J_{RKKY} \sim J^2$ (for weak J) [35–37]. The unconventional physics in this situation has been worked out recently [39] and is briefly sketched here.

Already for a *single* magnetic impurity, i.e. for the "Kondo-box" problem [40], there is a Δ-vs.-T_K competition: If $J = J_\Delta$, defined as the coupling strength where $\Delta \sim T_K$, logarithmic Kondo correlations are cut, and the extension of the Kondo screening cloud $\sim T_K^{-1}$ is actually given by the system size. There is eventually only a single conduction-electron state within the Kondo scale T_K around the Fermi energy which is available to form the "Kondo" singlet. Estimating Δ and comparing with typical Kondo temperatures, one finds that this type of physics takes place for systems with extensions on the nanometer scale.

The problem is even more interesting for the multi-impurity case where there is a *spatially dependent* competition of the above-described Kondo singlet with the RKKY interaction. We consider the model, (10.1), for an even total number $N + M$ of electrons and spins such that a Fermi-liquid state with a total spin singlet can be reached for $L \to \infty$. For two impurity spins and even N, the Fermi energy ε_F then lies in a finite-size gap of the single-conduction-electron spectrum ε_k ("off-resonant case") since all spin-degenerate ε_k are doubly occupied. There is a crossover from local Kondo-singlet formation at strong J to nonlocal RKKY coupling at $J = J_D$, defined by $T_K \sim J_{RKKY}$. The Kondo effect, however, is absent for $J \to 0$ (for $T_K < \Delta$): The spins cannot dynamically couple to the electron system as the Fermi sea is non-degenerate and thus a finite energy $\sim \Delta$ would be necessary for screening. Hence, the low-energy sector is covered by the RKKY two-spin model $H_{RKKY} = -J_{RKKY} \boldsymbol{S}_1 \boldsymbol{S}_2$ with $J_{RKKY} \propto (-1)^{|i_1 - i_2|} J^2 / |i_1 - i_2|$.

The quantum box with three impurities is qualitatively different: As N is odd, the highest one-particle eigenenergy ε_{k_F} is singly occupied. Hence, the ground state of the conduction-electron system is two-fold Kramers degenerate ("resonant case"), and screening of a spin is possible for $J \to 0$ but competes with the RKKY exchange. Perturbation theory for the Kondo problem is regularized due to the finite-size gap $\Delta > 0$ and predicts that, *if an impurity spin dynamically couples to the conduction-electron system* this happens on a linear-in-J scale. For sufficiently weak but finite J, this is larger than the RKKY scale $\propto J^2$ and we thus expect formation of a "Kondo" spin singlet involving the conduction-electron spins. *Whether or not there is a perturbative coupling of the impurity spin $\boldsymbol{S}_m$*, depends on the one-particle Fermi wave function U_{ik_F} at site $i = i_m$. Here, at $k_F = \pi/2$, we have $U_{ik_F} = \sqrt{2/(L+1)} \sin(i\, k_F) \neq 0$ for $i = 1, 3, \ldots, L-2, L$. These sites are denoted as "good", while $U_{ik_F} = 0$ for "bad" sites $i = 2, 4, \ldots, L-1$.

Consider a chain with $L = 4n + 1$ and integer n. Two spins $\boldsymbol{S}_1$ and $\boldsymbol{S}_3$ couple to bad sites neighboring the central site. The central site is good. Therefore, for small J and small L we have $T_K^{(\text{bulk})} < \Delta$, the perturbative arguments given above apply, the Kondo scale is linear in J, and $\boldsymbol{S}_2$ is screened. The weaker (ferromagnetic) RKKY interaction then couples $\boldsymbol{S}_1$ and $\boldsymbol{S}_3$ to a nonlocal spin triplet, and thus $S_{tot} = 1$.

What happens for finite Δ as J increases? This is nicely reflected in the different ground-state spin-correlation functions obtained by DMRG shown in Fig. 10.3 (right): For $L = 9$ at $J \to 0$, we find $\langle \boldsymbol{S}_1 \boldsymbol{S}_3 \rangle \to 1/4$ while $\langle \boldsymbol{S}_1 \boldsymbol{S}_2 \rangle \to 0$ and $\langle \boldsymbol{S}_2 \boldsymbol{S}_3 \rangle \to 0$ as well. The local correlations vanish $\langle \boldsymbol{S}_1 \boldsymbol{s}_{i_1} \rangle \to 0$ at the bad sites

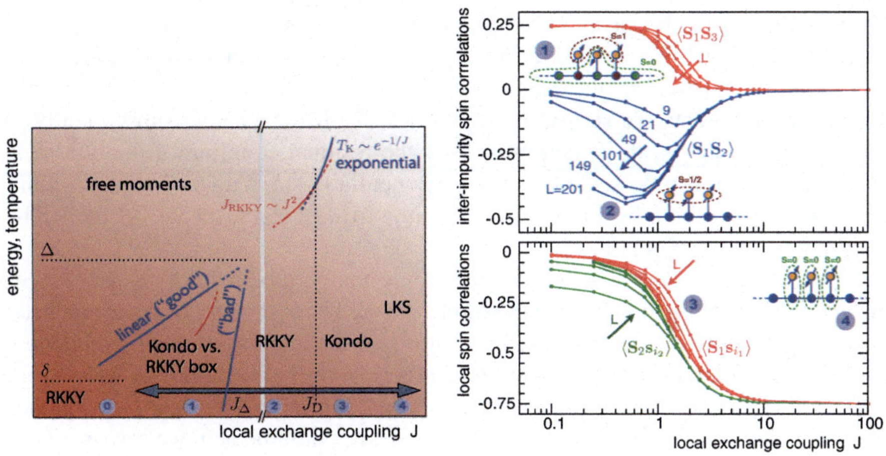

Fig. 10.3 (adapted from [39]). *Left*: Revised Doniach diagram for a nanosystem with competition between Kondo screening and RKKY interaction. *Right*: DMRG results for the inter-impurity and local spin correlations of the $M = 3$ model ("bad-good-bad") with increasing (see arrows) system size up to $L = 201$ sites. With decreasing J, the different regimes (1–4) marked in the Doniach diagram are found

but for the good one $\langle \boldsymbol{S}_2 \boldsymbol{s}_{i_2} \rangle$ remains finite as $J \to 0$. The strong-J limit is also easily understood: The distinction between good and bad sites becomes irrelevant for $J > J_\Delta$, and as $J \to \infty$ all local spin correlations approach $-3/4$. This *local* Kondo-singlet formation is basically independent of L.

At intermediate J, i.e. $J_\Delta < J < J_\mathrm{D}$, the conventional Kondo-vs.-RKKY competition should be recovered (since $J_\Delta < J$), and the RKKY interaction should win (since $J < J_\mathrm{D}$). However, this decisively depends on the system size: For small systems, we rather have $J_\Delta > J_\mathrm{D}$, implying that there is no intermediate-J regime. This can be seen in $\langle \boldsymbol{S}_1 \boldsymbol{S}_2 \rangle$ for $L = 9$ or $L = 21$, for example. Here, the spin correlation stays close to zero in the entire J regime, and there is a shallow minimum around $J = 1$ only.

Only if the system is sufficiently large, namely if $J_\Delta < J_\mathrm{D}$ (i.e. $\Delta < J_\mathrm{RKKY}$), do RKKY correlations develop among the impurity spins. In fact, as is seen in Fig. 10.3 (right) for $L = 201$, the correlation $\langle \boldsymbol{S}_1 \boldsymbol{S}_2 \rangle$ approaches $-1/2$ in the intermediate-J regime and $\langle \boldsymbol{S}_1 \boldsymbol{S}_3 \rangle$ is close to $1/4$. Note that these are exactly the correlations of a three-spin system with ferromagnetic coupling between $\boldsymbol{S}_1$ and $\boldsymbol{S}_3$ and antiferromagnetic ones between $\boldsymbol{S}_2$ and $\boldsymbol{S}_1$, $\boldsymbol{S}_3$. With decreasing J, the system eventually crosses over to the perturbative regime at a crossover coupling strength J_Δ. This strongly decreases with increasing L.

The physics is summarized with the "revised Doniach diagram" in Fig. 10.3 (left). Local Kondo singlets (LKS) are formed for extremely strong J (regime 4). With decreasing J, the energy to break up a Kondo singlet becomes exponentially small (regime 3), and below J_D, RKKY coupling is dominant (regime 2). Finite-size effects set in for $J < J_\Delta$, and if the system is sufficiently large, we have $J_\Delta < J_\mathrm{D}$. The singlet-formation energy is linear in J for "good" and vanishes for "bad" sites (regime 1). This implies that for sites where the k_F conduction-electron wave function has a finite (vanishing) weight, the Kondo effect "wins" (RKKY wins). The remaining unscreened moments are subjected to nonlocal RKKY exchange subsequently. At a still lower energy scale δ, determined by the residual coupling of the box to the environment, the Kondo scale becomes exponentially small again with a screening cloud leaking out into the environment (regime 0) [41]. Here, for $J < J_\delta$, the RKKY coupling is dominant again [42, 43].

Concluding, for nanosystems there is a subtle competition between Kondo screening and RKKY interaction even in the limit $J \to 0$. The symmetry of the ground state decisively depends on the geometrical positions of the impurities. It is predictable by perturbative techniques and is accessible experimentally by studies of field dependencies [11]. For possible applications, e.g., construction of nano-spintronic devices bottom-up [10], confinement normal to the surface could be achieved by insulating spacers [44], and lateral confinement will lead to strong variations in the local density of states as is known from quantum corrals [45], for example, but also from nonmagnetic adatoms, step edges etc. In cases where the confinement is not perfect but where there is a strong spatial dependence of the Kondo temperature, one can still utilize the Kondo-vs.-RKKY quantum-box physics.

A serious treatment of small residual couplings to the environment, however, represents the main challenge as concerns the further development of the theory. In this case, perturbative approaches are no longer applicable. Furthermore, generalizations to more impurities, systems off half-filling and multi-orbital systems are worth being explored.

10.4 Underscreening and Overscreening

One of the several interesting routes along which the quantum-box physics can be extended, is to consider systems with (i) a symmetry-induced degeneracy of one-particle energies at the Fermi level and (ii) localized spins with nonlocal couplings to the local conduction-electron spins in addition to the local one via J. This allows us to study the physics of (i) underscreening and (ii) overscreening in a finite (nano) system and, addressing the $L \to \infty$ limit, provides a new theoretical approach towards the very complex field of non-Fermi-liquid Kondo systems [46, 47]. To give an impression, we concentrate on the case of underscreening here and briefly highlight one example. A detailed overview and in fact a comprehensive theory of underscreening in a nanosystem has been presented in [48].

We apply degenerate perturbation theory of first order in J to the model, (10.1), for a lattice geometry where the N-electron ground state $|\text{FS}, \gamma\rangle$ (the Fermi sea) of the noninteracting tight-binding part of H, obtained by occupying all one-particle levels below the Fermi energy, $\varepsilon_k \leq \varepsilon_F$, is Γ-fold degenerate ($\gamma = 1, \ldots, \Gamma$). At weak coupling $J \to 0$, the finite-size Kondo effect (see previous section) always wins against RKKY coupling in the on-resonance case. But if several local spins are

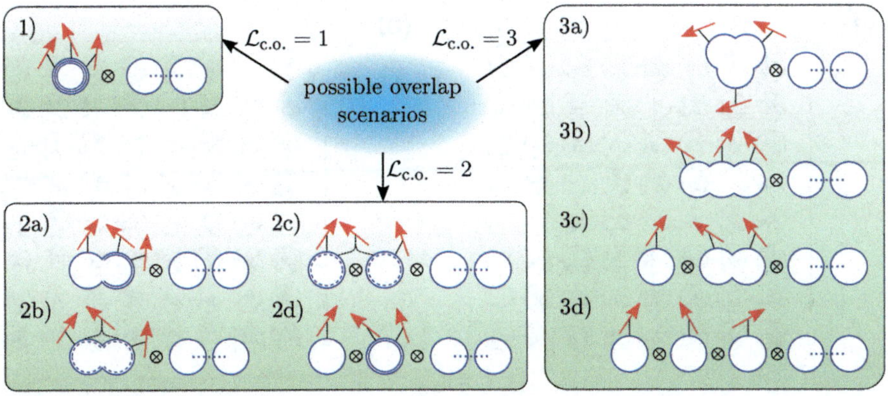

Fig. 10.4 Classification of different scenarios for perfect screening or underscreening of $M = 3$ local spins coupled to a conduction-electron system with a Γ-fold degenerate Fermi sea

present ($M > 1$), can they all be screened by finite-size Kondo effects? If not, does underscreening lead to degenerate ground states? How many screening channels are available and what defines a "channel"? Those (and more) questions can be answered convincingly by the effective generalized central-spin model emerging at weak coupling [48, 49].

All spins at the different sites in real space couple to the same small number of conduction-electron states at ε_F. Several scenarios can arise, depending on the nanostructure geometry and its electronic structure. It turns out that it is entirely possible for each spin $\boldsymbol{S}_m$ to be coupled to its own Fermi orbital, $|\text{F}, i_m, \sigma\rangle$, allowing exact finite-size-Kondo screening of all spins, and an overall spin-singlet ground state. In other situations, there are insufficient degrees of freedom at ε_F to screen all of the spins. The remaining unscreened spins are then coupled together via their mutual RKKY interaction. Depending on details, this could either lead to a degenerate ground state or a singlet ground state, with both finite-size-Kondo and RKKY mechanisms acting simultaneously on different spins.

A complete classification is possible. Figure 10.4 gives an example for $M = 3$. Here, $\mathcal{L}_{\text{c.o.}}$ is the dimension of the subspace spanned by the three coupling orbitals $|\text{F}, i_m, \sigma\rangle$, where $m = 1, 2, 3$. For $\mathcal{L}_{\text{c.o.}} = 1$, we have three identical coupling Fermi orbitals. Similarly, double circles with solid lines in scenarios 2a and 2d stand for two identical coupling Fermi orbitals. Dashed circles, see cases 2b and 2c, represent a single coupling Fermi orbital, which is a linear combination of the other two coupling Fermi orbitals, contrary to case 3b, where none of the coupling Fermi orbitals can be written as a linear combination of the others. The tensor product symbol indicates orthogonality between Fermi orbitals. Which ground state is realized crucially depends on the electron filling of the orbitals and can easily be determined following simple rules [49].

The theory can easily be applied to large nanostructures. As an example, Fig. 10.5 shows results for an 8×8 lattice (open boundaries), $N = 57$ electrons and an eight-

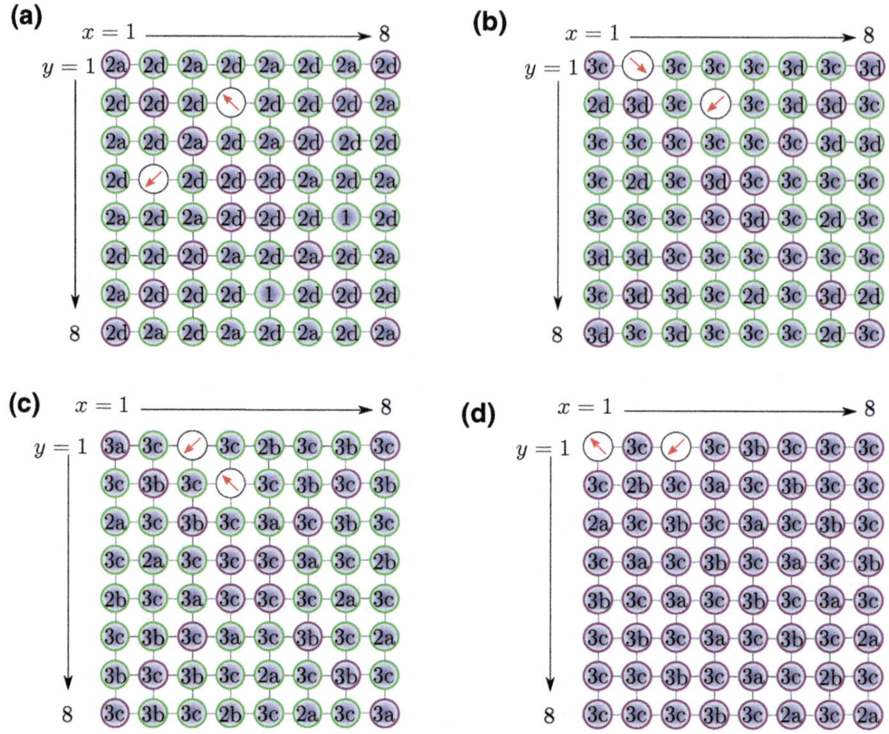

Fig. 10.5 Realization of the scenarios for 3 local spins on an 8 × 8 nanostructure, see text for discussion

fold degenerate highest occupied one-particle state. For (a)–(d), two spins are fixed at different sites (red circles with arrows), while the third is placed on one of the remaining sites. The resulting scenario is indicated. For example, in (a) two spins are located at positions (2, 4) and (4, 2). If the third one is placed at (1, 7) we find case (2a) of Fig. 10.4. Green circles around lattice sites represent cases where all three coupling strengths in the effective central-spin model are equal, while purple borders indicate that at least one effective coupling strength differs from the others.

10.5 Inverse Indirect Magnetic Exchange

So far, we have mainly discussed the weak-J regime. In special cases, however, the strong-J limit is particularly interesting, namely if there is confinement of conduction electrons due to scattering at almost local Kondo singlets. This may lead to local-moment formation at a priori uncorrelated sites and to couplings of these moments via virtual excitations of the Kondo singlets, a situation, which is "inverse" compared to

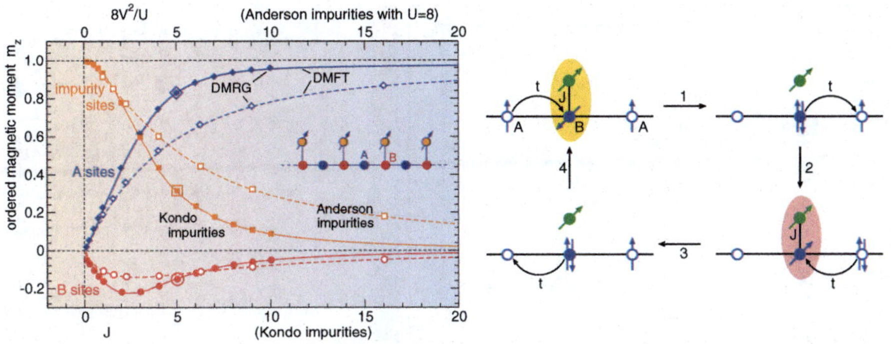

Fig. 10.6 (adapted from [50, 52]). *Left*: Crossover from the RKKY regime at weak coupling to the IIME regime at strong coupling. Ordered magnetic moments on A, B and impurity sites of a tight-binding chain with spin-1/2 Kondo impurities as functions of J (solid lines, filled symbols) and with Anderson impurities as functions of $8\,V^2/U$ at Hubbard $U = 8$ (dashed, open). Symbols: DMRG. Lines: DMFT. *Right*: A typical fourth-order process mediating the IIME

the conventional RKKY coupling at weak J, and which is called an "inverse indirect magnetic exchange" (IIME) mechanism [50–53].

Figure 10.6 (left) demonstrates the effect of the IIME for a one-dimensional system with $M = L/2$ spins at distances $d = 2$, see inset. DMRG and DMFT calculations for the Kondo and the related Anderson model show a clear crossover: In the weak-J limit there is ferromagnetic long-range order of the local spins in the ground state due to the RKKY coupling. With increasing J, however, the local moments are screened by electrons on the B sites, while the confinement of electrons on A sites leads to A-site local-moment formation.

The IIME mechanism favors ferromagnetic alignment of the A-site moments and, eventually, ferromagnetic order. Figure 10.6 (right) displays a typical fourth-order process where (1) a Kondo singlet at a B site is broken by excitation to a state where the site is either empty or doubly occupied, followed (2) by further excitation to a local triplet state at the same site, until (3) the system returns to a broken Kondo singlet state and (4) the local Kondo singlet is restored. Fourth-order perturbation theory results in an effective IIME model:

$$H_{\mathrm{IIME}} = -\frac{Z^2}{3}\alpha \sum_{m=1}^{Z} \sum_{i \in B_m} \boldsymbol{S}^2_{i,\mathrm{bonding}} \,. \tag{10.3}$$

where $\alpha = 64t^4/3J^3$ is the IIME coupling constant. Z is the number of sub-sublattices, see Fig. 10.7, right, for the case of a square lattice. Here, $\boldsymbol{S}_{i,\mathrm{bonding}}$ is the spin of the symmetric bonding orbital on a plaquette.

With Fig. 10.7, left, we demonstrate the effects of the IIME for strongly coupled magnetic adatoms in an artificial geometry on a two-dimensional conduction-electron system. Electrons in the chain of A-sites 1, 2, ..., 8 (see labels in the figure) with

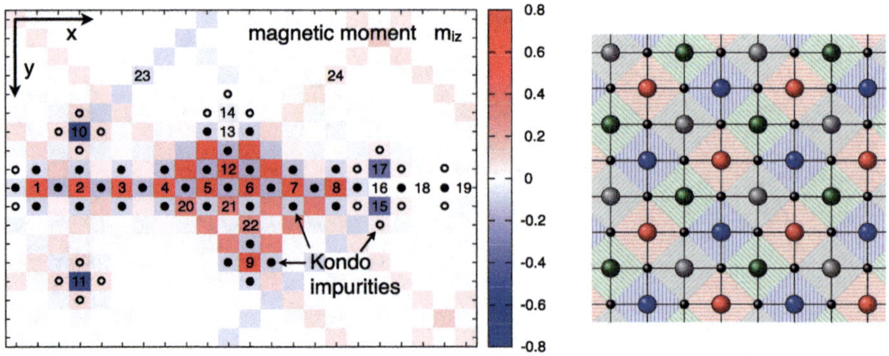

Fig. 10.7 (adapted from [50, 52]). *Left*: Magnetic structure of an IIME-coupled system in an artificial two-dimensional geometry. $M = 57$ local spins (filled and open dots) are locally strongly coupled ($J = 5$) to the electrons on an $L = 22 \times 18$ array of uncorrelated sites. R-DMFT results for the ordered magnetic moments (color code) at sites in the layer. Labels: see text. *Right*: $\mathcal{H}_{\text{eff}}$ for a square lattice. Small black circles: A-sublattice sites. Large circles: B-sublattice sites. For each B-sublattice site i, the spin $\boldsymbol{S}_{i,\text{bonding}}$ operates on basis states on the plaquette of neighboring A sites. The B-sublattice is thus subdivided into $Z = 4$ different and non-overlapping sub-sublattices represented by four different colors/patterns. Each sub-sublattice group fully covers the A-sublattice. Spins at neighboring B sites do not commute. Spins belonging to the same group do commute

relative distance $\delta = 2$ are confined and their moments couple ferromagnetically via IIME. The IIME, similar to RKKY, is oscillatory and is decreasing with distance, see A-site 9 (distance $\delta = 4$, ferromagnetic) and sites 10 and 11 ($\delta = 3$ and $\delta = 5$, antiferromagnetic coupling to the chain). Local moments at 10 and 11 are formed by confinement due to surrounding local Kondo singlets. The weak coupling to the rest of the system makes them extremely susceptible, such that $|m_\text{B}|$ increases with increasing distance from the chain. Neighboring A-sites with higher effective coordination mutually support magnetic polarization, see the slightly enhanced m_A at and around A-site 12. Confinement of an *odd* number of electrons is important, see 13 and 14 (almost no moment) and 15, 16, 17 (polarized). Electrons at 18 or 19 are not confined, and no polarization is found. Furthermore, there is a proximity effect (20, 21, or 22) and interference patterns from spin-dependent multiple scattering of conduction electrons at the magnetic structures (23, 24) are visible. Summing up, in the strong-J regime, the IIME, similar to RKKY [9–11], can be utilized to construct nanostructures with tailored magnetic properties.

10.6 Frustrated Quantum Magnetism

In many situations, in particular for non-bipartite lattices, the indirect (RKKY) magnetic exchange is frustrated. For the model (10.1) on the two-dimensional triangular lattice, a mechanism of partial Kondo screening (PKS) [54] has been suggested and

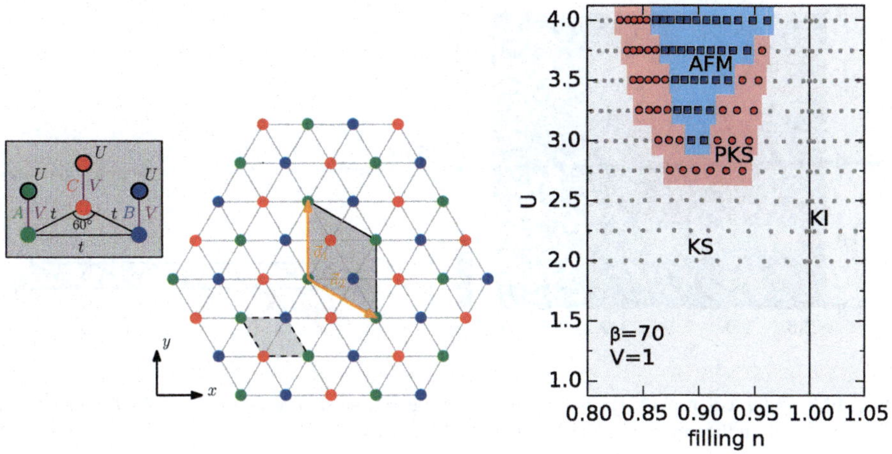

Fig. 10.8 (adapted from [57]). *Left*: Parameters and primitive and conventional unit cell of the periodic Anderson model on the triangular lattice. *Right*: R-DMFT phase diagram, obtained with the continuous-time quantum Monte–Carlo technique as impurity solver

has been studied with various approximate methods [54–56]: Starting from strong J in a spin-liquid state with dominating local Kondo correlations, the transition to an RKKY-induced magnetic state at weaker J is preempted by a PKS phase, where frustration is avoided via a site-selective Kondo-singlet formation on one of the three sites in the (conventional) unit cell such that the remnant local spins can order magnetically via the RKKY coupling. It would be highly interesting to check if this PKS state is realized in a magnetic nanostructure where one can profit from the much higher control over the geometrical degrees of freedom offered by STM techniques.

As a first orientation, we have recently employed "site-selective DMFT" to study the system (the Anderson lattice variant) as function of the filling $n = N/L$ [57]. This is equivalent with R-DMFT but assuming only the three orbitals in the conventional unit cell as inequivalent. The resulting rather complex magnetic phase diagram in the J-n plane (see Fig. 10.8) comprises the paramagnetic Kondo insulator (KI) at $n = 1$, a paramagnetic heavy-fermion metallic state (KS) and antiferro- (AFM) and ferromagnetic phases (not shown), and in fact an extended parameter region with PKS. As expected, this phase is located at the border between the nonmagnetic and the magnetically ordered phase. However, the insight gained by DMFT and also by the previous approaches must be questioned seriously, since the competition between magnetic coupling, Kondo screening and geometrical frustration in a two-dimensional strongly correlated system is one of the most complex problems one can think of.

We have therefore proceeded along a different route [58]. For the Kondo lattice on the zig-zag ladder, i.e., for an essentially one-dimensional model, numerically exact results are in reach so that the PKS hypothesis can be tested. Figure 10.9 (right) displays the lattice geometry and the model parameters. Opposed to the one-

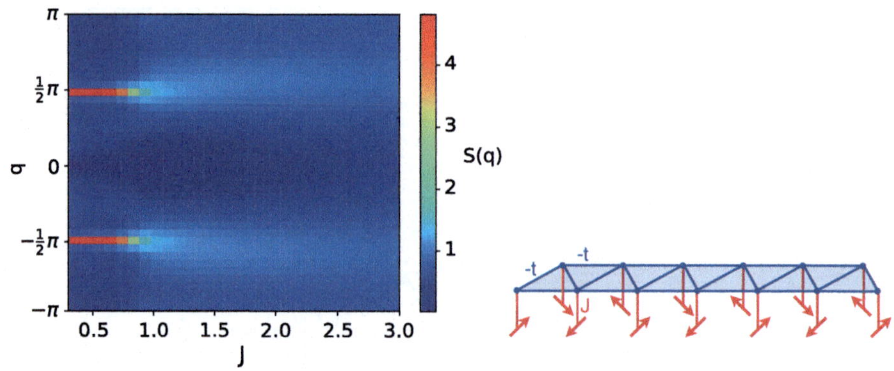

Fig. 10.9 (adapted from [58]). J-dependence of the spin-structure factor $S(q)$ for the half-filled Kondo model on the zig-zag ladder, as obtained by DMRG

dimensional Kondo model on a conventional chain [59], the Kondo lattice on the zig-zag ladder, with equal hopping on the legs and rungs, does show a magnetic phase at weak J already at half-filling $n = 1$ [58]. This is reminiscent of the quantum-critical point of the model on the (bipartite) square lattice [60].

Figure 10.9 (left) displays the J-dependence of the spin-structure factor $S(q) = L^{-1} \sum_{ij} e^{iq(R_i - R_j)} \langle \boldsymbol{S}_i \boldsymbol{S}_j \rangle$ as obtained by DMRG. We find a logarithmic-in-L divergence of $S(q)$ at $q = Q = \pi/2$ for $J < J_\mathrm{m} \approx 0.84$ which indicates magnetic quasi-long-range order with algebraically decaying correlations, $\langle \boldsymbol{S}_i \boldsymbol{S}_j \rangle \propto e^{-iQ(i-j)}/|i - j|$, i.e., a magnetic 90°-spin-spiral phase resulting as a compromise between predominantly antiferromagnetic indirect interactions and the non-bipartite lattice geometry.

Interestingly, the transition to the magnetic state is again preempted, at $J = J_\mathrm{d} \approx 0.90$, by a phase, similar to but different from PKS, where the frustration is alleviated due to a spontaneous breaking of the translational symmetries of the lattice. Namely, for $J < J_\mathrm{d}$ we find a non-zero order parameter O_d, defined as the asymmetry of spin correlations on geometrically equivalent rungs: $O_\mathrm{d} = \langle \boldsymbol{S}_i \boldsymbol{S}_j \rangle_{(i,j)=\diagdown} - \langle \boldsymbol{S}_i \boldsymbol{S}_j \rangle_{(i,j)=\diagup}$. Given the strong *ferromagnetic* short-range correlations, for $J < J_\mathrm{d}$, on one of the two sets of rungs, the remaining antiferromagnetic correlations are no longer frustrated. This means that spin dimerization helps to make the system "more bipartite" – on the level of the electronic rather than the geometrical structure. Like PKS, dimerization could play an important role for geometrically frustrated nanostructures. One should note that the quasi-long-range 90° spin spiral with dimerized short-range correlations represents a quantum state with no classical analog. An extended discussion of spin dimerization, of the detailed phase diagram and the underlying physical mechanisms is given in [58].

10.7 Conclusions

The multi-impurity Kondo model can provide a sketch of the complicated physics of magnetic nanostructures on metallic surfaces only. Due to its many-body character, however, already the model itself contains highly interesting physics which, in most cases, results from the competition of different interactions or mechanisms on the same energy scale. The uniqueness of nanostructures consists in the low dimensionality of the systems and the correspondingly strong role played by correlations, the additional energy scale given by the finite-size gap, which substantially modifies the theoretical description, and the extremely high degree of control of the system parameters on the experimental side.

A couple of predictions made in the recent past and reviewed here, at least partially, still await a verification or test. This may trigger further activities on the experimental side. On the other hand, there are also theoretical challenges. Theoretical modeling must get more refined as there are important aspects that have not been considered so far, such as all questions related to orbital physics, effects of magnetic anisotropies which, just for nanostructures, are expected to be particularly strong, or the explicit consideration of the coupling to the environment, i.e., to the bulk electronic structure of the substrate, to give some examples. For some of these topics one must proceed in a direction where many-body and conventional band theory are merging, i.e., towards a big LDA+DMFT-like machinery [61, 62]. For others, the phenomenological side is more pressing, and one thus has to adopt a more coarse-grained point of view which, however, is able to tackle larger and more inhomogeneous systems and to include more and longer-ranged interactions, for example. Both approaches are necessary and important complements of the quantum many-body approach presented here.

Acknowledgements This work has been made possible by the generous financial support of the Deutsche Forschungsgemeinschaft within the Sonderforschungsbereich 668 (project A14). We would like to thank the speaker R. Wiesendanger and all members of the SFB for the fruitful collaborations and intense discussions over the past years.

References

1. M. Donath, P.A. Dowben, W. Nolting (ed.), *Magnetism and Electronic Correlations in Local-moment Systems: Rare-Earth Elements and Compounds* (World Scientific, Singapore, 1998)
2. K. Baberschke, M. Donath, W. Nolting (ed.), *Band-Ferromagnetism* (Springer, Berlin, 2001)
3. E. Gull, A. Millis, A. Lichtenstein, A. Rubtsov, M. Troyer, P. Werner, Rev. Mod. Phys. **83**, 349 (2011)
4. A. Georges, G. Kotliar, W. Krauth, M.J. Rozenberg, Rev. Mod. Phys. **68**, 13 (1996)
5. T. Maier, M. Jarrell, T. Pruschke, M.H. Hettler, Rev. Mod. Phys. **77**, 1027 (2005)
6. M. Potthoff, in *AIP Conference Proceedings*, vol. 1419 (2011), p. 199
7. M. Potthoff, in *Strongly Correlated Systems: Theoretical Methods*, ed. by A. Avella, F. Mancini. Springer Series in Solid-State Sciences, vol. 171 (Springer, Berlin, 2012), p. 303
8. R. Wiesendanger, Rev. Mod. Phys. **81**, 1495 (2009)
9. L. Zhou, J. Wiebe, S. Lounis, E. Vedmedenko, F. Meier, S. Blügel, P. Dederichs, R. Wiesendanger, Nat. Phys. **6**, 187 (2010)

10. A.A. Khajetoorians, J. Wiebe, B. Chilian, R. Wiesendanger, Science **332**, 1062 (2011)
11. A.A. Khajetoorians, J. Wiebe, B. Chilian, S. Lounis, S. Blügel, R. Wiesendanger, Nat. Phys. **8**, 497 (2012)
12. D.C. Ralph, C.T. Black, M. Tinkham, Phys. Rev. Lett. **74**, 3241 (1995)
13. D.C. Ralph, C.T. Black, M. Tinkham, Phys. Rev. Lett. **78**, 4087 (1997)
14. C.H. Booth, M.D. Walter, M. Daniel, W.W. Lukens, R.A. Andersen, Phys. Rev. Lett. **95**, 267202 (2005)
15. R. Kaul, G. Zarand, S. Chandrasekharan, D. Ullmo, H. Baranger, Phys. Rev. Lett. **96**, 176802 (2006)
16. T.W. Odom, J.L. Huang, C. Cheung, C.M. Lieber, Science **290**, 1549 (2000)
17. Y. Bomze, I. Borzenets, H. Mebrahtu, A. Makarovski, H.U. Baranger, G. Finkelstein, Phys. Rev. B **82**, 161411(R) (2010)
18. F. Scazza, C. Hofrichter, M. Höfer, P.C.D. Groot, I. Bloch, S. Fölling, Nat. Phys. **10**, 779 (2014)
19. G. Cappellini, M. Mancini, G. Pagano, P. Lombardi, L. Livi, M.S. de Cumis, P. Cancio, M. Pizzocaro, D. Calonico, F. Levi, C. Sias, J. Catani, M. Inguscio, L. Fallani, Phys. Rev. Lett. **113**, 120402 (2014)
20. M. Balzer, W. Hanke, M. Potthoff, Phys. Rev. B **77**, 045133 (2008)
21. G. Li, W. Hanke, A.N. Rubtsov, S. Bäse, M. Potthoff, Phys. Rev. B **80**, 195118 (2009)
22. M. Balzer, B. Kyung, D. Sénéchal, A.M.S. Tremblay, M. Potthoff, Europhys. Lett. **85**, 17002 (2009)
23. M. Balzer, M. Potthoff, Phys. Rev. B **82**, 174441 (2010)
24. M. Balzer, M. Potthoff, Phys. Rev. B **83**, 195132 (2011)
25. E. Vedmedenko, N. Mikuszeit, T. Stapelfeldt, R. Wieser, M. Potthoff, A. Lichtenstein, R. Wiesendanger, Eur. Phys. J. B **80**, 331 (2011)
26. J.R. Schrieffer, P.A. Wolff, Phys. Rev. **149**, 491 (1966)
27. P. Sinjukow, W. Nolting, Phys. Rev. B **65**, 212303 (2002)
28. S. Doniach, Physica B+C **91**, 231 (1977)
29. S.R. White, Phys. Rev. Lett. **69**, 2863 (1992)
30. U. Schollwöck, Ann. Phys. **326**, 96 (2011)
31. M. Potthoff, W. Nolting, Phys. Rev. B **55**, 2741 (1997)
32. M. Caffarel, W. Krauth, Phys. Rev. Lett. **72**, 1545 (1994)
33. I. Titvinidze, A. Schwabe, N. Rother, M. Potthoff, Phys. Rev. B **86**, 075141 (2012)
34. M. Potthoff, Phys. Rev. B **64**, 165114 (2001)
35. M.A. Ruderman, C. Kittel, Phys. Rev. **96**, 99 (1954)
36. T. Kasuya, Prog. Theor. Phys. **16**(kk), 45 (1956)
37. K. Yosida, Phys. Rev. **106**, 893 (1957)
38. A.C. Hewson, *The Kondo Problem to Heavy Fermions* (Cambridge University Press, Cambridge, 1993)
39. A. Schwabe, D. Gütersloh, M. Potthoff, Phys. Rev. Lett. **109**, 257202 (2012)
40. W. Thimm, J. Kroha, J. von Delft, Phys. Rev. Lett. **82**, 2143 (1999)
41. P. Simon, I. Affleck, Phys. Rev. Lett. **89**, 206602 (2002)
42. S. Galkin, C.A. Balseiro, M. Avignon, Eur. Phys. J. B **38**, 519 (2004)
43. P. Simon, Phys. Rev. B **71**, 155319 (2005)
44. S.J. Tang, T. Miller, T.C. Chiang, Phys. Rev. Lett. **96**, 036802 (2006)
45. M.F. Crommie, C.P. Lutz, D.M. Eigler, E.J. Heller, Surf. Rev. Lett. **2**, 127 (1995)
46. P. Nozières, A. Blandin, J. de Phys. **41**, 193 (1980)
47. H. Kusunose, J. Phys. Soc. Jpn. **85**, 113701 (2016)
48. A. Schwabe, M. Hänsel, M. Potthoff, A.K. Mitchell, Phys. Rev. B **92**, 155104 (2015)
49. M. Hänsel, Theory of screening in the multi-channel multi-impurity Kondo box, Ph.D. thesis, University of Hamburg, 2017
50. A. Schwabe, I. Titvinidze, M. Potthoff, Phys. Rev. B **88**, 121107(R) (2013)
51. A. Schwabe, M. Hänsel, M. Potthoff, Phys. Rev. A **90**, 033615 (2014)
52. I. Titvinidze, A. Schwabe, M. Potthoff, Eur. Phys. J. B **88**, 42 (2015)
53. M.W. Aulbach, I. Titvinidze, M. Potthoff, Phys. Rev. B **91**, 174420 (2015)

54. Y. Motome, K. Nakamikawa, Y. Yamaji, M. Udagawa, Phys. Rev. Lett. **105**, 036403 (2010)
55. S. Hayami, M. Udagawa, Y. Motome, J. Phys. Soc. Jpn. **80**, 073704 (2011)
56. H. Ishizuka, Y. Motome, Phys. Rev. Lett. **108** (2012)
57. M.W. Aulbach, F.F. Assaad, M. Potthoff, Phys. Rev. B **92**, 235131 (2015)
58. M. Peschke, R. Rausch, M. Potthoff, Phys. Rev. B **97**, 115124 (2018)
59. H. Tsunetsugu, M. Sigrist, K. Ueda, Rev. Mod. Phys. **69**, 809 (1997)
60. F.F. Assaad, Phys. Rev. Lett. **83**, 796 (1999)
61. E. Pavarini, E. Koch, D. Vollhardt, A. Lichtenstein (ed.), *The LDA+DMFT Approach to Strongly Correlated Materials* (Forschungszentrum Jülich, Jülich, 2011)
62. F. Lechermann, A.I. Lichtenstein, M. Potthoff, Eur. Phys. J. Spec. Top. **226**, 2591 (2017)

Part II
Spin Dynamics and Transport in Nanostructures

Chapter 11
Magnetization Dynamics on the Atomic Scale

Stefan Krause and Roland Wiesendanger

Abstract Spin-polarized scanning tunneling microscopy is used to study the magnetization reversal of individual quasiclassical and quantum atomic-scale magnets. Modifications of the dynamics arising from the presence of the biased magnetic probe tip are identified in terms of spin-transfer torque, Joule heating, Oersted field, magneto-electric coupling and spin-polarized field emission. Observing the switching behavior over a wide temperature range allows for a detailed study of nucleation, propagation and annihilation of domain walls in quasi-classical nanomagnets. Quantum magnets are found to be intrinsically stable at ultra-low temperature, and magnetization dynamics can be controlled by the spin-polarized tunnel current injection from the magnetic tip.

Our today's world is taking advantage of digital information that is processed, transferred and stored at high speed and reliability. Understanding the microscopic processes of magnetization dynamics plays a crucial role for the development of future spintronic applications. Especially, the miniaturization of magnetic bit size is a prerequisite for the vast growth of storage density in magnetic media. However, thermal agitation leads to a magnetization reversal of the grains that represent a binary information unit (bit) below a critical size, thereby destroying stored information. Understanding the underlying physical processes that favor or hinder magnetization reversal inside small magnets is therefore highly relevant. Especially, for magnets consisting of only a few atoms the classical models for magnetization dynamics have to be adjusted to the quantum world.

Spin-polarized scanning tunneling microscopy (SP-STM) is an established and powerful tool for the investigation of magnetic nanostructures and the imaging of spin structures on magnetic surfaces with atomic resolution. Here, a biased magnetic probe tip is moved at atomic distance above the surface of a sample, and

S. Krause · R. Wiesendanger (✉)
Department of Physics, University of Hamburg, Jungiusstrasse 11A,
20355 Hamburg, Germany
e-mail: wiesendanger@physnet.uni-hamburg.de

S. Krause
e-mail: skrause@physnet.uni-hamburg.de

R. Wiesendanger (ed.), *Atomic- and Nanoscale Magnetism*, NanoScience and Technology, https://doi.org/10.1007/978-3-319-99558-8_11

information about the local magnetization configuration on the sample is extracted from the spin-polarized tunnel current. In contrast to lithographically fabricated devices, the atomically sharp magnetic probe tip of an SP-STM realizes a point-like spin-polarized tunneling contact. Imperfections and leakage channels, that are typical for planar devices, do not occur in the vacuum separating both electrodes. Whereas SP-STM is conventionally used to *read* magnetic information, the very same magnetic probe tip can be used to modify the magnetic state of a nanomagnet and thereby *write* magnetic information on the atomic scale. For nanomagnets exhibiting a stable magnetization, an energy barrier hinders magnetization reversals. To write magnetic information a reversal event has to be triggered, either by a spin-polarized tunnel current or by an external electric or magnetic field. Here, a threshold has to be overcome, otherwise the nanomagnet remains in its magnetic state. Nanomagnets that intrinsically reverse their magnetization frequently are perfect detectors for manyfold interactions with their environment. Here, no threshold has to be overcome, and tiny changes in terms of electro-magnetic fields, spin-polarized tunnel currents or temperature already drastically change their switching behavior.

Positioning the probe tip of an SP-STM above a nanomagnet allows for the observation and manipulation of the magnetization dynamics by means of telegraphic noise experiments and pump-probe techniques. With SP-STM, various parameters of the magneto-tunnel junction, like the current amplitude, the electro-magnetic field, the electron energy and the point of current injection can precisely be controlled. These unique capabilities are used to study the intrinsic dynamics of atomic-scale magnets as well as their interactions with spin-polarized electrons and electro-magnetic fields. Moreover, SP-STM can be combined with atomic manipulation by moving individual atoms, allowing for the tailoring and investigation of smallest magnets that are constructed atom by atom.

In the following, the basic concepts of time-resolved telegraphic noise experiments using SP-STM are presented (Sect. 11.1). High spin-polarized tunnel currents result in current-induced magnetization switching effects (Sect. 11.2). The combined action of Joule heating and spin-transfer torque is used to develop a spin-based pump-probe scheme that increases the temporal resolution to the nanosecond regime (Sect. 11.3). Various interactions between a high spin-polarized tunnel current and a quasiclassical nanomagnet are investigated in detail, namely in terms of Oersted field (Sect. 11.4), electric field (Sect. 11.5) and spin-polarized field emission (Sect. 11.6). In the second part of this chapter, time-resolved SP-STM experiments on the magnetization dynamics are performed in the quasiclassical as well as in the quantum regime. For quasiclassical magnets, an Arrhenius switching behavior is observed when changing the temperature (Sect. 11.7). Building magnetic clusters from only five individual iron atoms results in quantum magnets that are found to be intrinsically stable at ultra-low temperature, and magnetization dynamics is purely driven by the spin-polarized current generated by the magnetic SP-STM tip (Sect. 11.8).

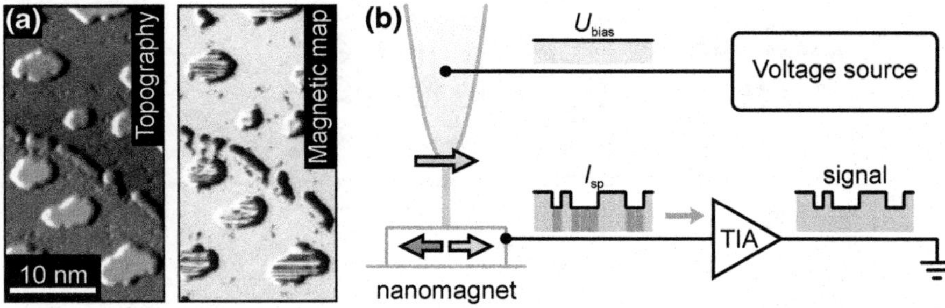

Fig. 11.1 **a** Topography (left) and simultaneously recorded magnetic map (right) of iron nanomagnets on W(110). The magnetic signal on the nanomagnets reveals frequent magnetization reversal events, as indicated by the striped or sprinkled pattern. **b** Schematics of point-mode SP-STM on an individual nanomagnet that thermally reverses magnetization. The spin-polarized tunnel current and the differential conductance changes according to the change of magneto-tunnel resistance, resulting in a telegraphic noise signal (TIA: transimpedance amplifier)

11.1 Telegraphic Noise Experiments on Nanomagnets

SP-STM is not only a powerful tool to investigate static magnetic surface structures on the atomic scale, it also provides resolution in the time domain. In Fig. 11.1a, an SP-STM image of a stepped W(110) surface decorated with atomic-scale iron magnets is shown. Whereas the z channel at closed feedback loop shows the topography, the magnetic map reveals that the nanomagnets exhibit a striped or sprinkled pattern that changes from image to image. It is known that the magnetic signal scales with $\cos\alpha$, where α is the angle between the magnetization directions of the tip and sample [1]. As the monolayer of Fe/W(110) is known to be ferromagnetic [2, 3], the nanomagnets are expected to be in a ferromagnetic monodomain state at low temperature. Consequently, they are expected to exhibit a homogeneous high or low magnetic signal, depending on the relative orientation of the magnetization between each nanomagnet and the tip. However, a two-state noise is observed. It is the manifestation of thermally induced switching events, thereby changing the magneto-tunneling configuration between each individual nanomagnet and the tip with fixed magnetization. As a consequence, the magnetic signal varies on a time scale that is given by the lifetime of the ferromagnetic ground state of each nanomagnet.

In this frame mode of SP-STM, many nanomagnets are imaged, but the magnetic signal on an individual nanomagnet is recorded only when the tip scans over it. Consequently, the temporal resolution in this mode is limited due to the waiting time until the same nanomagnet is imaged again. In order to increase the temporal resolution, the SP-STM tip is positioned stationary above a nanomagnet under investigation, and the evolution of the magnetic signal is recorded. Now the tunnel magneto-resistance between tip and sample changes as function of time, resulting in a frequent variation of the spin-polarized tunnel current and conductance, although a fixed bias voltage is applied. The signal is scaled by a transimpedance amplifier, resulting in a two-state

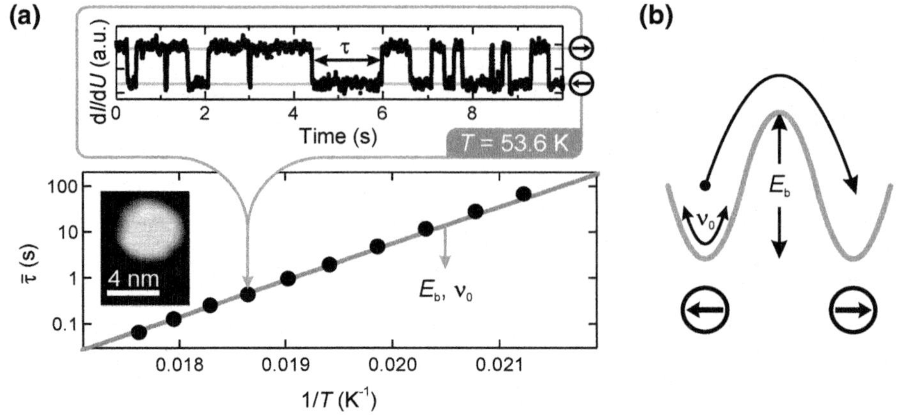

Fig. 11.2 **a** Telegraphic noise signal (top) and temperature dependent lifetime (bottom) for the nanomagnet shown in the inset. An Arrhenius law can be fitted to the data, yielding the effective activation energy barrier and the attempt frequency. **b** Schematics of the macrospin model for a nanomagnet with uniaxial magnetic anisotropy. For magnetization reversal, an activation energy barrier has to be overcome

telegraphic noise, as depicted in Fig. 11.1b. Note that this technique allows for the determination of lifetimes between consecutive switching events only. The details of the spin dynamics during individual magnetization reversal events are not resolved due to the limited temporal resolution of conventional transimpedance amplifiers.

Variable-temperature SP-STM allows for the investigation of the switching behavior of the same individual nanomagnet at different temperature [4]. Here, the telegraphic noise is recorded for every temperature setpoint and analysed for the mean lifetime between two consecutive switching events. The experiments reveal that the Fe/W(110) nanomagnets can be modelled as individual macrospins which have to overcome an effective activation energy barrier for reversal, resulting in an Arrhenius law for the switching behavior.

The switching probability for monodomain particles of uniaxial anisotropy has been calculated by Néel [5] and Brown [6] in early theoretical approaches. The mean lifetime $\bar{\tau}$ between consecutive magnetic switching events as a function of temperature T is characterized by its activation barrier E_b and attempt frequency ν_0,

$$\bar{\tau} = \nu_0^{-1} \exp\left(\frac{E_\mathrm{b}}{\mathrm{k_B} T}\right), \tag{11.1}$$

with k_B being the Boltzmann constant. In this model, a coherent rotation of all magnetic moments inside the particle is considered for the magnetization reversal, with E_b given by the total magnetic anisotropy of the particle. The prefactor ν_0 is commonly related to Larmor precession, with $\nu_0 \approx 10^{10}\,\mathrm{Hz}$ [6]. A systematic lifetime study is exemplarily shown in Fig. 11.2a. At a given temperature setpoint, the telegraphic noise has been recorded on the nanomagnet shown in the inset, revealing over 1000 switching events and yielding the mean lifetime $\bar{\tau}$. This procedure has been repeated for numerous temperature setpoints. As can be seen from the Arrhenius plot, the data perfectly agrees with the model given by (11.1), and fitting yields E_b and ν_0 for the given nanomagnet. Consequently, the magnetization switching behavior can be described by a macrospin model as depicted in Fig. 11.2b. The nanomagnet can be either found in one of the two monodomain ground states. Since the mean lifetime exponentially depends on the temperature T, small changes in temperature already result in a considerable change of the switching rate.

The dynamic system of a thermally switching nanomagnet can be used to study a huge variety of interactions of tiny magnets with the environment or passing electrons. Only in the undisturbed case the switching rate is given by the intrinsic activation energy barrier and attempt frequency of the nanomagnet. These parameters may strongly depend on the actual size and shape of the nanomagnet, as will be shown in Sect. 11.7. In a system of uniaxial magneto-anisotropy, a perfect symmetric bistable system is realized, since there is no preferred switching direction into one of the two ground states. However, tiny interactions can have a drastic influence on the switching behavior by lifting the activation barrier degeneracy or effectively increasing the temperature of the nanomagnet.

The detailed state-resolved lifetime analysis of telegraphic noise experiments provides a unique opportunity to study various interactions with (spin-polarized) electrons generated by a magnetic SP-STM tip or hosted by the substrate. Here, changes of E_b, ν_0 and of the effective temperature of the nanomagnet can be quantified on individual nanomagnets, yielding new insights into various microscopic interactions.

11.2 Current-Induced Magnetization Switching

As has been shown in the previous section, SP-STM is a powerful tool to investigate the magnetization dynamics with a spatial resolution down to the atomic scale. The same SP-STM tip that is used to read out the magnetization information at low tunneling currents may also serve as a highly localized source or drain of spin-polarized electrons at high tunneling currents, thereby manipulating magnetism and triggering magnetization dynamics on the atomic scale. [7]. Here, the magnitude of the spin-polarized tunneling current is tuned by distance variations between the magnetic probe tip and an individual nanomagnet.

In Fig. 11.3a, three traces of the magnetic signal at different setpoints for the spin-polarized tunneling current I are shown. They have been recorded with the tip positioned above the central region of a small Fe nanomagnet consisting of about 100 Fe atoms on a W(110) surface. At low tunneling current, the typical telegraphic noise is visible, with the signal statistically switching between two discrete levels. The histogram of the full traces, each containing more than 1000 switching events, is shown in Fig. 11.3b. For low tunnel current, both states occur with the same probability, as it is expected for the intrinsic superparamagnetic switching of a nanomagnet with two degenerate magnetic ground states. As the tunneling current increases, an

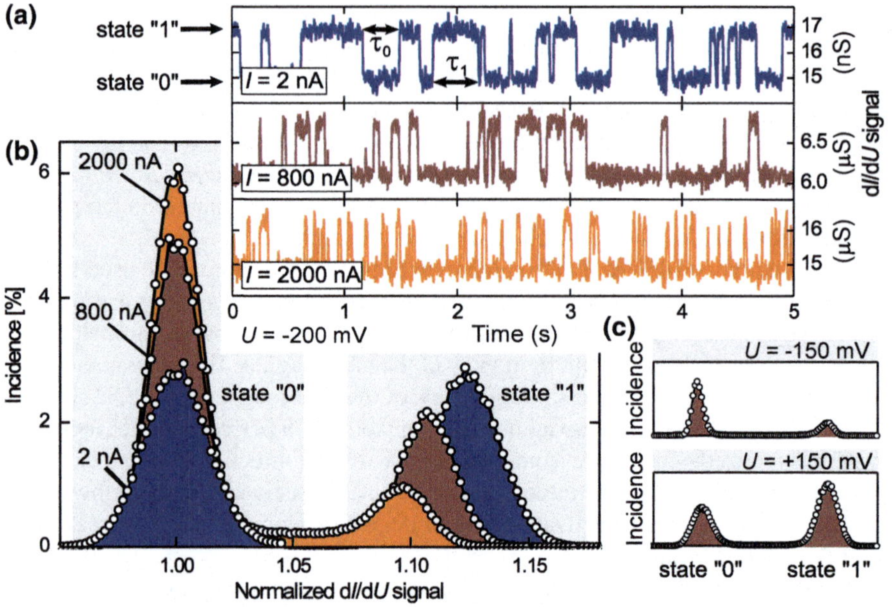

Fig. 11.3 **a** Trace section of the time-dependent magnetic dI/dU signal of one particular nanoisland (area: 5.7 nm^2) recorded with a Cr-coated probe tip at different tunneling currents I ($T = 48.4$ K, $U = -200$ mV). **b** Histogram of the overall magnetic signal normalized with respect to the state "0" level at different tunneling currents. While state "1" and state"0" are equally populated at low currents a significant asymmetry towards state "0" can be recognized at high currents. **c** Histogram of the normalized magnetic signal for $I = 600$ nA, recorded on a different nanomagnet. Inversion of U results in an inversion of the state population asymmetry. Figure adapted from [7]

imbalance between the two states builds up until one state clearly dominates. For $I = 2000$ nA, the magnetic signal is for most of the time on the lower level, indicating the nanomagnet being in the state "0", and magnetization reversals into the state"1" are immeadiately followed by another reversal back into state "0". A tunneling current flowing in the opposite direction results in an inversion of the asymmetry, as shown in Fig. 11.3c. Consequently, the interaction of the spin-polarized tunneling current with the nanomagnet leads to a directed switching. The scaling with the tunneling current is the first experimental demonstration of the spin-transfer torque generated by spin-polarized electrons that are tunneling through a vacuum barrier.

Quantification of Joule Heating and Spin-Transfer Torque

In order to elaborate the detailed influence of a high spin-polarized tunneling current on the magnetization dynamics, a systematic state-resolved lifetime analysis has been performed as a function of I on the nanomagnet shown in Fig. 11.4a. For each tunneling current setpoint I the telegraphic noise signal has been recorded, resulting in the respective state-resolved mean lifetimes $\bar{\tau}_0$ and $\bar{\tau}_1$ shown in Fig. 11.4b. At low tunneling current, $\bar{\tau}_0$ and $\bar{\tau}_1$ are very similar. At high tunneling currents the mean

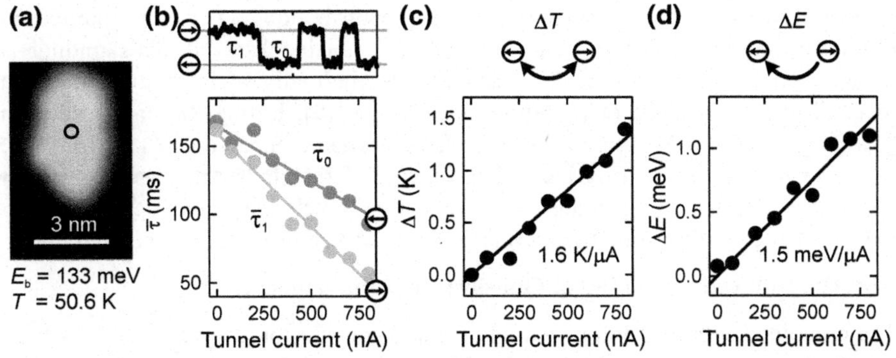

Fig. 11.4 **a** Topography image of an individual iron nanomagnet on W(110), with the effective energy barrier E_b and system temperature T indicated. **b** Telegraphic noise signal recorded when parking the SP-STM probe tip stationary above the nanomagnet at the position indicated by a circle in (**a**) (top) and state-resolved mean lifetime as a function of spin-polarized tunnel current (bottom). **c** Effective increase of the temperature of the nanomagnet due to Joule heating, ΔT, and **d** effective modification of E_b due to spin-transfer torque switching, ΔE, as function of spin-polarized tunnel current. A linear dependence is observed for both processes

lifetime strongly depends on the relative magnetization direction between the tip and the sample. Whereas the lifetimes are equal at low currents, a significant imbalance at high currents can clearly be recognized. Moreover, the mean lifetime of both states decreases with increasing current. This effect is attributed to Joule's heating of the high tunneling current. Hence, two mechanisms affect the switching behavior of the Fe/W(110) nanomagnets: Joule heating effectively increases the temperature by ΔT [8], and the spin-transfer torque modifies the activation energy barrier E_b by $\pm\Delta E$ [9], resulting in an asymmetry of the state lifetimes. To account for both, Joule heating and spin-transfer torque, the macrospin model is expanded to

$$\bar{\tau}_{0,1}(I) = \nu_0^{-1} \exp\left(\frac{E_b \pm \Delta E(I)}{k_B[T + \Delta T(I)]}\right), \tag{11.2}$$

where $\bar{\tau}_0(I)$ and $\bar{\tau}_1(I)$ are the mean lifetimes of state **0** and state **1**, respectively. Both contributions, ΔT and ΔE can be calculated from the state-resolved mean lifetimes [10]. The results as function of tunnel current setpoint I are shown in Fig. 11.4b, c, respectively. A linear increase of the effective temperature of the nanomagnet is observed, which is consistent with experimental studies of heat generation between an STM tip and a metallic sample [8]. For the nanomagnet an increase of the effective temperature by 1.6 K is found for 1 μA. Consequently, its thermal energy is increased by up to 3% of $k_B T$. This tiny temperature increase already considerably reduces the mean lifetimes by a factor of two due to the Arrhenius-like switching behavior. The Fe nanomagnet has a base area of about 5.5 nm^2, resulting in a temperature increase of 44 K/(mWμm^{-2}). This value is in accordance with STM experiments performed on Co/Cu(111) nanoislands, where the temperature increase has been roughly estimated to be 30–300 K/(mWμm^{-2}) [11].

The results for the spin-transfer torque contribution $\Delta E(I)$ are plotted in Fig. 11.4d. A linear scaling behavior with I is again observed. This finding is in accordance with theoretical studies on the thermally assisted magnetization reversal in the presence of a spin-transfer torque [9]. Fitting the data results in $\frac{d}{dI}\Delta E = 1.5\,\text{meV}/\mu\text{A}$. Again, this value is tiny when compared to $E_b = 133\,\text{meV}$. However, a high spin-polarized tunnel current of several hundred nA has a significant influence on the nanomagnet due to the Arrhenius switching behavior, clearly lifting the degeneracy of the ground states and resulting in a significant lifetime asymmetry.

Current-Induced Switching of Quasistable Nanomagnets

The combined action of Joule heating and spin-transfer torque generated by a high spin-polarized tunnel current can not only be used to modify the switching behavior of individual nanomagnets. It can also trigger a directed magnetization switching event for nanomagnets that exhibit a stable magnetization over more than half an hour [12]. At low temperature, the nanomagnets remain in one of their two degenerate magnetic ground states. In our experiments we increase the spin-polarized tunnel current by slowly approaching the magnetic tip to an individual nanomagnet while simultaneously recording the magnetic signal, as shown in Fig. 11.5a. When

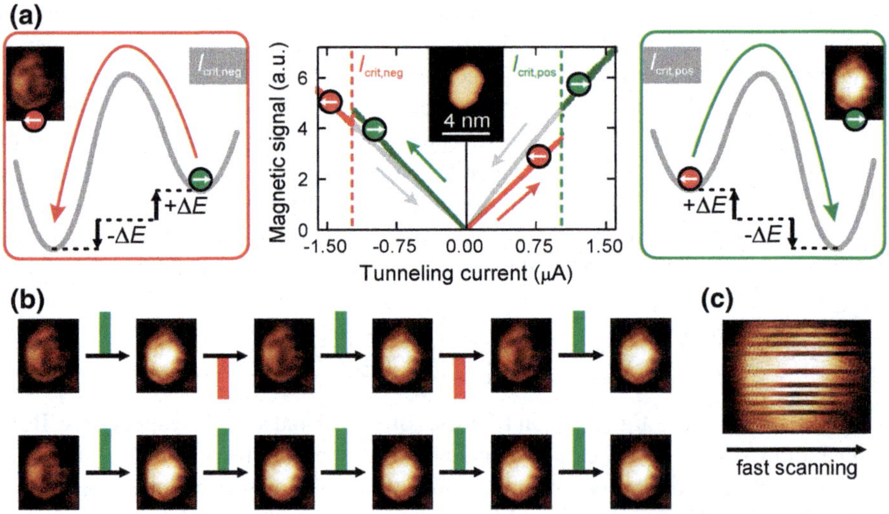

Fig. 11.5 **a** Switching a Fe/W(110) nanomagnet by increasing the spin-polarized tunnel current ($U = \pm 180\,\text{mV}$). Ramping the spin-polarized tunnel current results in a directed switching when exceeding the threshold. Decreasing the current from high values back to zero leads to no reversal (gray line). **b** Current pulse experiments on a Fe nanomagnet. Applying short high spin-polarized tunnel current pulses initiates magnetization reversal. Depending on the bias polarity, the nanomagnet can be switched into the desired state. Lower panel: Counter experiment without changing the bias polarity. Just the first pulse results in a magnetization reversal, but no back-switching events are observed. **c** Magnetic map of the nanomagnet. During scanning high current pulses have been applied to the center of the nanomagnet to reverse the magnetization, resulting in an artificially created stripe pattern. Figure adapted from [12]

exceeding a certain threshold current the nanomagnet switches, as indicated by a sudden jump of the magnetic signal. Depending on the bias polarity, the magnetization is switched into the parallel or antiparallel configuration with respect to the fixed tip magnetization and remains in this state even for higher currents. This finding clearly shows that the nanomagnet is not only destabilized by the Joule heating resulting from the high current. If that would be the case the nanomagnet is expected to undergo numerous additional switching events when further increasing the tunnel current. However, only one event is observed. The spin-transfer torque induces a directed switching and stabilizes the nanomagnet after switching, as depicted by the energy schematics in Fig. 11.5a. When decreasing the current back to low values, no further switching event is observed. Consequently, ramping the spin-polarized tunnel current can be used to selectively switch the magnetization of a single atomic-scale magnet into one of its two ground states.

An alternative approach to selectively switch the magnetization of a quasistable nanomagnet is the application of short high spin-polarized tunnel current pulses. In Fig. 11.5b the magnetic maps (taken at a low tunnel current setpoint) of the nanomagnet in a) are shown before and after application of high spin-polarized tunnel current pulses. As can be seen from the series, alternating the bias polarity ($U = \pm 180\,\mathrm{mV}$) results in a reversal event triggered by the pulse (length of 200 ms). For positive bias (green), the nanomagnet switches into the high conductance state, whereas it switches into the low conductance state for negative bias (red). In the lower panel, a series of magnetic maps is shown for applying high current pulses of fixed bias polarity. Here, only the first pulse results in a reversal, whereas the nanomagnet remains in the state of high conductance for the other pulses. This finding shows that the nanomagnet undergoes a directed switching during the application of the pulses, which is a consequence of the spin-transfer torque exerted by the spin-polarized tunnel current. In Fig. 11.5c the same nanomagnet is imaged using conventional SP-STM at low spin-polarized tunnel current. The scanning process has been interrupted every 20 lines for the injection of a short spin-polarized tunnel current pulse right at the center of the nanomagnet. As a result, the magnetic map exhibits an artificial stripe pattern perpendicular to the fast scanning direction. These stripes are the manifestation of the frequent magnetization reversals triggered by the pulse applications.

Vision for Future Data Storage Devices

As the same magnetic tip of an SP-STM can be used to read out and write magnetic information into individual nanomagnets, it is straightforward to envision a future data storage device that allows for a much higher data density than in today's hard disk drives. Conventionally, reading and writing information is realized by local magnetic fields that are sensed or generated by a combined read/write head with an integrated electric circuit. In the vision, this complex read/write head design is replaced by a single magnetic needle, as illustrated in Fig. 11.6. It is electrically connected to a pulse generator that has not to be part of the moving probe. The needle scans above atomic-scale magnets with uniaxial anisotropy, and the digital information stored in an individual nanomagnet is read out by applying a short low bias pulse, resulting in a spin-polarized tunnel current that depends on the magnetic orientation of the

Fig. 11.6 Artificial illustration: Vision of a hard disk drive with data readout and writing realized by a low or high spin-polarized tunnel current between a magnetic needle and individual nanomagnets. Each of the nanomagnets represents a magnetic bit, either being in state "0" or"1"

nanomagnet in terms of magneto-resistance. For writing, a high bias pulse of matching polarity is applied to the magnetic needle, thereby switching the nanomagnet into the desired magnetic state. Simultaneously, the success of writing is verified by measuring the magneto-resistance. As SP-STM allows for atomic resolution, this concept is scalable down to the single-atom limit, thereby realizing the ultimate data storage density of one bit per atom.

11.3 Spin Transfer-Torque Based Pump-Probe Experiments

The current-induced magnetization reversal triggered by SP-STM is a powerful method to control the magnetic state of an individual nanomagnet. The application of high spin-polarized tunnel current pulses forces the magnetization of individual nanomagnets into one preferred state. This concept can be used for controlled spin-pumping to drastically increase the temporal resolution of SP-STM [13]. Here, pump-probe schemes are applied, with a pump pulse driving the nanomagnet into one of its two degenerate ground states and afterwards probing its mean magnetic state by a low-current probe pulse as a function of pump-probe delay.

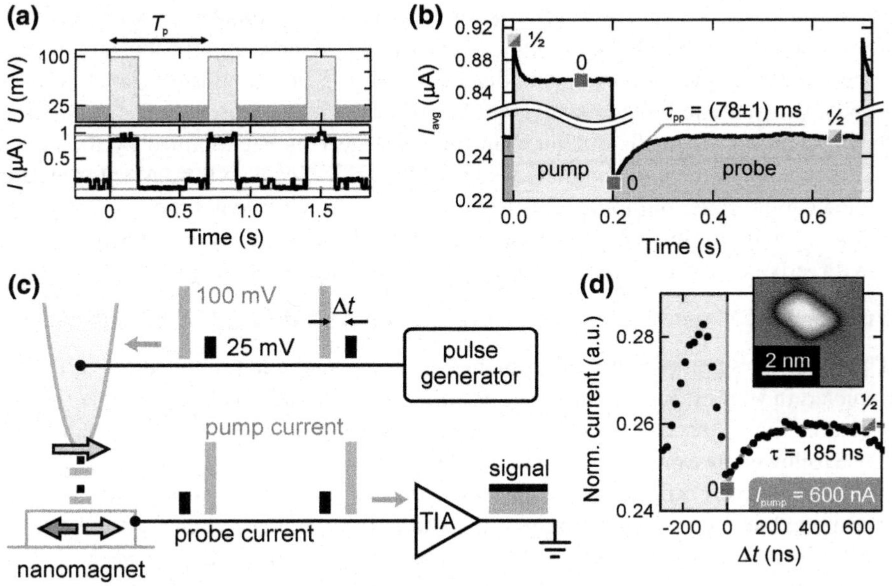

Fig. 11.7 **a** Top: Periodical sequence of high and low tunnel bias. Bottom: Resulting telegraphic noise on a thermally switching nanomagnet at constant tip-sample distance. **b** Averaged noise signal during one cycle, calculated from $I(t)$ in (**a**). **c** Concept of spin-transfer torque based pump-probe experiments using SP-STM. **d** Results of the pump-probe experiment on the nanomagnet shown in the inset. From the evolution of the probe signal the mean lifetime can be determined. Figure adapted from [13]

Real-Time Observations During Pump-Probe Experiments

The concept of spin-transfer torque based pump-probe SP-STM studies is illustrated in an experiment, where the evolution of the magnetic signal is recorded during and after pumping a nanomagnet with a high spin-polarized tunnel current pulse. A periodically modulated bias voltage U (with period T_p) is applied between the nanomagnet and the magnetic SP-STM tip being held at constant distance to record the telegraphic noise during and after the injection of a high spin-polarized tunnel current. Consequently, the spin-polarized tunnel current periodically changes between low and high values. The results of such an experiment are shown in Fig. 11.7a. Within the low-current and high-current sequences, a telegraphic noise is observed in the spin-polarized tunnel current, reflecting the switching behavior of the nanomagnet.

Averaging $I(t)$ over numerous cycles results in $I_{avg}(t)$ shown in Fig. 11.7b. The telegraphic noise is not present any more, and an exponential decay is found in $I_{avg}(t)$ after every change in U for pumping and probing. In each cycle, the magnetization is aligned when changing from the low to the high I regime, and it thermally relaxes when switching back into the low I regime. Initially ($t = 0$), the nanomagnet is in thermal equilibrium (state 1/2). During the high spin-polarized current pulse it is driven into the magnetic state **0**, as indicated by the significant decrease of I_{avg} during the pulse. When the pulse ends, the nanomagnet thermally relaxes back into the state $^1/_2$, as can be seen from the temporal evolution of I_{avg}. Consequently, the nanomagnet is pumped into one preferred state during the high I pulse, and its relaxation back into its thermal equilibrium state is probed after the pulse at low I. The decay time during pumping is much smaller than for probing. This is explained by the higher switching rate of the nanomagnet at high I due to significant spin-transfer torque and Joule heating. After pumping, the characteristic magnetic decay time is given by the intrinsic mean lifetime of the nanomagnet, divided by two. Hence, the mean lifetime can directly be determined by measuring the evolution of the magnetic probe signal.

Probing Fast Magnetization Dynamics with Pump-Probe SP-STM Experiments

The setup for a pump-probe experiment with spin-polarized tunnel current pulses is depicted in Fig. 11.7c. The feedback loop is switched off and U is set to zero after adjusting the distance between tip and sample. Consecutively, a sequence of a pump pulse followed by a probe pulse of low I at a given delay time Δt is generated, and this pump-probe cycle is repeatedly applied to the tunnel junction. As the bandwidth of the transimpedance amplifier is limited, the short spin-polarized tunnel current pulses are not resolved individually, and the average during the cycle is measured. This averaged spin-polarized tunnel current during the probe pulse provides a measure of the alignment between tip and sample magnetization, averaged over the (fixed) probe pulse length. The experiment is repeated for different Δt, thereby mapping the evolution of the magnetic state after pumping.

An exemplary pump-probe measurement on a nanomagnet is shown in Fig. 11.7d. The variation in I was recorded as a function of Δt between pump and probe spin-polarized tunnel current pulses that were injected into the nanomagnet. A characteristic triangular-shaped feature is observed for overlapping pump and probe pulses, resulting from the non-linearity in the $I(U)$ curve of the tunnel junction [14, 15]. For $\Delta t > 0$, no overlap is detected, and the tunnel current variation is given by the average current during the probe pulse. After the pump pulse, an exponential decay is observed. Obviously, the nanomagnet decays back into thermal equilibrium (mean state $^1/_2$) for $\Delta t > 0$. This is a consequence of the combined action of spin-transfer torque and Joule heating during the pump pulse that forces the nanomagnet into the state **0**. For the given nanomagnet, fitting the data with an exponential reveals $\tau = 185$ ns. Such short lifetimes cannot be resolved in conventional telegraphic noise experiments. Consequently, pump-probe schemes allow for a considerable extension of the temporal resolution in SP-STM experiments. The limit for the shortest resolvable lifetimes is only given by the rise time of the bias pulses and the high frequency capabilities of the cabling to the SP-STM probe tip.

In contrast to conventional pump-probe experiments where a system is pumped into an excited state and the relaxation back into the ground state is probed, in our experiments the ground state dynamics is observed without excitation into higher-energy states. Instead, the system is pumped into a state of lower entropy, and the relaxation back into the high entropy state is observed. Hence, using the spin-transfer torque based pump-probe technique allows for the quantification of state-resolved lifetimes for nanomagnets exhibiting magnetization dynamics between degenerate ground states. A lifting of the energy degeneracy of the ground states by an external magnetic field is not needed in this approach.

11.4 The Oersted Field Induced by a Tunnel Current

Using an SP-STM tip as the source/drain for spin-polarized electrons allows spatially resolved current-induced magnetization switching experiments to be performed by moving the tip to different sites of one particular nanomagnet while observing its magnetization dynamics. Thereby, information of site-specific properties can be gained which cannot be obtained in spatially averaging experiments performed with nanopillars [7].

While scanning a Fe nanomagnet consisting of about 100 atoms on W(110) the magnetic telegraphic noise is recorded for a duration of 12 s on each of the pixels. From the noise the site-specific lifetime asymmetry a_{H} is calculated on the basis of the corresponding datapoint histograms. The results are shown in a grayscale-coded representation in Fig. 11.8a. The data reveal a clear gradient along the [001] direction. For a quantitative analysis a_{H} is averaged column- and row-wise on the nanomagnet. When moving the tip along the $[1\bar{1}0]$ direction, $\bar{a}_{\mathrm{H}}$ is constant within the error, but it clearly reduces from the left to the right side of the nanomagnet.

The electrons will pass to the substrate, forming a conventional electric current after tunneling into the nanomagnet. This current gives rise to an Oersted field that surrounds the axis of current flow on a plane parallel to the sample surface. As

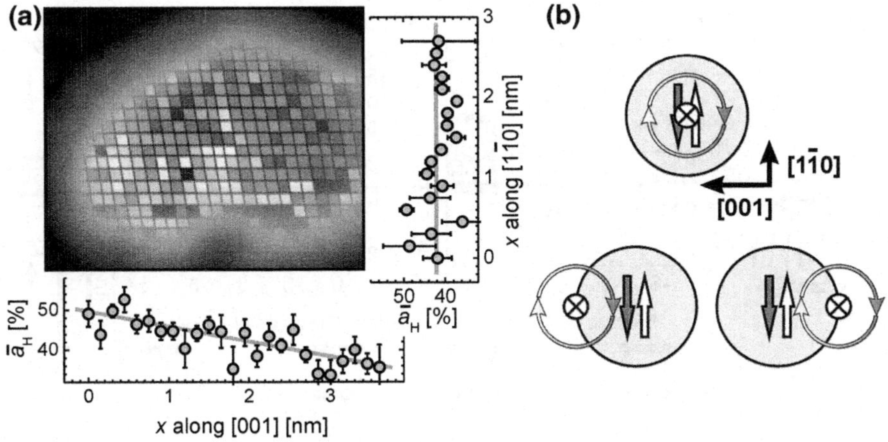

Fig. 11.8 **a** Map of the current-induced lifetime asymmetry a_{H} as measured with a Cr-coated probe tip at $I = 600\,\mathrm{nA}$ ($T = 55.0\,\mathrm{K}$). The plots show a_{H} averaged in rows and columns. **b** The schematic top view drawings illustrate the influence of the Oersted field with the tip positioned in the center and at the charge-free sides of the nanomagnet. In the latter case, the Oersted field results in a directed switching of the magnetization. Figure adapted from [7]

the experiments indicate, this Oersted field influences the switching behavior of a nanomagnet, depending on the site of current injection by the magnetic probe tip. In Fig. 11.8b, this influence of the lateral tip position on the magnetization dynamics is sketched schematically. The influence of the Oersted field cancels if the tip is positioned above a high symmetry point, and the nanomagnet experiences pure spin-current induced switching. When moving the tip from the center along the [001] direction one magnetic state is favoured over the other by the Oersted field. In this case the tip position affects the magnetic switching behavior by means of the Oersted field. As detailed analysis of the data shows, the Oersted field modifies the activation energy barrier of up to ± 0.7 meV when moving along the [001] direction, whereas the pure spin-transfer torque results in a modification of the barrier by -2.4 meV at the center of the nanomagnet. This finding indicates that the magnetization switching is dominated by the spin-transfer torque induced by the spin-polarized current. The influence of the Oersted field remains small, but is clearly detectable in the experiments.

11.5 Electric Field-Induced Magnetoelectric Coupling

In conventional STM experiments, a biased tip is approached into tunnel contact with the sample under investigation. Naturally, this is accompanied by an electric field that builds up between the tip and the sample, as depicted by Fig. 11.9a. As the distance between tip and sample in STM experiments is very small, a considerable

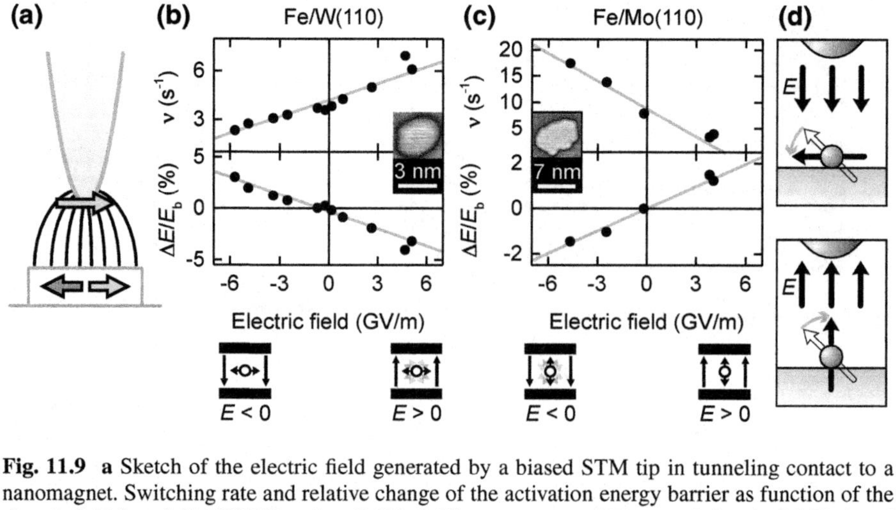

Fig. 11.9 **a** Sketch of the electric field generated by a biased STM tip in tunneling contact to a nanomagnet. Switching rate and relative change of the activation energy barrier as function of the electric field for a **b** Fe/W(110) and a **c** Fe/Mo(110) nanomagnet. **d** Picture of electric field-induced change of magnetic anisotropy. Depending on the polarity, an out-of plane or in-plane magnetic anisotropy is generated due to magneto-electric coupling. Figure adapted from [19]

electric field on the order of GV/m is expected to be present during the experiment. However, effects of the electric field on the properties of the sample are generally neglected in STM experiments. On the other hand manipulating magnetic properties with an electric field is in the focus of numerous ongoing research activities in the field of spintronics. A promising approach to solve the fundamental dilemma in data storage applications is tuning the magnetic anisotropy by an electric field: Whereas a large anisotropy helps to stabilize a magnetic bit against thermal agitation, a low anisotropy is desired during magnetization reversal when writing information. The magnetic anisotropy can be maximized or decreased by an electric field [16–18] for stabilizing a magnetic bit for long-term storage or facilitating magnetization reversal when writing information. Such devices could benefit from smaller power consumption compared to conventional magnetic field or spin-transfer torque based devices, as virtually no current is needed for this electric-field based scheme. In SP-STM experiments a well-defined, atomically sharp magnet-to-vacuum interface is realized to address single, defect-free magnets on the atomic scale. It can be used for a systematic study of electric field induced manipulation of magnetic properties [19].

Depositing iron onto W(110) and Mo(110) substrates leads to the pseudomorphic growth of atomic-scale monolayer nanomagnets. On W(110) they exhibit an in-plane uniaxial magnetic anisotropy with the easy axis of magnetization lying in the $[1\bar{1}0]$ direction [20, 21]. In contrast, when prepared on Mo(110), the easy axis is pointing out of the surface plane [22].

The switching behavior of individual Fe nanomagnets grown on W(110) and Mo(110) substrates as a function of applied bias has been recorded using time-resolved SP-STM. Using the simple model of a parallel plate capacitor, the electric field E is calculated to $E(U) = U/d(U)$, where U is the bias voltage and $d(U)$ is the (bias-dependent) distance between the tip and the sample at a constant tunnel current. As d only slightly changes with U, the electric field already changes dramatically up to $E = \pm 6\,\text{GV/m}$ at $U = \pm 5\,\text{V}$. From the telegraphic noise the mean lifetime τ is extracted for every value of U (and respective E). The results are shown in the upper panels of Fig. 11.9b and c for a nanomagnet placed on W(110) and Mo(110), respectively.

Surprisingly, the switching rate increases for positive electric fields for the nanomagnet on W(110), but decreases for Mo(110). This finding is not in agreement with current-induced effects since the applied spin-polarized tunnel current is low. Hence, spin-transfer torque and Joule heating are negligible. Consequently, the observed modification of the switching rate is attributed to the coupling of the nanomagnet to the electric field E [19]. From the E-dependent mean lifetime the respective relative change $\Delta E/E_b$ of the activation energy barrier can be calculated. It is shown in the lower panels of Fig. 11.9b, c. Obviously, the energy barrier is decreased at positive electric fields for the in-plane system, but increases for the out-of-plane system. The electric field results in a charge redistribution at the surface of the nanomagnet. Whereas positive fields lead to a depletion, negative fields result in an accumulation of electrons. This in turn causes a shift of the Fermi level and a modification of the occupation of the 3d bands which are responsible for magnetism [23]. In the

SP-STM experiment the electric field modifies the effective activation energy barrier E_b for magnetization reversal. An additional magnetic anisotropy is induced by the electric field, thereby changing E_b. In a simple model it superimposes onto the intrinsic anisotropy of the nanomagnet. It is reasonable to assume that the mechanism of magneto-electric coupling is the same on Fe/W(110) and Fe/Mo(110), as the lattice constant, chemical properties and thus the electronic structure of molybdenum are similar compared to tungsten. The magnetoelectric coupling favors in-plane magnetization for $E < 0$, whereas an out-of-plane magnetization is preferred for $E > 0$, as illustrated in Fig. 11.9d. Consequently, a distinct electric-field induced modification of the switching behavior is observed when changing between the two systems.

The experiments show that the electric field generated by a biased STM probe tip influences the magnetic properties of nanomagnets due to magneto-electric coupling, and magnetization dynamics can change significantly. Only at vanishing electric field the intrinsic magnetic properties of the system under investigation are observable.

11.6 Spin-Polarized Field Emission

Numerous spin-electronic applications rely on the interaction of nonequilibrium hot-electron spins with magnetic solids [24–26]. A magnetic SP-STM probe tip can be used to generate spin-polarized field emission for the imaging of the magnetic surface of bulk-like samples via their spin-split field-emission resonance (FER) states [27]. The question arises whether magnetic structures ultimately being only one atomic layer in height can be imaged as well and how they will be affected by the spin-polarized field emission current.

The basic idea of resonantly tunneling from a scanning probe tip into a FER of a nanomagnet is illustrated in Fig. 11.10a. The FER is located outside the magnet *in vacuo*, being up to several nanometers away from the surface. Consequently, electrons that are emitted from the tip are not directly injected into the sample but tunnel resonantly into the FER and consecutively relax into the sample. The FER of a nanomagnet is spin-split, as shown in the energy schematics in Fig. 11.10b. The resonance condition for spin-up electrons is fulfilled at an energy that differs from that for spin-down electrons. Depending on the relative orientation between the tip and sample magnetization, this lifting of the energy degeneracy of the spin-up and spin-down channel results in a variation of the spin-polarized field emission current. In Fig. 11.10c, a typical field emission spectrum is shown for a Fe/W(110) nanomagnet [28]. Here, the emission current feedback loop is closed in order to realize an almost constant electric field between tip and sample. As can be seen from the data, the tip is retracted by more than 3 nm when increasing the bias from 0 to 10 V. Steps are observed whenever a peak is detected in the simultaneously obtained channel of differential conductance dI/dU. These features are attributed to the FER above the nanomagnet: It adds a channel of conductance to the junction at distinct energy levels, whenever the resonance condition is fulfilled. A close-up of the first FER is shown in Fig. 11.10d. Numerous spectra are plotted on top of

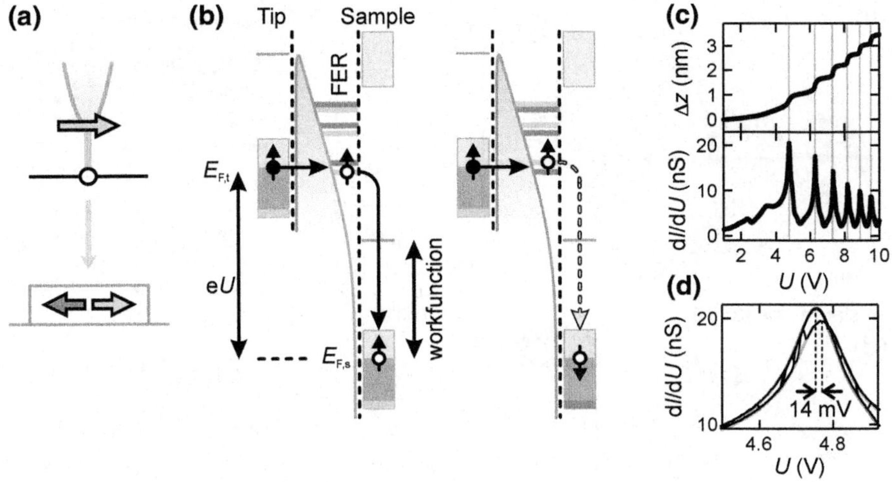

Fig. 11.10 **a** Sketch of spin-polarized field emission of a magnetic probe tip into field emission resonances located above a nanomagnet. **b** Physical picture of resonantly injecting spin-polarized field-emitted electrons for parallel (left) and antiparallel (right) alignment of tip and sample magnetization. **c** Spectroscopy data on a Fe nanomagnet, revealing the first six field emission resonances. **d** First FER peak. It changes as the nanomagnet switches its magnetization. Figure adapted from [28]

each other, revealing a frequent switching between two curves that are shifted by about 14 meV with respect to each other. This finding is attributed to the thermally switching nanomagnet, resulting in an inversion of the FER spin channels with every reversal event. Consequently, the magnetic state of the nanomagnet can directly be read out by measuring the differential conductance of the spin-polarized field emission from the tip to the nanomagnet at enlarged tip-sample distances compared to the tunneling regime. Note that the spin-up and spin-down FER states overlap due to their considerable width, resulting in a non-zero conductance for both the spin-up and spin-down electrons even at energies that are off-resonant to the FER states.

In Fig. 11.11a, an ensemble of thermally switching Fe/W(110) nanomagnets is observed using spin-polarized field-emitted electrons at $U = 4.6\,\text{V}$. The topography is textured with the simultaneously recorded magnetic map of differential emission conductance. On the nanomagnets, a telegraphic noise is observed due to the thermally induced magnetization reversals, resulting in a characteristic stripe pattern. Parking the magnetic probe tip above one individual nanomagnet, the mean lifetime of each magnetic state can be determined from the telegraphic noise [28]. In Fig. 11.11b, the results are shown for different values of the emission current. When increasing I from 2 to 140 nA, the mean lifetime drops by about three orders of magnitude. Obviously, this experimental finding indicates a considerable Joule heating generated by the field-emitted electrons. Additionally, an increasing lifetime asymmetry $a_\tau(I) = (\bar{\tau}_1 - \bar{\tau}_0)/(\bar{\tau}_1 + \bar{\tau}_0)$ of up to 40 % is observed with increasing current. Consequently, one magnetic orientation of the nanomagnet is favored at the cost of

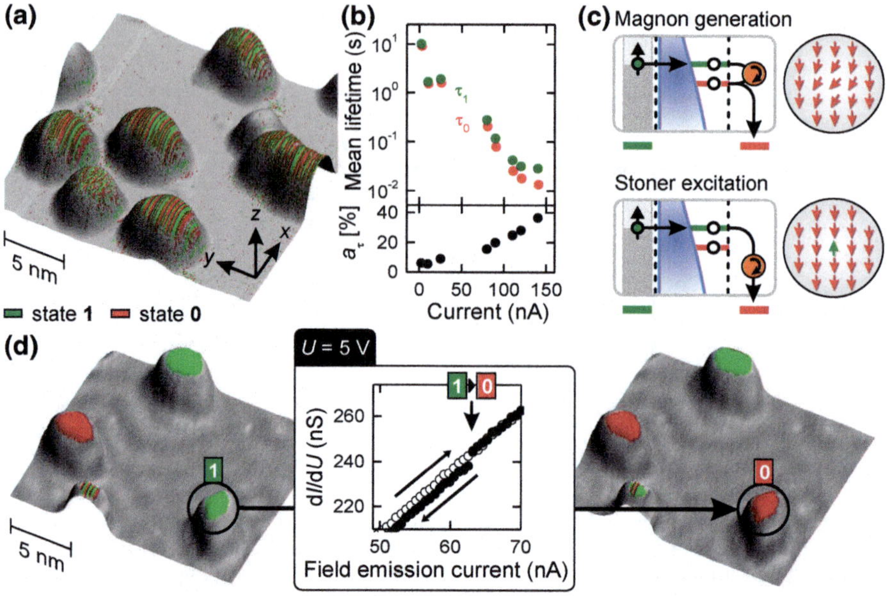

Fig. 11.11 Spin-polarized scanning field emission microscopy of nanomagnets: **a** Topography textured with the magnetic map. **b** State-resolved mean lifetime of an individual nanomagnet affected by a spin-polarized field emission current (top) and evolving lifetime asymmetry (bottom). **c** Inelastic spin-flip processes of a spin-polarized electron. Top: When flipping spin within a FER, a low energy magnon is excited in the magnet. Bottom: When flipping spin while relaxing into the sample, a high energy Stoner excitation is generated. **d** Switching the magnetization of a nanomagnet using a spin-polarized field emission current. SP-STM images before (left) and after (right) ramping the emission current reveal the switching event of the nanomagnet under investigation (circle). Figure adapted from [28]

the other. This purely magnetic feature is a result of the spin-transfer torque generated by spin-polarized field emission. The microscopic processes of the spin-transfer torque observed in field emission differ from those in the tunneling regime, the latter predominantly creating magnons [29]. For field-emitted spin-polarized electrons a two-step process is expected. First they are injected into the spin-split FER and subsequently relax into the sample. Consequently, for a simple model with an assumed spin polarization of one, the electron has to flip its spin in order to relax into the sample when the magnetizations of the tip and the sample are in the antiparallel configuration. This is accomplished by interactions with the sample, either by relaxing within the FER spin states and generating low-energy spin waves (magnons) [30], or when relaxing from the FER state into the sample, resulting in higher-energetic single-site spin flips (Stoner excitations) [31], as shown in the schematics of Fig. 11.11c. The sample magnetization is destabilized by both processes, and ultimately a reversal is triggered, resulting in a directed switching accompanied by an asymmetry of the lifetimes.

The high impact of field-emitted electrons onto the switching behavior of nanomagnets can be used to trigger magnetization reversal of nanomagnets that are otherwise thermally stable, as shown in Fig. 11.11d. The SP-STM image shows four nanomagnets, with three of them exhibiting a stable magnetization. The tip is positioned above a nanomagnet being in the magnetic state **1**. From a long-term observation with low-energy tunneling electrons the intrinsic lifetime on this nanomagnet is found to be on the order of an hour. The spin-polarized field emission current is then ramped up and down at moderate values between 2 and 70 nA while simultaneously recording the magnetic signal. As can be seen from the data, with increasing emission current the magnetic signal increases, reflecting the enhancement of conductance when decreasing the tip-sample distance. Within the curve, a sudden jump is observed, indicating a magnetic switching event of the nanomagnet. A subsequently taken SP-STM image on the same area reveals that the nanomagnet in fact has switched to the magnetic state **0**, whereas the surrounding remained unchanged [28]. Consequently, spin-polarized field emission severely decreases the state lifetime of the nanomagnet, ultimately resulting in the magnetization becoming unstable and flipping its orientation.

11.7 Magnetization Dynamics of Quasiclassical Magnets

Spin-polarized scanning tunneling microscopy is a powerful tool to investigate the magnetization dynamics on the very local, atomic scale. Especially, it can be used to observe the switching behavior of individual nanomagnets over a very wide temperature and switching rate regime. For the system of Fe/W(110) nanomagnets, variable temperature SP-STM is applied to gain a detailed understanding of the microscopic processes involved in the magnetization reversal on the atomic scale [4, 13].

Evaporating 0.14 atomic layers of iron onto a W(110) substrate held at room temperature leads to the formation of pseudomorphically grown nanomagnets which have a typical diameter between 2 nm and 6 nm, thereby consisting of about 30–150 atoms. In Fig. 11.12a, a topography patchwork of eleven nanomagnets of different size and shape is shown. Positioning the magnetic probe tip on top of each nanomagnet results in a characteristic telegraphic noise. Recording more than a thousand switching events allows the determination of the mean lifetime. As discussed in Sect. 11.1, a temperature dependent study of the telegraphic noise yields the effective activation energy barrier E_b and prefactor ν_0. These parameters have been determined for each individual nanomagnet. They are summarized in Fig. 11.12b. Apparently, there is a huge variation both in E_b and ν_0 with no obvious scaling behavior. However, when plotting E_b as a function of the nanomagnet length along the $[1\bar{1}0]$ direction $N_{[1\bar{1}0]}$, a clear linear dependence is visible, as shown in Fig. 11.12c. This scaling behavior indicates that reversal takes place via the nucleation and propagation of a domain wall rather than via a collinear rotation of all spins. The domain wall aligns along the $[1\bar{1}0]$ direction and consequently moves along the [001] direction. As $N_{[1\bar{1}0]}$ increases, the respective length of the domain wall (and therefore its energy) increases accordingly.

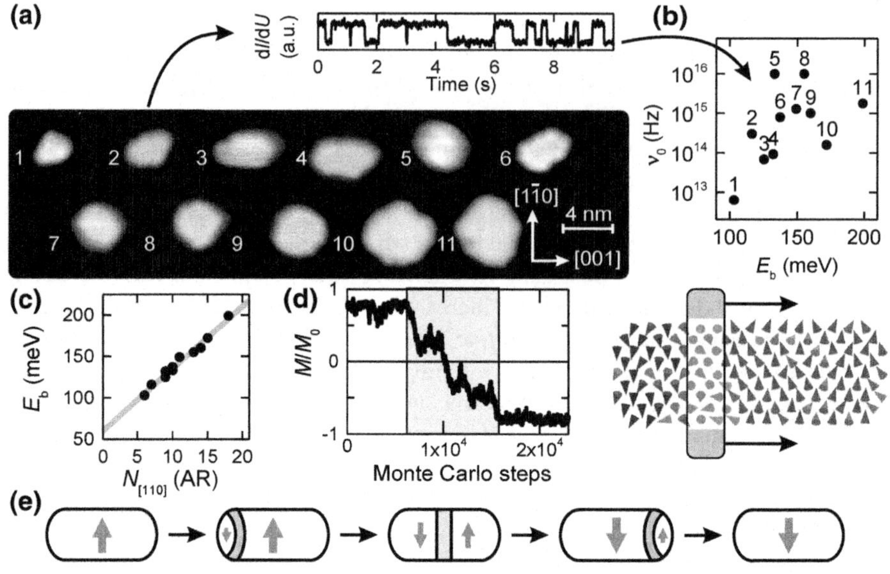

Fig. 11.12 **a** Topography maps of eleven Fe/W(110) nanomagnets. On each nanomagnet, a telegraphic magnetic noise is observed. **b** Effective activation energy barrier E_b and attempt frequency ν_0, resulting from a temperature-dependent study. **c** E_b as function of the nanomagnet length along the [1$\bar{1}$0] direction, in atomic rows (AR). **d** Monte Carlo simulation results: Evolution of magnetization during a switching event. Right panel: Snapshot of the spin configuration during the reversal. **e** Model of magnetization reversal in a nanomagnet, realized via nucleation and propagation of a domain wall. Figure adapted from [4]

As the magnetization reversal process is much faster than it can be resolved by SP-STM, the microscopic details of the spin dynamics during a switching event are not experimentally resolved. However, Monte Carlo simulations of the thermally activated magnetization reversal in Fe/W(110) nanomagnets have been performed, thereby resolving the detailed magnetization reversal processes at any given simulation time, as shown in Fig. 11.12d. The snapshot in the right panel of Fig. 11.12d shows the magnetic configuration of the simulated nanomagnet during reversal. Here, a domain wall separates a spin-up domain from a spin-down domain. Composing a movie from many consecutive snapshots reveals that the domain wall propagates from one end of the nanomagnet to the other, thereby resulting in a reversal of the magnetization [4].

In Fig. 11.12e, the microscopic processes for magnetization reversal in a nanomagnet are depicted. The nanomagnet is in a monodomain state initially, and the reversal is triggered by the coherent rotation of several magnetic moments within a small nucleus. This nucleus is separated from the monodomain by a domain wall. It can propagate along the [001] direction through the whole nanomagnet, following Boltzmann probability. A domain wall propagation from one end of the nanomagnet to the other leads to a net magnetization reversal of the nanomagnet. Likewise, for a

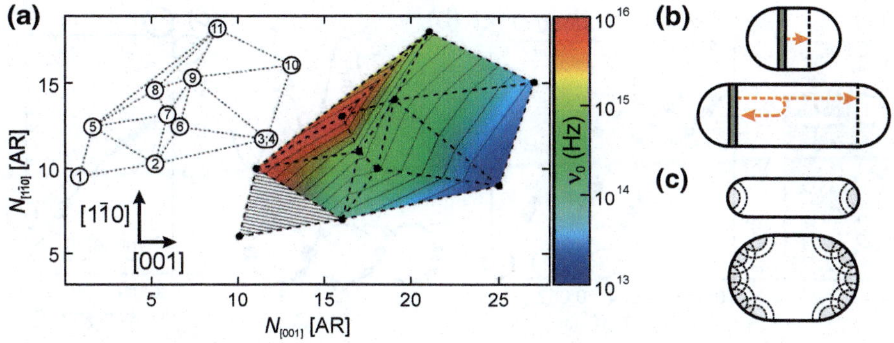

Fig. 11.13 **a** Experimentally determined prefactor of the nanomagnets under investigation, plotted as function of their lateral dimensions, and interpolated color-coded contour plot. See labeled mesh inset for nanomagnet correlation. (AR: atomic rows). Physical picture of the prefactor scaling. **b** The longer the distance between nucleation and annihilation site, the lower the prefactor. **c** The wider the ends of the nanomagnet, the higher the number of nucleation sites, resulting in an increase of the prefactor. Figure adapted from [4]

domain wall that annihilates at the nucleation site the initial magnetic configuration is restored, resulting in no reversal.

In Fig. 11.13a, the experimentally determined prefactor is plotted as function of the nanomagnet dimensions. Obviously, the prefactor decreases with increasing dimension along the [1$\bar{1}$0] direction, but increases with increasing dimension along the [001] direction $N_{[001]}$. In accordance to Monte Carlo simulations the following physical model is developed: The domain wall can be described as a quasiparticle propagating through the nanoisland. In the experiments no external magnetic field is applied. The absence of external forces implies that moving forward and backward is energetically degenerate. Random walk theory for a particle moving along a line with absorbing ends [32] is applied, yielding the probability for a successful propagation of a domain wall from one end of the nanomagnet to the other. It decreases with increasing $N_{[001]}$, as depicted in Fig. 11.13b. For very elongated nanomagnets it is likely that the domain wall returns to its nucleation site and annihilates there, with no net magnetization reversal. The number of nucleation sites for magnetization reversal increases with increasing $N_{[1\bar{1}0]}$, and every additional nucleation center increases the probability for a magnetization reversal. In summary, the prefactor ν_0 is determined by both, the probability of a domain wall to diffuse through the whole nanomagnet and the number of nucleation sites [4].

The temporal resolution of SP-STM can be drastically increased with pump-probe schemes, as described in Sect. 11.1. On the system of Fe/W(110) nanomagnets, nanosecond time resolution is easily achieved without any HF optimization of the STM cabling. The magnetic switching behavior can be investigated over a wide temperature and switching rate regime when combining telegraphic noise experiments and pump-probe schemes [13]. In Fig. 11.14a, the result of such an extensive study on three individual nanomagnets is shown in an Arrhenius plot. Here, for the

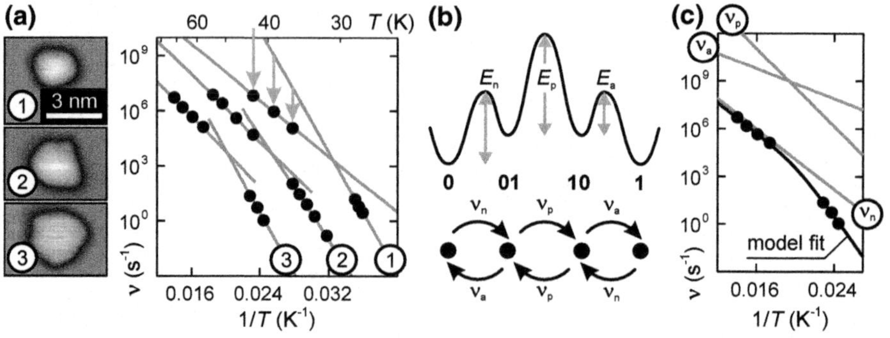

Fig. 11.14 **a** Switching rates of the three nanomagnets (left), as experimentally determined by SP-STM over a wide temperature and switching rate regime. At high temperatures, the switching rates are found to be by orders of magnitude lower than expected (indicated by arrows). **b** Four-state hopping model: Energy landscape (top) and hopping rates (bottom), modelling nucleation, annihilation and propagation of a domain wall in a nanomagnet. **c** Fitting the model to the experimental data: The effective switching rate is given by the temperature dependent rates for nucleation, annihilation and propagation. Figure adapted from [13]

low switching rate regime below $10^3\,\mathrm{s}^{-1}$ the telegraphic noise has been analyzed, whereas for the high switching rate regime up to $10^8\,\mathrm{s}^{-1}$ pump-probe schemes have been applied. As can be seen from the data, each of the nanomagnets exhibits a characteristic switching behavior, following two distinct Arrhenius laws for the low and high switching rate regimes. The effective activation energy barrier E_b for magnetization reversal is given by the slope of the lines in the Arrhenius plot. Consequently, the data indicate that E_b of every nanomagnet is constant within the low and the high T range, respectively. Between both ranges, a transition regime within a T window of about 10 K is observed. Interestingly, when extrapolating the Arrhenius behavior at low T to high T, ν is generally found to be by orders of magnitude lower than expected.

The microscopic processes which are involved in the magnetization reversal inside the nanomagnets are given by nucleation, annihilation and propagation [13]. An analytical four-state hopping model is developed that accounts for these processes. Its scheme is depicted in Fig. 11.14b. At any time, the nanomagnet can be found in one of four magnetic states: **0**, **01**, **10** and **1**. Here, **0** and **1** are the two monodomain ground states, and metastable states **01** and **10** correspond to situations with the magnetization within a nucleus volume being reversed with respect to the rest of the nanomagnet. Hopping between the states at intrinsic rates ν_n, ν_a and ν_p models the microscopic processes of nucleation, annihilation and propagation, respectively. Each hopping rate follows an Arrhenius law:

$$\nu_i = \nu_{0,i} \exp\left(-\frac{E_i}{k_\mathrm{B}T}\right) \qquad i = \mathrm{n, a, p}. \tag{11.3}$$

Here, $\nu_{0,\mathrm{n}}$, $\nu_{0,\mathrm{p}}$, $\nu_{0,\mathrm{a}}$, E_n, E_p and E_n are the prefactors and effective activation energy barriers. The effective switching rate ν_s between the two ground states **0** and **1** is calculated by solving the rate equations, yielding

$$\nu_\mathrm{s} = \left(2 + \frac{\nu_\mathrm{a}}{\nu_\mathrm{p}}\right)^{-1} \nu_\mathrm{n}. \tag{11.4}$$

Hence, ν_s is given by ν_n reduced by a factor that depends on the ratio $\nu_\mathrm{a}/\nu_\mathrm{p}$. In the asymptotic limits, a distinct Arrhenius behavior is found for the low and high T regime, in agreement with the experimental results. The experimental data obtained on an individual nanomagnet and the corresponding model fit results are shown in Fig. 11.14c for illustration. At high T, $\nu_\mathrm{p} \gg \nu_\mathrm{a}$. Here, a nucleation event is immediately followed by the propagation of a domain wall. The domain wall undergoes numerous backscattering events, because annihilation is hindered. Due to the uncertainty of the annihilation site, the probabilities for one or no net magnetization reversal after nucleation is one half, resulting in the effective switching rate $\nu_\mathrm{s} = \nu_\mathrm{n}/2$. At low T, $\nu_\mathrm{p} \ll \nu_\mathrm{a}$. As propagation is hindered, the annihilation of the nucleus at its nucleation site is very likely. Only in very rare events the domain wall propagates through the nanomagnet, resulting in a magnetization reversal. Consequently, $\nu_\mathrm{s} \ll \nu_\mathrm{n}$, but again ν_s follows an Arrhenius law, as can be seen from Fig. 11.14c.

The experiments reveal that nanomagnets consisting of less than a hundred atoms exhibit a very rich temperature-driven magnetization dynamics. Although the nanomagnets are ferromagnetic, a magnetization reversal is realized via nucleation, propagation and annihilation. Each of these microscopic processes is found to have a distinct Arrhenius behavior, resulting in an effective switching rate of the nanomagnet that is governed by their complex interplay.

11.8 Magnetization Dynamics of Quantum Magnets

When approaching the atomic size limit of magnets, classical approaches for the description of the magnetization dynamics may fail due to their emerging quantum nature. As long as small magnets are sufficiently decoupled from an electron bath, an adequate description of the spin dynamics is given by a generalized Anderson model taking into account the coupling of the electron bath to the quantum spin [33–36]. However, if a small magnet is directly adsorbed on a metallic substrate with a strong coupling of the moment to the substrate conduction electrons [37, 38], the exact role of the substrate conduction electrons on the spin dynamics of the magnet is not obvious [39].

In Fig. 11.15a, STM topography images of single Fe atoms on a Cu(111) surface and a five-atom Fe magnet are shown. Note that the five-atom magnet has been assembled from the individual Fe atoms by atom manipulation using the SP-STM probe tip, literally moving together the atoms [40]. Positioning the tip above the

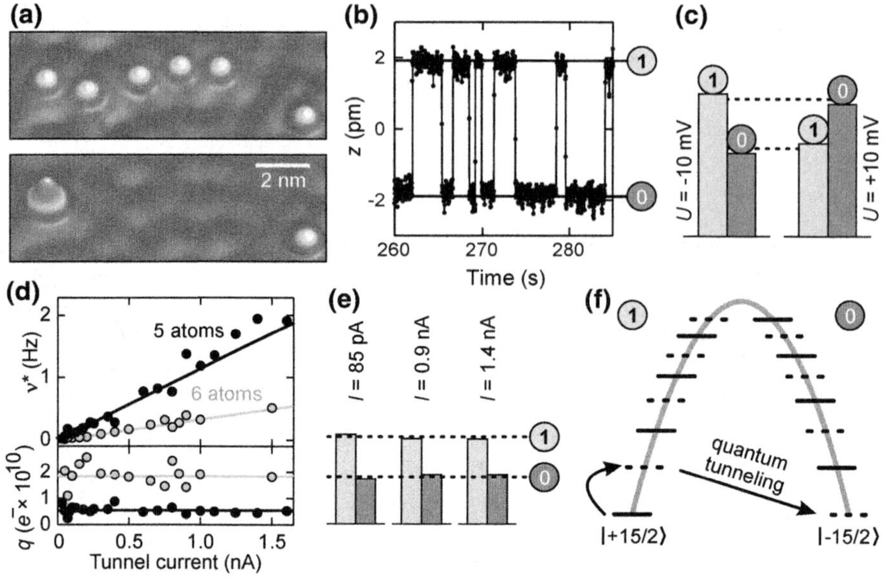

Fig. 11.15 **a** Topographic maps of six individual Fe atoms on Cu(111) (top). By using STM atom manipulation, five of them have been assembled to an iron cluster (bottom). **b** Magnetic telegraph noise, recorded on top of the cluster shown in a). **c** Histogram of the telegraphic noise, revealing a magnetic state asymmetry that changes sign with bias. **d** Tunnel current dependent switching rate of atomic clusters, consisting of five and six iron atoms, respectively. A constant number q of electrons have to be injected to trigger magnetization reversal (bottom). **e** Current dependent magnetic state asymmetry. A constant asymmetry is observed. **f** Quantum model for the cluster. Only discrete magnetic states are allowed, and direct ground state quantum tunneling of the magnetization is prohibited due to a non-vanishing magnetic anisotropy component parallel to the sample surface. Figure adapted from [40]

cluster at closed feedback results in a telegraphic noise in the z channel, as shown in Fig. 11.15b. Consequently, the cluster frequently switches between two ground states and exhibits a magnetization that is stable over several seconds (at $T = 300\,\text{mK}$). The determination of the state-dependent lifetimes results in the histograms that are shown in Fig. 11.15c. Here, a considerable asymmetry is observed, indicating a preferential state. A considerable spin-transfer torque is induced by the spin-polarized tunneling electrons, as indicated by the reversal of the asymmetry when reversing the bias-polarity. The results of a current-dependent study of the mean switching rate are shown in Fig. 11.15d for clusters consisting of five and six iron atoms, respectively. In contrast to SP-STM studies on quasiclassical magnets that are thermally switching, the switching rate scales linearly with the tunnel current. Moreover, the experiments indicate no switching events at vanishing tunnel current. Obviously, the reversals are purely triggered by a constant charge that has to be injected into the cluster to initiate a switching event. For a five-atom magnet, approximately 5×10^9 electrons tunneling into the cluster are needed to switch the magnetization. This finding is a hallmark of inelastic spin excitations, with each tunneling electron having

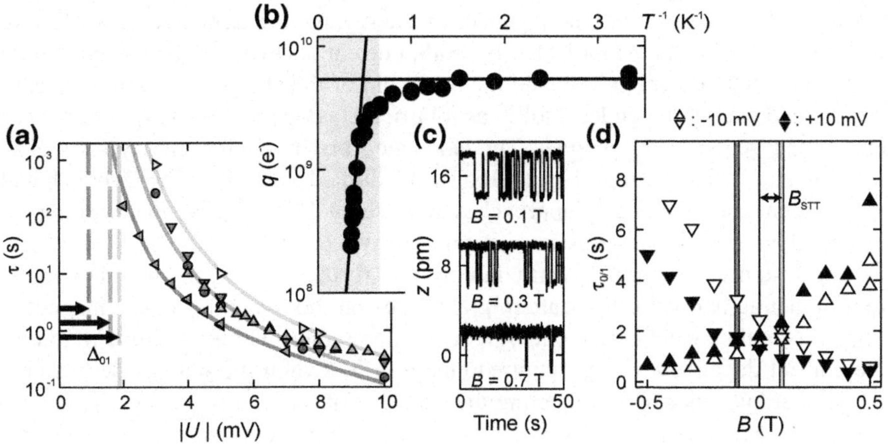

Fig. 11.16 **a** Experimentally determined magnetic lifetime as a function of applied bias voltage for selected five-atom magnets. With decreasing bias, the lifetime rapidly increases to very high values, and below a certain threshold (arrows) the quantum magnet exhibits a stable magnetic state. **b** Temperature dependence of the charge needed to reverse a five-atom magnet. A transition from an Arrhenius behavior (grey) to a quantum tunneling regime is observed. **c** Evolution of the spin-dependent telegraphic noise recorded on a five-atom magnet under the influence of an external magnetic field B. **d** State-resolved lifetime of a five-atom magnet for negative and positive bias polarity. The bias is indicated by the empty and filled triangles. The vertical lines indicate the strength of the magnetic field B_{STT} needed to compensate the spin-transfer torque effect. Figure adapted from [40]

a fixed probability of exciting the spin eigenstates of the magnet [41]. As shown in Fig. 11.15e, the observed asymmetry between the state-dependent lifetimes, resulting from spin-transfer torque, does not depend on the magnitude of the tunnel current I. It has been demonstrated on larger scale Fe islands that increasing I reduces the spin-transfer torque induced asymmetry, which is a consequence of significant Joule heating that symmetrically reverses the magnetization [7]. For the small Fe clusters, Joule heating can be ignored in the chosen range of both I and U in the experiment. The comparatively small bias voltage U is too small for a strong spin-phonon coupling [42].

From density functional theory a total magnetic moment of $\approx 3\,\mu_B$ per atom was found for such five-atom Fe magnets on a copper surface, which is close to the magnetic moment of a single Fe adatom ($\approx 3.5\,\mu_B$) [37, 43]. This results from a strong ferromagnetic exchange coupling between the constituent Fe atoms to the total angular momentum [44]. Consequently, the magnet is treated as a single total angular momentum, J [38], which is related to the total magnetic moment via $m_J = g\mu_B J$. In the following the total angular momentum will be referred to the spin for short. A schematic level diagram of the spin states is shown in Fig. 11.15f.

In Fig. 11.16a, the dependence of the lifetime τ on the applied bias U is shown for several selected five-atom magnets. They exhibit a stronger than exponential increase in τ when lowering the energy of the tunneling electrons. In the absence of current or for very small tunneling electron energy, these magnets remain stable for extremely long times, as indicated by both, current and voltage dependent studies of τ. The magnetic anisotropy and the total spin of the system determine the exact current and energy dependence of τ, as it intricately depends on the spin-dependent energy landscape of the magnet. The calculated first transition energy for a given magnet, Δ_{01}, is indicated in Fig. 11.16a by the dashed lines. The divergent behavior of τ is seen when the energy approaches Δ_{01}. Below $|U| < 2\,\text{mV}$, no switching was observed up to maximum observation times of two hours.

The system temperature is varied in order to further investigate the role of quantum tunneling and spin-flip scattering electrons on the dynamics of the magnet. In Fig. 11.16b, the switching charge $q(T)$ is plotted rather than the lifetime to remove the current dependency. The required charge q to switch the magnet decreases as the temperature increases, reflecting the decrease of the lifetime. From the log plot in Fig. 11.16b the Arrhenius behavior for $T \geq 2\,\text{K}$ is evident, whereas a plateau-like regime is observed for $T \leq 1.5\,\text{K}$. Such plateaus are attributed to quantum tunneling of the magnetization through the anisotropy barrier [36]. However, a significant deviation from a constant transition rate in the low temperature regime is observed, indicating that quantum tunneling alone cannot account for the temperature dependence of τ in the experiment [40].

A quantum impurity model with axial and transverse magnetic anisotropy is used to link the temperature dependence of $q(T)$ to the different spin transition processes. The model reveals that the temperature dependence results from two spin transition mechanisms which contribute to the switching of the magnet in addition to sequential transitions driven by tunneling electrons. At higher T, sequential transitions of the magnet induced by spin-flip scattering of thermally excited substrate electron-hole pairs generated by the broadening of the Fermi function dominate the switching behavior. Additionally, a non-vanishing transverse anisotropy results in non-equilibrium quantum tunneling. The equilibrium quantum tunneling, which fully reverses the magnetic state from one ground state to the other, without additional spin-flip processes [45], is blocked: for a half-integer spin the states across the barrier, which have the same $|J_z|$, have zero overlap. However, a two-step process where an electron changes J_z can lead to quantum tunneling in non-equilibrium, resulting from a finite overlap of states across the barrier for particular values of $\Delta J_z \neq 0$, as indicated by the arrows in Fig. 11.15f. The T-dependent interplay between the two transition effects creates a shoulder in $q(T)$ (not shown here) [40]. Its shape of the shoulder depends on the size of the transverse anisotropy E. With increasing T, the Arrhenius behavior is recovered. The underlying mechanism of how the magnetization is fully reversed is revealed within the quantum impurity model when accounting for both, the substrate conduction electrons and non-equilibrium induced quantum tunneling [40].

The asymmetry in the spin noise increases when an external out-of-plane magnetic field B is applied, as shown in Fig. 11.16c. This effect is a result of the Zeeman energy that favors the ground state of the magnet which has a magnetization pointing parallel to B. As shown in Fig. 11.16d, the respective lifetimes show a monotonic increase or decrease with increasing B. Note that the balance of lifetimes occurs at a nonzero magnetic field, $|B_{STT}| \approx 0.1\,T$. At $B = 0\,T$, the observed asymmetry is a result of the spin-polarization of the tunneling current generated by the magnetic tip. The value of B_{STT} changes sign when reversing the direction of tunneling by changing the bias polarity. This finding is a proof of the spin-transfer torque effect of the spin-polarized current acting on the cluster magnet. It favors transitions over the barrier in one particular direction as opposed to the other [34].

The relaxation of the Fe/Cu(111) cluster magnets is far from purely quantum because the quantum phase is destroyed. However, despite the strong hybridized spin coupling to an electron bath, quantum effects are important to fully describe the magnetization dynamics of the system.

References

1. A. Wachowiak, J. Wiebe, M. Bode, O. Pietzsch, M. Morgenstern, R. Wiesendanger, Science **298**, 577 (2002)
2. M. Pratzer, H.J. Elmers, M. Bode, O. Pietzsch, A. Kubetzka, R. Wiesendanger, Phys. Rev. Lett. **87**, 127201 (2001)
3. M. Pratzer, H.J. Elmers, Phys. Rev. B **67**, 094416 (2003)
4. S. Krause, G. Herzog, T. Stapelfeldt, L. Berbil-Bautista, M. Bode, E.Y. Vedmedenko, R. Wiesendanger, Phys. Rev. Lett. **103**, 127202 (2009)
5. M.L. Néel, Ann. Géophys. **5**, 99 (1949)
6. W.F. Brown, Phys. Rev. **130**, 1677 (1963)
7. S. Krause, L. Berbil-Bautista, G. Herzog, M. Bode, R. Wiesendanger, Science **317**, 1537 (2007)
8. I. Bat'ko, M. Bat'ková, Eur. Phys. J. Appl. Phys. **31**, 191 (2005)
9. Z. Li, S. Zhang, Phys. Rev. B **69**, 134416 (2004)
10. S. Krause, G. Herzog, A. Schlenhoff, A. Sonntag, R. Wiesendanger, Phys. Rev. Lett. **107**, 186601 (2011)
11. N. Néel, J. Kröger, R. Berndt, Appl. Phys. Lett. **95**, 203103 (2009)
12. G. Herzog, S. Krause, R. Wiesendanger, Appl. Phys. Lett. **96**, 102505 (2010)
13. S. Krause, A. Sonntag, J. Hermenau, J. Friedlein, R. Wiesendanger, Phys. Rev. B **93**, 064407 (2016)
14. G. Nunes, M.R. Freeman, Science **262**, 1029 (1993)
15. S. Loth, M. Etzkorn, C.P. Lutz, D.M. Eigler, A.J. Heinrich, Science **329**, 1628 (2010)
16. K. Nakamura, R. Shimabukuro, Y. Fujiwara, T. Akiyama, T. Ito, A.J. Freeman, Phys. Rev. Lett. **102**, 187201 (2009)
17. M. Tsujikawa, T. Oda, Phys. Rev. Lett. **102**, 247203 (2009)
18. C.-G. Duan, J.P. Velev, R.F. Sabirianov, Z. Zhu, J. Chu, S.S. Jaswal, E.Y. Tsymbal, Phys. Rev. Lett. **101**, 137201 (2008)
19. A. Sonntag, J. Hermenau, A. Schlenhoff, J. Friedlein, S. Krause, R. Wiesendanger. Phys. Rev. Lett. **112**, 017204 (2014)
20. U. Gradmann, G. Liu, H.J. Elmers, M. Przybylski, Hyperfine Interact. **57**, 1845 (1990)
21. H.J. Elmers, J. Hauschild, H. Höche, U. Gradmann, H. Bethge, D. Heuer, U. Köhler, Phys. Rev. Lett. **73**, 898 (1994)
22. M. Bode, O. Pietzsch, A. Kubetzka, R. Wiesendanger, Phys. Rev. Lett. **92**, 067201 (2004)
23. T. Maruyama, Y. Shiota, T. Nozaki, K. Ohta, N. Toda, M. Mizuguchi, A.A. Tulapurkar, T. Shinjo, M. Shiraishi, S. Mizukami, Y. Ando, Y. Suzuki, Nat. Nanotechnol. **4**, 158 (2009)
24. S.S.P. Parkin, X. Jiang, C. Kaiser, A. Panchula, K. Roche, M. Samant, Proc. IEEE **91**, 661 (2003)

25. R. Jansen, J. Phys. D: Appl. Phys. **36**, R289 (2003)
26. S.A. Wolf, D.D. Awschalom, R.A. Buhrman, J.M. Daughton, S. von Molnár, M.L. Roukes, A.Y. Chtchelkanova, D.M. Treger, Science **294**, 1488 (2001)
27. A. Kubetzka, M. Bode, R. Wiesendanger, Appl. Phys. Lett. **91**, 012508 (2007)
28. A. Schlenhoff, S. Krause, A. Sonntag, R. Wiesendanger, Phys. Rev. Lett. **109**, 097602 (2012)
29. T. Balashov, A.F. Takács, W. Wulfhekel, J. Kirschner, Phys. Rev. Lett. **97**, 187201 (2006)
30. A.B. Schmidt, M. Pickel, M. Donath, P. Buczek, A. Ernst, V.P. Zhukov, P.M. Echenique, L.M. Sandratskii, E.V. Chulkov, M. Weinelt, Phys. Rev. Lett. **105**, 197401 (2010)
31. J. Hong, D.L. Mills, Phys. Rev. B **62**, 5589 (2000)
32. G. Ehrlich, J. Chem. Phys. **44**, 1050 (1965)
33. F. Delgado, J.J. Palacios, J. Fernández-Rossier, Phys. Rev. Lett. **104**, 026601 (2010)
34. S. Loth, K. von Bergmann, M. Ternes, A.F. Otte, C. Lutz, A.J. Heinrich, Nat. Phys. **6**, 340 (2010)
35. M. Misiorny, J. Barnaś, Phys. Rev. B **76**, 054448 (2007)
36. S. Loth, S. Baumann, C.P. Lutz, D.M. Eigler, A.J. Heinrich, Science **335**, 196 (2012)
37. A.A. Khajetoorians, S. Lounis, B. Chilian, A.T. Costa, L. Zhou, D.L. Mills, J. Wiebe, R. Wiesendanger, Phys. Rev. Lett. **106**, 037205 (2011)
38. T. Schuh, T. Balashov, T. Miyamachi, S.-Y. Wu, C.-C. Kuo, A. Ernst, J. Henk, W. Wulfhekel, Phys. Rev. B **84**, 104401 (2011)
39. D.L. Mills, P. Lederer, Phys. Rev. **160**, 590–599 (1967)
40. A.A. Khajetoorians, B. Baxevanis, C. Hbner, T. Schlenk, S. Krause, T.O. Wehling, S. Lounis, A. Lichtenstein, D. Pfannkuche, J. Wiebe, R. Wiesendanger, Science **339**, 55 (2013)
41. N. Lorente, J.-P. Gauyacq, Phys. Rev. Lett. **103**, 176601 (2009)
42. H. Gawronski, M. Mehlhorn, K. Morgenstern, Science **319**, 930 (2008)
43. A.A. Khajetoorians, J. Wiebe, B. Chilian, S. Lounis, S. Blügel, R. Wiesendanger, Nat. Phys. **8**, 497 (2012)
44. A. Bergman, L. Nordström, A. Burlamaqui Klautau, S. Frota-Pessôa, O. Eriksson, Phys. Rev. B **75**, 224425 (2007)
45. W. Wernsdorfer, R. Sessoli, Science **284**, 133 (1999)

Chapter 12
Magnetic Behavior of Single Nanostructures and Their Mutual Interactions in Small Ensembles

Stefan Freercks, Simon Hesse, Alexander Neumann, Philipp Staeck, Carsten Thönnissen, Eva-Sophie Wilhelm and Hans Peter Oepen

Abstract A method is introduced that allows for the investigation of the magnetic behavior of single nanostructures in a wide range of temperatures and external magnetic fields. The technique is based on the anomalous Hall Effect utilizing nano-scaled Hall crosses. The nano-devices as well as the nanostructures are created out of sandwich films made from Pt/Co/Pt via ion milling. For the single nanostructures we identify a much too large attempt frequency which is shown to be partially caused by a temperature dependence of the magnetic anisotropy. It is shown that the magnetic behavior of and the mutual interaction in small numbers of nanostructures can be resolved. The interactions cause a strong change of the magnetic behavior of the individual particle particularly in case the system is affected by thermal energy. The latter happens even on a reasonable large scale of particle separation.

12.1 Introduction

The physics of systems of reduced dimensions is a fascinating research field as it addresses, on the one hand, questions of the impact of limited size on material properties and/or the bridging between atomic and solid state behavior. On the other hand, it gains more and more importance in technology as the trend in many device architectures goes towards the nanoscale [1, 2]. In magnetic storage technology for example a promising concept, which makes highest storage densities possible, is the **B**it **P**atterned **M**edia (BPM) in which single-domain ferromagnetic nanostructures are used as bits [3, 4]. The main reason for proceeding towards single-domain magnetic nanostructures is to prevent the continuous transition of magnetization between oppositely magnetized bits. The so called domain walls limit the storage density due to their spatial extent and act as a nucleation center for easier reversal via wall

S. Freercks · S. Hesse · A. Neumann · P. Staeck · C. Thönnissen · E.-S. Wilhelm · H. P. Oepen (✉)
Department of Physics, University of Hamburg, Hamburg, Germany
e-mail: oepen@physnet.uni-hamburg.de

S. Freercks
e-mail: sfreerck@physnet.uni-hamburg.de

R. Wiesendanger (ed.), *Atomic- and Nanoscale Magnetism*, NanoScience and Technology, https://doi.org/10.1007/978-3-319-99558-8_12

movement. In the research and development of BPM a key issue is the switching fields which have for technical reasons to be very similar for every nanostructure in a large array [5]. To get hands on the switching much effort is put into determining the internal switching field and separate the latter from external effects (like magnetostatic coupling) [4]. The second challenge in the field of BPM is the lower limit of sizes, which is given by thermal excitations and called the superparamagnetic limit [6].

Investigations of single particle behavior and the interaction of individual nanostructures can help to sort out the influence of the different parameters. In basic research there is a fundamental interest in the understanding of the impact of reduced dimension on the properties of nanostructures. Thus, there is an overlap of interests from both of the sides the research on and development of nanoscale magnets. A long-standing research field is that of the superparamagnetic behavior of large ensembles of nanoparticles. Nowadays, the very successful research merges into the inherent problem of ensemble research where statements can be made only on averaged quantities with a certain spread of quantities, like distribution of sizes, of separations and of magnetic properties [7, 8]. Although theoretical models for the behavior of a particle ensemble are elaborated, it is difficult to extract the correct single particle properties. Many parameters, more or less justified, are introduced to get the appropriate quality of fitting. From that respect the investigation of the behavior of the single nanostructure is a very important back up for the interpretation of the ensemble behavior. Moreover, understanding the mutual interactions of individual nanostructures would be even more desirable. In recent times the investigation of single nanostructures has been successfully tackled. In principle two approaches are used. The first is the technique based on nano-scaled SQUIDs [9, 10] which is limited to low-temperature applications. The second is based on scanning probe techniques where the spin-polarized scanning tunneling microscope has been demonstrated to be best suited for that purpose [11, 12].

Here, the development of a new technique with sufficient sensitivity for investigating single nanostructures is presented which has the promise for a high flexibility regarding the experimental parameter range. The method is based on the anomalous Hall effect and gives the possibility to perform experiments in a large temperature range from very low up to room temperature with no limits concerning the applied magnetic field strength. Although the main idea sounds intriguingly simple and easy to realize, much effort had to be put into the fine tuning of sensitivity. The successful approach and realization is explained in the first part of the paper. The method also allows for the investigation of small ensembles of nanostructures and in particular to study their mutual interactions. Most importantly, it has to be noted that fingerprints of interactions are frequently found and has sometimes strong impact on the magnetic behavior of the nanostructures. A brief survey of published results and some examples of new, recent findings are given in the second paragraph.

12.2 Experimental Details

(a) Preparation and Fabrication

The nanostructures are carved out of a Pt/Co/Pt sandwich that exhibits perpendicular magnetic anisotropy in a reasonable wide range of Co thicknesses. A seed layer of platinum (thickness of at least 4 nm) is fabricated on SiO_2 substrate via ion beam sputtering utilizing an electron cyclotron source. The ion beam sputtering has been shown to yield a polycrystalline film with a smooth seed layer surface and high texture independent of the substrate [13, 14]. Via DC magnetron sputtering another 1 nm Pt layer is deposited followed by Co and covered with a 3 nm Pt cap layer. The thickness, roughness and texture are controlled via X-ray reflectometry and –diffraction [14]. The X-ray analysis reveals a preferential [111] orientation of the fcc Pt crystallites. The Co film has no influence on the texture at small Co thicknesses. On the Pt(111) the Co grows in the fcc modification which is deduced from the thickness dependence of the anisotropy values. (X-ray analysis does not allow for distinguishing between fcc and hcp for small Co thicknesses [14]). The Co films show a thickness dependent change of lattice constant that merges into the Co bulk value at a Co thickness of about 5 nm [15]. It is found that the interfaces are very smooth with a roughness of about one monolayer ($\sim$0.2 nm) and a zone of intermixing on the average of 2 atomic layers ($\sim$0.4 nm). Due to the high interface quality the multilayers exhibit a perpendicular magnetic anisotropy with a spin reorientation transition at about $t_{Co} \sim 1.3$ nm. The lateral grain size is not depending on Co thickness and was determined via high resolution scanning electron microscopy. The lateral grain diameter is $\Phi \sim 11$ nm [16]. The grains show a Gaussian distribution of crystal orientations around the [111] direction with a full width at half maximum of $\sim 23°$.

The nanostructures are fabricated via ion milling with nano-sized etching masks on the surface of the sandwich. Two different approaches are utilized. (A) The first approach is based on diblock copolymer micelles filled with porous SiO_2. The micelles are deposited via dip or spin coating. They form a close packed mono-micellar layer on the multilayer film [17]. In the next step the polymer shell of the micelles is removed in oxygen plasma. The SiO_2 cores remain on top of the film [18]. The structure of the particle array is transferred into the magnetic film via Ar ion milling $<$500 eV. The sputtering takes off the film material between the particles while the SiO_2 cores protect the underlying film from being removed. The ion dose determines the amount of material which is taken off. The milling has to be stopped when the magnetic material is totally removed and before the material below the SiO_2 cores is affected. The net outcome is an array of nanostructures with dot sizes that is determined by the core dimensions and dot separation that is determined by the diameter of the micelles [19]. Perpendicular magnetized nanostructures with diameters in the range of $\Phi \sim 10$–30 nm can be created via the micelle route [20, 21]. (B) The second approach uses electron beam lithography to create the nano-sized etching mask utilizing a negative resist. The smallest achievable sizes of particle diameter are in the range of $\Phi \sim 20$ nm. The advantage here is that the location of the

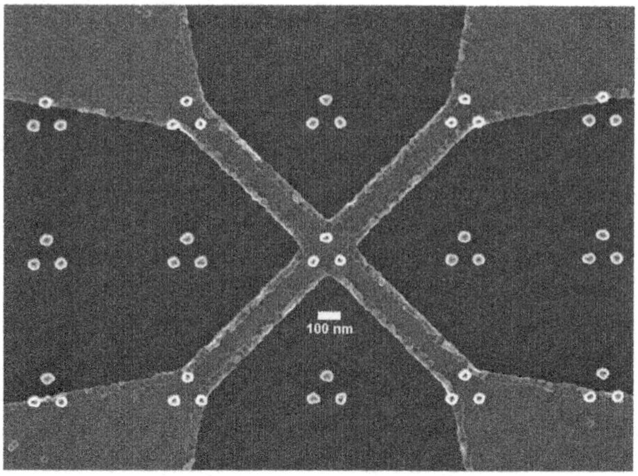

Fig. 12.1 Hall cross created in the Pt seed layer with three nanostructures in the sensitive area. The contacting of the device is made via wire bonding utilizing the four large pads

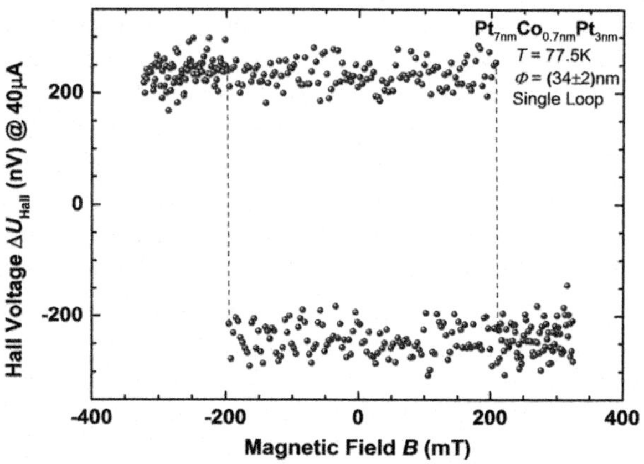

Fig. 12.2 Hysteresis loop of a single nanostructure. The magnetization curve has been obtained in a single field sweep. Plotted is the measured Hall voltage versus applied magnetic field. The field is perpendicular to the plane of the Hall cross

nanostructure and geometrical arrangement of a limited number of nanostructures is at the hand of the researcher.

In any of the preparation methods it is important that the sputtering is stopped within the Pt seed layer. The arrangement of nanostructures in a matrix of conducting material is essential for the creation of the in-situ measuring device.

(b) Measuring Procedure

The magnetic states of nanostructures are investigated via the anomalous Hall effect (AHE). A Hall cross is structured via e-beam lithography into the remaining seed layer that contains the nanostructures (on top). A SEM micrograph of such a Hall structure with three nanostructures in the center of the cross is shown in Fig. 12.1. The measurement goes as follows: A current is applied between opposite leads and the transverse voltage is measured. Due to the AHE the voltage is proportional to the component of magnetization that is normal to the Hall cross. In Fig. 12.2 the Hall voltage (transverse voltage) is plotted versus the magnetic field applied perpendicularly to the Hall cross that contains a single nanostructure. The dot switches between two stable perpendicular orientations of magnetization causing the sign change of voltage. AHE has been used before to study large arrays of small nanostructures or small ensembles of structures of considerable larger size [22–25]. We have been successful to demonstrate the single particle sensitivity by appropriate design of a nano-scaled Hall cross as measuring tool. The sensitivity was estimated to be equivalent to detecting the switching of a Co sphere of radius $r = 3.3$ nm [21]. A further improvement of sensitivity was set aside as sensitivity was sufficient to tackle the first fundamental questions/problems. The critical ingredient to obtain the high sensitivity is to tune the size of Hall bar and the nanostructure to achieve an as high as possible filling of the cross with the magnetic structure [23, 26]. The second tuning knob is the balance of the resistances of the ferromagnetic material and the leads while the quality of the interface should be as good as possible. The ferromagnet should have a lower resistivity than the lead so that the electrons will penetrate the ferromagnetic material while passing through the non-magnetic material beneath the nanostructure. In the present design this criterion is fulfilled as Pt has a roughly three times higher resistance than Co [26]. Any interface pollution is prevented as the different layers are grown in the same chamber under UHV conditions. The resistance of the leads, however, should not be too high as they will easily be destroyed due to Ohm heating.

An important question appears when an ensemble of a few nanostructures on the Hall cross (see Fig. 12.1) is investigated. It is the question about the allocation of a measured magnetic response to an individual dot and, correlated, the question about the dependence of signal height on position. Within the framework of our research project the latter problem was attacked via simulations utilizing the COMSOL software package [27]. The results reveal that the sensitivity is almost constant within the crossing range independent of the actual lateral scale [26]. A similar result was also obtained for systems which were much larger [28]. Hence, the signal height is not a good quantity for the identification of nanostructures. For the allocation of different magnetic signals to individual nanostructures a new procedure was developed. Starting from the simulation it was found that when using the Hall cross in an unusual way (applying the current between adjacent leads) a position sensitive sign signature appears [21, 26]. This idea was checked and experimentally elaborated which is now routinely used to distinguish the magnetic behavior of different nanostructures [21].

As our method to investigate the magnetic states of nanostructures via a nano-sized Hall bar is not limited in temperature and magnetic field strength (apart from a small ordinary Hall effect signal, which can be eliminated) it is well suited to investigate the two phases, i.e. blocked and superparamagnetic state, of the same nanostructure and particularly to study the influence of the magnetic environment on the transition.

(c) Magnetic Properties

The general idea of our approach was to separate tuning of magnetic properties and dimensions of the nanostructures. Magnetic sandwich systems have been chosen as they allow for manipulating the magnetic anisotropy and magnetization easily via the thickness of the Co film. In principle that means that at least in a certain range the size dependence of magnetic properties can be decoupled from the pure magnetic properties. The reality turned out to be more complex. At first, experiments reveal that the magnetic properties of film and nanostructure vary considerably in many cases. The reason is that a film property, particular the magnetic anisotropy, is a quantity that is obtained as an average over a large area. In other words the averaging includes a large number of grains which results in an extraordinary axis that is perpendicular to the film for symmetry reasons (textured film). When a small fraction of the film, however, is dissected only a few grains contribute and the mean over all orientations of the (111) grains is no longer necessarily vertical. The latter is the good and the bad news for the measurement on very small ensembles. On the one hand it allows for separating apparently identical nanostructures via the AHE signal due to different coercive fields. On the other hand the analysis of data becomes difficult because some angle uncertainties enter in an unknown magnitude, which even changes with strength of external magnetic fields. The canted easy axis is also responsible for the strong deviations from an ideal Stoner–Wohlfarth asteroid that we find in our nanostructures [29]. The unpredictable direction of the dots' easy axis is the first effect of the reduced size. A further consequence of the lateral limitation of size is a considerable change of the effective magnetic anisotropy via shape anisotropy. Erroneously, it was expected to have not a strong impact as the lateral dimensions are still large compared to the thickness (height of the nanostructure). Although the change of shape anisotropy is small it is in many cases comparable to the contributions of interface and magneto-crystalline anisotropy. The shape anisotropy is included in all of the following measurements and analysis if necessary. Fortunately, an exact calculation of the shape anisotropies of cylinders is possible which has been derived some years ago [30].

A third effect of the lateral confinement is a more efficient strain relief and an additional surface tension caused by the new cylindrical surface. It has been demonstrated that the structuring causes changes of the magnetic anisotropy with regard to the film values, after correcting for shape effects. At low Co thicknesses the anisotropy is reduced in the nanostructures, which is due to an efficient strain relief in the structures of small extent. For nanostructures with larger Co thickness a strong increase of the shape-corrected anisotropy is attributed to a diffusionless transformation from fcc to hcp within the Co nanostructures. The martensitic transformation is driven by additional surface stress that changes the overall energy balance in the magnetic system of limited dimensions [31].

It has to be mentioned that the ion milling might also give small changes of the anisotropy which is assumed to be small and which has not been further investigated systematically. The milling, however, causes a redeposition of sputtered material which can be resolved in SEM micrographs (Fig. 12.3). The size distribution of nanomasks, which are used for the creation of nanostructures, and the size distribution of the resulting nanostructures are shown. The cloning of nanostructures via ion milling results in a larger diameter of the metallic structures which is caused by

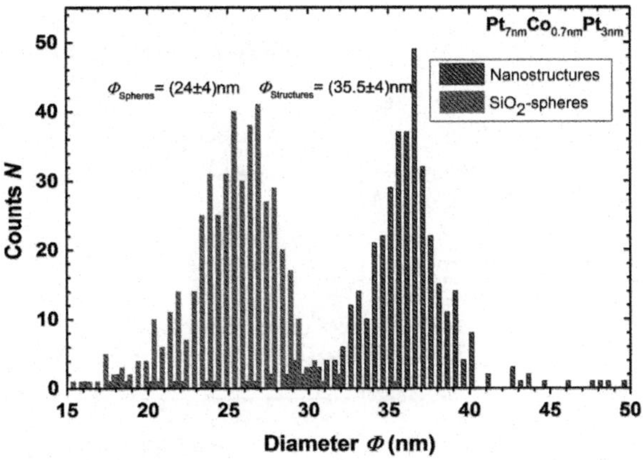

Fig. 12.3 Size distribution of nanomasks and -structure. In two steps the size distribution of nanomasks and nanostructures are evaluated via SEM. The size distribution of the SiO_2 cores reveals a smaller diameter than the nanostructures which have been fabricated utilizing the same nanomasks

redeposited material. The composition of the redeposited material is magnetically not active which has been experimentally verified.

12.3 The Physics of Single Nanostructures and Small Ensembles of Nanostructures

(a) Interactions

The big advantage of the new method is that it gives direct access to the investigation of coupling phenomena in comparison to the isolated single particle behavior. To demonstrate the latter option we aimed at investigating an arrangement of nanostructures in a so-called frustrated geometry (three particles on an equilateral triangle). Figure 12.4 depicts the switching of the three particles in a magnetic field that is applied perpendicular to the plane of the Hall cross. Each loop is a single loop, i.e. obtained in one magnetic field cycle. In the plot the magnetization behavior is shown for sweeps with different scan width (full loop and minor loops). At first one can recognize that the particles switch at different fields and that the signal height is not the same although the particles appear to be very similar. As mentioned above, this is most likely due to the grain distribution and composition of the individual nanostructures. Hence, effects of frustration cannot be investigated as the particles differ too much in their magnetic properties. On the other hand, the varying coercive fields and Hall voltages come as an advantage because they give the chance to identify all the different particles in their magnetic behavior. The major loop reveals only two switching events which apparently seem to indicate that one particle is not effective. In the minor loops (bright and greyish), however, it turns out that the first large step comes from two nanostructures switching at the same time, the candidates are particle A and C. By means of the minor loop one can recognize that the particle A is the

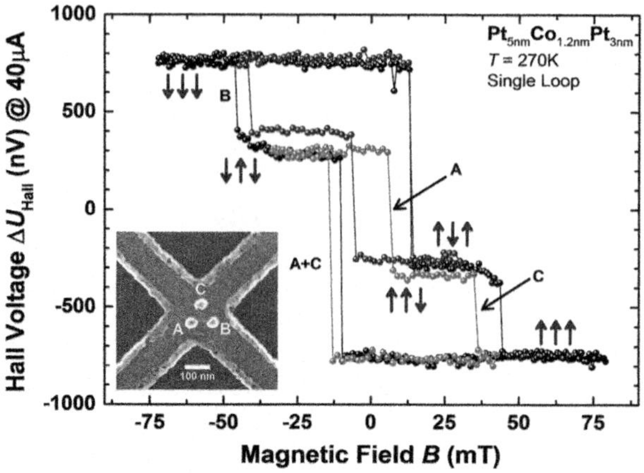

Fig. 12.4 Hysteresis loops of three nanoparticles. The hysteresis loops are all single field sweeps. The dark dots give a full field scan while the bright and greyish dots are obtained by partially sweeping the field which are so called minor loops. The switching events are indicated by letters which represent the dots shown in the micrograph (inset). The large arrows represent the orientation of magnetization of the individual nanostructures (from left to right, A, B, C). The minor loops reveal that the large step about 10 mT comes from two nanostructures that switch almost simultaneously

one that has the smallest switching field. In the major loop the two particles switch simultaneously because of magneto-static coupling. Apparently this is a dynamic process in which the change of total energy landscape opens up a certain pathway for the second particle. Or in other words the total energy landscape has a small barrier for the simultaneously switching of particle A and C. The loop does not reveal how strong the switching fields of the two particles differ because the switching field distribution is broad, due to the stochastic origin of the switching. In Fig. 12.4, for example, the different switching events at about -10mT are due to the spread of switching fields. A decent investigation of the switching field distribution gives a typical width of about ± 5 mT for the nanostructures shown in Fig. 12.4 [29]. The coercive field and the width of the switching field distribution changes drastically when the surrounding particles change their orientation of magnetization [29]. Note that the non-symmetric appearance of the total hysteresis is due to the same effect.

Concluding, it turns out that effects of coupling can have dramatic impact on the magnetic behavior even on scales where it is usually considered as a minor effect [32]. To make the point: For the use of bit patterned media (BPM) the results mean that the local symmetry around a bit can have strong impact on the switching field. Investigations to determine SFD, as found in literature [33, 34], where a whole array is switched cannot reveal a correct picture of the interactions that appear if individual particles have to be switched with arbitrary orientations of surrounding magnetic moments. In experimental approaches, where the behavior of only one particle is studied and the interference with others is not detectable [35, 36], a large

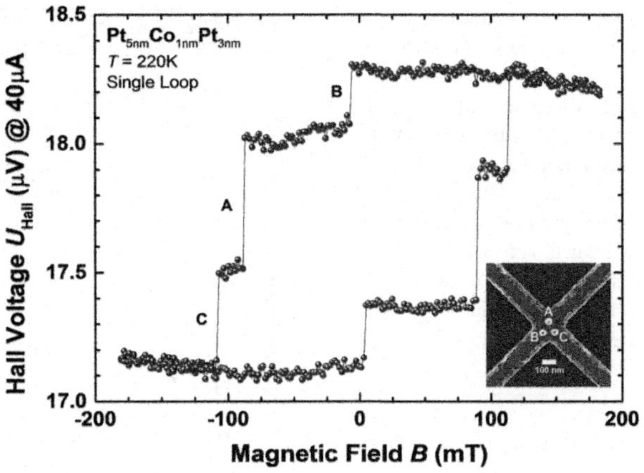

Fig. 12.5 Hysteresis loops of three nanostructures. The reversal of the nanostructures happens at well separated switching fields (coercivity). The nanostructure B has a very small coercivity which makes it susceptible to thermal excitations. The diameter of the structures are $\Phi_B = 51.8\,\text{nm}$ and $\Phi_{A/C} = 52.8\,\text{nm}$. The separation (center - center) are A–B = 89 nm, B – C = 84 nm and C – A = 85.5 nm

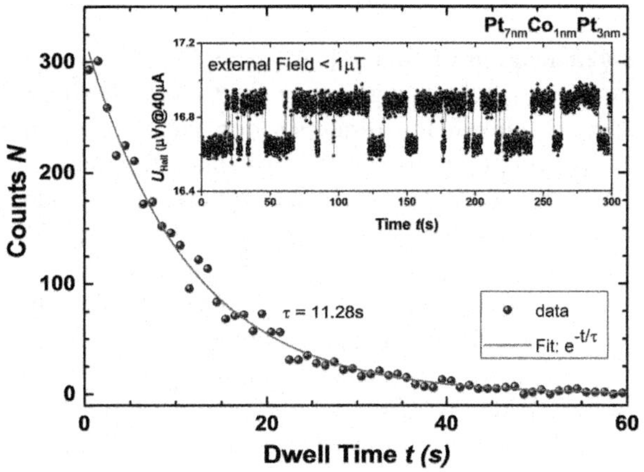

Fig. 12.6 Thermal excitation of nanostructure B shown in Fig. 12.5. In the inset telegraph noise in zero external field (compensated) is shown. The plot gives the frequency of the time a particle stays in one of the two states versus the stay or dwell time. An exponential decrease is expected which gives a mean stay time of 11.28 s

spread of switching fields for particle A and/or C will be found apparently depending erratically on magnetic field treatment.

The investigation of coupling was continued utilizing another ensemble with three particles on an equilateral triangle (Fig. 12.5). This time three well separated switching events are found for the particles. The nanostructure B has a low coercive field (smaller barrier height) which makes it susceptible to thermal excitations. The temporal behavior of nanostructure B has been investigated via the fluctuations of the AHE voltage [37, 38]. The inset of Fig. 12.6 depicts the telegraph noise that has been obtained at the same temperature while residual external fields were compensated (residual field $< 1\,\mu\text{T}$). Two voltage levels are found while each level itself reveals a broad distribution of random (white) noise. In Fig. 12.6 the analysis of the telegraph noise is depicted as a plot of the frequency that a state is occupied for a given time (dwell time) versus the time of occupation. Fitting an exponential behavior [39] one obtains the averaged dwell time for staying in the one or other

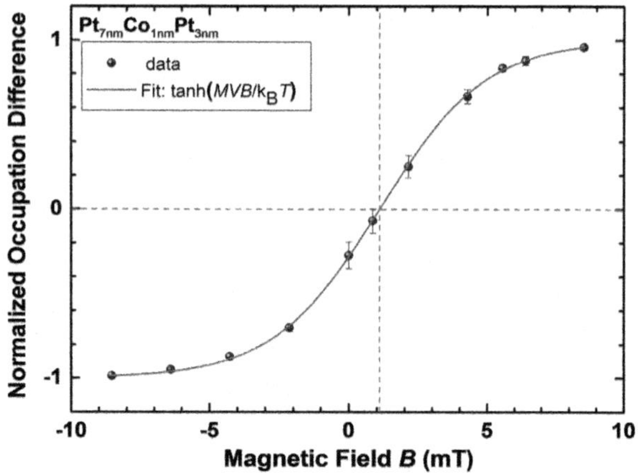

Fig. 12.7 Normalized difference of occupation of one state is shown as a function of an applied field. From a tanh-fit the moment of the nanostructure can be determined. The even population (zero) does not appear at zero field which indicates that the two remaining particles create a field at B

magnetization orientation, i.e. $\tau = 11.28$ s. The dwell time is large compared to the rate of field change in the field loop measurement, so that the switching level appears stable in the hysteresis loop. Next, when applying a field along the easy axis an uneven population of the two states is generated. The occupation of the energy state with orientation of magnetization parallel to the field direction increases while the opposite is less frequent occupied. In Fig. 12.7 the normalized occupation difference $\tau_{\text{diff}} = \frac{\tau_+ - \tau_-}{\tau_+ + \tau_-}$ is plotted versus the external field [38] which is equivalent to the average magnetization of a large ensemble of such identical particles or the expectation value for the magnetic moment of one particle (averaged over a long time). The intriguing result is that the even population ($\tau_{\text{diff}} = 0$) does not happen to appear at zero field. An offset of 1.1 mT is found which is caused by the stray field of the two other nanostructures. Hence, the experiment gives access to the magnetic field that is caused by the adjacent particles. The latter analysis (τ_{diff}) is correct only in case the attempt frequency is identical for the particle in states with orientation of magnetization parallel (+) and antiparallel (−) to the external field. The data points fit well to a hyperbolic tangent-function. The distribution function of a two-state system is a quite good approximation here. The description of the reversal utilizing Brown's expression [40] for the attempt frequency gives a relative deviation of $<2.6\%$ while for the most exact description of Coffey and Kalmykov [41] a deviation of $<3.4\%$ is found [29]. Hence it is reasonably accurate to use the Ising like approximation here. The fit gives a value of $7.64 \cdot 10^{-19}$ Am^2 for the total magnetic moment of the particle. The magnetic moment can be used to cross-check the magnetically effective size of the particle. Taking for the saturation magnetization $M_S = 1.44$ MA/m [42] a diameter $\Phi = 26$ nm results for the nanostructure. The value is drastically smaller than determined via SEM (Fig. 12.5) and closer to the size of the etching mask ($\Phi \sim 40$ nm). A reduced "activation" volume has also been found in previous investigations of particle ensembles [24, 43, 44] and in studies on single particles [36, 37]. It is important to note that the activation volume found here is

Table 12.1 Measured effective field at the position of nanostructure B as a function of different magnetic configurations of the nanostructures A and C. The arrows give the orientation of magnetization of nanostructure A and C

A	C	Resulting field (mT)
↑	↑	-0.68 ± 0.44
↑	↓	-0.65 ± 0.29
↓	↑	1.26 ± 0.26
↓	↓	1.15 ± 0.36

larger than the grain size in the Pt/Co/Pt film. The grain size is sometimes considered as nucleation/activation volume.

By means of the same procedure it is possible to determine the stray fields for various magnetization arrangements of the nanostructures in the vicinity. The telegraph noise of particle B is always investigated, while in a pre-step a minor loop is run to switch one of the other particles. The result is listed in Table. 12.1. What is essential is the fact that upon total reversal of the two particles A and C the effective field along the easy axis of particle B is not just opposite. Similarly, the statement is also true for every reversal of a single particle. From symmetry consideration the latter result is only understandable if the easy axis of particle B is tilted against the vertical direction. In the framework of a simple model, i.e. three spins on an equilateral triangle, it follows for symmetry reasons that at least one of the two particles that cause the field (at nanostructure B) has to have a tilted easy axis as well. Obviously, the orientation of the easy axis of magnetization differ which has an impact on the occupation (field along the easy axis) as well as on the effective barrier height (field component perpendicular to the easy axis) of the individual nanostructure. Besides the impact of the possible variation of the [111] axis of the grains, their number and particularly the number of grains at the rim, which might be more or less affected by the ion milling, can considerably vary. An important but still open question concerning the influence of the grain composition is that of the strength of exchange coupling between the grains. If the inter-grain coupling is small the grains at the rim can become (super)paramagnetic depending on the extent of size reduction by ion milling. The latter can be responsible for the reduction of the active magnetic volume and thus the signal change.

Although the magnetic field that is caused by the nanostructure in the surrounding is apparently not very large, the magneto-static interactions have a big influence on the dynamical behavior when the particles are susceptible to thermal excitation [45]. The investigations have been made by artificially destabilizing particles of an array of four nanostructures via magnetic fields that are applied close to the hard axis of the particles (here parallel to the plane of the Hall cross). The particles had diameters $\Phi \sim 40$ nm and separations (center to center) of about 100 nm [45]. At room temperature one of the particles is superparamagnetic while three particles are blocked. When applying a field along the hard axis one of the three blocked particles started switching and telegraph noise can be measured. At 150 K all particles are blocked. The telegraph noise, however, turns out to be drastically changed. On field increase to destabilize the particle that was superparamagnetic at room temperature (lowest coercive field), multiple level switching appears. Three of the particles show correlated switching, which means that the switching of particles does not happen independently as before. After the event of simultaneous switching of the three particles a frequent simultaneous switching of two of the three particles is initiated. Obviously, in the latter situation the antiparallel orientation of the two particles is energetically degenerated with a small barrier in between and the two particle switch back and forth between the two states of anti-parallel alignment. The third particle is only switchable in one of the degenerate states of the other two particles. When comparing the measured frequencies with the calculated ones (from independently measured magnetic anisotropies) drastic changes of the time scales are found. The two particles that switch simultaneously between the two antiparallel arrangements should behave almost superparamagnetic ($f_{1,2} = 10^9/10^{11}$ Hz) while the third particle should not switch within 27 h under the given conditions. Apparently the single particle switching cannot describe the dynamics of the experiment that appear on the time scale of seconds to minutes. The explanation is that the dynamics are determined by multiple particle behavior. The fluctuations of the total energy landscape allows for different pathways with different transition times for correlated switching in comparison to the single particle behavior. In [45] the dynamical correlation function

$$C_{\text{dyn}} = 1/T \int_{-\infty}^{\infty} dt\ S_{z,i}(t) S_{z,j}(t+s)$$

between pairs of magnetic moments i and j is defined, where S_z is the z-component of magnetization of moment i and j at times t and $t+s$, respectively. This function gives the information whether a given state is still correlated after a delay time s. Hence, $C_{\text{dyn}} \to 1$ corresponds to the correlated, in-phase switching, while $C_{\text{dyn}} \to 0$ to stochastic noise.

It has been shown that the dynamic correlations have a larger range of impact than the dipolar interaction which is the physical effect behind the correlations.

The main message of our studies on interactions is that the surrounding of a particle has a strong influence on the single particle behavior. In the quasi static mode it was demonstrated that the field at the position of the particle can be measured for varying magnetic configurations in the surrounding. While it was demonstrated that the fields cause a change of switching fields it can be more generally be expected that also the barrier height and thus the switching frequency of the particle can be changed. When the particles are susceptible to thermal energies the influence on the switching behavior and frequency can be even more dramatic. In the latter case the fluctuations give a dynamic variation of the total energy landscape which causes correlated events to take place.

(b) Attempt Frequency

As already mentioned above, our method allows for the investigations of nanoparticles in a large temperature range. This feature is used to study the properties and behavior in the superparamagnetic and the blocked regime and compare the num-

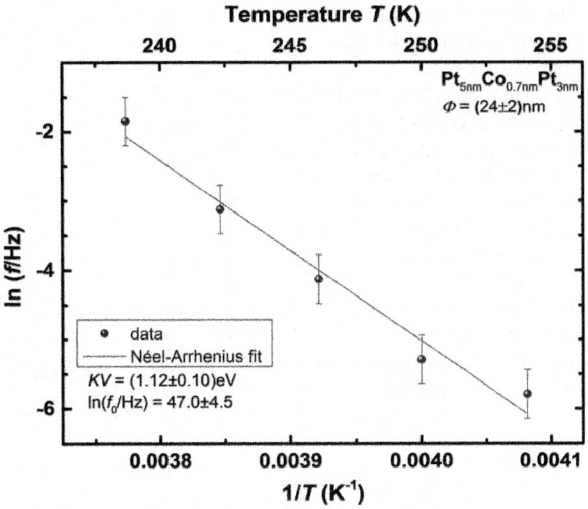

Fig. 12.8 Switching frequencies versus 1/T for a nanoparticle in the superparamagnetic regime. The fit represents a Néel-Arrhenius dependence. The fitting parameters are the total anisotropy of the particle and the natural logarithm of the attempt frequency in Hz

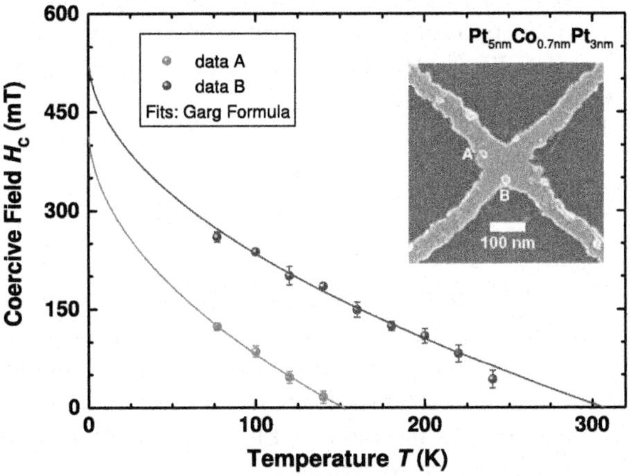

Fig. 12.9 Dependence of switching field on temperature for nanoparticles in the blocked state. The particle B is the same as the one investigated in the superparamagnetic state (Fig. 12.8). The fit is according to the formula derived by Garg [46]. The two nanostructures which were localized on the same Hall cross give similar attempt frequencies

bers. One example is the following: In the superparamagnetic state the dependence of switching rate on temperature is evaluated (Fig. 12.8). The data can be perfectly fitted by a Néel–Arrhenius law. In a semi-log plot versus 1/T a straight line is obtained which gives a number for the total magnetic anisotropy and from the intercept with the ordinate a number for the attempt frequency f_0. The values are KV = 1.12 eV and f_0 is in the range of 10^{19}–10^{23} Hz. The range of f_0 is incredibly high and in no accordance with any expectable value that fits into the span 10^9–10^{13} Hz [40, 41, 47]. To make an independent proof the dependence of switching field on temperature was measured for the same nanoparticle in the blocked state (Fig. 12.9). The data have been fitted utilizing the formula by Garg and the anisotropy determined from the former analysis [46]. Again a high value for the attempt frequency $f_0 = 10^{18}$ Hz (for the second particle in Fig. 12.9, $f_0 = 10^{18}$–10^{19} Hz) is obtained [48]. Similarly,

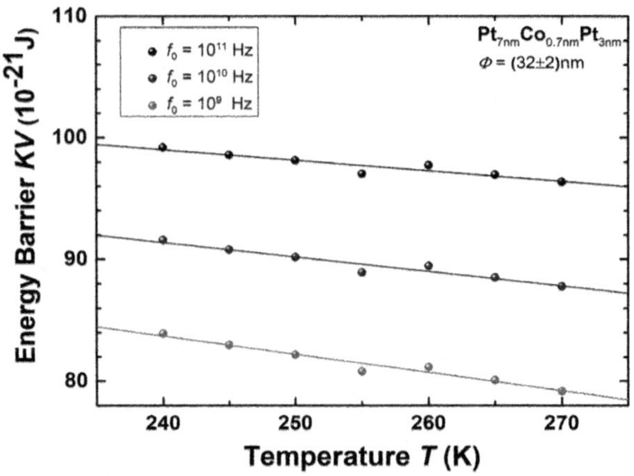

Fig. 12.10 Plot of KV versus temperature for a nanoparticle in the superparamagnetic state. The nanostructure is not the same as used for the analysis shown in Figs. 12.8 and 12.9. For the different attempt frequencies in the range that is expected from theory a linear temperature dependence of KV is found which indicate that the anisotropy is not constant but changes with temperature

extremely high attempt frequencies are also reported for other systems in literature [35, 49]. Despite existing theories that have considered domain wall nucleation and propagation [35], spin wave excitations [50] or laterally varying anisotropies [51, 52] as explanation for the large pre-factor our results give some hints to other effects that can partially explain the finding. The main idea becomes evident when re-writing the Néel–Arrhenius law

$$KV = -k_B T \ln\left(\frac{f}{f_0}\right),$$

with f the measured frequency and k_B the Boltzmann constant, and plotting KV versus temperature for assumed f_0 in the expected range (Fig. 12.10). The plot gives for all attempt frequencies a linear decrease. As the volume is constant it follows that apparently the anisotropy is temperature dependent and not constant as usually conjectured. As the plots in Fig. 12.10 show a linear temperature dependence the anisotropy can be written as

$$K(T) = K_0 + K_T^* T,$$

with K_T^* as the change of anisotropy per Kelvin and the Néel–Arrhenius law can be reformulated to

$$f = f_0 \exp\left(\frac{K_T^* \mathrm{V}}{k_B}\right) \exp\left(-\frac{K_0 V}{k_B T}\right).$$

As K_T^* enters the formula in the exponent a strong offset of the actual value of f_0 can be the consequence. Effects of a changing magnetic anisotropy with temperature on the superparamagnetic behavior have been mentioned before [53, 54].

Note: The linearity of the temperature dependence is only proven in the small range of temperatures that is investigated (which is determined by the lower time limit needed to accurately detect the AHE voltage of a switching and the upper time limit needed to capture a statistically relevant number of switches).

Currently that ansatz is tested quantitatively. The aim of the investigation is to determine independently the anisotropy of the nanostructures in the superparamagnetic and blocked state. To determine the magnetic anisotropy of a nanostructure is a formidable task and time consuming. Hopefully, exact temperature dependences of K(T)V can be determined that can be put into the modified Néel–Arrhenius formula. It is our goal to quantify the impact of the conjectured mechanism on the attempt frequency.

The focus of the corresponding theoretical investigation [55] is on the general impact of temperature dependent magnetic anisotropies on the analysis of the superparamagnetic behavior via its switching characteristics. It is demonstrated that the temperature interval (which is usually somewhat limited in experiments) in which a finite number of data points are fitted to an Arrhenius law has to be selected with care. It may determine very sensitively the resulting parameters of the Arrhenius plot, being the slope (anisotropy) and the intercept (attempt frequency). In general, besides the blocking temperature T_B in superparamagnetism, a second temperature scale becomes relevant in the macrospin description, i.e., the Curie temperature T_C, being decisive for the magnetic ordering and the variation of the magnetic anisotropy in the vicinity of T_C. Thus, experimentally determined pre-factors have to be carefully interpreted, particularly, but not exclusively, when the temperature range, in which the switching measurements are performed, is comparable to the magnetic ordering temperature. The mere numbers that come out of such an analysis via a Néel–Arrhenius fit are correct on their own, but the assignment to specific material properties has to be done with care. In particular, precise knowledge of the temperature dependence of the exponent in the Néel–Arrhenius law is necessary for reliable extractions of material parameters. Likewise, a straightforward extrapolation of the Arrhenius plot to $T = 0$ in order to derive the pre-factor F_0 as a constant "attempt frequency" has to be done with caution [55]. The consideration of our qualitative treatment has the bigger impact the smaller the nanoparticle is, as finite size effects can dramatically reduce the Curie temperature even for materials which are known to have high T_C values. As the latter condition is often fulfilled in experiments on superparamagnetism, our qualitative contribution is of high importance in that field.

12.4 Conclusions

A method for the investigation of the magnetic behavior of single nanostructures has been successfully developed. Much effort has been put into the fabrication of well characterized nanostructures and the nano-scale measuring device based on the anomalous Hall effect. It could be demonstrated that the magnetic behavior of a small number of nanostructures is simultaneously accessible in our approach. In the latter studies mutual interactions between individual nanostructures have been identified and separated. It turns out that particularly in the regime where the nanostructures are susceptible to thermal excitations the interactions gain a strong impact and can even determine the behavior at large. The studies on single particles reveal that the attempt frequencies obtained via a Néel-Arrhenius plot have to be considered with care. It could be demonstrated that even small temperature variations of the anisotropy can cause a change of the apparent attempt frequency by orders of magnitude.

Acknowledgements It is our pleasure to thank Dr. A. Philippi-Kobs, Dr. G. Winkler, A. Frauen, J. Jacobsohn, A. Meyer, and R. Frömter for valuable discussions and many experimental contributions to our research. Financial support by the Deutsche Forschungsgemeinschaft for our project (B14) within the SFB668 is gratefully acknowledged.

References

1. B. Hoefflinger (ed.), *Chips 2020: A Guide to the Future of Nanoelectronics* (Springer, Heidelberg, 2012)
2. T.W. McDaniel, J. Appl. Phys. **112**, 093920 (2012)
3. D. Szabados, Inside IT Storage (Seagate Enterprise) (2013)
4. B. Terris, T. Thomson, G. Hu, Microsyst. Technol. **13**, 189 (2007)
5. T. Albrecht et al., IEEE Trans. Magn. **49**, 773 (2013)
6. D. Weller, A. Moser, I.E.E.E. Trans, Magnetics **35**, 4423 (1999)
7. S. Morup, M. Hansen, C. Frandsen, *Magnetic Nanoparticles in Comprehensive Nanoscience and Technology*, vol. 1 (Elsevier, Amsterdam, 2011)
8. J.L. Dormann, D. Fiorani, E. Tronc, Adv. Chem. Phys. **98**, 283 (2007)
9. W. Wernsdorfer et al., J. Appl. Phys. **78**, 7192 (1995)
10. W. Wernsdorfer, Supercond. Sci. Technol. **22**, 064013 (2009)
11. M. Bode, O. Pietzsch, A. Kubetzka, R. Wiesendanger, Phys. Rev. Lett. **92**, 067201 (2004)
12. S. Ouazi, S. Wedekind, G. Rodary, H. Oka, D. Sander, J. Kirschner, Phys. Rev. Lett. **108**, 107206 (2012)
13. M. Wellhöfer, M. Weißenborn, R. Anton, S. Pütter, H.P. Oepen, J. Magn. Magn. Mater. **292**, 345–358 (2005)
14. G. Winkler, A. Kobs, A. Chuvilin, D. Lott, A. Schreyer, H.P. Oepen, J. Appl. Phys. **117**, 105306 (2015)
15. G. Winkler, Korrelation zwischen strukturellen und magnetogalvanischen Eigenschaften von Pt/Co/Pt- und Pd/Co/Pd-Schichtsystemen, Thesis University Hamburg, 2015
16. A. Kobs, Magnetogalvanic effects in ferromagnets of reduced dimensions, Thesis University Hamburg, 2013
17. A. Frömsdorf, A. Kornowski, S. Pütter, H. Stillrich, L.T. Lee, Small **3**, 880 (2007)
18. A. Neumann, N. Franz, G. Hoffmann, A. Meyer, H.P. Oepen, Open Surf. Sci. J. **4**, 49 (2012)
19. A. Meyer, N. Franz, H.P. Oepen, J. Perlich, G. Carbone, T.H. Metzger, Nano Res. **10**, 456 (2017)
20. H. Stillrich, A. Frömsdorf, S. Pütter, S. Förster, H.P. Oepen, Adv. Funct. Mater. **18**, 76 (2008)
21. A. Neumann, C. Thoennissen, A. Frauen, S. Hesse, A. Meyer, H.P. Oepen, Nano Lett. **13**, 2199 (2013)
22. J.B.C. Engelen, M. Delalande, A.J. le Febre, T. Bolhuis, T. Shimatsu, N. Kikuchi, L. Abelmann, J.C. Lodder, Nanotechnology **21**, 035703 (2010)
23. N. Kikuchi, S. Okamoto, O. Kitakami, Y. Shimada, K. Fukamichi, Appl. Phys. Lett. **82**, 4313 (2003)
24. N. Kikuchi, R. Murillo, J.C. Lodder, J. Appl. Phys. **97**, 10J713 (2005)
25. T. Shimatsu, H. Kataoka, K. Mitsuzuka, H. Aoi, N. Kikuchi, O.J. Kitakami, J. Appl. Phys. **111**, 07B908 (2012)

26. C. Thönnißen, Untersuchung der Schichtdickenabhängigkeit und Zuordnung der Hall-Signale einzelner Co/Pt-Nanodots, Diploma Thesis University Hamburg, 2012
27. COMSOL Multiphysics®, www.comsol.de
28. M. Alexandrou, P.W. Nutter, M. Delalande, J. de Vries, E.W. Hill, F. Schedin, L. Abelmann, T. Thomson, J. Appl. Phys. **108**, 043920 (2010)
29. P. Staeck, Untersuchung zur Wechselwirkung von magnetischen Nanostrukturen mittels Magnetotransport, Thesis University Hamburg, 2017
30. Y.T. Millev, E.Y. Vedmedenko, H.P. Oepen, J. Phys. D: Appl. Phys. **36**, 2945 (2003)
31. A. Neumann, A. Frauen, J. Vollmers, A. Meyer, H.P. Oepen, Phys. Rev. B **94**, 094430 (2016)
32. O. Fruchart, J.P. Nozieres, W. Wernsdorfer, D. Givord, F. Rousseaux, D. Decanini, Phys. Rev. Lett. **82**, 1305 (1999)
33. O. Hellwig, A. Berger, T. Thomson, E. Dobisz, Z.Z. Bandic, H. Yang, D.S. Kercher, E.E. Fullerton, Appl. Phys. Lett. **90**, 162516 (2007)
34. J.W. Lau, X. Liu, J. Appl. Phys. **112**, 093918 (2012)
35. S. Krause, G. Herzog, T. Stapelfeldt, L. Berbil-Bautista, M. Bode, E.Y. Vedmedenko, R. Wiesendanger, Phys. Rev. Lett. **103**, 127202 (2009)
36. S. Wedekind, G. Rodary, J. Borme, S. Ouazi, Y. Nahas, M. Corbetta, H. Oka, D. Sander, J. Kirschner, I.E.E.E. Trans, Magnetics **47**, 3351 (2011)
37. W. Wernsdorfer, E.B. Orozco, K. Hasselbach, A. Benoit, B. Barbara, N. Demoncy, A. Loiseau, H. Pascard, D. Mailly, Phys. Rev. Lett. **78**, 1797 (1997)
38. O. Sinwani, J.W. Reimer, L. Klein, Phys. Rev. B **89**, 020404R (2014)
39. M. Lederman, S. Schultz, M. Ozaki, Phys. Rev. Lett. **73**, 1986 (1994)
40. W.F. Brown, Phys. Rev. **130**, 1677 (1963)
41. W.T. Coffey, Y.P. Kalmykov, J. Appl. Phys. **112**, 121301 (2012)
42. M.B. Stearns, *Phase Diagrams, Lattice Constants and Elastic Moduli in Landolt-Börnstein - Group III Condensed Matter 19a* (Springer, Berlin, 1986)
43. N. Kikuchi, K. Mitsuzuka, T. Shimatsu, O. Kitakami, H. Aoi, J. Phys.: Conf. Ser. **200**, 102003 (2010)
44. S. Rusponi, T. Cren, N. Weiss, M. Epple, P. Buluschek, L. Claude, H. Brune, Nat. Mater. **2**, 546 (2003)
45. A. Neumann, D. Altwein, C. Thönnißen, R. Wieser, A. Berger, A. Meyer, E. Vedmedenko, H.P. Oepen, New J. Phys. **16**, 083012 (2014)
46. A. Garg, Phys. Rev. B. **51**, 15592 (1995)
47. C.P. Bean, J.D. Livingston, J. Appl. Phys. **30**, 120S (1959)
48. A. Neumann, Magnetisierungsverhalten einzelner ferromagnetischer Nanostrukturen, Thesis University Hamburg, 2014
49. J.L. Dormann, D. Fiorani, E. Tronc, J. Magn. Magn. Mater. **202**, 251 (1999)
50. S. Rohart, P. Campiglio, V. Repain, Y. Nahas, C. Chacon, Y. Girard, J. Lagoute, A. Thiaville, S. Rousset, Phys. Rev. Lett. **104**, 137202 (2010)
51. J. Lee, C. Brombacher, J. Fidler, B. Dymerska, D. Suess, M. Albrecht, Appl. Phys. Lett. **99**, 062505 (2011)
52. P. Campiglio et al., Phys. Rev. B **84**, 235443 (2011)
53. F. Wiekhorst, E. Shevchenko, H. Weller, J. Kötzler, Phys. Rev. B **67**, 224416 (2003)
54. C. de Julian Fernandez, Phys. Rev. B **72**, 054438 (2005)
55. M. Stier, A. Neumann, A. Philippi-Kobs, H.P. Oepen, M. Thorwart, J. Magn. Magn. Mater. **447**, 96 (2018)

Chapter 13
Fluctuations and Dynamics of Magnetic Nanoparticles

Elena Vedmedenko and Michael Potthoff

Abstract The stability of magnetic moments in a nanostructure against thermal and quantum fluctuations and the real-time dynamics of strongly excited nanosystems on metallic surfaces are studied theoretically on the basis of microscopic models addressing the degrees of freedom on the atomic level. To this end, different theoretical approaches and computational tools are employed and developed, such as classical Monte-Carlo simulations, quantum-classical hybrid dynamics and time-dependent density-matrix renormalization group.

13.1 Introduction

Fluctuations and real-time dynamics of a magnetic moment in an infinite magnetic film, in an isolated nanostructure or in a nanostructure coupled to a nonmagnetic metallic surface can be very different from fluctuations of the spin degrees of freedom in a magnetic bulk system. In addition, the magnetic properties of real nanomagnetic samples are measured during a finite observation time [1]. This fact and the strong influence of the substrate electrons may give rise to a couple of sometimes rather fundamental questions. For magnetic nanoparticles and atomic clusters it also has considerable physical and technical consequences. Namely, the recent investigations on magnetic properties of nanoislands [2] very much rely on the fact that information, encoded in the magnetic state of a small unit representing a single bit, is sufficiently stable over the respective observation time. This stability is important for applications in magnetic data storage technology [3] and intimately related to the above-mentioned dynamical properties.

E. Vedmedenko (✉)
Institute of Nanostructure and Solid-State Physics, University of Hamburg, Hamburg, Germany
e-mail: vedmeden@physnet.uni-hamburg.de

M. Potthoff (✉)
I. Institute of Theoretical Physics, University of Hamburg, Hamburg, Germany
e-mail: michael.potthoff@physik.uni-hamburg.de

R. Wiesendanger (ed.), *Atomic- and Nanoscale Magnetism*, NanoScience and Technology, https://doi.org/10.1007/978-3-319-99558-8_13

The purpose of this review is threefold: (i) We first discuss, on a rather fundamental level, different theoretical approaches to the real-time dynamics of a single spin coupled to a metallic substrate or host. In particular, this discussion serves to place the famous Landau–Lifschitz–Gilbert (LLG) equation in a broader context. This is important as the LLG equation serves as the basis for extended and very successful simulations of dynamical properties of magnetic systems when formulated for several spins and when including various magnetic interactions and anisotropies.

(ii) In a second step, we move from dynamics to thermodynamics and consider the equilibrium thermal properties of a magnetic nanoparticle. In particular, we discuss the role of finite lateral dimensions of magnetic nanoobjects for the definition of various crossover and critical temperatures.

(iii) In the third part we finally address a couple of highlights of the application of the spin-dynamics theory and review the study of life-times of polarizations of magnetic adatoms placed on semiconducting and metallic substrates and perturbed by the tip of a spin-polarized scanning tunneling microscope (SP-STM) and discuss a theoretical proposal of the manipulation of magnetic domain walls by the SP-STM tip.

The chapter is organized as follows: Sect. 13.2 gives an introduction and some overview of the dynamics of spins coupled to conduction electrons. Spin dynamics in the prototypical Kondo-impurity model is discussed in Sect. 13.3, followed by linear-response theory in Sect. 13.4. Effects of electron correlations on the spin dynamics are addressed in Sect. 13.5. Section 13.6 is devoted to the theoretical description of static and dynamic correlation functions in nanomagnets of finite lateral extensions, while theoretical concepts of microscopic manipulation of magnetic domain walls at the nanoscale are reviewed in the last Sect. 13.7.

13.2 Dynamics of Spins Coupled to Conduction Electrons

The real-time dynamics of a classical spin $\boldsymbol{S}$ in an external magnetic field is determined by the Laudau–Lifschitz (LL) [4] equation

$$\dot{\boldsymbol{S}} = \boldsymbol{S} \times \boldsymbol{B} \, , \tag{13.1}$$

where we have absorbed constants, such as the g-factor, Bohr's magneton μ_{B} and Planck's constant $\hbar$ in the definition of the field $\boldsymbol{B}$. Considering as a prototypical system a magnetic atom on a metallic surface, the LL equation does not realistically model the dynamics of the magnetic moment since there is a coupling to the substrate. This coupling will result in a damping of the spin dynamics. If, for example, the field direction is suddenly flipped from $-z$ to $+z$-direction, the spin will align to the new field direction, i.e., the system will reach its ground state in the course of time. This is well described by the Landau-Lifschitz-Gilbert (LLG) equation [4, 5],

$$\dot{\boldsymbol{S}} = \boldsymbol{S} \times \boldsymbol{B} - \alpha \boldsymbol{S} \times \dot{\boldsymbol{S}} \, , \tag{13.2}$$

where α is the famous Gilbert-damping constant.

There is a huge number of studies on spin-dynamics based on the LLG equation and of its various extensions [6, 7] covering many-spin systems, direct and indirect magnetic couplings between the spins, nonlocal Gilbert damping, and anisotropic interactions of different kind. Furthermore, the LLG approach is extremely successful in practice, even on an atomistic level [8]. On the other hand, from a fundamental perspective, the LLG equation (13.2) has been introduced phenomenologically. Apparently, it also violates energy and spin conservation. We will therefore consider a different theoretical route here and study the dynamics of a classical spin in contact with a large Fermi sea of conduction electrons by treating the electron degrees of freedom explicitly.

Figure 13.1 gives an overview: The paradigmatic model for a single spin coupled to a Fermi sea is the Kondo-impurity model. In case of a classical spin, the model is easily amenable to an exact numerical solution which is reminiscent of Ehrenfest dynamics in the context of molecular dynamics (see, e.g., [9] for an overview), where one also treats the nuclear degrees of freedom classically while the electron system is (effectively) noninteracting, i.e., moves in an effective single-electron potential. An important simplification is linear-response theory which applies to the weak-coupling limit. This approach is quite attractive since it represents a spin-only theory and, as the spin dynamics is typically much slower than the electron dynamics, allows us to access long time scales. The linear-response approach still keeps the full memory effects. This is not always necessary, and exploiting the largely different time scales of spins and electrons explicitly, one can re-derive the LLG theory with the help of a Markov approximation. The physics of the classical-spin Kondo model is very rich and has been studied in various ways and with various model extensions in recent years [10–13]. To some extent it will be reviewed here. Beyond that, there are quantum-spin effects, either on the exact level of the quantum-spin Kondo model or within a time-dependent hybridization mean-field theory [14], which become important for antiferromagnetic coupling, small spin quantum numbers and at extremely low temperatures. Below the Kondo temperature, time-dependent screening effects are worth being explored, see [12, 15, 16].

13.3 Tight-Binding Spin Dynamics

The prototypical model is the Kondo-impurity model of a classical spin coupled to a Fermi sea of conduction electrons:

$$H = \sum_{ij\sigma} T_{ij} c^{\dagger}_{i\sigma} c_{j\sigma} + J \boldsymbol{s}_{i_0} \boldsymbol{S} - \boldsymbol{B}\boldsymbol{S} \,. \tag{13.3}$$

The inset of Fig. 13.2 gives a sketch of the Hamiltonian. Here, a classical spin $\boldsymbol{S}$ with fixed length S is antiferromagnetically ($J > 0$) exchange-coupled to a system of conduction electrons hopping with an amplitude T_{ij} over the sites i, j of a lattice.

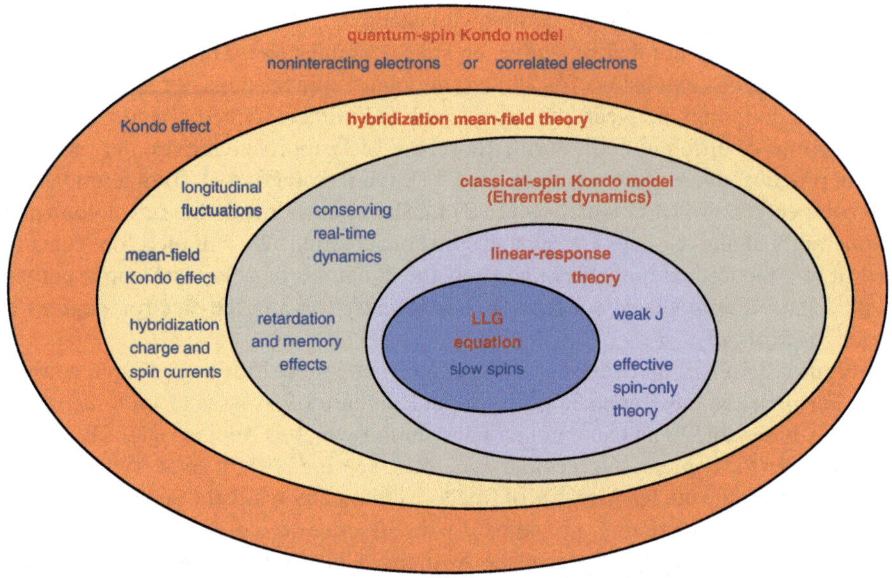

Fig. 13.1 Various approaches to real-time spin dynamics. See text for discussion

For simplicity, we assume $T_{ij} = -T$ for nearest neighbors on a one-dimensional chain of length L. Setting $T = 1$ fixes the energy and (with $\hbar \equiv 1$) the time scale. We further consider a half-filled system, i.e., if N is the total electron number, the electron density $n = \sum_{\sigma=\uparrow,\downarrow} \langle c_{i\sigma}^\dagger c_{i\sigma} \rangle$ is fixed to $n = N/L = 1$. Real-time dynamicsis initiated by suddenly flipping the field direction from $-z$ to z-direction at time $t = 0$, while the spin initially points in the $-z$-direction (with a slight distortion to break the symmetry).

We will start the discussion with the classical-spin Kondo (or s-d) model, i.e., with the numerically exact solution of the model (13.3). The equation of motion for $\boldsymbol{S}$ derives from the classical Hamilton function $H_{\text{class}} = \langle \Psi(t)|H|\Psi(t) \rangle$ and is of the LL form but with an additional Weiss field produced by the magnetic moment $\langle \boldsymbol{s}_{i_0} \rangle$ of conduction electrons at the site i_0 where the classical spin couples to. With the help of the Pauli matrices $\boldsymbol{\tau}$, this can be expressed in terms of the one-particle reduced density matrix $\rho = \langle c^\dagger c \rangle$ as $\langle \boldsymbol{s}_{i_0} \rangle = (1/2) \sum_{\sigma\sigma'} \rho_{i_0\sigma, i_0\sigma'} \boldsymbol{\tau}_{\sigma'\sigma}$. The equation of motion for ρ is of the von Neumann type and, besides the hopping matrix $\boldsymbol{T}$, involves a contribution from the Weiss field produced by $\boldsymbol{S}$. Therewith, one obtains a closed nonlinear system of coupled ordinary differential equations which can be solved by means of a high-order Runge-Kutta method, for example. System sizes of $L = 10^3$ are easily accessible in this way.

Figure 13.2 shows a typical result. After the sudden flip of the field, the x- and y-components of $\boldsymbol{S}$ show oscillations reflecting the precessional motion with Larmor frequency $\omega = B$. In addition, visible in the z-component, there is spin damping. The spin aligns to the new field direction in a few hundred inverse hoppings. This behavior

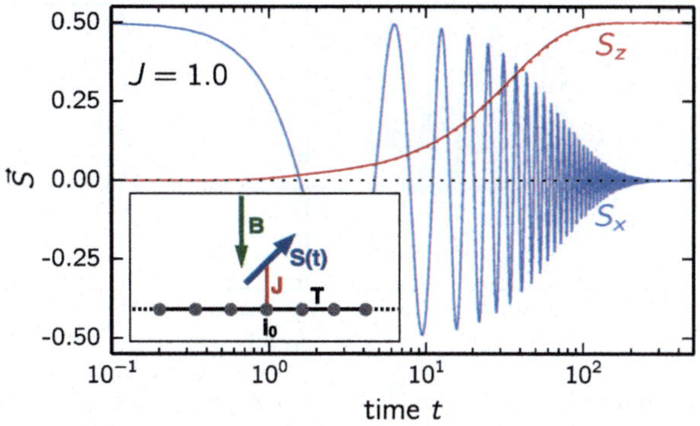

Fig. 13.2 (Adapted from [10]). Real-time dynamics of a classical spin (x and z components) within the model (13.3). Parameters: $S = 1, J = 1, L = 1001, i_0 = 501$, chain geometry with open boundaries. Energy and time units are chosen by fixing the nearest-neighbor hopping to $T = 1$. At time $t = 0$, the field direction is suddenly flipped from the $-z$ to the z direction. The inset provides a sketch of the system studied

is very much like the spin dynamics that could have been derived within the LLG framework, and one could easily extract a value of the Gilbert damping by fitting the numerical results. However, the spin dynamics shown in the figure emerges from a non-phenomenological, microscopic setup where, e.g., all macroscopic conservation laws resulting from the symmetries of H are fully satisfied. One should note that the energy of the local excitation of the system, $2BS$, is dissipated into the bulk such that *locally*, at i_0, the system approaches its ground state. The calculation stops at a maximum propagation time beyond which one would get (unphysical) distortions of the spin due to reflections from the open boundaries of the system. This also implies that the accessible time scale is approximately given by twice the half length of the chain divided by the Fermi velocity $v_F = 2T$. In our example, this amounts to about 500 inverse hoppings. Also the spin changes locally and thus, besides the energy, also spin must be transported into the bulk of the system. A detailed analysis [10] shows that this takes place in form of a spinful wave packet that is emitted from the core region, propagating with v_F and broadening slightly due to dispersion.

13.4 Linear-Response Theory

Typically, local exchange couplings are weak as compared to the energy scale set by the electron hopping and, therefore, a perturbative approach for the weak-J limit is reasonable (see "linear-response theory" in Fig. 13.1). Via standard time-dependent first-order perturbation theory in J, namely via the Kubo formula, $\langle \boldsymbol{s}_{i_0} \rangle_t = J \int_0^t dt' \chi_{\text{loc}}(t - t') \boldsymbol{S}(t')$, the response of the conduction-electron magnetic moment at i_0 due to the spin that stirs in the Fermi sea can be expressed in terms of the local retarded magnetic susceptibility $\chi_{\text{loc}}(t)$ at i_0. The latter is an equilibrium

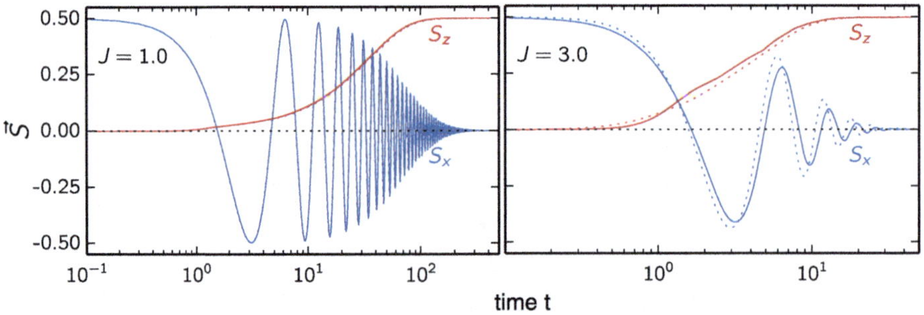

Fig. 13.3 (Adapted from [10]). Spin dynamics after a sudden flip of the field from x- to z-direction, as obtained by linear-response theory, see (13.4), for two different coupling constants J (solid lines). Dashed lines: results of the full tight-binding spin dynamics for comparison

quantity which can easily be calculated for a system of noninteracting electrons. Using this result for $\langle s_{i_0} \rangle_t$ the classical-spin equation of motion yields an effective spin-only theory, i.e., an equation of motion for an open quantum system with the full memory effect [10, 17, 18]:

$$\dot{\boldsymbol{S}}(t) = \boldsymbol{S}(t) \times \boldsymbol{B} - J^2 \boldsymbol{S}(t) \times \int_0^t dt' \chi_{\text{loc}}(t-t') \boldsymbol{S}(t') \,. \tag{13.4}$$

Figure 13.3 shows corresponding results for $J = 1$ and $J = 3$. The linear-response theory provides extremely accurate results for rather strong couplings up to $J = 1$ as can be seen in the figure by comparing with the exact nonperturbative result, and only at $J = 3$ do we find significant deviations. This may appear surprising as the relevant dimensionless parameter tJ is not "small"; one should note, however, that its effects are rather moderate as only the non-adiabatic terms $\sim \boldsymbol{S}(t) \times \boldsymbol{S}(t')$ with $t \neq t'$ contribute, see (13.4). For $J = 3$ (see figure), visible artifacts of the linear-response theory already appear on a time scale which is smaller by almost two orders of magnitude as compared to the $J = 1$ case. This is attributed to the strong enhancement of retardation effects with increasing J, which lead to a much more effective perturbation.

The LLG equation (13.2) can be derived from (13.4) as a Redfield equation [19] using a Markov-type approximation [18, 20] which assumes that the electron dynamics is much faster than the spin dynamics. Formally, we exploit that the memory kernel $\chi_{\text{loc}}(t-t')$ in (13.4) is peaked at $t' \approx t$ and cut the expansion $\boldsymbol{S}(t') \approx \boldsymbol{S}(t) + (t'-t)\dot{\boldsymbol{S}}(t) + \cdots$ at first order under the integral in (13.4). Furthermore, the upper bound of the integral must be set to infinity to get a constant. This is usually again justified by the peak structure of the kernel. With this we arrive at $\dot{\boldsymbol{S}}(t) = \boldsymbol{S}(t) \times \boldsymbol{B} - \alpha(t)\boldsymbol{S}(t) \times \dot{\boldsymbol{S}}(t)$ where the damping constant is given by

$$\alpha = -J^2 \int_0^\infty d\tau \, \tau \, \chi_{\text{loc}}(\tau) \,. \tag{13.5}$$

Interestingly, for a one-dimensional system, this quantity is ill-defined as the integral diverges at the upper bound $t' \to \infty$ due to the slow long-time decay $\propto 1/t$ of the kernel (see [10] for a detailed discussion of this point). This implies that the LLG approach becomes questionable in one dimension. A pragmatic way out is to supplement the theory with an artificial cut-off for long times or by an *ad hoc* damping of van Hove singularities on the frequency axis. A physical regularization of the theory comes, for example, with the additional inclusion of electron-correlation effects.

13.5 Correlated Conduction Electrons

Electron correlations among the conduction electrons are expected to have a substantial impact on the spin dynamics. Within different models and using various approximations, this has been addressed in a few pioneering studies [21–23] but only via the effect of the Coulomb interaction on the Gilbert damping, i.e., in an indirect way. The emergence of new energy (and time) scales, however, is a hallmark of strongly correlated systems, and here the entire LLG concept is expected to break down. Studying these effects is important, e.g., for a microscopic understanding of the relaxation time scales emerging in modern nano-spintronics devices [24–26]. Here, we review our recent theoretical work [11] where the breakdown of the LLG theory is demonstrated by referring to the correlation-induced Mott insulator as a paradigmatic example [27].

We start from (13.4) assuming a weak (or moderate) coupling of the spin to the system of conduction electrons but replace the noninteracting Fermi sea by a one-dimensional Hubbard model with local Coulomb interaction U [28]. The time-dependent density-matrix renormalization-group approach (t-DMRG) [29–31] is used to compute the retarded local magnetic susceptibility of the Hubbard model, and the numerical solution of (13.4) provides us with the full memory and correlation effects. Results for different U are shown in Fig. 13.4.

At half-filling, the Hubbard model is a correlation-induced Mott insulator. For strong U, it perturbatively maps onto an antiferromagnetic Heisenberg model with Heisenberg coupling $J_{\mathrm{H}} = -4T^2/U$, i.e., there is a single energy scale only. Consequently, the susceptibility must show a scaling behavior according to $\chi_{\mathrm{loc}}(t) = F(tJ_{\mathrm{H}})$ with some function F. Inserting this into the expression for the Gilbert damping, (13.5), we find

$$\alpha = \frac{J^2}{J_{\mathrm{H}}^2} \int_0^\infty dx\, x\, F(x) = \frac{J^2}{J_{\mathrm{H}}^2} \alpha_0 = \frac{J^2 U^2}{16 T^4} \alpha_0 \,, \tag{13.6}$$

where α_0 is a universal dimensionless Gilbert damping constant of a Mott insulator. From the numerical data we find $\alpha_0 \approx 4.8$. Thus, for fixed J, T, we have $\alpha \propto U^2$ implying that increasing interactions lead to a shorter spin relaxation time.

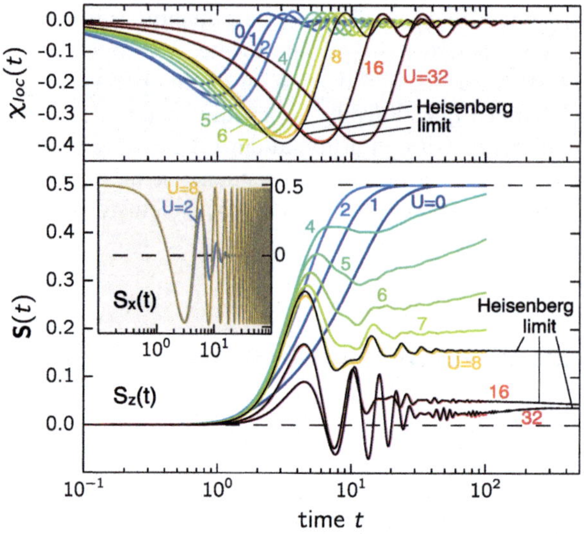

Fig. 13.4 (Adapted from [11]). *Upper panel:* Time dependence of the local spin susceptibility $\chi_{\text{loc}}(t)$ at $i_0 = 1$ for an open Hubbard chain with $L = 60$ sites. t-DMRG for half-filling and different U as indicated. The nearest-neighbor hopping $T = 1$ fixes energy and time scales. *Lower panel:* Resulting real-time dynamics of a classical spin $\boldsymbol{S}(t)$ with $|\boldsymbol{S}(t)| = \frac{1}{2}$ coupled at $i_0 = 1$ (main plot: $S_z(t)$, inset $S_x(t)$). Calculations using (13.4) for $J = 1$ and different U. The spin dynamics is initiated by switching the field direction at $t = 0$ from x to z where $B_{\text{final}} = 1$. Thin black lines: Heisenberg model with $J_{\text{H}} = \frac{4T^2}{U}$ ($L = 400$) and, for improved accuracy at $U = 8$, with n.n. and n.n.n. couplings $J_{\text{H}} = \frac{4T^2}{U} - \frac{16T^4}{U^3}$ and $J'_{\text{H}} = \frac{4T^4}{U^3}$ [32] ($L = 300$)

The actual physics, however, is completely different: Fig. 13.4 (top panel) demonstrates that magnetic excitations which provide the dissipation of energy and spin to the bulk of the system get more and more weight with increasing U. However, they also become active on a later and later time scale. For $U \to \infty$ this means that they will actually never be activated and, since these magnetic excitations are the only ones available for a Mott insulator, this implies that there is no spin damping at all in this limit.

This expectation is indeed verified by the spin-dynamics results (see lower panel). In the weak-U regime, there is a fast relaxation of the spin to the field direction with a relaxation time which decreases with increasing U, see results for $U = 0$ up to $U = 2$. This regime is followed by some intermediate-coupling regime with a less regular behavior of $\boldsymbol{S}$. For an interaction strength of $U = 8$ or stronger, however, the relaxation time appears to diverge; the spin, after some initial dynamics, does not fully relax at all. It is easily shown, however, that a constant z-component of the spin and an undamped oscillation of the x- and y-component (see inset) is not a solution of the equation of motion (13.4). In the extreme long-time limit, we therefore expect that the spin finally undergoes a full relaxation.

The incomplete relaxation of the spin should be understood as a "pre-relaxation", i.e., analogously to the concept of prethermalization which is well known for purely electronic systems [33–36]. In cases of close parametric distance to points or limits of integrability, those systems are trapped, during a substantial period of time, in a prethermalized state and do not thermalize directly. For the present quantum-classical hybrid system and referring to relaxation rather than thermalization, the analog of the integrable point is given by the limit $U \to \infty$. Namely, for every finite time t, the memory kernel $\chi_{\text{loc}}(t) \equiv 0$ in this limit, and thus (13.4) simplifies to the *linear* Landau-Lifschitz equation [4]. The situation is also similar to the case of quantum excitations which are metastable on a long time scale due to a very small phase space for decay. A prominent example is given by doublon excitations in the Hubbard model for interactions U much larger than the bandwidth [37–40]. Energy conservation requires the decay via a high-order process, such that the relaxation time diverges for $U \to \infty$.

13.6 Critical Properties and Magnetization Reversal in Nanosystems

Physical characteristics of the magnetization reversal are closely related to the critical properties of the object. While the critical properties of infinitely large magnets are well known, they have to be reconsidered for nanoparticles. One of the most important critical parameters is the Curie temperature. In the following, we describe some aspects of this critical temperature at the nanoscale and describe the magnetization switching in nanomagnetic ensembles.

13.6.1 Crossover Temperatures of Finite Magnets

From the experimental point of view, the critical temperature is a well defined quantity which can be measured. From the theoretical point of view, there is no Curie temperature as there is no phase transition in a system with a finite number of degrees of freedom and magnetic susceptibilities stay finite in the entire temperature regime. Nevertheless, they show enhancements at $T = T_{\text{C}}(L)$, which defines a "reduced Curie temperature". Typically, one finds $T_{\text{C}}(L) < T_{\text{C}}(\infty)$ for open boundaries.

Above $T_{\text{C}}(L)$, the magnetic state of the system is not stable temporally, but rather shows a superparamagnetic (SPM) behavior at temperatures $T_{\text{b}}(L) < T < T_{\text{C}}(L)$, which is neglected in finite-size scaling. For storage technology the blocking temperature $T_{\text{b}}(L)$ is even more relevant than $T_{\text{C}}(L)$ because it characterizes the crossover from the ferromagnetic (FM) state at low T to the SPM state, where the system changes its magnetization orientation between several energy minima determined by magnetic anisotropies. The $T_{\text{b}}(L)$ is not a pure property of the system, but rather a relative value, which depends on the observation time τ. For infinitely large τ, there is no temporal blocking of the magnetization because of quantum tunneling and thermal excitations, and thus $T_b(L) \to 0$. For $\tau \to 0$ (referring to e.g. laser-probe methods), $T_b(L) \to T_C(L)$ while for intermediate τ (like in spin-polarized scanning tunneling microscopy) $0 < T_b(L) < T_C(L)$. This discussion shows that theoretical studies based on a microscopic spin model have to be refined when dealing with the superparamagnetic regime. A recent publication [41] shows that additional to a Monte-Carlo (MC) approach, where with an increasing number of MC steps per temperature, the blocking temperature $T_b(L)$ decreases, a simple but reliable effective theory can cover the phenomenology of magnetic nanoparticles. A central quantity of this phenomenological theory is a two-point order-parameter correlation function defined between two spins $\mathbf{S}_i$ and $\mathbf{S}_j$ at sites $\mathbf{r}_i$ and $\mathbf{r}_j$ by

$$G(r) = \frac{1}{n(r)} \sum_{\substack{i<j \\ |r_i - r_j| = r}} \langle \mathbf{S}_i \mathbf{S}_j \rangle \,, \tag{13.7}$$

and a connected correlation function

$$\tilde{G}(r) = G(r) - M^2 = \frac{1}{n(r)} \sum_{\substack{i<j \\ |r_i - r_j| = r}} \langle S_i S_j \rangle - M^2 \,, \tag{13.8}$$

which explicitly takes account of the finite system size. Here, the sum in the first term runs over all $n(r)$ pairs separated by the distance r. The second term involves the magnetization $M = |\sum_i \langle S_i \rangle|/N$ with N being the number of sites. It turns out that the quantities defined above can be used for unambiguous definitions of $T_C(L)$ and $T_b(L)$, and moreover, lead to a new method to determine critical temperatures for infinite systems.

To get insight into the new method of determining the critical temperatures the above-defined correlation function $\tilde{G}(r)$ is evaluated and presented in Fig. 13.5 for a linear chain of Ising spins (see Fig. 13.5a) and for a 5×5 Ising system on a square lattice (Fig. 13.5b). Analyzing the data of Fig. 13.5 in detail we identify three specific temperatures: (i) For any finite T, the magnetization is $M = 0$, while for $T_1 = 0$ the magnetization jumps to $M = 1$. Therefore, the function $\tilde{G}(r)$ jumps from unity to zero as the temperature approaches $T = 0$ as seen in Fig. 13.5a, b. (ii) At the temperature T_2 ($1.7J$ in Fig. 13.5b) the curvature of $\tilde{G}(r)$ and $G(\mathbf{r})$ changes its sign. (iii) At the temperature $T_3 = 2.3J$ the trend of $\tilde{G}(r)$ changes from algebraic convex to exponential.

The first characteristic temperature T_1 is evidently equivalent to $T_b(L)$. The mostly remarkable feature, however, is the change in the curvature of $\tilde{G}(r)$ from concave to convex at some temperature which cannot be expected from previously published results.

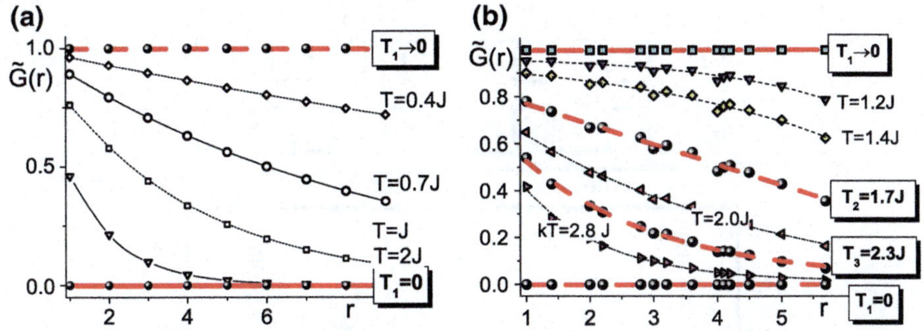

Fig. 13.5 (Adapted from [41]). **a** Two-point correlation functions $\tilde{G}(r)$ (symbols) calculated for an Ising chain with open ends and consisting of 10 atomic sites; **b** Two-point correlation functions for a 5×5 square lattice of Ising spins at $T = 0.5J \ll T_C(\infty)$. The lines in (**a**) and (**b**) are fits of the numerical data by the model correlation function given in (13.9). Three thick lines correspond to three critical temperatures: $T_b(L)$ - the blocking temperature, $T_C(L)$ the reduced Curie temperature, and $T_C(\infty)$ the Curie temperature of an infinite sample

To identify the physical meaning of T_2 and T_3, a model function $G(r)$ has been constructed.

$$G(r) = G(r, T) \approx B(T)e^{-r/\varepsilon(T)} + y(T) \,, \tag{13.9}$$

The form of $G(r)$ is reminiscent of the Ornstein-Zernicke theory. Note, however, that (13.9) refers to a finite system and that it applies to all distances r up to the system boundary. In particular, there are no constraints on the sign of the r-independent constants $\varepsilon(T)$, $B(T)$ and $y(T)$.

Using the above mentioned model, the numerically exact expectation values for the 1D, 2D and 3D finite Ising systems (Fig. 13.5) can be used to fit the parameters $\varepsilon(T)$, $B(T)$ and $y(T)$. It has been found that the quality of the fit is very good, see lines in (a) and (b) and it excellently describes the data in the entire temperature range. From this fit (see details in [41]) the temperature T_3 can be interpreted as the Curie temperature at a thermodynamic limit (infinite samples): $T_3 \approx T_C(\infty)$, and the temperature $T_2 = T_C(L)$ corresponding to the change in the curvature of $\tilde{G}(r)$ gives the reduced Curie temperature of a finite magnet. Thus, the described method yields a good estimate of $T_C(\infty)$, $T_C(L)$ and $T_b(L)$ from a single calculation without finite-size scaling. Analysis presented in [41] shows that $T_C(L)$ as well as $T_b(L)$ satisfy the finite-size scaling law $(T_C(\infty) - T(L))/T_C(\infty) = (L/L_0)^{-1/\nu}$ [42].

Hence, a simple analytical form for the two-point magnetic correlation function suggested in [41] for magnetic nanoparticles leads to an excellent numerical agreement with exact Ising, Heisenberg and Monte-Carlo data of finite anisotropic spin models and gives accurate definitions of crossover temperatures for finite systems.

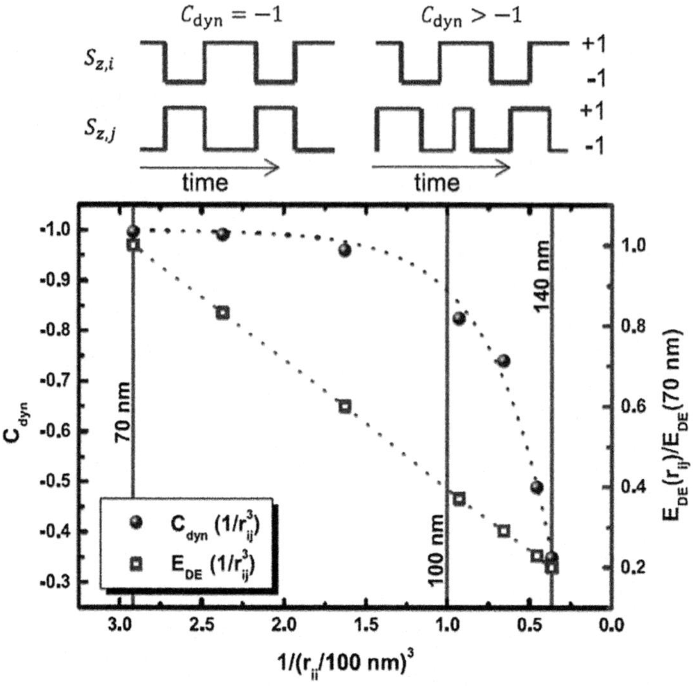

Fig. 13.6 Dynamical correlation-function C_{dyn} (spheres) and normalized dipolar energy C (circles) as a function of $1/r_{ij}^3$ normalized to 100 nm. It can be seen that C_{dyn} decreases more slowly than the C, which is responsible for the antiparallel correlation. The damping parameter $\alpha = 0.2$ has been considered in the calculations. The form of $C_{\text{dyn}}(1/r_{ij}^3)$, however, is independent of α

13.6.2 Switching of Nanoparticles in Systems with Long-Range Interactions

While the described spin-spin correlations define averaged, time independent properties of nanosystems, the understanding of time-dependent correlations in nanomagnetic systems is also of great interest for fundamental science as well as for applications. Particularly crucial is the dynamical spin-spin correlation function for certain magnetic states used as bits of information, because the ultimate goal for any storage media is to create a highest possible packing density. This requires a small distance between islands or grains and, at the same time, one needs to switch the bits individually. So the question arises whether the spin-spin correlations remain the same in static (non-switching) and dynamical (switching) systems of identical geometry?

To investigate this exciting question theoretical and experimental investigations of magnetization reversal of ferromagnetic nanodots and their switching field distributions have recently been performed in [43–45]. In these investigations the dynamical correlation function $C_{\text{dyn}} = \int_{-\infty}^{\infty} dt S_{z,i}(t) S_{z,j}(t+s)$ between pairs of magnetic moments i and j has been considered. Hereby, S_z is the vertical component of moment i or j at times t and $t+s$, respectively. The proposed expression gives information on time correlations in a given state. For $C_{\text{dyn}} \to 1$ the switching is correlated, while $C_{\text{dyn}} \to 0$ corresponds to stochastic noise. The dynamical correlation function was calculated numerically for $s = 0$ and variable distances r_{ij} to check for the degree of correlation during the switching of different ensembles of dipolar nanoparticles coupled by the dipolar interaction [43]. The dependence of the correlation function on the interdot distance is shown in Fig. 13.6.

Figure 13.6 demonstrates that the time-dependent correlations decrease slower than the dipolar coupling itself. Even the dots at rather large separations ($r_{ij} > 140$ nm) show strong correlations (see Fig. 13.6), which are responsible for the in-phase switching of magnetization, although the dipolar energy is very weak at these large distances. The physical essence of this dynamical phenomenon is the time-averaged minimization of the potential energy of the ensemble of dots, which manifests itself in the many-body long-range correlations and prohibits dephasing of individual magnetic moments. Hence, the many-body dynamical effects described here, correspond to a minimization of a spin dynamical version of the action, rather than to a mere minimization of static dipolar energy. The most intriguing outcome of presented investigation is the long-ranged order of dynamical correlations and, thus, their influence on the collective states close to the thermal instability. The importance of these long-range order interactions and correlations for thermally assisted switching is supported by experiments described in [43].

13.7 Control of Ferro- and Antiferromagnetic Domain Walls with Spin Currents

There are two main concepts of the magnetic data storage: the first one relies on magnetic domains or other two-dimensional magnetic objects as bits of information; the second one uses one-dimensional transition regions between two domains; that is domain walls, as bits of information. Previous chapters were devoted to the switching of magnetic domains between two binary states corresponding to the first concept. Within the second concept towards new storage and logic devices the information is written or deleted by the current- and field-driven motion of magnetic domain walls (DWs) [46–48]. There are several proposals for reading or writing devices. Generally, an object corresponding to a bit of information has to be moved to the reading or writing device or the device has to be moved towards the bit. Nevertheless, it is difficult to address each DW individually in both cases, because a current often moves neighboring bits (DWs) in the same direction, whilst a magnetic field requires a movement in opposite directions. Domain walls can be individually addressed by the stray field coming from a tip of a magnetic force microscope (MFM) [49]. A particular challenge is, however, the manipulation of narrow DWs like those in monolayer thick nanowires of Fe/W(111). The width of these walls is of the order of 2 nm, which is at least one order of magnitude smaller than the resolution of MFM. Therefore, new concepts for individual manipulations of such narrow DWs are required. One possible theoretical concept for such manipulations will be reviewed below [50–54].

In [50–52, 54] the manipulation of a magnetic DW was proposed to be performed by the tip of a Spin-Polarized Scanning Tunneling Microscope (SP-STM). The influence of an SP-STM tip on a DW has been studied by means of quantum- and classical atomistic spin dynamics as well as by Monte-Carlo (MC) simulations. Different modes, systems and time regimes were studied. The investigations have shown that this setup might be suitable for manipulation of ferromagnetic [50] as well as antiferromagnetic [51] domain walls. The theoretically proposed setup is depicted in Fig. 13.7.

An example is given by a ferromagnetic monolayer stripe of dimensions up to $L_x \times L_y = 40\,a \times 70\,a$ (15 × 26 nm (sc(001) or bcc(110) stacking) with the lattice constant a and classical Heisenberg moment $\mathbf{S_i} = (S_i{}^x, S_i{}^y, S_i{}^z)$ of unit length $\boldsymbol{\mu}_i/\mu_s$ at each lattice point. The magnetic properties are given by the following Hamiltonian:

$$\mathcal{H} = -J\sum_{\langle ij\rangle} \mathbf{S_i}\mathbf{S_j} - D_x\sum_i (S_i^x)^2 + D_z\sum_i (S_i^z)^2, \tag{13.10}$$

with $J > 0$ being the ferromagnetic exchange coupling between nearest neighbors, $D_x > 0$ an easy-axis and $D_z > 0$ a hard-axis anisotropy, respectively. The Heisenberg magnetic rotors were confined to the xy-plane by the anisotropies. The dipolar interaction of such an in-plane system is typically vanishing. Material parameters for Fe/W(110) or Co/Pt(111) monolayers ($J = 10...13$ meV, $D_x = 0...5$ meV and $D_z = 0...2.5$ meV) have been used in the calculations.

To describe the motion of DWs the generalized Landau-Lifshitz-Gilbert equation has been utilized:

$$\begin{aligned}\frac{\partial \mathbf{S}_i}{\partial t} = &- \frac{\gamma}{\left(1+\alpha^2\right)\mu_S}\mathbf{S}_i \times [\mathbf{H}_i + \alpha\,(\mathbf{S}_i \times \mathbf{H}_i)] \\ &+ \mathcal{C}\mathbf{S}_i \times \mathcal{T}_i + \mathcal{D}\mathbf{S}_i \times (\mathbf{S}_i \times \mathcal{T}_i)\ ,\end{aligned} \tag{13.11}$$

where γ is the gyromagnetic ratio, $\alpha = 0.025$ is the Gilbert damping, $\mathbf{H}_i = -\partial\mathcal{H}/\partial\mathbf{S}_i$ is the internal field, and $\mathcal{T}_\mathbf{i}$ is the polarized spin current. The last two terms are the contributions corresponding to the precession and relaxation terms of the torque from the tunnel current. The concept similar to the case of a spin valve $\mathcal{C} = 0$ and $\mathcal{D} = 1$ has been used. In the numerical MC calculations, the sd-model has been used to account for the spin torque of tunneling electrons:

$$\mathcal{H}_\mathcal{T} = -g\sum_i \mathcal{T}_i \cdot \mathbf{S}_i, \tag{13.12}$$

with the coupling constant $g = 1$. Here, the complete information on the current is provided by $\mathcal{T}_\mathbf{i}$.

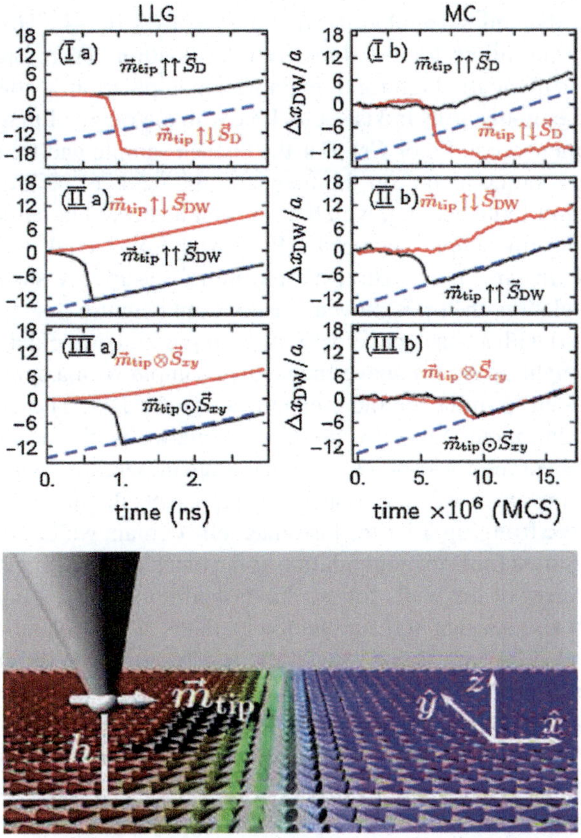

Fig. 13.7 Scheme of the initial state of the studied sample: two domains are separated by a DW elongated in y-direction, while the magnetic moments are pointing in $-x$ (blue) and $+x$ (red) directions. A tip with a magnetization $\boldsymbol{m}_{\text{tip}}$ and distance h from the sample moves towards the DW on the indicated track (white dashed line). Displacement of DW Δx_{DW} as a function of time obtained in MC simulations (left) and spin dynamics (LLG, right) are shown in the top panel. The tip moves with constant velocity marked by the dashed line. (I a) and (I b) correspond to a tip magnetization parallel and antiparallel to the initial domain ($\boldsymbol{m}_{\text{tip}}|| \pm x$), (II a) and (II b) parallel and antiparallel to the DW orientation ($\boldsymbol{m}_{\text{tip}}|| \pm y$), and (III a) and (III b) pointing into or out-of-plane ($\boldsymbol{m}_{\text{tip}}|| \pm z$) [50]

The local strength and the orientation of the tunneling current can often be described by the Tersoff-Hamann model:

$$\mathcal{T}_{\mathbf{i}} = -I_0 \cdot \mathrm{e}^{-2\kappa\sqrt{(x_i - x_{\text{tip}})^2 + (y_i - y_{\text{tip}})^2 + h^2}} \cdot P \cdot \boldsymbol{m}_{\text{tip}}, \tag{13.13}$$

where P is the polarization of the SP-STM tip with magnetization $\boldsymbol{m}_{\text{tip}}$, κ is the inverse decay length of the wave function in vacuum, the time dependent tip and atom positions are $\mathbf{r}_{\text{tip}} = (x_{\text{tip}}, y_{\text{tip}}, h, t)$, $\mathbf{r}_i = (x_i, y_i, 0, t)$, and the current density

I_0. In the spin dynamics simulation we set $I_0 = 1.0 \cdot 10^7 \frac{\mu_S}{\gamma t J}$, However, the MC procedure does not allow for the direct time evaluation. Therefore, we choose a current density sufficiently high for domain wall manipulation. κ contains the work-function ϕ of the tip material and is set to 4.5 eV, in the range of magnetic materials used in experiments (e.g. Fe or Cr). For the chosen sample and tip parameters, we assume the spin torque acting on the magnetic moments to be large compared to Oersted-fields and Joule heating, which have been neglected in our simulations.

The following tip-sample geometries have been considered: (i) $\boldsymbol{m}_{\text{tip}}|| \pm x$; (ii) $\boldsymbol{m}_{\text{tip}}|| \pm y$, and (iii) $\boldsymbol{m}_{\text{tip}}|| \pm z$. The explored initial set-ups are shown in Fig. 13.7. During the simulation, the tip is moved at a constant height $h \; || + z$ along the stripe (in $+x$-direction) with a constant velocity v_{tip}. All presented calculations correspond to a constant height of the tip above the studied sample with a spin current, which is sufficiently high to influence the magnetization of a DW. The time dependence of the DW displacement Δx_{DW} for the three scenarios is plotted in Fig. 13.7. The black/gray solid curves correspond to the parallel/antiparallel orientation of the tip to a corresponding axis while the dashed line represents the tip displacement.

As one can see from Fig. 13.7 for ferromagnetic domain walls, almost all studied geometries are suited for DW manipulation as the black and gray lines corresponding to the displacement of the walls follow the dashed line representing the motion of the tip. One can also see that both simulation methods give identical results with one exception revealed for $\boldsymbol{m}_{\text{tip}} \otimes \mathbf{S}_{\text{d}}$. This effect is a result of the different simulation models and disappears in the regime of strong currents, when neglecting the hard-axis anisotropy, and also in the regime of strong damping. The detailed analysis given in [50–52] has demonstrated that there are at least three different modes suitable for DW manipulation, but the investigation of costs and benefits reveals the $\boldsymbol{m}_{\text{tip}} \uparrow\uparrow \mathbf{S}_{\text{DW}}$ as the optimal one. Similar calculations have been performed to study the possibility of manipulation of antiferromagnetic domain walls, which possess zero net magnetization. It has been concluded in [51, 52] that in order to manipulate such domain walls, the $\boldsymbol{m}_{\text{tip}}$ has to be othogonal to the magnetization of the domain as well as the domain walls. The directionality of the domain wall motion is identical for $\boldsymbol{m}_{\text{tip}}|| \pm z$. To change the direction of the domain wall motion one has to change the orientation of the tip polarization to the opposite one [51, 52].

13.8 Conclusions

On the fundamental level, the studies that have been reviewed here can be extended in various ways: Most interesting is the generalization of the theory to the case of the Kondo lattice model, i.e., to several spins which represent, for example, a chain of magnetic atoms on a metallic surface. The real-time dynamics of such magnetic chains is crucially determined by the underlying electronic system, and there is not much known about such setups on the theoretical side. While long-time dynamics in the quantum-spin Kondo lattice is probably out of reach with the presently available numerical methods (also including the one-dimensional case), studying the classical-spin Kondo lattice appears quite promising and first computations are in progress. Particularly, two-dimensional systems featuring the Kosterlitz-Thouless transition, systems with superconducting substrates (modeled by extensions of standard BCS theory to the inhomogeneous and time-dependent case) and systems in the extreme adiabatic limit, where the electron dynamics can be assumed to follow instantaneously the trajectories of classical spin configurations, represent highly interesting avenues for future research in this field. On the level of applications, the developed methods can be extended to describe the non-equilibrium dynamics of systems with complex magnetic structures and magnetic quasiparticles for their successful manipulation.

Acknowledgements We would like to thank David Altwein, Mario Krizanac, Roman Rausch, Mohammad Sayad, Thim Stapelfeldt, Kolja Them, Robert Wieser for the intense and fruitful cooperation over the years. Financial support of this work by the Deutsche Forschungsgemeinschaft through the Sonderforschungsbereich 668 (project B3) is gratefully acknowledged.

References

1. K. Binder, Rep. Prog. Phys. **60**, 487 (1997)
2. S. Krause, G. Herzog, T. Stapelfeldt, L. Berbil-Baitusta, M. Bode, E.Y. Vedmedenko, R. Wiesendanger, Phys. Rev. Lett. **103**, 127202 (2009)
3. E. Tsymbal, I. Zutic (eds.), *Handbook of Spin Transport and Magnetism* (Chapman and Hall/CRC, Florida, 2011)
4. L.D. Landau, E.M. Lifshitz, Physik. Zeits. Sowjetunion **8**, 153 (1935)
5. T. Gilbert, Phys. Rev. **100**, 1243 (1955); T. Gilbert, Magnetics, IEEE Trans. **40**, 3443 (2004)
6. G. Tatara, H. Kohno, J. Shibata, Phys. Rep. **468**, 213 (2008)
7. G. Bertotti, I.D. Mayergoyz, C. Serpico, *Nonlinear Magnetization Dynamics in Nanosystemes* (Elsevier, Amsterdam, 2009)
8. B. Skubic, J. Hellsvik, L. Nordström, O. Eriksson, J. Phys.: Condens. Matter **20**, 315203 (2008)
9. D. Marx, J. Hutter, Ab initio molecular dynamics: theory and implementation,in *Modern Methods and Algorithms of Quantum Chemistry*. NIC Series, Vol. 1, Ed. by J. Grotendorst, p. 301 (John von Neumann Institute for Computing, Jülich, 2000)
10. M. Sayad, M. Potthoff, New J. Phys. **17**, 113058 (2015)
11. M. Sayad, R. Rausch, M. Potthoff, Phys. Rev. Lett. **117**, 127201 (2016)
12. M. Sayad, R. Rausch, M. Potthoff, Europhys. Lett. **116**, 17001 (2016)
13. C. Stahl, M. Potthoff, Phys. Rev. Lett. **119**, 227203 (2017).
14. M. Sayad, L.M. Woelk, R. Rausch, M. Potthoff, unpublished (2017)
15. M. Medvedyeva, A. Hoffmann, S. Kehrein, Phys. Rev. B **88**, 094306 (2013)
16. M. Nuss, M. Ganahl, E. Arrigoni, W. von der Linden, H.G. Evertz, Phys. Rev. B **91**, 085127 (2015)
17. A. Sakuma, J. Phys. Soc. Jpn. **81**, 084701 (2012)
18. S. Bhattacharjee, L. Nordström, J. Fransson, Phys. Rev. Lett. **108**, 057204 (2012)
19. H.P. Breuer, F. Petruccione, *The Theory of Open Quantum Systems* (Oxford University Press, New York, 2002)
20. J. Fransson, Nanotechnology **19**, 285714 (2008)
21. E.M. Hankiewicz, G. Vignale, Y. Tserkovnyak, Phys. Rev. B **78**, 020404(R) (2008)
22. I. Garate, A. MacDonald, Phys. Rev. B **79**, 064404 (2009)
23. K.M.D. Hals, K. Flensberg, M.S. Rudner, Phys. Rev. B **92**, 094403 (2015)
24. M. Morgenstern, Science **329**, 1609 (2010)

25. S. Loth, M. Etzkorn, C.P. Lutz, D.M. Eigler, A.J. Heinrich, Science **329**, 1628 (2010)
26. A.A. Khajetoorians, J. Wiebe, B. Chilian, R. Wiesendanger, Science **332**, 1062 (2011)
27. F. Gebhard, *The Mott Metal-Insulator Transition* (Springer, Berlin, 1997)
28. J. Hubbard, Proc. R. Soc. Lond. A **276**, 238 (1963)
29. U. Schollwöck, Ann. Phys. (N.Y.) **326**, 96 (2011)
30. J. Haegeman, J.I. Cirac, T.J. Osborne, I. Pižorn, H. Verschelde, F. Verstraete, Phys. Rev. Lett. **107**, 070601 (2011)
31. J. Haegeman, C. Lubich, I. Oseledets, B. Vandereycken, F. Verstraete, Phys. Rev. B **94**, 165116 (2016)
32. L.N. Buleavskiĭ, Sov. J. Exp. Theor. Phys. **24**, 154 (1967)
33. M. Moeckel, S. Kehrein, Phys. Rev. Lett. **100**, 175702 (2008)
34. M. Moeckel, S. Kehrein, New J. Phys. **12**, 055016 (2010)
35. M. Kollar, F.A. Wolf, M. Eckstein, Phys. Rev. B **84**, 054304 (2011)
36. M. Marcuzzi, J. Marino, A. Gambassi, A. Silva, Phys. Rev. Lett. **111**, 197203 (2013)
37. A. Rosch, D. Rasch, B. Binz, M. Vojta, Phys. Rev. Lett. **101**, 265301 (2008)
38. N. Strohmaier, D. Greif, R. Jördens, L. Tarruell, H. Moritz, T. Esslinger, Phys. Rev. Lett. **104**, 080401 (2010)
39. F. Hofmann, M. Potthoff, Phys. Rev. B **85**, 205127 (2012)
40. R. Rausch, M. Potthoff, New J. Phys. **18**, 023033 (2016)
41. E. Vedmedenko, N. Mikuszeit, T. Stapelfeldt, R. Wieser, A.L.M. Potthoff, R. Wiesendanger, Eur. Phys. J. B **80**, 331 (2011)
42. O. Iglesias, A. Labarta, Phys. Rev. B **63**, 184416 (2001)
43. A. Neumann, D. Altwein, C.T. En, R. Wieser, A. Berger, A. Meyer, E.Y. Vedmedenko, H.P. Oepen, New J. Phys. **16**, 083012 (2014)
44. M. Krizanac, E.Y. Vedmedenko, R. Wiesendanger, New J. Phys. **18**, 033029 (2016)
45. M. Krizanac, E.Y. Vedmedenko, R. Wiesendanger, New J. Phys. **19**, 013032 (2017)
46. M.K. et al., J. Magn. Magn. Mater. **14**, 53 (2009)
47. R. Wieser, E.Y. Vedmedenko, R. Wiesendanger, Phys. Rev. B **79**, 144412 (2009)
48. R. Wieser, E.Y. Vedmedenko, P. Weinberger, R. Wiesendanger, Phys. Rev. B **82**, 144430 (2010)
49. T. Yamaoka, K. Watanabe, Y. Shirakawabe, K. Chinone, E. Saitoh, H. Miyajima, Jpn. J. Appl. Phys. **45**, 2230 (2006)
50. T. Stapelfeldt, R. Wieser, E.Y. Vedmedenko, R. Wiesendanger, Phys. Rev. Lett. **107**, 027203 (2011)
51. R. Wieser, E.Y. Vedmedenko, R. Wiesendanger, Phys. Rev. Lett. **106**, 067204 (2011)
52. R. Wieser, T. Stapelfeldt, E.Y. Vedmedenko, R. Wiesendanger, Eur. Phys. Lett. **97**, 17009 (2012)
53. P. Weinberger, E.Y. Vedmedenko, R. Wieser, R. Wiesendanger, Philoso. Magn. **91**, 2248 (2011)
54. K. Them, E.Y. Vedmedenko, K. Fredenhagen, R. Wiesendanger, J. Phys. A: Math. Theor. **48**, 075301 (2014)

Chapter 14
Picosecond Magnetization Dynamics of Nanostructures Imaged with Pump–Probe Techniques in the Visible and Soft X-Ray Spectral Range

Philipp Wessels and Markus Drescher

Abstract The most direct way of accessing and understanding fast dynamical processes in nature is by capturing the motion in real space with a high temporal and spatial resolution. This chapter details time-resolved imaging techniques for probing the transient evolution of the magnetization in small magnetic systems in the visible and soft X-ray spectral range. Optical methods using femtosecond laser pulses can follow ultrafast processes with an extreme temporal resolution. The spatial resolution, however, is limited by diffraction to a few hundred nanometers at visible wavelengths. The dynamics of smaller structures can be investigated using X-ray microscopy at synchrotron radiation sources. A resolution of a few ten nanometers can be achieved, however, the time-resolution is limited to a few hundred picoseconds due to the pulse duration of the synchrotron bunches. Spin-wave packets are captured by optical methods using a time-resolved confocal Kerr microscope where backward volume spin-wave packets with counterpropagating group- and phase velocity are observed directly. Time-resolved X-ray microscopy is used to monitor the destruction and emergence of equilibrium domain patterns out of uniformly magnetized states.

Magnetic systems have proven to be very efficient data storage elements so that nowadays the majority of information is recorded on magnetic hard-disk-drives. Information can be encoded in a persistent way by aligning the magnetization of ferromagnetic elements. In order to achieve a high data-density, the size of these magnetic bits has to decrease and has approached domain sizes around 20 nm at 1 Tbit/in^2 data densities in commercially available hard disks. The nature of magnetism in such nanoscale systems has to be understood very well in order to further diminish the domain size and the spacing between adjacent bits. Ideally, a direct

P. Wessels · M. Drescher (✉)
The Hamburg Centre for Ultrafast Imaging (CUI), Hamburg, Germany
e-mail: markus.drescher@physik.uni-hamburg.de

P. Wessels
e-mail: pwessels@physnet.uni-hamburg.de

M. Drescher
Institut Für Experimentalphysik, University of Hamburg, Hamburg, Germany

R. Wiesendanger (ed.), *Atomic- and Nanoscale Magnetism*, NanoScience and Technology, https://doi.org/10.1007/978-3-319-99558-8_14

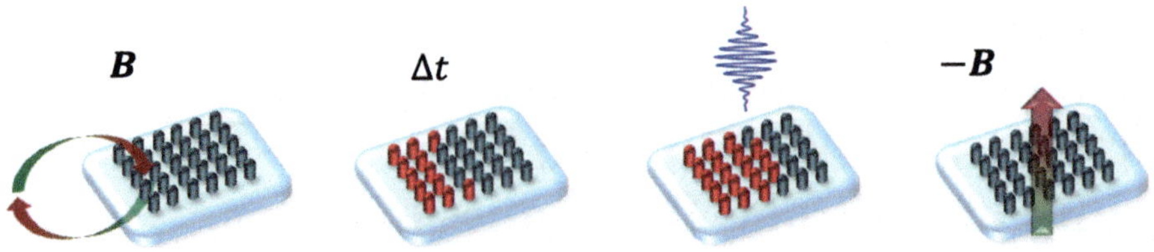

Fig. 14.1 Pump–probe/reset sequence for studying magnetization dynamics: A magnetic sample can be excited via a local magnetic field pulse $\boldsymbol{B}$ so that the magnetization evolves in time. Here, single-domain elements in a nanomagnetic array start reversing their magnetization (*red dots*). A snapshot of the magnetization pattern after a certain time Δt can be imaged by a light pulse. If the dynamic process is non-reversible, the magnetization has to be reset by a global field pulse $-\boldsymbol{B}$. Repetition of this process for a varying time delay Δt leads to a movie sequence of reproducible events. Adapted from P. Wessels, Ph.D. thesis (Verlag Dr. Hut, München, 2014) [1]

real-space observation is favored to investigate magnetic systems at such small length scales.

Magnetism, moreover, is no static phenomenon and domains in a magnetic system can reorganize and evolve in time. Adjacent magnetic elements, for example, may see the stray-field of neighboring elements and trigger dynamics that might delete stored information when decreasing the bit distance in storage media. A locally excited coupled spin-system will transfer such an excitation due to interactions so that spin-waves emerge. These dynamic processes occurring in micro- and nanoscale elements evolve on very fast time-scales in the nano- to femtosecond range so that an instrument capable of following magnetization dynamics in these elements does not only need a high spatial but also a high temporal resolution.

This can be achieved with ultrashort light pulses providing extremely fast imaging flashes for the observation of magnetization dynamics by employing the pump–probe technique introduced in Fig. 14.1. A sample is excited by a short magnetic field pulse and a snapshot of the magnetization is captured by a laser pulse after a variable time Δt. By repeating this procedure for several time delays Δt, a movie of the reproducible dynamics can be reconstructed. However, structures below 200 nm size can not be resolved with visible light due to the fundamental diffraction limit at optical wavelengths. The use of soft X-ray radiation available for example at synchrotron radiation sources extends the observable structural size down to a few nanometers where the resolution is limited by the fabrication capabilities of X-ray optics. However, the time-resolution usually is not competitive with laser-based instruments since the light pulses in a storage ring are typically a few hundred picoseconds long so that faster processes can not be followed.

In Sect. 14.1 we present how magnetization dynamics can be explored in real-space and real-time using optical techniques with a high temporal but a medium spatial resolution where the investigation is focused on spin-wave packets in magnetic micro-stripes. An opto-electronic high peak-current source enables the excitation and direct observation of backward volume spin-wave packets in permalloy ($Ni_{80}Fe_{20}$). These modes feature the unusual property of a negative group velocity where the phase fronts move in the opposite direction of the propagating wave packet.

Soft X-ray microscopy discussed in Sect. 14.2, allows for imaging with a higher spatial but a lower temporal resolution so that magnetization processes in nano- and microstructures can be followed where domain pattern destruction and recovery has been observed. For this purpose a full-field imaging X-ray microscope has been set up at the 3rd generation synchrotron source PETRA III that is equipped with a gated detector so that time-resolved pump–probe experiments can be performed. Here, an intense magnetic field pulse saturates the magnetization of a permalloy microsquare and subsequently the build-up of the equilibrium domain pattern can be followed by the X-ray microscope. This allows direct access to the domain formation process and sheds light on the question how (fast) magnetic domains form.

Recent experiments have shown alternative approaches for time-resolved imaging of small magnetic samples. For example an investigation of spin-wave solitons has been performed using X-ray microscopy [2]. Another experiment has demonstrated the time-resolved capabilities of scanning electron microscopes with polarization analysis [3].

14.1 Direct Observation of Spin-Wave Packets in Permalloy

Besides the routine use of ferromagnetic elements in data storage, magnetic structures also have the potential to solve a problem of increasing importance when building smaller and faster miniaturized circuits: In conventional electronic devices information is transported and processed by moving electrons steered by their electric charge. This involves friction so that heat is generated which limits the clock rate in such circuits, especially in mobile devices where active cooling is not an option.

Instead of using the charge for information transport and processing also the magnetic moment - or spin - of electrons can be utilized to complete these tasks. The advantage of magnetic transport is the negligible heat load since particle movement is not necessary because the information is transferred purely by interactions of neighboring spins. The research field of spintronics [4–6] is dedicated to building magnetic devices for signal transport and processing which may extend or even replace conventional electronic circuits. Magnetic moments can be driven into a precessional motion and the gyration frequency in metallic ferromagnetic structures such as permalloy ($Ni_{80}Fe_{20}$) is compatible with current clock rates of several GHz in conventional electronic microprocessors. This motion can be transferred in confined magnetic structures so that spin-waves emerge (compare also Fig. 14.3a). The wavelength of spin-waves in permalloy is in the μm to nm range also matching the size of logical units in conventional electronics.

Extensive research has been carried out to study the properties of spin-waves and their application to logic devices [7–9]. Most experiments focus on continuously excited spin-waves but to implement a scheme for encoding information the wave's amplitude or phase has to be modified dynamically. In this work we address spin-wave packets and investigate the dynamical properties such as phase-and group velocities as well as the dispersion of these wave packets.

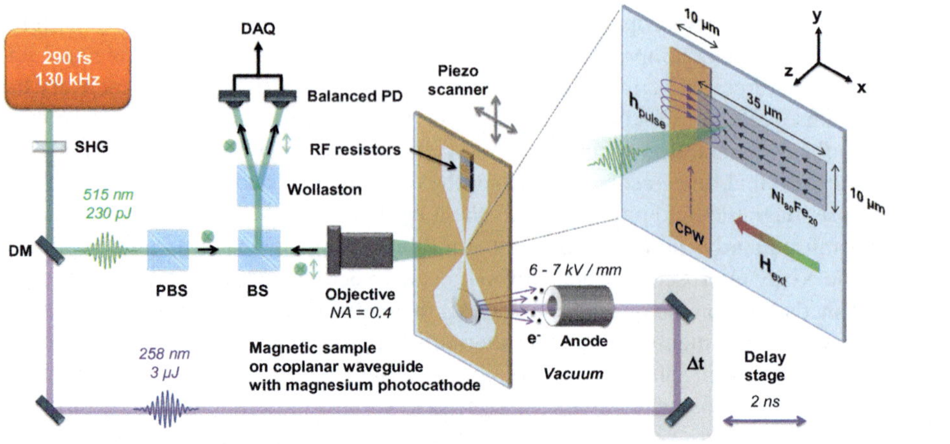

Fig. 14.2 Time-resolved scanning Kerr microscope: Ultrashort visible laser pulses of 290 fs pulse duration are converted into the ultra-violett (UV) spectral range via second harmonic generation (SHG). The delayable UV pulses trigger a magnesium photocathode on a coplanar waveguide (CPW) supplied by a static extraction field. The generated current pulses propagate through the waveguide and are terminated in a series of radio-frequency (RF) resistors. The co-propagating magnetic field $\mathbf{h}_{\text{pulse}}$ is able to excite spin-wave packets in a permalloy ($Ni_{80}Fe_{20}$) layer in the center of the waveguide. The residual visible light pulses are separated by a dichroic mirror (DM), poralization filtered by a polarizing beamsplitter (PBS) and focused onto the magnetic sample. The polarization change due to the magneto-optical Kerr effect (MOKE) is analyzed behind an intensity beamsplitter (BS) and Wollaston prism in a balanced photodetector (PD). This observable is proportional to the z component of the magnetization. By scanning the sample through the laser focus and variation of the pump–probe delay Δt, a movie of the spin-wave packets can be captured. Adapted from P. Wessels et al., Sci. Rep. **6**, 22117 (2016) [10]

The experimental setup for exciting and capturing spin-wave packets is presented in Fig. 14.2 and described in more detail in [1, 10]. Briefly, a confocal scanning microscope analyzes the polarization of a laser pulse in the visible spectral range reflected by a magnetic sample. The polarization change after reflection encodes information on the out-of-plane component of the sample magnetization via the magneto-optical Kerr effect (MOKE). Synchronized ultraviolet (UV) laser pulses release electrons via the photoeffect from a metallic photocathode fabricated onto the sample. The emerging current pulse propagates towards the magnetic sample in a tapered coplanar wave-guide (CPW) where the copropagating magnetic field is able to trigger magnetization dynamics in the vicinity of the waveguide conductor. The magnetic moments starts to gyrate around the axis of an external magnetic field and due to interactions a spin-wave packet is launched which propagates along the permalloy microstripe.

Spin-waves in confined magnetic structures can be excited in different configurations. Most popular are Damon-Eshbach (DE) modes where an external bias field is aligned perpendicular to the propagation direction of the spin-wave. These modes are easy to excite and feature high gyration frequencies and small wavelengths in fer-

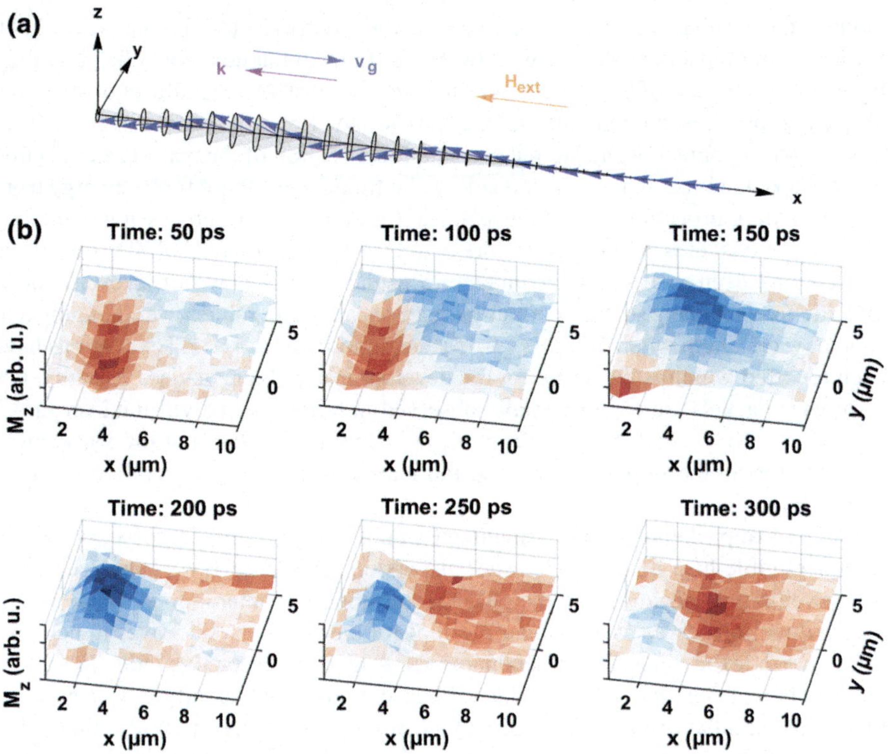

Fig. 14.3 Observation of a backward volume spin-wave packet: **a** Geometry of the precessional motion. An external bias field $\boldsymbol{H}_{\rm ext}$ aligned parallel to the wave vector **k** features backward volume spin-wave packets with a negative group velocity $\boldsymbol{v}_{\rm g}$ so that the phasefronts propagate into the opposite direction of the wave packet. **b** Snapshots of the magnetization for six delays Δt. The wave packet is excited on the left around $x = 2\,\mu$m and propagates to the right along the positive x axis. The phase-fronts clearly approach the source region from the right side indicating that phase- and group velocity are counteraligned. Reproduced from P. Wessels et al., Sci. Rep. **6**, 22117 (2016) [10]

romagnetic layers. Thus, experimental efforts have focused on continuously excited DE spin-waves via a radio-frequency (rf) current coupled into a waveguide. For information processing, however, an impulsive excitation is favorable so that information encoded into spin-wave bursts can be transported and processed. Damon-Eshbach spin-wave packets are directional, meaning that a certain propagation direction is favored. Similar to most coupled systems the oscillating phase-fronts propagate into the same direction as the wave packet envelope so that wave vector **k** and group velocity $\mathbf{v}_{\rm g}$ point into the same direction. For detailed studies of directly captured DE spin-wave packets in permalloy microstripes with the time-resolved MOKE microscope the reader is referred to [10].

By changing the direction of the bias field parallel to the propagation direction of the phase-fronts (compare Fig. 14.3a), another mode called backward volume (BV) mode is favored which is usually harder to excite compared to DE spin-waves [11]. We have captured and analyzed such BV modes in ferromagnetic metallic permalloy layers by making use of a high peak current source in form of a metallic photocathode. The results are presented in Fig. 14.3b. Characteristic for such modes is a phase front motion that is counteraligned to the propagation direction of the wave packet. While the wave packet is excited on the left side of the frame, new phase fronts emerge from the right side approaching the source region as time elapses. The precession frequency of the wave packet is 3.5 GHz and more details are discussed in [10]. BV spin-wave packets have been observed in ferrimagnetic insulators [12], however, an excitation in metallic ferromagnets such as permalloy features a considerable lower damping and smaller wavelength where micro- [13–15] and nanoscale [2, 16, 17] excitations have been reported which is favorable for the development of logic devices. Usually, BV spin-wave packets are experimentally observed in ferrimagnetic yttrium iron garnet $Y_3Fe_5O_{12}$ (YIG) which benefits from a low damping constant but the wavelength typically lies in the mm range and the fabrication of the monocrystalline layers is demanding.

Spin-wave packets not only implement a data transport and processing protocol with negligible heat load. The unusual dispersion relation of BV modes where the group velocity is negative and the phase front motion counteraligned to the propagation direction serves as a simulator toolkit for electromagnetic waves in negative refractive index media such as metamaterials. A negative refractive index ($n < 1$) is equivalent to a negative scalar product of phase- and group velocity ($\boldsymbol{v}_{\mathrm{ph}} \times \boldsymbol{v}_{\mathrm{g}} < 1$). Often only simulations provide access to the phase-resolved electromagnetic field in such materials. A sample with an engineered BV dispersion relation can serve as a model system for studying more complex spin-wave dynamics such as interference, reflection etc. for designing advanced metamaterials.

14.2 Time-Resolved Imaging of Domain Pattern Destruction and Recovery

Soft X-ray radiation as delivered by synchrotron radiation sources is a well suited tool for magnetic imaging. The short wavelength enables a high spatial resolution and via the X-ray magnetic circular dichroism (XMCD) the magnetization of a sample can be mapped with element specificity. Due to the pulsed character of synchrotron radiation also time-resolved experiments can be performed so that magnetization dynamics can be captured with a typical time resolution of about 100 ps.

X-ray microscopy has provided tremendous insight into magnetic nanostructures and their dynamics [18, 19]. Due to the high absorption of soft X-ray radiation in almost all materials the use of conventional refractive optics in such microscopes is not possible. Instead, one relies on holographic techniques or refractive optics such as zone plates where the outermost zone width determines the achievable resolution. Usually, these endstations operate as transmission microscopes where a thin sample

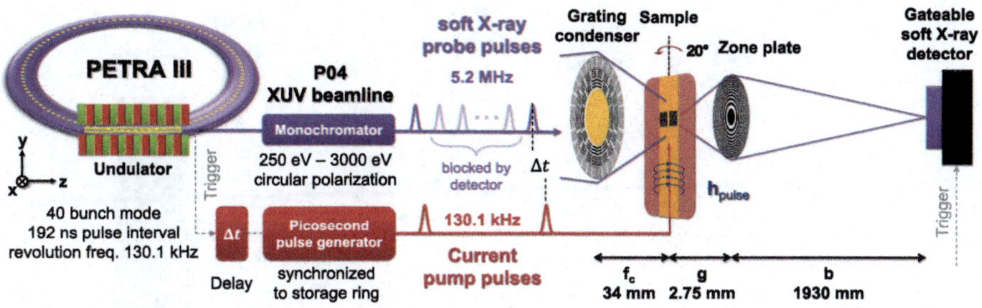

Fig. 14.4 Magnetic full-field X-ray microscope with time-resolution: Soft X-ray flashes generated by the storage ring PETRA III are monochromatized in the P04 beamline and focused onto a magnetic sample located on the conductor of a coplanar waveguide (CPW) by a grating condenser. The transmitted light is imaged by a zone plate onto a gateable detector. Magnetization dynamics can be triggered by injecting picosecond current pulses into the waveguide where the copropagating magnetic field $\mathbf{h}_{\text{pulse}}$ excites the sample. The dynamics can be followed by taking snapshots for several delays Δt between pump and probe pulse. Adapted with permission from P. Wessels et al., Phys. Rev. B **90**, 184417 (2014) [23]. Copyrighted by the American Physical Society

layer is deposited onto a silicon nitride membrane with a high transmission for soft X-ray radiation. For time-resolved measurements most instruments are operated in scanning mode where the X-rays are focused onto the sample by a zone plate and the transmitted radiation is detected by a fast photodiode. We have demonstrated the first realization of a full-field X-ray microscope with time-resolution where the sample is illuminated by a condenser zone plate and imaged by a micro zone plate onto a gated two–dimensional detector [20–23].

The experimental setup is depicted in Fig. 14.4. In contrast to many other X-ray microscopes, the developed instrument is not a fixed endstation but a mobile unit that can be installed for example at the 3rd generation synchrotron radiation source PETRA III at DESY in Hamburg. The soft X-ray undulator beamline P04 [24] delivers monochromatized circularly polarized photons in the energy range of 250–3000 eV. In time-resolved mode the storage ring is filled with 40 bunches circulating at a repetition rate of 130.1 kHz.

The synchrotron light is collected by a grating condenser [25] that creates a homogeneously illuminated spot on the magnetic sample. The transmitted light is imaged onto a two–dimensional detector by a micro zone plate with an outermost zone width of 50 nm so that the spatial resolution is limited to 65 nm according to the Rayleigh criterion.

The magnetic sample is excited by intense current pulses at a repetition rate of 130.1 kHz synchronized to the storage ring round trip frequency. The pulses are injected into a tapered coplanar waveguide and the copropagating magnetic field $\mathbf{h}_{\text{pulse}}$ is able to initiate magnetization dynamics in the sample located on the conductor of the waveguide. The sample is tilted by 20° to be sensitive to in-plane magnetization components.

The full-field microscope directly images the sample plane onto a two–dimensional detector. For time-resolved studies a single out of 40 bunches of the storage ring has

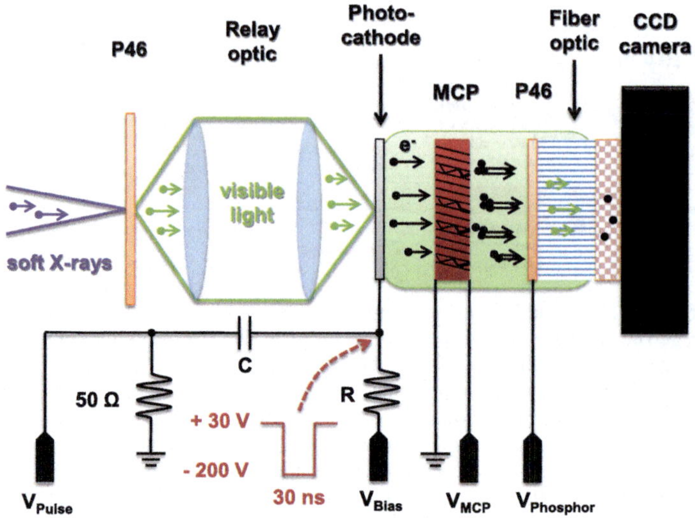

Fig. 14.5 Gateable two–dimensional X-ray detector: The sample image is projected onto a P46 phosphor screen and the emitted visible light transferred via a relay optic onto the photocathode of an image intensifier assembly. By applying a negative voltage pulse to the photocathode the emitted electrons are passed onto a microchannel plane (MCP) where the signal is amplified and converted into visible light by another P46 phosphor screen fiber-tapered onto a CCD camera. To block light from a bunch, a positive bias voltage is applied to the photocathode so that the electrons can not reach the MCP. Reproduced from J. Ewald et al., AIP Conf. Proc. **1696**, 020005 (2016) [22], with the permission of AIP Publishing

to be isolated. Available cameras cannot continuously be read out at the necessary 130 kHz rate, so that a gating technique has to be applied. Figure 14.5 depicts the layout of the gateable X-ray detector. The micro zone plate projects the sample plane onto a P46 phosphor screen mounted on a window inside the vacuum chamber. The emitted visible light pulse is transferred by two fast camera lenses in a relay configuration onto the photocathode of a commercial image intensifier. The photocathode is biased with a positive voltage so that the electrons by default can not escape the cathode surface. By applying a short negative gate pulse of 30 ns and −200 V amplitude the electrons are passed on to the front side of a microchannel plate (MCP) where the signal is amplified and accelerated onto another P46 phosphor screen that converts the electronic signal into visible light. The output port of the phosphor screen is fiber-tapered onto the CCD chip of a cooled camera where the image is recorded.

Due to the intense excitation pulse, the sample can only be excited at the round trip frequency of the storage ring of 130.1 kHz while the ring is filled with 40 bunches. This means that only one bunch contains a dynamic signal so that images generated by the remaining 39 bunches have to be filtered out by the gateable detector. The gate pulse duration of 30 ns and the decay time of the P46 phosphor screens ($\approx$200 ns) are fast enough to isolate light from single bunches in the detector (compare also [1, 21, 22]). For a fixed delay Δt the signal can be integrated on the camera chip until a sufficient signal-to-noise ratio has been reached.

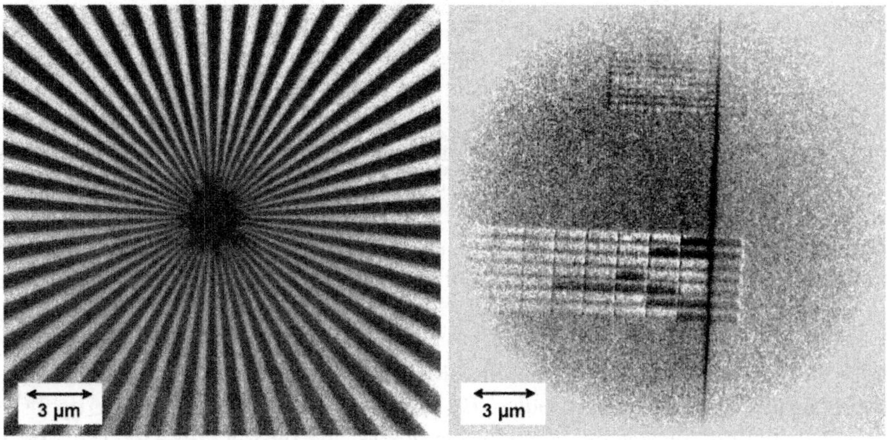

Fig. 14.6 Siemens star and array of magnetic elements: The Siemens star with 50 spokes is imaged at a magnification of $M = 700$. The central part of the XMCD image on the right shows the magnetization configuration of an array of $1.3 \times 0.3\,\mu\text{m}^2$ permalloy elements of 10 nm thickness separated by ≈40 nm. The magnetization of some elements points to the right side of the image (dark contrast) while other elements are aligned to the left side (white contrast). Adapted from P. Wessels, Ph.D. thesis (Verlag Dr. Hut, München, 2014) [1]

To demonstrate the imaging capabilities of the X-ray microscope, a Siemens star with 50 spokes is recorded at a magnification of $M = 700$ in Fig. 14.6 on the left. The Siemens star supports a resolution where features below 75 nm are discernible. The inner part of the Siemens star is obscured by lithography remnants so that no statement on a possibly higher resolution can be made here.

The (static) magnetic imaging capabilities are demonstrated on the right image in Fig. 14.6. An array of $1.3 \times 0.3\,\mu\text{m}^2$ permalloy elements of 10 nm thickness separated by 40 nm are presented here. The separation is still resolvable with the X-ray microscope. For XMCD measurements, the instrument is operated at the nickel L_3 edge at a photon energy of 852.7 eV with a bandpass of 1 eV. An XMCD image is obtained by capturing two images with circular polarized light at opposite photon helicities. The helicitiy can be changed from σ^+ to σ^- by shifting the magnetic structures in the beamline undulator. The difference of the two images directly maps the magnetic orientation in the sample. Elements aligned homogeneously along the x axis (dark contrast) as well as elements aligned into the opposite direction (white contrast) are visible. Dark contrast in the XMCD images corresponds to a magnetization aligned along the x or z axis while bright contrast indicated an alignment along the $-x$ or $-z$ axis. Due to the sample alignment with respect to the X-ray propagation direction (z axis) the instrument is not sensitive to magnetization components along the y axis. The magnetic signal is expected to originate mainly from the M_x component because of the sample geometry.

A larger homogeneously magnetized sample in general will decay into domains of constant magnetization pointing into different directions to minimize its internal energy. The equilibrium domain patterns for a large variety of structures and materials have been observed with static magnetic imaging techniques. Time-resolved approaches have also captured the dynamics followed by a weak transient perturbation of the equilibrium configuration [26–29]. However, observation of the essential process how (fast) an equilibrium domain pattern evolves out of a homogeneously magnetized sample was lacking so far. We have captured the destruction and recovery of the magnetization configuration in small $2 \times 2\,\mu m^2$ permalloy squares of 30 nm thickness. These elements will relax into a Landau pattern consisting of four triangular-shaped domains with a magnetization circling around a central vortex core (e.g. in Fig. 14.7 at a delay of 42.37 ns).

An intense current pulse launched into the coplanar waveguide on the sample can align the magnetization in such samples along the direction of the magnetic field that copropagates with the current pulse. By acquiring XMCD images using opposite photon helicities for several pump–probe delays, the destruction and recovery of the equilibrium Landau pattern can be monitored as a movie sequence with the time-resolved full-field X-ray microscope. Figure 14.7 shows the measured image sequence of the magnetization dynamics in two permalloy squares (I + II) located on the inner conductor of a CPW.

The indicated pump–probe delay is calibrated such that the peak amplitude of the magnetic field pulse of 438 ps FWHM duration is reached at 0 ps. The field of the excitation pulse is directed along the x axis. On arrival of the pulse the magnetization follows the excitation field and changes its direction on a fast timescale so that it points along the x axis (dark contrast) until a maximal alignment is reached at a delay of $\Delta t = 370$ ps with no detectable substructure. A closer look into the previous frames reveals that the vortex core in the center leaves the magnetic sample during destruction of the domain pattern (for a more elaborate discussion see [23]). Previous experiments have shown that the vortex core can be driven into an oscillatory motion by an external oscillating field and that it reverses its polarity at core velocities exceeding $\approx$300 m/s [30]. Due to the strong excitation field available here, we can study the vortex core motion in a new regime and find that the vortex core exceeds a velocity of 1 km/s and has no time to complete an oscillation period before it is ejected out of the sample.

After the pulse has vanished, the energetically unfavorable homogeneous magnetization configuration decays back into the equilibrium Landau pattern. The dynamics is similar for both squares (I + II) and up to 1 ns the M_x component (visible as dark contrast) is reduced homogeneously without any detectable substructure. A quantitative analysis of the gray values together with micromagnetic simulations [23] have shown that the majority of the magnetic polarization along the x axis is removed in this first fast phase when the x component decays from $M_x = 0.9M_S$ down to $M_x = 0.3M_S$ where M_S corresponds to the saturation magnetization. Subsequently, a zigzag domain starts building up on the top and another small M_x domain appears at the bottom until 3.37 ns where the decay of the polarization along the x axis slows down and reaches a value of $M_x = 0.2M_S$ at 3 ns. Although the M_x polarization has almost vanished the dynamics still continues and during the next 10 ns the magnetization reorganizes into the final Landau state with an imbalanced domain size.

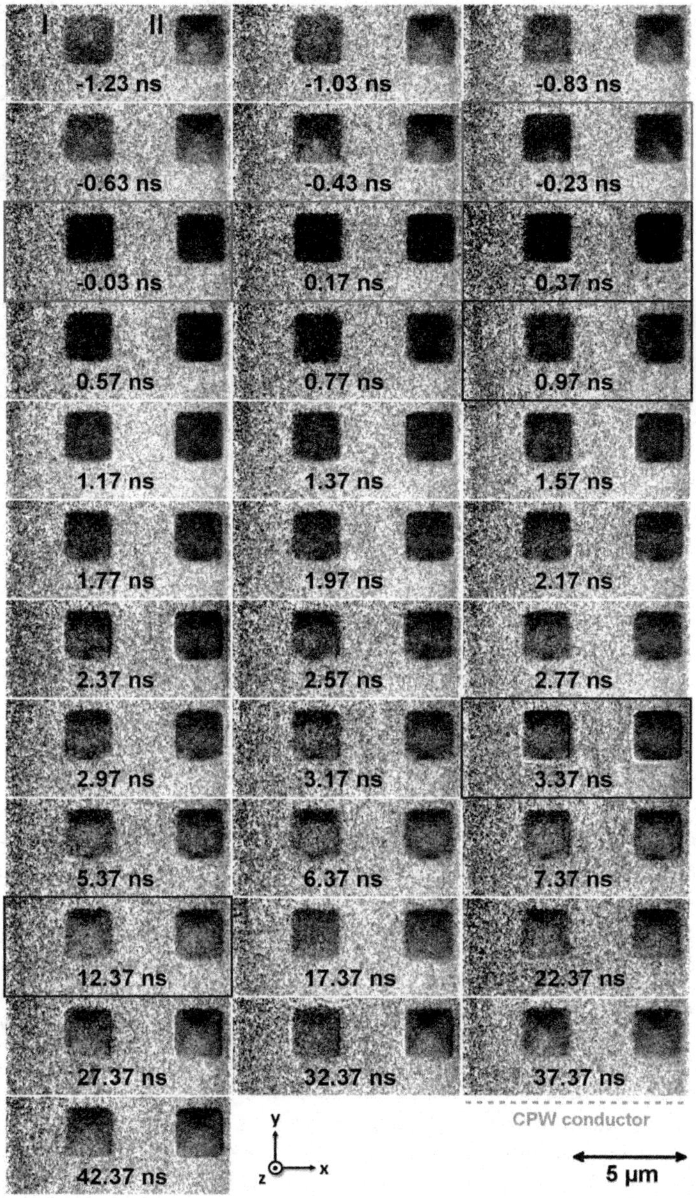

Fig. 14.7 Domain pattern destruction and recovery: Image sequence of the magnetization dynamics in two $2 \times 2\,\mu m^2$ permalloy squares (I + II) of 30 nm thickness exposed to a short magnetic field pulse (438 ps FWHM centered at 0 ps). Before arrival of the excitation pulse, the squares feature an equilibrium Landau domain pattern that consists of four domains. The pulse destroys this domain pattern and aligns the magnetization homogeneously along the x axis. After the pulse has vanishes the sample decays back to its Landau configuration. Adapted with permission from P. Wessels et al., Phys. Rev. B **90**, 184417 (2014) [23]. Copyrighted by the American Physical Society

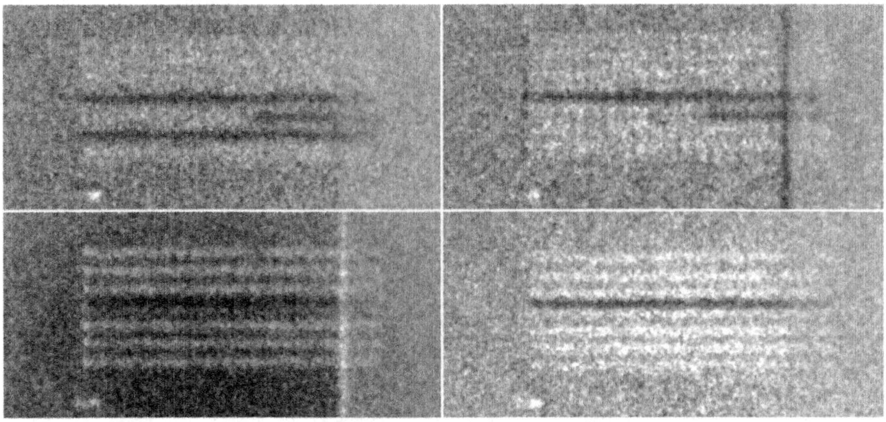

Fig. 14.8 Stochastic switching in a pump–probe / reset sequence: Dynamic magnetization of an array consisting of $400 \times 100\,\text{nm}^2$ elements separated by 50 nm horizontally and 250 nm vertically. The array partially overlaps with the waveguide conductor on the right and the XMCD images show the magnetization configuration at a fixed pump–probe delay Δt and constant reset pulse parameters after integrating the signal from several pump-probe cycles. The different magnetic orientation indicates a stochastic switching characteristic. Reproduced from P. Wessels, Ph.D. thesis (Verlag Dr. Hut, München, 2014) [1]

Between 12 ns and 42 ns the top M_x domain (black contrast) grows until the vortex core is positioned in the center of the square.

The previously discussed permalloy squares with a Landau magnetization pattern decay back into the ground state on their own. More complex magnetic systems such as arrays consisting of single domain particles coupled via their stray field can end up in a final stable state different from the initial state so that an active reset pulse is necessary to initialize the sample before the next measurement can begin (compare Fig. 14.1). Figure 14.8 shows XMCD images of an array of $400 \times 100\,\text{nm}^2$ elements separated by 50 nm horizontally and 250 nm vertically. For a pump–probe / reset cycle the magnetization of the whole array would be aligned along the $-x$ direction and then the elements on the waveguide switch their magnetization into the $+x$ direction. Due to coupling via the stray-field the chain reverses its magnetization direction element by element over time.

The images in Fig. 14.8 were taken by integrating the signal from several pump–probe cycles at the same delay and yet the dynamic magnetization configuration is different in all images indicating an unstable switching process so that a dynamic measurement was not possible since the pump–probe/reset approach relies on the reproducibility of the dynamics. In order to ensure a high reproducibility of the switching dynamics the indistinguishability of the elements and the spacings have to be improved. Moreover, external magnetic fields have to be compensated. Given a perfect sample preparation, simulations have shown that the switching dynamics should be observable with the instrument. Also arrays of lateral magnetic elements such as Co/Pt multilayer systems can be investigated and were imaged with the soft X-ray microscope [20].

14.3 Conclusion

In conclusion, two time-resolved instruments for imaging the dynamics of small magnetic structures have been set up. A laser based scanning Kerr microscope allows for the investigation of magnetic transport phenomena using visible light. An opto-electronic current source enables a direct observation of Damon-Eshbach as well as backward volume spin-wave packets with unusual dispersion relations so that the phase-fronts approach the excitation region while the wave packet propagate into the opposite direction. For experiments demanding a higher spatial resolution the first full-field transmission X-ray microscope has been set up that allows time-resolved measurements. The fundamental process of the destruction and recovery of equilibrium domain patterns out of uniform magnetization states can be followed with this instrument. Moreover, the vortex core motion in a strongly driven system has been studied where the core can not complete a round trip at its eigenfrequency and where the core velocity exceeds 300 m/s despite the fact that other experiments have observed a switching of the polarity at these velocities. First studies on non-reversible dynamics have been initiated.

Acknowledgements The authors would like to thank Andreas Vogel and Guido Meier for contributions to all sample preparation processes and Jan-Niklas Tödt for supporting the spin-wave data analysis and interpretation. The group of Thomas Wilhein is acknowledged for support in developing and running the X-ray microscope, especially during beamtime shifts attended by Johannes Ewald, Thomas Nisius, Gennaro Abbati, Stefan Baumbach, and Johannes Overbuschmann. We also thank Jens Viefhaus, Leif Glaser, Jörn Seltmann, and Frank Scholz for assistance in using beamline P04 as well as Pambos Charalambous (ZonePlates.com) for manufacturing of zone plate and grating condenser. Alexander Neumann and Hans Peter Oepen are acknowledged for providing Co/Pt multilayer structures. The authors are grateful to Benjamin Krüger and Daniela Pfannkuche for calculations of non-reversible switching dynamics in permalloy elements. Moreover, the authors would like to thank Marek Wieland for supporting these projects. Parts of this research were carried out at the light source PETRA III at DESY, a member of the Helmholtz Association (HGF). Financial support of the Deutsche Forschungsgemeinschaft via the SFB 668 (project B5) is gratefully acknowledged.

References

1. P. Wessels, Time-resolved imaging of magnetic nanostructures in the visible and soft x-ray spectral range. Ph.D. thesis, University of Hamburg (2014)
2. S. Bonetti, R. Kukreja, Z. Chen, F. Macià, J.M. Hernàndez, A. Eklund, D. Backes, J. Frisch, J. Katine, G. Malm, S. Urazhdin, A.D. Kent, J. Stöhr, H. Ohldag, H.A. Dürr, Nat. Commun. **6**, 8889 (2015)
3. R. Frömter, F. Kloodt, S. Rößler, A. Frauen, P. Staeck, D.R. Cavicchia, L. Bocklage, V. Röbisch, E. Quandt, H.P. Oepen, Appl. Phys. Lett. **108**, 142401 (2016)
4. S.A. Wolf, D.D. Awschalom, R.A. Buhrman, J.M. Daughton, S. von Molnár, M.L. Roukes, A.Y. Chtchelkanova, D.M. Treger, Science **294**, 1488 (2001)
5. A.V. Chumak, V.I. Vasyuchka, A.A. Serga, B. Hillebrands, Nat. Phys. **11**, 453 (2015)
6. C. Chappert, A. Fert, F.N. Van Dau, Nat. Mater. **6**, 813 (2007)

7. M.P. Kostylev, A.A. Serga, T. Schneider, B. Leven, B. Hillebrands, Appl. Phys. Lett. **87**, 153501 (2005)
8. M. Jamali, J.H. Kwon, S.M. Seo, K.J. Lee, H. Yang, Sci. Rep. **3**, 3160 (2013)
9. K. Vogt, F.Y. Fradin, J.E. Pearson, T. Sebastian, S.D. Bader, B. Hillebrands, A. Hoffmann, H. Schultheiss, Nat. Commun. **5**, 3727 (2014)
10. P. Wessels, A. Vogel, J.N. Tödt, M. Wieland, G. Meier, M. Drescher, Sci. Rep. **6**, 22117 (2016)
11. H.J.J. Liu, G.A. Riley, K.S. Buchanan, IEEE. Magn. Lett. **6**, 4000304 (2015)
12. T. Satoh, Y. Terui, R. Moriya, B.A. Ivanov, K. Ando, E. Saitoh, T. Shimura, K. Kuroda, Nat. Photonics **6**, 662 (2012)
13. Z. Liu, F. Giesen, X. Zhu, R.D. Sydora, M.R. Freeman, Phys. Rev. Lett. **98**, 087201 (2007)
14. K. Perzlmaier, G. Woltersdorf, C.H. Back, Phys. Rev. B **77**, 054425 (2008)
15. V.E. Demidov, S.O. Demokritov, K. Rott, P. Krzysteczko, G. Reiss, Phys. Rev. B **77**, 064406 (2008)
16. V.E. Demidov, S. Urazhdin, S.O. Demokritov, Nat. Mater. **9**, 984 (2010)
17. M. Madami, S. Bonetti, G. Consolo, S. Tacchi, G. Carlotti, G. Gubbiotti, F.B. Mancoff, M.A. Yar, J. Akerman, Nat. Nanotechnol. **6**, 635 (2011)
18. P. Fischer, IEEE. Trans. Magn. **51**, 0800131 (2015)
19. S. Bonetti, J. Phys.: Condens. Matter **29**, 133004 (2017)
20. P. Wessels, M. Schlie, M. Wieland, J. Ewald, G. Abbati, S. Baumbach, J. Overbuschmann, T. Nisius, A. Vogel, A. Neumann, A. Meents, J. Viefhaus, H.P. Oepen, G. Meier, T. Wilhein, M. Drescher, J. Phys.: Conf. Ser. **463**, 012023 (2013)
21. P. Wessels, J. Ewald, M. Wieland, T. Nisius, G. Abbati, S. Baumbach, J. Overbuschmann, A. Vogel, A. Neumann, J. Viefhaus, H.P. Oepen, G. Meier, T. Wilhein, M. Drescher, J. Phys.: Conf. Ser. **499**, 012009 (2014)
22. J. Ewald, P. Wessels, M. Wieland, T. Nisius, A. Vogel, G. Abbati, S. Baumbach, J. Overbuschmann, J. Viefhaus, G. Meier, T. Wilhein, M. Drescher, AIP. Conf. Proc. **1696**, 020005 (2016)
23. P. Wessels, J. Ewald, M. Wieland, T. Nisius, A. Vogel, J. Viefhaus, G. Meier, T. Wilhein, M. Drescher, Phys. Rev. B **90**, 184417 (2014)
24. J. Viefhaus, F. Scholz, S. Deinert, L. Glaser, M. Ilchen, J. Seltmann, P. Walter, F. Siewert, Nucl. Instrum. Methods Phys. Res., Sect. A **710**, 151 (2013)
25. U. Vogt, M. Lindblom, P. Charalambous, B. Kaulich, T. Wilhein, Opt. Lett. **31**, 1465 (2006)
26. S.B. Choe, Y. Acremann, A. Scholl, A. Bauer, A. Doran, J. Stöhr, HA. Padmore. Science **304**, 420 (2004)
27. B. Van Waeyenberge, A. Puzic, H. Stoll, K.W. Chou, T. Tyliszczak, R. Hertel, M. Fähnle, H. Brückl, K. Rott, G. Reiss, I. Neudecker, D. Weiss, C.H. Back, G. Schütz, Nature **444**, 461 (2006)
28. A. Vogel, T. Kamionka, M. Martens, A. Drews, K.W. Chou, T. Tyliszczak, H. Stoll, B. Van Waeyenberge, G. Meier, Phys. Rev. Lett. **106**, 137201 (2011)
29. C. Behncke, M. Hänze, C.F. Adolff, M. Weigand, G. Meier, Phys. Rev. B **91**, 224417 (2015)
30. K.S. Lee, S.K. Kim, Y.S. Yu, Y.S. Choi, K.Y. Guslienko, H. Jung, P. Fischer, Phys. Rev. Lett. **101**, 267206 (2008)

Chapter 15
Magnetic Antivortices

Matthias Pues and Guido Meier

Abstract We investigate the dynamics of magnetic antivortices by time-resolved magnetic X-ray microscopy and high-frequency absorption spectroscopy. A method for the reliable generation of isolated magnetic antivortices is devised using specifically formed microstructures and two-dimensional sequences of magnetic fields. For antivortices the measured resonance frequency is lower than for comparable vortices. The deflection of antivortices by external fields reveals strong deviations from a harmonic potential as well as rather low annihilation fields. Spectroscopy yields a characteristic absorption signal for antivortices for strong excitation that indicates a continuous switching of the antivortex core.

15.1 Introduction

Magnetic domains are regions with a homogeneous magnetisation. In soft magnetic materials like permalloy ($Ni_{80}Fe_{20}$) the magnetization in such domains lies in the film plane. A magnetisation pointing out of the plane would give rise to a large stray field and is prohibited by shape anisotropy. The coercive field of an extended film lies in the range of only a few mT. The in-plane magnetisation can be guided by the lateral confinement, as the shape anisotropy also tries to minimise the stray field on the sides, see Fig. 15.1. The energetically favoured state is a flux-closed magnetisation pattern, see Fig. 15.1c, d. The regions between the domains, the domain walls, can be separated into two main types: *Bloch walls* and *Néel walls*. In thin films, the

M. Pues
Institut für Angewandte Physik und Zentrum für Mikrostrukturforschung,
Jungiusstr. 11, 20355 Hamburg, Germany

G. Meier (✉)
Max-Planck Institute for the Structure and Dynamics of Matter,
Luruper Chaussee 149, 22761 Hamburg, Germany
e-mail: gmeier@mpsd.mpg.de

R. Wiesendanger (ed.), *Atomic- and Nanoscale Magnetism*, NanoScience and Technology, https://doi.org/10.1007/978-3-319-99558-8_15

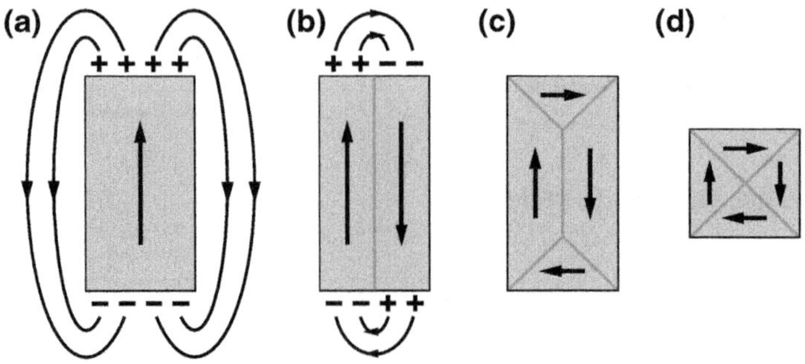

Fig. 15.1 Magnetic domains in in-plane magnetised thin film elements: **a** homogeneous magnetisation. **b** Two domains separated by a 180° domain wall. The stray field is reduced. **c** Triangular end domains close the inner flux. The so-called *Landau pattern* appears. **d** In equilateral polygons, like a square, the inner 180° domain wall becomes point-like

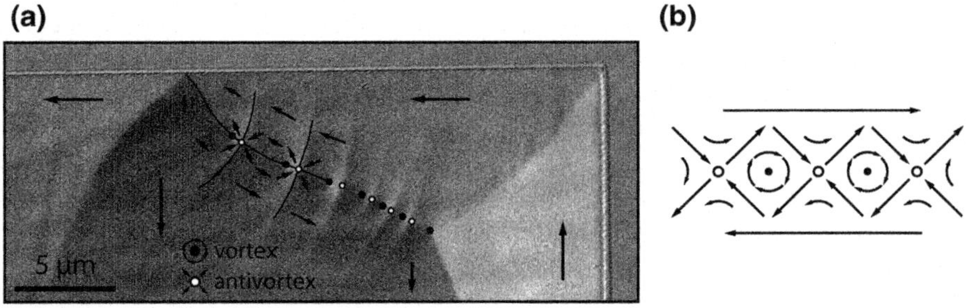

Fig. 15.2 Edge of a large permalloy rectangle observed by Kerr microscopy. The domains align along the edges of the microstructure. The chain of *vortices* and *antivortices* forming a *cross-tie wall* in the middle is partially marked. **b** Schematic of the magnetisation pattern of a 180° *cross-tie wall*

Bloch wall is characterised by a rotation of the magnetisation out of the plane of the substrate, whereas the rotation lies in the plane for Néel walls. For the investigated film thicknesses of up to 60 nm, Néel walls are the dominant type, but domain walls with a more complex inner structure also occur [1]. Figure 15.2a shows a larger permalloy rectangle, with an in-plane magnetisation, and a preferred orientation of the domains along the boundaries of the structure, as expected for this material. Some domain walls show no visible internal structure at this resolution, but the marked inner wall is a so-called *cross-tie wall*, a complex magnetisation pattern shown in Fig. 15.2b. This type of domain wall is a chain of special magnetisation points, where the opposing magnetic moments point antiparallel to each other. These magnetic quasi singularities are addressed as vortex and antivortex and are of special interest.

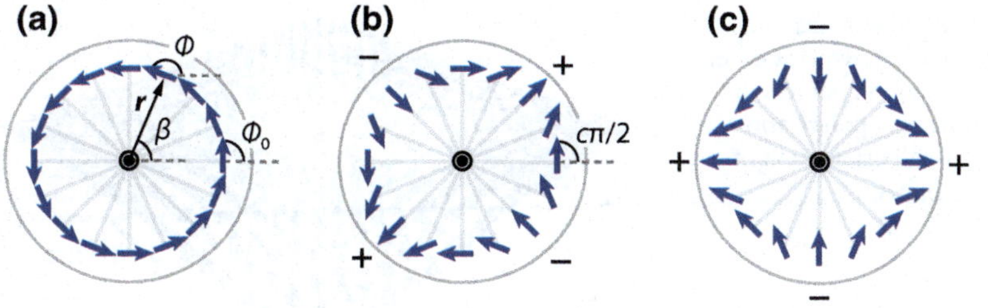

Fig. 15.3 Schematic magnetisation pattern of **a** a vortex with $c = +1$, and an antivortex with **b** $c = +1$ and **c** $c = 0$. The magnetic flux is not closed for the antivortex, indicated by the magnetic charges representing magnetic poles

15.2 Magnetic Singularities – Antivortices

The crossing points of the complex magnetisation pattern found in cross-tie walls are characterised by a specific winding of the magnetisation around the centre of those singularities [2]:

$$\phi(\beta) = n\beta + \phi_0 \quad \text{and} \quad \phi_0 = \frac{c\pi}{2}, \tag{15.1}$$

where ϕ is the angle of the local magnetisation vector with respect to the x-axis, β is the angle of the position vector $\boldsymbol{r}$ to the respective local magnetisation, see Fig. 15.3. The curling of the magnetisation is defined by the so-called winding number n, the orientation of the pattern by the offset angle ϕ_0. For $n = +1$, the magnetisation curls around the centre in a closed loop, see Fig. 15.3a, thus for $\phi = c\pi/2$ only two values are possible. The defining value $c = \pm 1$ is called circularity, as it determines the rotation direction of the magnetisation. For antivortices, with $n = -1$, the magnetisation curls in the opposite sense around the centre, see Fig. 15.3b. This pattern is not chiral, thus c can have any value in the interval $c \in [-2, 2[$ and is simply called the *c-value* or the orientation of the antivortex.

The antiparallel alignment of opposite moments in the centre of these magnetisation patterns is energetically unfavourable due to high exchange energy contributions for the large angles between neighbouring moments. This is overcome by a rotation out of the film plane of the centre magnetisation, see Fig. 15.4. The 180° angle is reduced to 90°. The orientation of the vortex or antivortex core up or down with respect to the z-axis is described by the polarisation p:

$$p = \operatorname{sgn} M_z(\boldsymbol{r} = 0) = \pm 1. \tag{15.2}$$

When considering the influence of external fields on both magnetic singularities, the quasi-static case and the dynamic case need to be discussed separately. In the quasi-static case, the change of the external field is slow compared to the intrinsic time scales of magnetization dynamics. Here, the domain pattern is shifted into the new

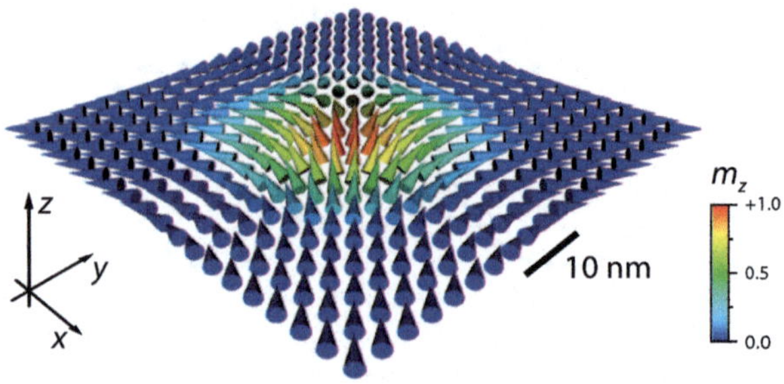

Fig. 15.4 Magnetisation pattern of the antivortex core. The inner region points out of the plane. A polarisation of $p = 1$ and orientation of $c = 1$ is shown

energetically favoured state according to the change in Zeeman energy induced by the external field. In a qualitative approach to estimate the new pattern, the domain with a magnetisation parallel to the applied field grows, while the one with an antiparallel magnetisation shrinks. For the dynamic case the Landau–Lifshitz–Gilbert equation (LLG) for the micromagnetic model can be used [3, 4]

$$\frac{d\boldsymbol{M}}{dt} = -\gamma \boldsymbol{M} \times \boldsymbol{H}_{\text{eff}} + \frac{\alpha}{M_{\text{s}}} \boldsymbol{M} \times \frac{d\boldsymbol{M}}{dt}, \tag{15.3}$$

where the effective field is the sum of the exchange field, the Zeeman field and the demagnetisation field

$$\boldsymbol{H}_{\text{eff}} = \boldsymbol{H}_{\text{ex}} + \boldsymbol{H}_{\text{Zeeman}} + \boldsymbol{H}_{\text{demag}}. \tag{15.4}$$

The gyromagnetic ratio in micromagnetism is defined as $\gamma = \mu_0 g e / 2m_e$. M_S is the saturation magnetisation and α a phenomenological damping term introduced by Gilbert [3]. The first term of this differential equation describes the Larmor precession of the magnetic moment $\boldsymbol{M}$ around the effective field $\boldsymbol{H}_{\text{eff}}$. For the complex magnetisation pattern of a vortex or an antivortex the displacement of the equilibrium position by an external field yields a spiral-like motion [5]. This can be illustrated for the starting point of the core of the singularity, as the effective field is parallel to the applied field for this particular moment, see Fig. 15.5. For a vortex, the infinitesimal right hand rotation caused by the Larmor precession results in an effective core motion antiparallel to the effective field, see Fig. 15.5c. The antivortex however has a more complex structure, see Fig. 15.5b, d. Depending on the orientation, the situation is similar to the vortex as for the magnetisation of an intersection line along $c = \pm 1$. The magnetisation rotates to an axis parallel to the intersection, a so-called *Bloch-line* [6]. For an intersection at $c = 0$, the magnetisation rotates around an axis perpendicular to the intersection line, the Larmor precession yields a perpendicular displacement of the magnetisation pattern, i.e. the antivortex core, see Fig. 15.5d, f. A detailed analytical description of the vortex and antivortex dynamics can be found in [7].

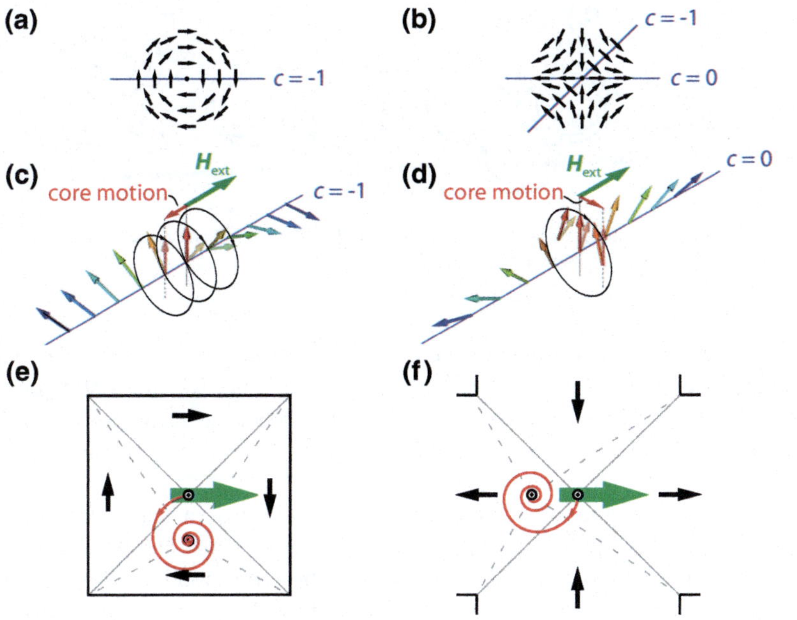

Fig. 15.5 Magnetisation pattern of **a** a vortex and **b** an antivortex with a polarisation $p = +1$. The magnetisation along certain intersection lines corresponds to **c** a *Bloch-line* for $c = \pm 1$ and **d** a *Néel-line* for $c = 0$. The Larmor precession induced by a magnetic field results in different effective core displacement directions. Gyrational motion of **e** a vortex core and **f** an antivortex core to the new equilibrium position caused by an external field

15.3 Antivortex Generation

The process of antivortex generation is understood by means of micromagnetic simulations and is experimentally demonstrated by X-ray transmission microscopy and magnetic force microscopy. The probability of antivortex nucleation is optimised by the variation of the structure geometry and the nucleation field strength. In contrast to the closed magnetisation of a vortex, which can be stabilised in thin-film structures with proper thickness and width that resemble a disc [8, 9], the in-plane magnetisation of an antivortex features four alternating poles, see Fig. 15.3b. The magnetic flux is not closed, but needs to be guided away from the centre region, thus the antivortex is metastable in a disc-like structure [10]. In the literature, different shapes have been shown to be suitable to stabilise antivortices [2, 11–14]. For micromagnetic simulations a star-like shape, whose contour is shaped by four adjacent discs, has been used [14], see Fig. 15.6a. The four apexes of the structure pin the magnetisation, pointing towards and away from the centre of the antivortex. The magnetic state can be predefined in the simulations, but to find the antiparallel alignment of the domains in real structures is very unlikely. Instead, in experiments this star-shape is used as a centre in a clover-shaped sample, see Fig. 15.6b, where wire loops close the magnetic flux

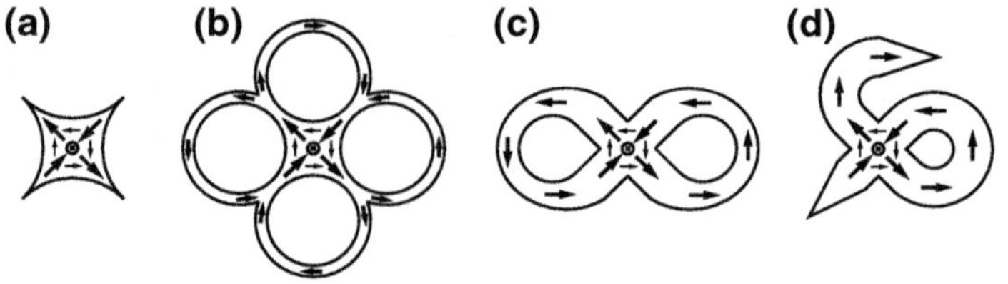

Fig. 15.6 Shapes of structures used to stabilise an antivortex: **a** star-like shape, **b** clover shape, **c** "infinity" structure, and **d** φ-shaped structure

[12]. After saturation of the magnetisation in such a sample, a successful regeneration of an antivortex with homogeneous external magnetic fields could not be observed. When altering the shape to a lemniscate, see Fig. 15.6c, the antivortex was found in some structures as the as-grown state. This so-called "infinity"-structure reduces the number of wire loops from four to two.

In principle the antivortex state could be recovered in such structures by a demagnetisation sequence of alternating fields of decreasing magnitude to annihilate possible multi-domain patterns that occur after the saturation [15]. Nonetheless the probability of the antivortex generation in these structures is too low to allow ensemble measurements. The reliable generation process proposed and demonstrated in this section makes use of a special shape of the structure and a two-dimensional bias field sequence to generate the antivortex. In addition, the field sequence is exactly specified and physically motivated so that it allows for high reproducibility and adaptability for other sample geometries. The φ-shaped design, see Fig. 15.6d, features three main characteristics that are crucial for the formation process: (i) tapered wire ends to pin the magnetisation along the wire and to prevent a magnetisation reversal after the saturation of the structure, (ii) curved segments to reproducibly create domain walls when saturated radially, and (iii) a wire junction in which the antivortex will be stabilised.[1]

The idea of the remagnetisation process is illustrated in Fig. 15.7a. The process starts with the saturation of the magnetisation in the φ-shaped structure in-plane with the saturation field H_s along the x-axis, see Fig. 15.7a step I. The coordinate system is defined by Fig. 15.7b. When the external field is lowered, the magnetisation aligns along the wires due to the now dominating shape anisotropy and two domain walls form in the curved segments, see Fig. 15.7a step II. This field amplitude is denoted as the nucleation field H_{nuc}. By a quasi-static counter-clockwise rotation of the in-plane field with a constant amplitude the domain walls are moved through the curved segments, see Fig. 15.7a II–IV. When both walls reach the wire junction, the antivortex nucleates, see Fig. 15.7a IV. The external field is then set to zero. The characterising parameters of such a remagnetisation, i.e. the saturation field H_s, the nucleation field H_{nuc}, and the nucleation field angle θ, cf. Fig. 15.7b, have to be

[1]Figures 15.6, 15.7, 15.9, 15.10, and 15.11 reprinted from M. Pues, M. Martens, T. Kamionka, and G. Meier, Applied Physics Letters **116**, 162404 (2012), with the permission of AIP Publishing.

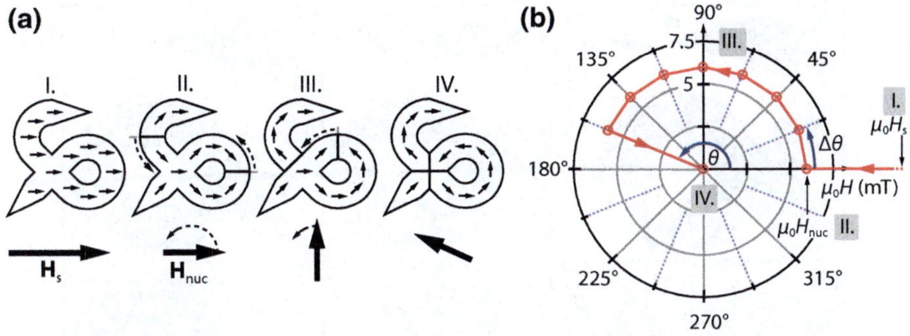

Fig. 15.7 **a** Schematic remagnetisation process of the φ-shaped structure: I. Saturation, II. domain-wall formation, III. domain-wall propagation driven by a quasi-static rotating magnetic field, IV. antivortex nucleation in the wire junction. **b** Polar graph of external in-plane fields used. An exemplary sequence of field steps for the remagnetisation process is shown

adjusted depending on the thickness, the wire width, and the edge roughness of each investigated structure.

The remagnetisation process is tested by means of micromagnetic simulations, as well as full field X-ray transmission microscopy performed at the Advanced Light Source (XM-1, beamline 6.1.2.) in Berkeley, CA, USA. In the simulation, the Gilbert damping is set to $\alpha = 0.5$. The high damping value is used as only the relaxed magnetisation states for the different bias fields are of interest, here. This and choosing a structure with a wire width of 0.5 mm, which is the size of the smallest experimentally analysed structures, reduces the computing time significantly. The height of the structure is 50 nm and the mesh cell size is $5 \times 5 \times 12.5\ \text{nm}^3$. The simulation starts from a parametrised homogeneous magnetisation parallel to the x-axis and an external field of 80 mT in the same direction. Subsequently, the bias field is reduced to 20 mT in four steps and then in nine steps to the nucleation field $\mu_0 H_{\text{nuc}} = 8\,\text{mT}$. The nucleation field is then rotated in steps of $\Delta\theta = 10°$. For the microscopy investigations, structures with two curved arms were used, see transmission X-ray micrographs in Fig. 15.8, for a more symmetric environment of the wire junction and to be able to investigate more domain walls within one microstructure. The elements have a wire width of 1 mm and a film thickness of 30 nm. To be able to achieve an in-plane contrast, the samples are tilted by 30° with respect to the X-ray beam path. The elements are tilted by $-22.5°$ in the plane with respect to the simulation in order to identify all four domains of the antivortex unequivocally when it is nucleated, cf. Fig. 15.8f, l. To depict the remagnetisation process the external field is set to the nucleation field strength after saturation at 100 mT in the x-direction and then rotated in-plane by steps of $\Delta\theta = 22.5°$. When decreasing the external field from saturation to the nucleation field, vortex domain walls form in the curved wire segments, see Fig. 15.8a, g. With the rotation of the external field, the domain walls move along the curved wire segments as expected. Different rotation angles are necessary in the simulation and the experimental microstructure, which can be explained by pinning

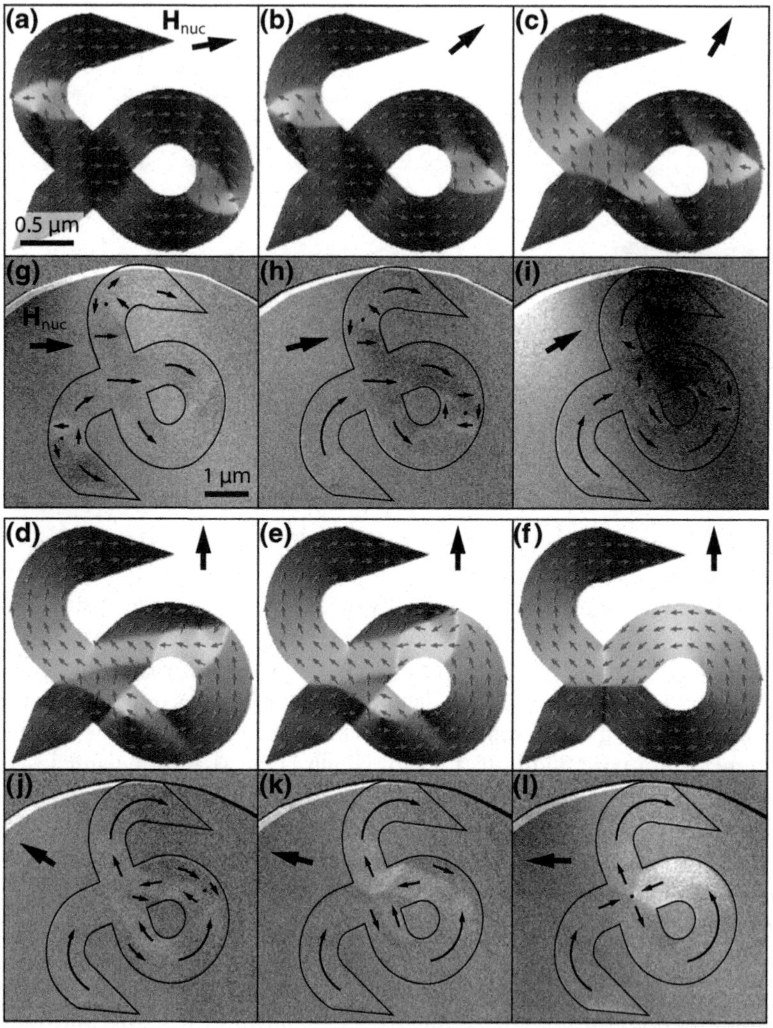

Fig. 15.8 Comparison of the simulated remagnetisation process (**a**)–(**f**) and transmission X-ray micrographs (**g**)–(**l**). Snapshots of the nucleation process for selected field angles are shown. The last images (*f*) and (**l**) show the stabilised antivortex. The nucleation fields are $\mu_0 H_{\mathrm{nuc}} = 8$ and 3 mT, respectively. The greyscale represents the x magnetisation in both image sets. The element is tilted deliberately by $-22.5°$ for the transmission microscopy samples to identify the antivortex unequivocally

of the domain walls at edge roughnesses on the wire walls and impurities within the wire. The vortex wall in the upper left arm moves into the junction, see Fig. 15.8c and (i), and forms an elongated 180° domain wall. From this wall, the antivortex nucleates, see Fig. 15.8d, and with the other vortex wall being pushed into the junction, the remaining vortices move towards the edges of the junction and annihilate.

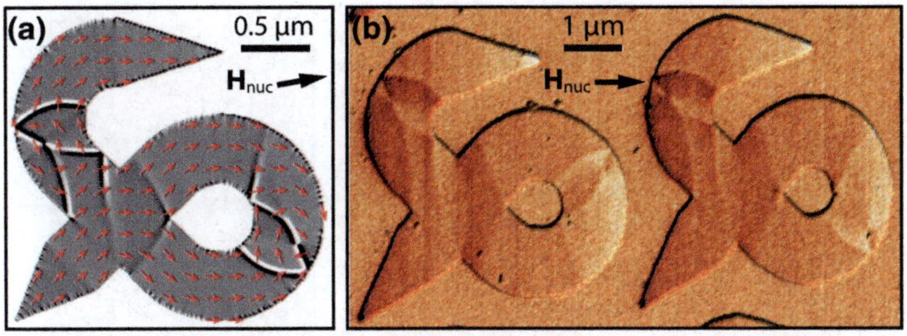

Fig. 15.9 Depiction of vortex domain walls in the curved wire segments by **a** micromagnetic simulation and **b** magnetic force microscopy. The greyscale in **a** represents the divergence of the magnetisation to be comparable with the MFM image. The nucleation fields are $\mu_0 H_{\text{nuc}} = 8$ mT and 5 mT, respectively

Finally in Fig. 15.8f, l the relaxed magnetisation constitutes an isolated antivortex that is stabilised in the wire junction.

The vortex walls can have either chiralities, showing no effect on the success of the remagnetisation process, which can be seen in the direct comparison of the simulated state and a magnetic force micrograph, see Fig. 15.9a, b. The magnetic force micrograph is recorded in an external field of 5 mT along the x-direction. Note that this and the following MFM images represent the first scan on that position after the saturation or remagnetisation. Thus the absence of characteristic lines in the micrographs due to a change of the magnetisation during the scan indicates a negligible influence of the magnetic tip on the magnetisation.

For a statistical analysis of the success of the remagnetisation process, arrays of identical structures were investigated with magnetic force microscopy. The array of structures in Fig. 15.10a are scanned after a remagnetisation with the parameters $\mu_0 H_s = 100$ mT and $\mu_0 H_{\text{nuc}} = 5$ mT. The maximum rotation angle is $\theta = 135°$. The external magnetic field is generated with a pair of Helmholtz coils surrounding the microscope and a manual rotation of the sample. The micrograph shows that an antivortex could be generated in all 25 structures. Since the remagnetisation process determines the direction of the magnetisation in the wires of the junction, thus dictating the orientation or c-value [2] of the antivortex, a remagnetisation with a negative saturation and nucleation field leads to the opposite orientation of the generated antivortex. With the fixed orientation of the structure to the chosen coordinate system, a remagnetisation with positive field amplitudes leads to $c = -1$, see Fig. 15.10b, and with negative field amplitudes to $c = +1$, see Fig. 15.10c. If the wire width w of the structure is varied, different nucleation fields are required to ensure a successful antivortex generationx . Detailed MFM analyses like depicted in Fig. 15.10 of further arrays of 25 to 36 identical structures with wire widths from 0.5 to 1.4 μm reveal the expected influence of w and H_{nuc} on the probability of successful antivortex generation, see Fig. 15.10d, which can be understood as follows. With higher nucleation fields the probability increases because the field is then strong enough to overcome a pinning due to edge roughness and move the domain walls

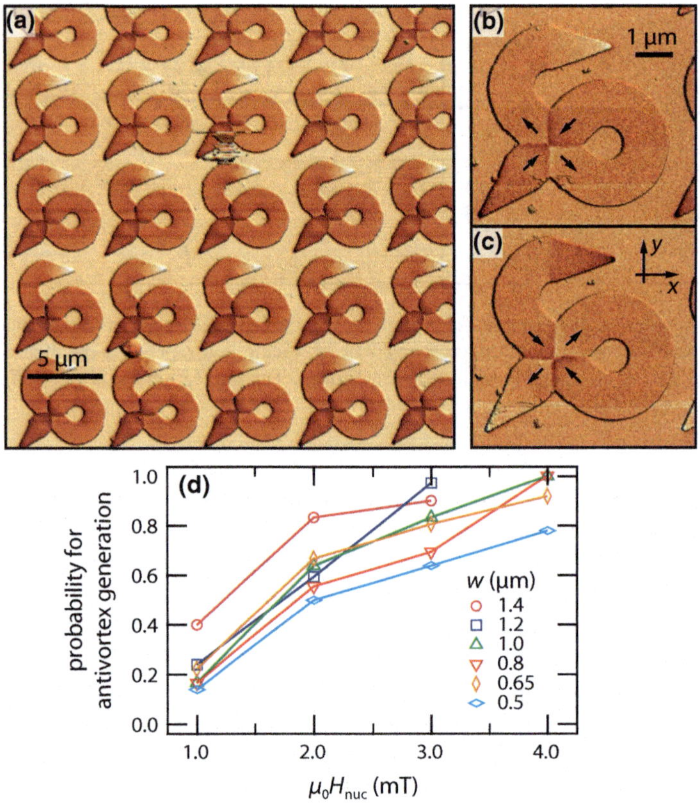

Fig. 15.10 **a** Magnetic force micrograph of an array of 25 φ-shaped structures after a remagnetisation process. All structures are in the antivortex state. Selected structures with antivortices of opposite orientations: **b** $c = -1$ and **c** $c = +1$. **d** Probability of a successful remagnetisation process for different nucleation fields and wire widths for an element height of 50 nm. Lines are guides to the eye

along the curved segments. The edge roughness is independent of w. Therefore, the ratio of the edge roughness to wire width becomes higher for smaller w. This results in a stronger pinning, explaining the higher fields necessary for successful antivortex generation for structures with smaller w. Note that the antivortex remagnetisation process is no longer successful for nucleation field strengths exceeding circa 10 mT, depending on the wire width.

The antivortex nucleation process has been investigated in detail showing how the manipulation of transient states of domain walls and vortices results in the desired antivortex state with a negative winding number of $n = -1$. The investigations reveal the reliability of the method for antivortex generation in φ-shaped structures with two-dimensional homogeneous in-plane fields. The antivortex-nucleation routine can now be used for spectroscopy, transport and X-ray microscopy measurements.

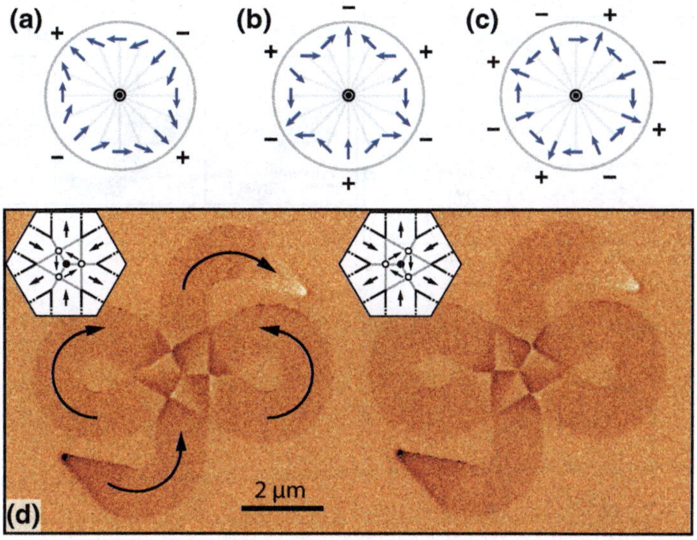

Fig. 15.11 Vector fields with winding numbers **a** $n = -1$, **b** $n = -2$, and **c** $n = -3$, all having an orientation of $c = -1$. **d** Magnetic force micrograph of two structures stabilising a magnetisation configuration in the junction with a total winding number of $n = -2$. The insets show the domain patterns within the junctions. Antivortices are indicated by circles, vortices by dots

15.4 Higher Winding Numbers

The method to control the magnetisation in curved nanowires with rotating in-plane fields can also be used to create even more complex domain patterns than antivortices. Magnetisation vector fields with decreasing winding numbers n exhibit increasing numbers of alternating magnetic poles around the centre, see Fig. 15.11a–c. Two structures designed so as to stabilise a hypothetical singularity with $n = -2$ are depicted in Fig. 15.11d. Before scanning the structures with a film thickness of 50 nm and a wire width of $w = 1.0\,\mu\text{m}$, they were remagnetised with a similar field sequence as the φ-structures with the parameters $\mu_0 H_s = 100\,\text{mT}$, $\mu_0 H_{nuc} = 5\,\text{mT}$, and $\theta = 360°$. The micrograph shows that the total winding number of the wire junction is in fact $n_{tot} = -2$. This is realised by the known stable singularities, by a vortex with $n = +1$, surrounded by three antivortices with $n = -1$. The chirality of the centered vortex is different in the two structures, resulting in a mirrored domain pattern, but not affecting the overall orientation.

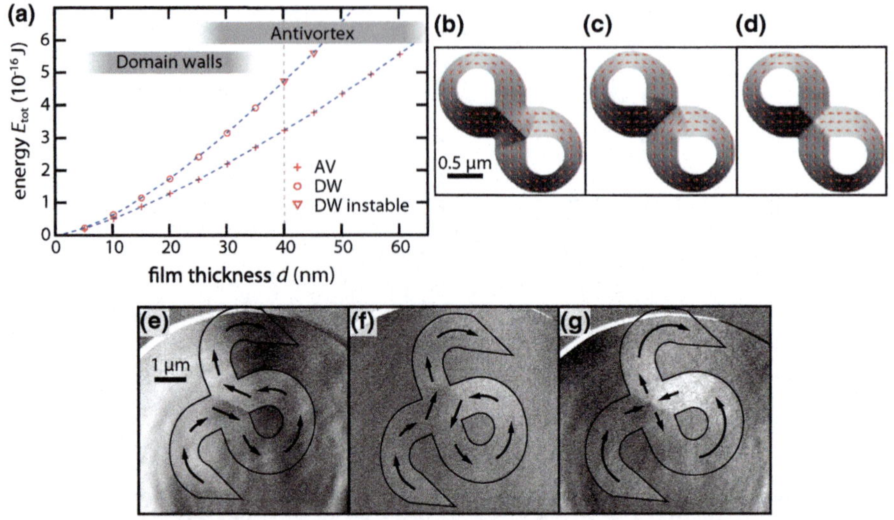

Fig. 15.12 **a** Total energy of antivortex or domain wall states obtained by micromagnetic simulations in infinity structures of different thickness. The two domain-wall states and the antivortex: **b–d** Simulated in a 50 nm thick infinity structure, and **e–g** measured by X-ray transmission microscopy in 30 nm thick symmetric φ-structures. Contrast shows the magnetisation in x-direction

15.5 Thickness Dependence

With the remagnetisation method discussed above an investigation of the thickness dependence of the antivortex and the opposing magnetic configuration, a 180° domain wall, is possible. A comparison of the total energy of the states obtained by micromagnetic simulations of infinity structures for film thicknesses of 5–60 nm shows that the domain wall can only be stabilised in elements with a thickness of up to 35 nm and transforms into an antivortex for a thicker element. The two domain wall configurations, see Fig. 15.12b, c, are nearly equal in energy, whereas the antivortex is always lower in energy, see Fig. 15.12a. As the simulations test only the static stability of predefined magnetic configurations, no direct conclusion can be drawn from the energy comparison to the probability of finding any of the different states in real elements, e.g. in the investigations by Bocklage et al. in [16], reporting an observation of a domain wall in a 20 nm thick infinity structure. Therefore, samples of symmetric φ-elements with thicknesses of 10, 20, 30, and 40 nm are investigated by full field transmission X-ray microscopy. The remagnetisation process is analysed for different nucleation fields, as well as the resulting magnetic states for the desired arm magnetisation. For the 10 and 20 nm samples only domain walls of both orientations can be observed, whereas for 30 nm, all three states occur, see Fig. 15.12e–g, cf. with (b)–(d). For 40 nm, only antivortices are generated by the remagnetisation sequence. This means that the domain wall is the dominant state for the used generation process for film heights up to 30 nm although it is higher in energy. Domain wall

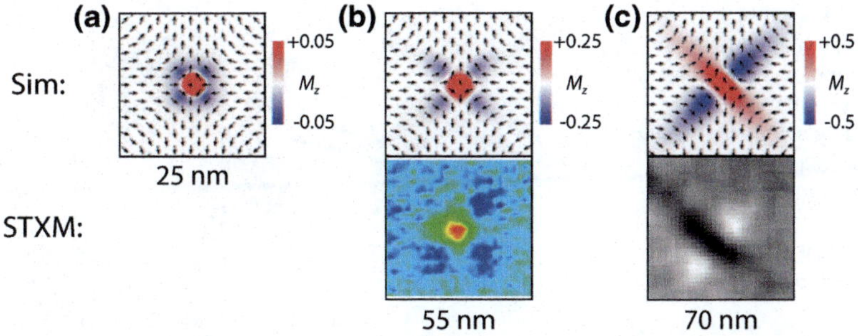

Fig. 15.13 Simulated antivortices for thicknesses **a** 25 nm, **b** 55 nm, and **c** 70 nm. Since the z-component is small for the thinner samples the colour scales are adjusted to 5, 25, and 50% M_z. The lower row shows scanning transmission X-ray micrographs of a 55 nm and a 70 nm thick structure at normal incidence

states cannot be found for thicker samples, as also indicated by the micromagnetic simulations, cf. shaded areas in Fig. 15.12a.

The upper thickness limit for antivortices is reached at 70 nm, where the four 90° domain walls from the kinks of the wire junction to the antivortex residing in the centre change from pure Néel walls to more complex walls. This can be seen in the change of the magnetisation pattern surrounding the antivortex core, simulated for different film heights, see Fig. 15.13. For thin films, the core is surrounded by a nearly circular region that points in an antiparallel direction to the core magnetisation out of the film plane. This ring divides into four distinguishable dips for 55 nm thick samples, as measured by scanning transmission X-ray microscopy at the MAXYMUS microscope of the BESSY II synchrotron in Berlin, Germany, see Fig. 15.13b. In direct proximity to the antivortex core, the 90° domain walls become Bloch like. For film heights above 65 nm the symmetry changes drastically, see Fig. 15.13c. The out-of-plane rotation of the domain walls becomes much more prominent and also two opposing domain walls change the sign of their z-magnetisation and fuse with the now parallel antivortex core. In this case the magnetic microstructure will no longer behave like a single 2D oscillator, as the antivortex or the vortex. No gyrational mode can be excited, but instead the antivortex performs a much more complex trajectory, since each domain wall itself is excited by the external fields. Moreover, the radius of the trajectory is decreased significantly. Consequently this analysis yields an optimal film thickness for isolated antivortices in permalloy from 40 to 60 nm.

15.6 Antivortices Influenced by Static and Dynamic External Magnetic Fields

With the method for antivortex generation proven to be reliable, the dynamics of isolated antivortices can be investigated spectroscopically by scalar network analysis. The influence of varying the exciting- and the static bias field on the gyrotropic

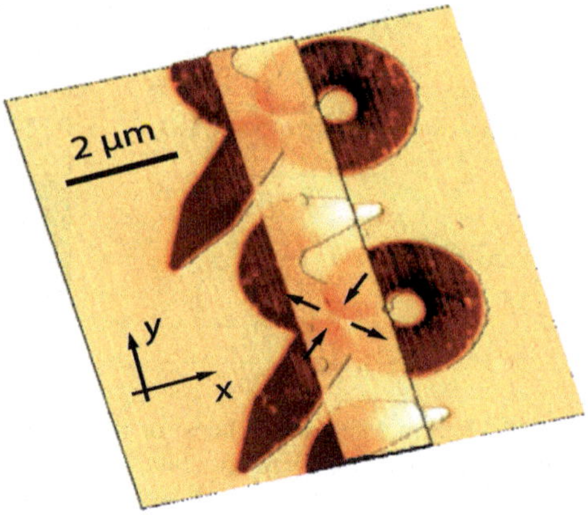

Fig. 15.14 Combined atomic and magnetic force micrograph of two φ-shaped permalloy microstructures overlaid by a copper stripline for high-frequency absorption measurements. The magnetic information shows an antivortex state in each structure

eigenmode of the antivortex is presented and discussed in the following. The spectroscopic measurements are complemented by magneto-resistance measurements and micromagnetic simulations [17].[2]

The resonance frequency of the gyrotropic eigenmode of isolated antivortices is determined by means of absorption spectroscopy. Figure 15.14 shows two φ structures of the ensemble, containing typically hundred antivortices, and the overlaying stripline, which generates the exciting Oersted field. A typical Lorentzian absorption spectrum for the antivortex ensemble is shown in Fig. 15.15b, with a resonance frequency of $f_{\mathrm{res}} = 169\,\mathrm{MHz}$. This frequency is approximately 40% lower than the resonance frequency of a vortex confined in a square-shaped element with the same film thickness and an edge length comparable to the wire width [20]. The lower frequency indicates a weaker confining potential if the antivortex is considered as a rigid quasiparticle as it is successfully done for vortices [21]. Absorption spectra for varying excitation power and thus for increasing Oersted field excitation are depicted in Fig. 15.15a. Three regions can be distinguished: (1) linear gyrotropic motion, (2) non-linear gyrotropic motion, and (3) continuous switching of the antivortex polarity. In regime (3) the cone-shaped absorption signal is caused by the continuous switching processes. For low excitation powers in regime (1), the resonance frequency is rather constant. The gyration of the antivortex can be described as a rigid quasiparticle in a nearly parabolic potential [19]. If the excitation power is increased, the resonance frequency drops significantly by about 20% in regime (2). In addition to the frequency drop, an asymmetric absorption curve can be observed in this non-linear regime, see Fig. 15.15c. This asymmetry has also been reported for numerical simulations and measurements of resonance curves of vortices by Drews et al. in [18]. The core switching regime (3) is reached at $\mu_0 H_{\mathrm{hf}} = 3\,\mathrm{mT}$, indicated by the characteristic cone shaped signal [22] above 3 mT. The Oersted field to induce antivortex core switching is about six times the field strength needed to induce core switching

[2]Figures 15.15, 15.16, 15.18, 15.19, 15.20, 15.21, and 15.22 reprinted from M. Pues, M. Martens, and G. Meier, Journal of Applied Physics **116**, 153903 (2014), with the permission of AIP Publishing

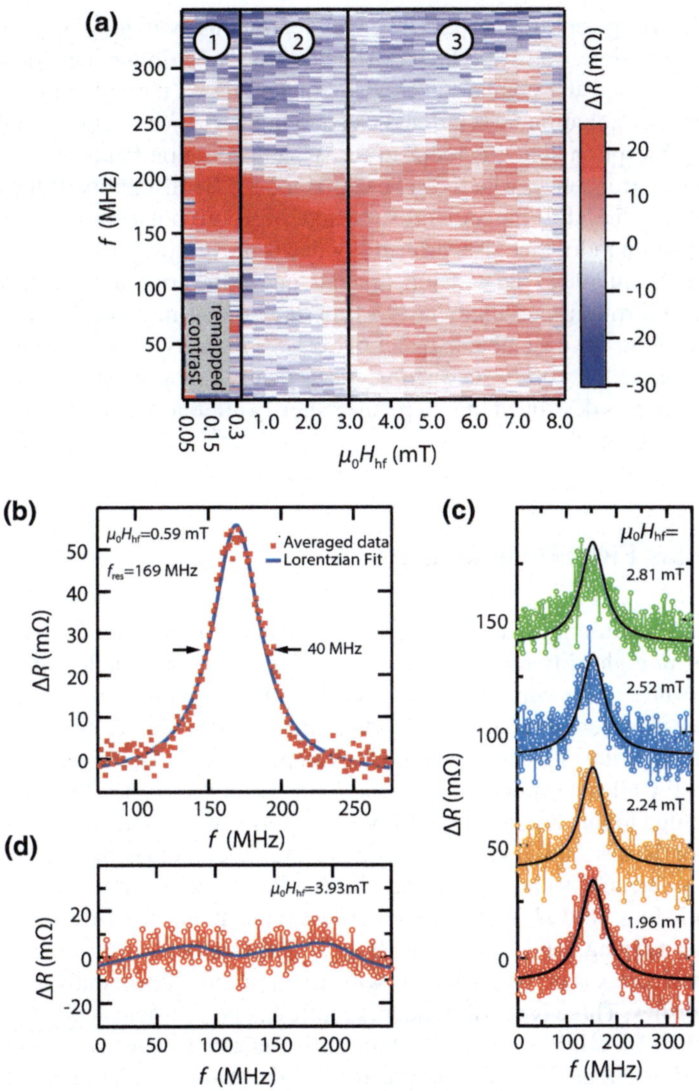

Fig. 15.15 Absorption spectra of antivortex generation. **a** Influence of the exciting field strength H_{hf}. The three regions are described below. **b** Frequency sweep in the linear regime. **c** Absorption signals for increasing excitation fields in the non-linear regime of the gyrotropic mode. The curves are offset successively by 50 mΩ. In order to show the increasing asymmetry and red shift of the absorption curves [18, 19], the same Lorentzian fit at the lowest excitation field strength of 1.96 mT is plotted as a solid line for all data curves. **d** Absorption signal from the core switching regime. The solid line is a guide to the eye

in a similar setup for vortices [20, 22]. The continuous switching of the polarity of the core is possible, since the linear excitation by the stripline can couple to both the clockwise and the counter-clockwise eigenmode of the antivortex. This is in contrast to the excitation by rotational Oersted fields [13] or rotational spin currents [15]. The absorption signal is reduced for strong excitation fields above 3 mT near the resonance frequency, which can be explained by the instability of the antivortex state, see Fig. 15.15d. If the antivortex switches its core polarisation multiple times at high-gyration radii, there is a probability that it moves away from the equilibrium position in the middle of the wire junction into one of the arms and is destroyed there. Since the absorption measurement is a time-integrating method, this process does not contribute to the ensemble signal. For vortices in discs, these considerations are irrelevant, since the vortex state is the energetically favoured state. Consequently, even if a vortex is destroyed, it will recover shortly after and will again contribute to the absorption signal.

15.7 Bias Field Dependence

To probe the confining potential of the antivortex, a static magnetic in-plane field is applied in eight different directions, $H_{\text{ext}}(\Theta)$, in order to deflect the antivortex from the centre of the wire junction. Since the generation process of the antivortices determines the orientation of all antivortices in the ensemble, see Sect. 15.3, a single deflection direction of the whole ensemble can be ensured. An orientation of $c = -1$ is chosen for all measurements. A detailed description of the shift of the energy minimum due to this Zeeman field for vortices has been presented by Langner et al. in [23]. For the absorption spectroscopy a low excitation field in the harmonic regime of $\mu_0 H_{\text{hf}} = 0.3$ mT is chosen, see Fig. 15.16. The resonance frequency dependence of the static bias field $H_{\text{ext}}(\Theta)$ shows a different behaviour for different field angles as well as a varying annihilation field $H_{\text{an}}(\Theta)$ of the antivortices. At the annihilation field the antivortex is pushed out of the wire junction and consequently the absorption signal vanishes. Three types of frequency shifts can be distinguished. The frequency shift labelled (I) exhibits a small drop in the resonance frequency of 6 MHz, which corresponds to about 4%, as well as a decrease in the absorption. The annihilation field is around 2 mT. The second type (II) is characterised by a similarly small annihilation field, but shows a frequency increase of about 33 MHz (20%) with an increase of the absorption signal for bias field values close to the annihilation field. The third type (III) has relatively high annihilation fields of up to 8 mT and the resonance frequency rises about 25 MHz (15%). The small annihilation fields of antivortices in comparison to annihilation fields of vortices of about 35 mT [20] indicate again the comparably shallow potential, which confines the antivortex within the wire junction. Moreover, the anisotropy of the annihilation fields and the frequency shifts show a strong influence of the wire arms and the asymmetry of the confining structure on the antivortex.

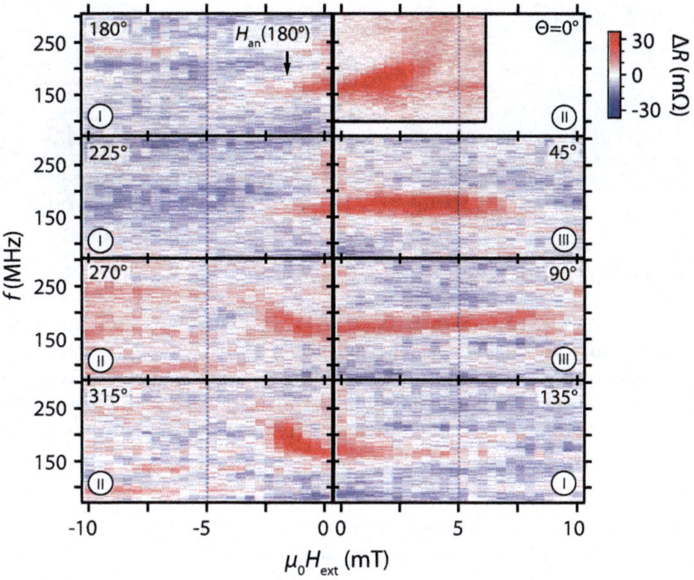

Fig. 15.16 Dependence of the antivortex resonance on a static bias field $H_{\text{ext}}(\Theta)$. The disappearance of the resonance signal for higher bias fields indicates the annihilation of the antivortex. The absorption spectra can be categorised into three types of frequency shifts (I, II, and III, compare Fig. 15.20)

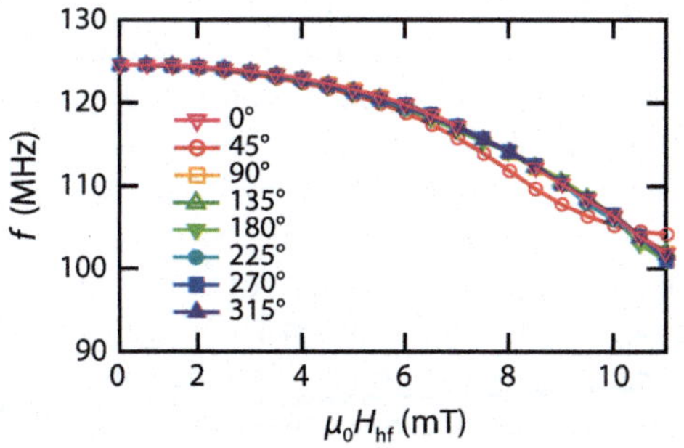

Fig. 15.17 Dependence of the free resonance frequency of the antivortex on a static in-plane bias field obtained by micromagnetic simulations. The frequency shift is nearly constant for every field angle, contrary to the experimental results. The change in frequency only deviates from the other curves for the field angle $\theta = 45°$ (red circles), when the antivortex is pushed into the straight arm

This complex resonance frequency dependence cannot be reproduced by micromagnetic simulations, cf. Figs. 15.16 and 15.17. To be able to achieve results for the eight different bias field angles within reasonable computation times, only the free gyration frequency is determined for the deflected antivortex. This may also give an explanation for the discrepancy to the experimental spectroscopy results. The

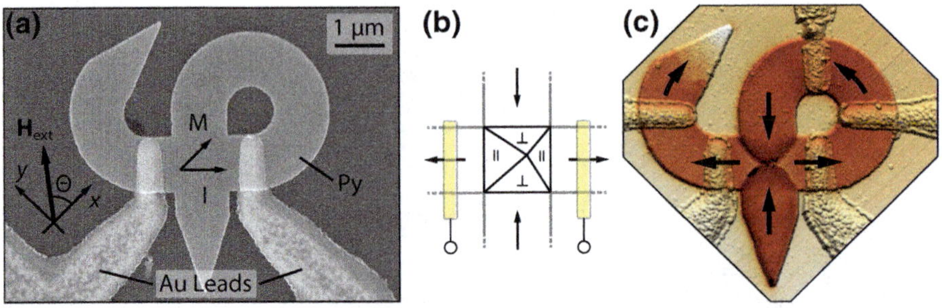

Fig. 15.18 **a** Scanning electron micrograph of a φ-shaped structure with Au contact leads for resistance measurements. **b** Scheme of a deflected antivortex. The triangularly shaped domains are denoted as parallel or perpendicular to the current flow. **c** Combined atomic and magnetic force micrograph of a contacted φ-structure containing an antivortex with an orientation of $c = -1$. For the MR measurements presented here, the structure with two contacts shown in **a** is used

simulations yield a uniform drop of the free gyration frequency for every bias field angle of about 23 MHz (18%). Only when pushed into the straight arm, the frequency shift differs slightly from the shifts caused by a bias field in the other directions, see Fig. 15.17.

To compare the drastic asymmetry of the annihilation field for gyrating antivortices with the annihilation fields for a static antivortex, magneto-resistance measurements are performed, see Fig. 15.19. The resistance of the wire junction of the φ-structure in saturation parallel to the current flow direction, indicated in Fig. 15.18a, is compared to the resistance after the antivortex generation and application of the bias in-plane field with angle Θ. This procedure is done for each data point, thus abrupt changes is the signal reveal the annihilation field of the antivortex for the corresponding bias field closest to zero fields in the MR signal. The following assumption leads to the identification of the antivortex state in the magneto-resistance signal: The magnetisation pattern of the antivortex consists of four triangularly shaped uniformly magnetised domains, see Fig. 15.18b, c. When a homogeneous current flows between the contact leads, a constant magnetisation within the wire arms, and no deformation of the 90° domain walls is assumed, the MR signal is expected to be constant to a good approximation for a deflected antivortex. Figure 15.19 shows this expected behaviour for the deflected antivortex state. The MR signal jumps from a nearly constant plateau to a MR signal indicating a magnetisation diagonal, perpendicular, or parallel to the current flow, see insets in Fig. 15.19. For some field angles the annihilation of the antivortex is followed by additional step-like transitions most likely due to sudden depinning processes of domain walls from the corners of the junction. As every data point represents a whole field sweep starting after an antivortex nucleation from zero field to either positive or negative field values, the smooth field dependence of the MR signal for each field angle indicates both a successful antivortex nucleation for every data point and a single path for the expulsion of the antivortex. The annihilation fields range from 7.2 to 12.7 mT and show a similar but much less distinct asymmetry compared to those derived from the absorption measurements, see Fig. 15.20a.

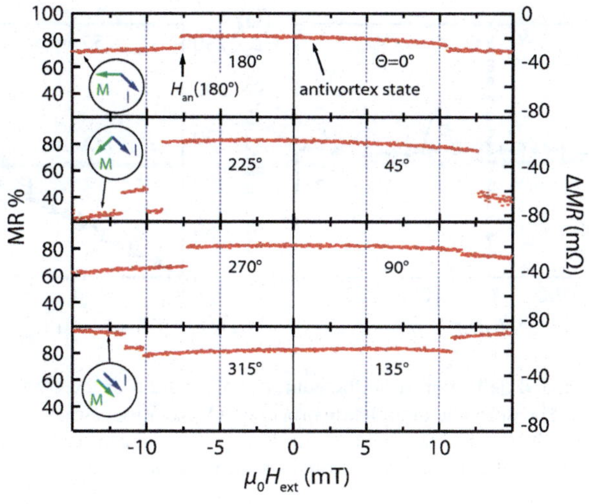

Fig. 15.19 Dependence of the magneto-resistance of an antivortex in the wire junction of a φ-shaped microstructure on a static bias field $H_{ext}(\Theta)$. The jumps in the signal indicate the annihilation of the antivortex

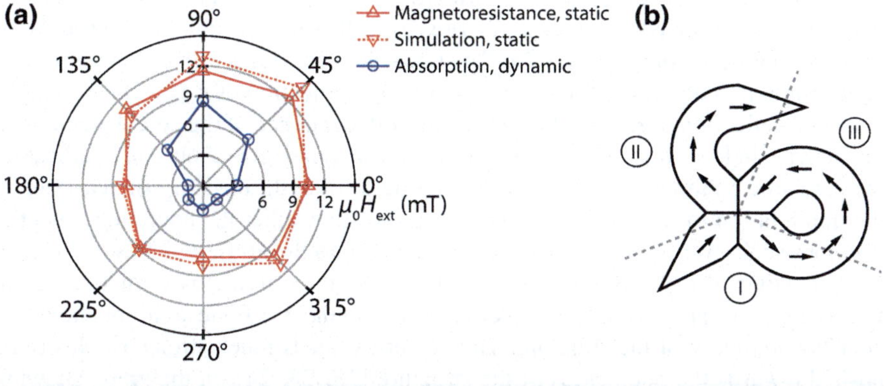

Fig. 15.20 **a** Comparison of the annihilation fields of the deflected antivortex for different in-plane field angles of the static field H_{ext}. Annihilation fields from magneto-resistance measurements, micromagnetic simulations, and high-frequency absorption measurements are shown. **b** Sketch of a φ-shaped structure. A deflection of the antivortex into one of the marked regions corresponds to the resonance frequency shift types from Fig. 15.16

15.8 Annihilation Process

Analogous to the MR measurements, micromagnetic simulations are performed, where the deflection of the antivortex by a quasi-static in-plane field $H_{ext}(\Theta)$ for eight different field angles as well as the free gyration frequency of the deflected antivortex is investigated. The simulated element is mapped from a scanning electron

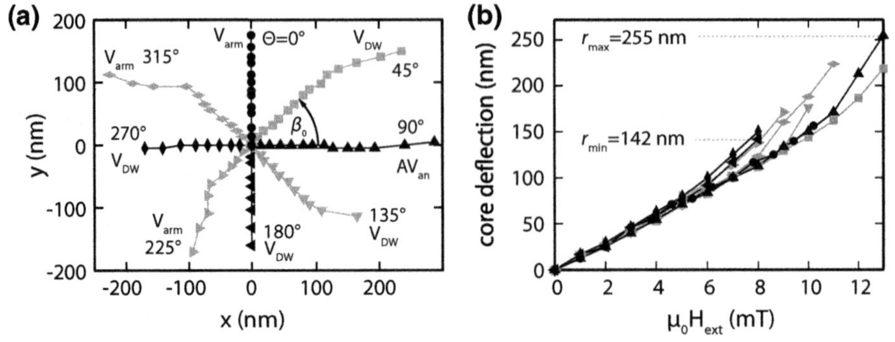

Fig. 15.21 **a** Simulated deflection from the equilibrium position of the core for bias fields with the indicated angle. The different annihilation processes (V_{arm}, V_{DW}, and AV_{an}) for each angle are explained in the text. **b** Distance from the equilibrium position at zero field of the antivortex core depending on the bias field obtained by micromagnetic simulations

micrograph of a real element, ensuring the same dimensions and edge roughness. The annihilation fields derived from these simulations match the ones determined by the MR measurements, see Fig. 15.20a. Furthermore, the simulations give a deep insight in the deflection behaviour of the antivortex, as well as in the annihilation process for different field directions. Figure 15.21b shows the core deflection distance from its equilibrium position for all fields with a minimal deflection of $r_{min} = 142$ nm and a maximal deflection of $r_{max} = 255$ nm. A linear dependence of the core deflection on the external field with a slope of 14.6 nm/mT can be found up to about 5 mT, indicating the parabolic confining potential. For higher field strengths the slope increases up to the annihilation of the antivortex. This increase shows that the confining potential becomes shallower once the antivortex is pushed into the arms, which is contrary to the behaviour of isolated vortices. A vortex is confined in a closed microstructure like a disc or a square, resulting in a deviation from the parabolic confining potential near the boundary of the structure. Thus a vortex needs much stronger fields to be pushed towards the boundaries of the structure [18, 23, 24]. In the graph showing the core position for all field strengths and angles, see Fig. 15.21a, another effect of the open junction can be seen. For field strengths in the linear regime up to 5 mT, cf. Fig. 15.21b, the displacement of the core follows a fixed angle. This azimuth angle β_0 to the new equilibrium position of the core can be generally derived by $\beta_0 = n(\Theta + \pi - \Phi_0) = n(\Theta + \pi - c\pi/2)$, taking a rotational symmetric potential and the Zeeman energy into account. Here, the antivortex with a winding number [25] of $n = -1$ and the fixed orientation of $c = -1$ yields $\beta_0 = -\Theta - 3\pi/2$. At certain field strengths the core deflection deviates from this direction, mostly for the deflection into the arms. This can be attributed to the depinning of the 90° domain walls from the corners of the wire junction, see Fig. 15.22d, and a deformation of the walls. Moreover, for high field strengths the magnetisation within the arms, whose magnetisation points antiparallel to the external field, starts to bend in a zig-zag fashion, leading to the deflection of the core away from the centre of the arm. This

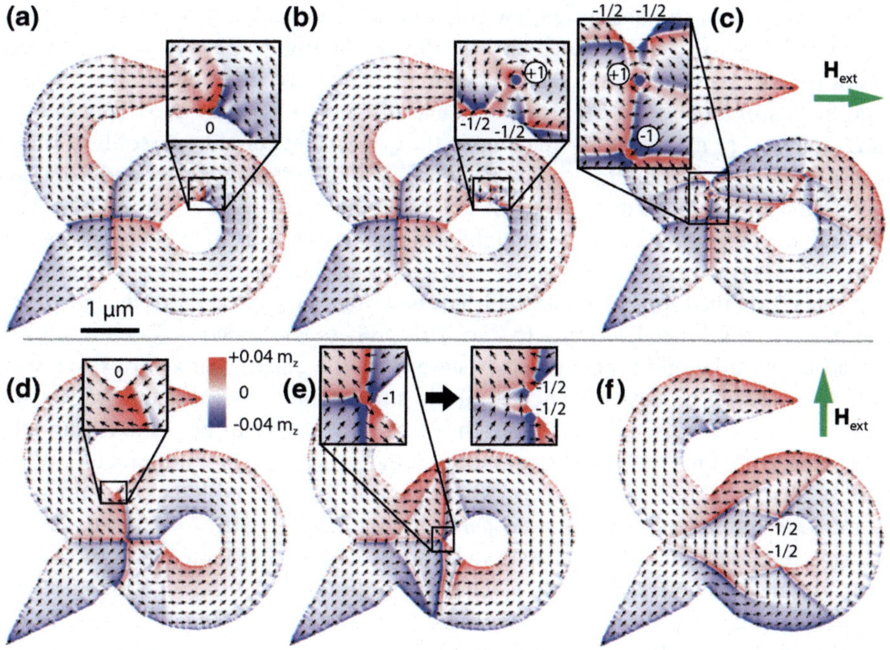

Fig. 15.22 **a–c** Sequence of the simulated annihilation process of the antivortex with an orientation of $c = -1$ by a static external field of $\mu_0 H_{\mathrm{ext}} = 11\,\mathrm{mT}$ in x-direction. **a** A 180° domain wall forms in a curved segment of the φ-shaped structure. **b** A vortex nucleates from the domain wall. **c** The upper 90° domain wall of the antivortex detaches from the corner of the wire junction forming another vortex. This vortex moves to the centre of the junction and both the vortex and the antivortex are annihilated. **d–f** Sequence of the annihilation of the antivortex at $\mu_0 H_{\mathrm{ext}} = 14\,\mathrm{mT}$ in y-direction. **d** The upper and lower 90° domain walls are no longer pinned at the respective corner and start moving into the arms. **e** The antivortex moves to the right corner and is annihilated there. Two edge defects are formed, that move further into the wire. The numbers denote the winding number of the respective magnetisation pattern

behaviour is not described by a simple rigid quasiparticle model of the antivortex, where the antivortex is confined in a parabolic potential. An extension of the parabolic potential to describe the deviation of the potential towards the boundaries of the structure has been done for vortices in squares [23], but for antivortices this approach poses several difficulties. For antivortices in φ-shaped structures the potential is highly anisotropic, whereas for vortices in squares only two different field directions have to be considered, i.e. towards the edge of the square and diagonally.

Similar shifts of the resonance frequency of the antivortex can be found for adjacent field angles and considering the deflection angle of the antivortex core caused by the external field reveals the influence of the φ-shaped structure on the confining potential, see Fig. 15.20b, even though the wire junction in which the antivortex is confined is completely rectangular. Thus a much more complex potential confines the antivortex.

An analysis of the antivortex annihilation for each field angle Θ reveals three different processes, as indicated in Fig. 15.21a. In Fig. 15.22 two of these annihilation processes are examplarily shown in the micromagnetic simulations for field angles of $\Theta = 0°$ in Fig. 15.22a–c, and 90° in (d)–(f). At an external field of $\mu_0 H_{ext}(0°) = 11\,mT$, the magnetisation in the upper right curved wire folds, forming a 180° domain wall, see Fig. 15.22a. From this wall a vortex nucleates and a domain that no longer points antiparallel to the external field is created, see Fig. 15.22b. This tilts the magnetisation in the upper right arm and destabilises the upper 90° domain wall of the antivortex. It detaches from the corner of the junction and another vortex nucleates from the former domain wall. The new vortex moves towards the antivortex and both are annihilated, see Fig. 15.22c. This process is labelled V_{arm} in Fig. 15.21a since the annihilation of the antivortex starts with the creation of a vortex in a wire arm. The second process, labelled V_{DW}, is similar, but a 90° domain wall of the antivortex detaches from the corner of the junction via a vortex nucleation without a preceding vortex nucleation in a curvature of the structure. The third process, AV_{an}, is only observed for a bias field in y-direction. The upper and lower domain walls of the antivortex depin from the corners, but are still attached to the boundary of the structure, see Fig. 15.22d. The domain walls and the antivortex start moving to the right until the antivortex core reaches the right corner of the wire junction, see Fig. 15.22e. Here the antivortex is annihilated and two edge defects [25] form, which move further along the inner boundary of the curved wire. Similarly, edge defects are generated when a vortex nucleates in the other processes described above. The magnetic texture of these edge defects can be described by half integer winding numbers $n = \pm 1/2$. When taking all magnetic defects that are generated or annihilated during the annihilation of the antivortex into account, it can be observed that the sum of the winding numbers of the whole structure remains $n_{sum} = -1$, the same as the initial antivortex. This holds for all bias field angles up to the maximum field strength of 15 mT in the simulation. The different annihilation processes exhibit no special symmetry that can be found in the φ-shaped structure, cf. Figs. 15.20b and 15.21a, but give a possible explanation of the anisotropy of the measured annihilation fields. For most cases, the antivortex is not pushed to the boundary of the wire junction and destroyed there, but its magnetic texture is distorted by changes of the magnetisation far from the antivortex. The annihilation processes for the real element may differ from the ones observed in the simulations, however a possible diversity in the annihilation mechanism is revealed. Thus it cannot be distinguished in the absorption and magneto-resistance measurements, if the annihilation field of the antivortex is measured or a field at which the magnetisation is reversed at some point in the φ-shaped structure by the external field as in the V_{arm} process.

The annihilation fields determined by MR measurements and micromagnetic simulations for a static antivortex are in good agreement, but the results from absorption spectroscopy, where the antivortex is gyrating at the resonance frequency, exhibit much smaller fields, see Fig. 15.20a. This discrepancy may not be attributed to high gyration radii and an expulsion of the antivortex from the decentred equilibrium position at a certain bias field. An estimation of the gyration radius r_{gyr} can be done by results from transmission X-ray microscopy on antivortices in elements with comparable dimensions [13, 15]. In these works maximal gyration radii of 95 to 115 nm at 140 and 120 MHz, respectively, are measured before core switching starts. Assuming a similar switching threshold for the structures presented here and a linear dependence of the gyration radius on the excitation field H_{hf}, the radius r_{gyr} can be estimated to be about 10 nm, a tenth of the maximal radius.

The displacement for the lowest annihilation field for the absorption measurements with $\mu_0 H_{\mathrm{an}}(180°) = 1.4\,\mathrm{mT}$ can be deduced from the simulations, where the constant deflection rate yields a displacement of the equilibrium position for this field of about 20 nm. The maximal displacement for a gyrating antivortex at this position can thus be estimated to be under 30 nm away from the centre of the wire junction, whereas it can be deflected up to 142 nm in the static case. Consequently, the reduced annihilation fields for the deflection of an excited antivortex cannot be attributed to the gyration radius of the antivortex, but rather to a reduction of the activation fields of the above described processes of domain wall depinning and vortex nucleation by the high-frequency field. A similar reduction of the switching field of a nanoparticle by radio-frequency field pulses is described by Thirion et al. in [26]. Thus the maximal gyration radius of about 100 nm before core switching starts, derived from the transmission X-ray microscopy results in [13, 15], and the minimal deflection distance of 142 nm for the static antivortex annihilation obtained by micromagnetic simulations (Fig. 15.21b) could explain the drastic decrease in the absorption signal for increasing excitation fields in the core switching regime, shown in Fig. 15.15a. It supports the above-mentioned hypothesis of an antivortex destruction for strong excitation fields and high gyration radii.[3]

Another comparison of the antivortex and the vortex can be made concerning the critical velocity needed for core switching. For vortices, a critical switching velocity $v_{\mathrm{crit}} = 2\pi r f_{res}$ was found by analytical and micromagnetic calculations [27] to be 320 m/s for vortices in discs or 250 m/s for vortices in squares by absorption measurements [22, 28]. The squares have comparable dimensions as the wire widths and thicknesses of the structures that stabilise the antivortex. The vortices in these squares exhibit a resonance frequency of 320 MHz and thus reach a gyration radius of about 124 nm at the critical velocity. The microscopy results for antivortices yield a much lower critical velocity of about 85 m/s at the comparable radii of 95 nm to 115 nm.

15.9 Conclusion

The gyrotropic eigenmode of isolated antivortices has been measured by high-frequency absorption spectroscopy for a varying excitation field strength in the linear, the non-linear and the core switching regime. The behaviour of isolated antivortices is similar to the one of excited vortices. When comparing antivortices with vortices

[3] In the transmission X-ray microscopy investigations in [13, 15] the disappearance of the antivortex core after several switching processes was occasionally observed.

of similar dimensions, the gyrotropic mode of the antivortex has a lower resonance frequency. To induce core switching, the antivortex needs to be exposed to a much stronger excitation field than the vortex. Deviations from a confining harmonic potential of the antivortex due to a static in-plane field have been demonstrated by the shift of the resonance frequency in the absorption measurements. To complement the anisotropy of the annihilation fields found in the absorption spectroscopy, magnetoresistance measurements as well as micromagnetic simulations have been discussed. While the annihilation fields determined from the MR measurements and the simulations without an excitation of the antivortex match quite well, the annihilation fields drop about 50% for the absorption spectroscopy, where the antivortex is excited. Moreover, a much more distinct anisotropy of the annihilation can be found for the excited state. It is demonstrated by means of micromagnetic simulations that for some field angles the annihilation of the antivortex is caused by a rotation of the magnetisation within an arm of the structure far away from the antivortex itself, thus destabilising the antivortex. These results show that a simple quasi-particle description for the antivortex is no longer applicable for the case of a strong excitation and an additional deflection of the equilibrium position by a static field.

Acknowledgements We are grateful to Ulrich Merkt for continuous support and fruitful discussions over many years. We thank Markus Weigand, Hermann Stoll, and Gisela Schütz, Max-Planck-Institute for Intelligent Systems, Stuttgart, Germany as well as Mi-Young Im and Peter Fischer, LBNL Berkeley, CA, USA for excellent, long-standing cooperation. We acknowledge financial support from the Deutsche Forschungsgemeinschaft via SFB 668 'Magnetism from the Single Atom to the Nanostructure', via Graduiertenkolleg 1286 'Functional Metal-Semiconductor Hybrid Systems', and via excellence cluster 'The Hamburg Centre for Ultrafast Imaging - Structure, Dynamics and Control of Matter on the Atomic Scale'.

References

1. S. Middelhoek, Domain walls in thin NiFe films. J. Appl. Phys. **34**, 1054 (1963)
2. A. Drews, B. Krüger, M. Bolte, G. Meier, Current- and field-driven magnetic antivortices. Phys. Rev. B **77**, 094413 (2008)
3. T.L. Gilbert, A Phenomenological Theory of Damping in Ferromagnetic Materials. IEEE Trans. Magn. **40**, 3443 (2004)
4. L. Landau, E. Lifshits, on the theory of the dispersion of magnetic permeability in ferromagnetic bodies. Phys. Zeitsch. der Sow. **169**, 14 (1935)
5. K.Y. Guslienko, B.A. Ivanov, V. Novosad, Y. Otani, H. Shima, K. Fukamichi, Eigenfrequencies of vortex state excitations in magnetic submicron-size disks. J. Appl. Phys. **91**, 8037 (2002)
6. J. Stöhr, H.C. Siegmann, *Magnetism: From Fundamentals to Nanoscale Dynamics* (Springer, Berlin, 2006)
7. M. Martens, T. Kamionka, A. Drews, B. Krüger, G. Meier, Influence of the winding number on field- and current driven dynamics of magnetic vortices and antivortices. J. Appl. Phys. **112**, 013917 (2012)
8. K.J. Kirk, S. McVitie, J.N. Chapman, C.D.W. Wilkinson, Imaging magnetic domain structure in sub-500 nm thin film elements. J. Appl. Phys. **89**, 7174 (2001)
9. T. Shinjo, Magnetic vortex core observation in circular dots of permalloy. Science **289**, 930 (2000)

10. A. Lyberatos, S. Komineas, N. Papanicolaou, Precessing vortices and antivortices in ferromagnetic elements. J. Appl. Phys. **109**, 023911 (2011)
11. A. Drews, B. Krüger, G. Meier, S. Bohlens, L. Bocklage, T. Matsuyama, M. Bolte, Current- and field-driven magnetic antivortices for nonvolatile data storage. Appl. Phys. Lett. **94**, 062504 (2009)
12. K. Shigeto, T. Okuno, K. Mibu, T. Shinjo, T. Ono, Magnetic force microscopy observation of antivortex core with perpendicular magnetization in patterned thin film of permalloy. Appl. Phys. Lett. **80**, 4190 (2002)
13. T. Kamionka, M. Martens, K.W. Chou, M. Curcic, A. Drews, G. Schütz, T. Tyliszczak, H. Stoll, B. Van Waeyenberge, G. Meier, Magnetic antivortex-core reversal by circular-rotational spin currents. Phys. Rev. Lett. **105**, 137204 (2010)
14. H. Wang, C. Campbell, Spin dynamics of a magnetic antivortex: micromagnetic simulations. Phys. Rev. B **76**, 220407 (2007)
15. T. Kamionka, M. Martens, K.W. Chou, A. Drews, T. Tyliszczak, H. Stoll, B. Van Waeyenberge, G. Meier, Magnetic antivortex-core reversal by rotating magnetic fields. Phys. Rev. B **83**, 224422 (2011)
16. L. Bocklage, B. Krüger, R. Eiselt, M. Bolte, P. Fischer, G. Meier, Time-resolved imaging of current-induced domain-wall oscillations. Phys. Rev. B **78**, 180405 (2008)
17. M. Pues, M. Martens, G. Meier, Absorption spectroscopy of isolated magnetic antivortices. J. Appl. Phys. **116**, 153903 (2014)
18. A. Drews, B. Krüger, G. Selke, T. Kamionka, A. Vogel, M. Martens, U. Merkt, D. Möller, G. Meier, Nonlinear magnetic vortex gyration. Phys. Rev. B **85**, 144417 (2012)
19. B. Krüger, A. Drews, M. Bolte, U. Merkt, D. Pfannkuche, G. Meier, Vortices and antivortices as harmonic oscillators. J. Appl. Phys. **103**, 07A501 (2008)
20. T. Kamionka, M. Martens, A. Drews, B. Krüger, O. Albrecht, G. Meier, Influence of temperature on the gyrotropic eigenmode of magnetic vortices. Phys. Rev. B **83**, 224424 (2011)
21. B. Krüger, A. Drews, M. Bolte, U. Merkt, D. Pfannkuche, G. Meier, Harmonic oscillator model for current- and field-driven magnetic vortices. Phys. Rev. B **76**, 224426 (2007)
22. M. Martens, T. Kamionka, M. Weigand, H. Stoll, T. Tyliszczak, G. Meier, Phase diagram for magnetic vortex core switching studied by ferromagnetic absorption spectroscopy and time-resolved transmission X-ray microscopy. Phys. Rev. B **87**, 054426 (2013)
23. H.H. Langner, T. Kamionka, M. Martens, M. Weigand, C.F. Adolff, U. Merkt, G. Meier, Vortex dynamics in nonparabolic potentials. Phys. Rev. B **85**, 174436 (2012)
24. J.-S. Kim, O. Boulle, S. Verstoep, L. Heyne, J. Rhensius, M. Kläui, L.J. Heyderman, F. Kronast, R. Mattheis, C. Ulysse, G. Faini, Currentinduced vortex dynamics and pinning potentials probed by homodyne detection. Phys. Rev. B **82**, 104427 (2010)
25. O. Tchernyshyov, G.-W. Chern, Fractional vortices and composite domain walls in flat nanomagnets. Phys. Rev. Lett. **95**, 197204 (2005)
26. C. Thirion, W. Wernsdorfer, D. Mailly, Switching of magnetization by nonlinear resonance studied in single nanoparticles. Nat. Mater. **2**, 524 (2003)
27. K. Guslienko, K.-S. Lee, S.-K. Kim, Dynamic origin of vortex core switching in soft magnetic nanodots. Phys. Rev. Lett. **100**, 027203 (2008)
28. A. Vansteenkiste, K.W. Chou, M. Weigand, M. Curcic, V. Sackmann, H. Stoll, T. Tyliszczak, G. Woltersdorf, C.H. Back, G. Schütz, B. Van Waeyenberge, X-ray imaging of the dynamic magnetic vortex core deformation. Nat. Phys. **5**, 332 (2009)

Chapter 16
Nonequilibrium Quantum Dynamics of Current-Driven Magnetic Domain Walls and Skyrmions

Martin Stier and Michael Thorwart

Abstract Magnetic textures in solid-state nanostructures can be manipulated by an applied electrical current and are thus promising candidates for new classes of electronic devices. The manipulation of a magnetization texture occurs via a spin torque exerted by the spin-polarized current on the local magnetic moments. We present different approaches how to calculate this spin torque under nonequilibrium conditions. In particular, we generalize the conventional approach to calculate the spin torque to treat magnetic structures which have a large spatial gradient (steep structures). We discuss this for the case of ferromagnetic domain walls and show how their chirality can be switched by an external spin-orbit torque. Besides domain walls, we also investigate the dynamics of two-dimensional ferromagnetic skyrmions and derive an equation of motion for the skyrmion's topological charge density. By this equation we are able to explain the current-induced creation of neutral skyrmion-antiskyrmion pairs which can be used for the production of stable skyrmions after the antiskyrmion partner has disappeared due to dissipation.

16.1 Introduction

Magnetism has been used in information technology for decades. A major step in miniaturization was often accompanied by the discovery of new physical phenomena as for example the giant magnetoresistance [1–3]. This allowed for a reduction of the size of the reading head in a hard drive. The required manipulation of the magnetic domain by a magnetic field, however, sets a physical limit to further reducing the size

M. Stier · M. Thorwart (✉)
I. Institut für Theoretische Physik, Universität Hamburg, Jungiusstraße 9,
20355 Hamburg, Germany
e-mail: michael.thorwart@physik.uni-hamburg.de

M. Stier
e-mail: mstier@physnet.uni-hamburg.de

R. Wiesendanger (ed.), *Atomic- and Nanoscale Magnetism*, NanoScience and Technology, https://doi.org/10.1007/978-3-319-99558-8_16

of the domains. To circumvent this principal limitation, applied electrical currents can also be used to manipulate the magnetization of a nanostructure, e.g., by moving magnetic domain walls (DWs). This concept also leads to current-based magnetic storage devices in the form of racetrack memories [4]. Since the electrons forming the charge current also carry a spin, they are able to exert a spin torque on the local magnetic moments, thus changing the overall magnetization. The dynamics of this process is commonly described by the Landau-Lifshitz-Gilbert equation (LLG)

$$\partial_t \mathbf{n} = -\gamma_0 \mathbf{n} \times \mathbf{B}_{\text{eff}} + \alpha \mathbf{n} \times \partial_t \mathbf{n} + \mathbf{T}, \tag{16.1}$$

which describes the temporal change of the normalized magnetization $\mathbf{n}(x, t)$ under the influence of an effective magnetic field $\mathbf{B}_{\text{eff}}$, the Gilbert damping $\propto \alpha$ and the aforementioned current-induced spin torque $\mathbf{T}$. Typical contributions to the spin torque are the adiabatic and the non-adiabatic spin-transfer torque (STT) [5, 6], but also include field-like terms which may be caused by spin-orbit interactions (SOI) in the itinerant electrons [7–9].

In particular, these field-like terms may be used to control the chirality of DWs, which defines how the magnetization winds within the DW along its longitudinal direction (cf. Fig. 16.1). In absence of chiral interactions, e.g., if the spin torque is only formed by the STT, the chirality is of minor importance as no special chirality is preferred and the DW dynamics is independent of the chirality. However, when a SOI is present, this symmetry is broken. In fact, with SOI we are able to change the DW's chirality [10] by an applied electron current and its dynamics.

Below, we discuss how contributions to the spin torque can be efficiently calculated analytically. Starting from an expansion in the spatial gradient of the magnetic texture, we extend the method to an exact calculation of the spin torque. This approach is particularly valuable for textures with a strong magnetic gradient ("steep" magnetic textures). We use these results to identify conditions under which the chirality of DWs can be switched all-electronically. Additionally, we show that non-local spin torques arise which change the dynamics of DWs qualitatively when the magnetic gradient becomes significant. Chiral (or, non-collinear) interactions are also important for ferromagnetic skyrmions which are another important class of magnetic objects. In particular the Dzyaloshinskii-Moriya interaction (DMI) can stabilize ferromagnetic skyrmions under certain conditions which are revealed below. Skyrmions

Fig. 16.1 Néel walls of different chirality: **a** negative chirality **b** positive chirality

have several advantages over DWs. For example, they are comparably stable and less susceptible to pinning. Consequently, the critical current which is necessary to move skyrmions is much lower than for DWs. Despite their stability, skyrmions can be efficiently produced by means of an electrical current [11]. In fact, this process occurs via the creation of a skyrmion-antiskyrmion pair. To show this, we derive a skyrmion equation of motion for the topological charge density directly. After the creation of this topologically neutral pair, the antiskyrmion is not stable against dissipation and vanishes, such that only the skyrmion survives. The skyrmion equation of motion clearly reveals the general prerequisites of current induced skyrmion creation.

16.2 Model and Equations of Motion

To discuss the principles of the current induced dynamics of magnetic textures, we investigate model systems which are sufficiently simple but still contain the essentials of real systems. The Hamiltonian describing such a system in general consists of three parts, i.e.,

$$H = H_d + H_s + H_{sd}, \tag{16.2}$$

where H_d refers to the part containing only localized electrons in, e.g., d-like bands, H_s only contains the itinerant, current carrying electrons in s-like bands and H_{sd} describes the interaction between both types of electrons (cf. Fig. 16.2). In this section we will mainly discuss the effect of the latter two parts as they directly generate the spin torque acting on the localized moments. The first part determines the effective field in the LLG (16.1) via the definition $\mathbf{B}_{\text{eff}} = -\frac{1}{\gamma_0 \hbar}\frac{\partial H_d}{\partial \mathbf{n}}$. To be specific, the effective field describing a ferromagnetic DW is given by

$$\mathbf{B}_{\text{eff}} = -J_{\text{IA}}\partial_x^2\mathbf{n} - K_{\|}n_{\|}\mathbf{e}_{\|} + K_{\perp}n_{\perp}\mathbf{e}_{\perp} \,. \tag{16.3}$$

It contains a Heisenberg interaction with interaction strength J_{IA} and two anisotropy terms referring to the hard ($K_{\perp}$) and the easy axis ($K_{\|}$) of the nanostructure.

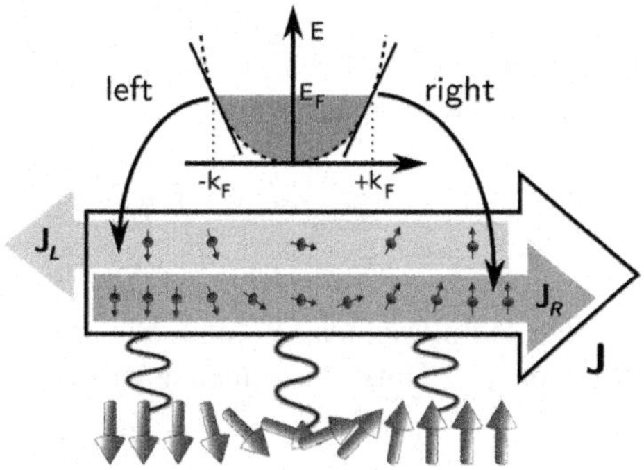

Fig. 16.2 Schematic picture of the 1D model system used to calculate the spin torques. Itinerant electrons may be separated into right and left movers which form the according spin currents and the total spin current density **J**. Interaction with the local magnetic moments (large arrows at the bottom) yield the spin torques

As we only consider one-dimensional DWs in the following, it is useful to describe the kinetic part H_s in terms of a gas of non-interacting electrons in one dimension. This is inspired by the Luttinger model which is exact in the low-energy regime and can be solved exactly even in the presence of electron-electron interaction [12, 13]. The key feature is that the electron dispersion is linearized in the vicinity of the Fermi edge, such that effectively two types of electrons arise, namely those moving solely in the right or solely in the left direction. The Hamiltonian reads

$$H_s = -i\hbar v \sum_{\sigma,p} \int dx c^{\dagger}_{p\sigma}(x) \partial_x c_{p\sigma}(x), \tag{16.4}$$

where $c^{(\dagger)}_{p\sigma}$ are the annihilators (creators) of electrons with spin $\sigma = \uparrow, \downarrow$ which are moving in the left or right direction ($p = L/R = -/+$) with the Fermi velocity v. It turns out that for our purpose of calculating the spin torque, it is beneficial to rewrite the Luttinger model in its Sugawara form [13] by introducing the spin density operators

$$\hat{\mathbf{J}}_p(x) = \frac{1}{2} : c^{\dagger}_{p\sigma}(x) \boldsymbol{\sigma}_{\sigma\sigma'} c_{p\sigma'}(x) : \tag{16.5}$$

with the Pauli matrices $\boldsymbol{\sigma}$ and the colons $: \cdots :$ denoting normal ordering. The Hamiltonian changes to

$$H_s = \hbar v \sum_p \int dx \; : \hat{\mathbf{J}}_p \cdot \hat{\mathbf{J}}_p : + H_{\text{charge}} \tag{16.6}$$

with an irrelevant charge part which will not contribute to the spin torque. A major advantage of this new formulation is the possibility of direct definitions of the total spin densities

$$\hat{\mathbf{s}} = \hat{\mathbf{J}}_R + \hat{\mathbf{J}}_L \,, \tag{16.7}$$

and more importantly of the spin current density

$$\hat{\mathbf{J}} = v(\hat{\mathbf{J}}_R - \hat{\mathbf{J}}_L) \,. \tag{16.8}$$

Due to the definition in (16.8) we do not need to introduce electric fields to create current flows which would add rather complicated terms to the calculations [14]. An additional advantage of the Sugawara formulation becomes apparent when we introduce the *sd* Hamiltonian

$$H_{sd} = \Delta_{\text{sd}} \int dx \, \hat{\mathbf{s}} \cdot \mathbf{n} = \Delta_{\text{sd}} \sum_p \int dx \, \hat{\mathbf{J}}_p \cdot \mathbf{n} \,, \tag{16.9}$$

as both Hamiltonians, H_s and H_{sd} are solely expressed in terms of spin density operators $\hat{\mathbf{J}}_p$. Thus, it is quite easy to set up the Heisenberg equation of motion

$$\partial_t \hat{\mathbf{J}}_p = -\frac{i}{\hbar}\left[\hat{\mathbf{J}}_p, H\right]_- , \tag{16.10}$$

when one keeps in mind that the spin density operators in the Sugawara formalism follow a slightly modified spin algebra [13]

$$\left[\hat{J}_p^\mu(x), \hat{J}_{p'}^\nu(x')\right]_- = i[p\partial_x + \epsilon^{\mu\nu\lambda}\hat{J}_p^\lambda]\delta_{pp'}\delta(x-x') , \tag{16.11}$$

which is the Kac-Moody algebra. A straight-forward calculation yields

$$(\partial_t + vp\partial_x)\hat{\mathbf{J}}_p = -\frac{\Delta_{\text{sd}}}{\hbar}[\hat{\mathbf{J}}_p \times \mathbf{n} + \beta(\hat{\mathbf{J}}_p - \hat{\mathbf{J}}_p^{\text{relax}})], \tag{16.12}$$

where we have added a relaxation term $\propto \beta = \hbar/(\Delta_{\text{sd}}\tau)$ with the relaxation time τ to incorporate non-adiabatic contributions [15] to the time evolution of the spin density operators. Equation (16.12) effectively represents a continuity equation for the spin density. The spatial derivative directly evolves from the derivative in the commutation relation of the Kac-Moody algebra in (16.11). Finding a solution of (16.12) is the main task, as we immediately obtain the spin torque

$$\mathbf{T} = -\frac{\Delta_{\text{sd}}}{\hbar}\mathbf{n} \times \langle\hat{\mathbf{s}}\rangle \tag{16.13}$$

from it. Strictly speaking, we need to calculate the expectation value of the operator $\hat{\mathbf{s}}$ as obviously the spin torque is a classical quantity in the LLG. Equation (16.12), however, is the same for $\langle\hat{\mathbf{J}}_p\rangle$ as for $\hat{\mathbf{J}}_p$ since the operator appears only linearly in this equation and for convenience, we will only write $\mathbf{J}_p \equiv \langle\hat{\mathbf{J}}_p\rangle$ from now on.

One-dimensional systems also offer the advantage that we can add a SOI in a straightforward way. We show this for the case of the Rashba SOI which is commonly expressed by the Hamiltonian

$$H_{\text{Rashba}} = \alpha'_R(\mathbf{k} \times \boldsymbol{\sigma}) \cdot \hat{\mathbf{z}} \tag{16.14}$$

with the Rashba coupling α'_R, the wave vector $\mathbf{k}$ and the vector of the Pauli matrices $\boldsymbol{\sigma}$. However, in our strict 1D Luttinger-type model we can replace $\mathbf{k} = (k_x, k_y, k_z)^T \to (k_x, 0, 0)^T$. As we have only left and right movers at the Fermi edge with $|\mathbf{k}| = k_F$ (cf. Fig. 16.2) the only possible values $k_x = \pm k_F$ remain and the Rashba Hamiltonian can be simplified to

$$H_{\text{Rashba}} = \Delta_{\text{R}}\sum_p p\int dx\,\mathbf{J}_p \cdot \hat{\mathbf{y}}, \quad \Delta_{\text{R}} = 2\tilde{\alpha}_R k_F. \tag{16.15}$$

This has almost the same form as H_{sd} and we can combine both of them to

$$H_{sd} + H_{\text{Rashba}} = \Delta_{\text{sd}} \sum_p p \int dx\, \mathbf{J}_p \cdot \mathbf{m}_p \tag{16.16}$$

with $\mathbf{m}_p = \mathbf{n} + p\alpha_R \hat{\mathbf{y}}$ and $\alpha_R = \Delta_{\text{R}}/\Delta_{\text{sd}}$. By this, we obtain the equation of motion

$$(\partial_t + vp\partial_x)\mathbf{J}_p = -\frac{\Delta_{\text{sd}}}{\hbar}[\mathbf{J}_p \times \mathbf{m}_p + \beta(\mathbf{J}_p - \mathbf{J}_p^{\text{relax}})], \tag{16.17}$$

which is only slightly different as compared to (16.12).

16.3 Ferromagnetic Chiral Domain Walls

Equipped with the equation of motion, we can solve (16.17) approximately in terms of a gradient expansion. This is done by the Ansatz

$$\mathbf{J}_p = \mathbf{J}_p^{(0)}(\mathbf{m}_p) + \mathbf{J}_p^{(1)}(\partial_x \mathbf{m}_p, \partial_t \mathbf{m}_p) + \cdots, \tag{16.18}$$

where $\partial_{x,t}\mathbf{m}_p = \partial_{x,t}\mathbf{n}$. When we insert (16.18) into (16.17) and arrange all the terms according to their order of the derivative, we get

$$\begin{aligned} 0 =& - \Delta_{\text{sd}}[\mathbf{J}_p^{(0)} \times \mathbf{m}_p - \beta(\mathbf{J}_p^{(0)} - \mathbf{J}_p^{\text{relax}})] \\ (\partial_t + vp\partial_x)\mathbf{J}_p^{(0)} =& - \Delta_{\text{sd}}[\mathbf{J}_p^{(1)} \times \mathbf{m}_p - \beta(\mathbf{J}_p^{(1)} - 0)] \\ \cdots =& \dots . \end{aligned} \tag{16.19}$$

This set of equations can be successively solved. To get explicit results, we have to define the state $\mathbf{J}_p^{\text{relax}}$ to which the spin density tries to relax. Typical choices would be the stationary state $\mathbf{J}_p^{\text{relax}} = \mathbf{J}_p^{\text{stat}}$ defined by $\partial_t \mathbf{J}_p^{\text{stat}} = 0$ or a relaxation towards the local magnetization, i.e., $\mathbf{J}_p^{\text{relax}} \propto \mathbf{n}$. Even though the explicit choice of $\mathbf{J}_p^{\text{relax}}$ has some influence on the results [10, 16], we will not discuss the details here and set $\mathbf{J}_p^{\text{relax}} = j_p \mathbf{n}$. Due to the definition in (16.8), we can fix the spin current density to $v_s = v(j_R - j_L)$. This can also be expressed in terms of experimental parameters as $v_s = PI_c/(2eM_s)$ with the spin polarity P, the charge current density I_c, the elementary charge E and the saturated magnetic moment M_s.

Equation (16.19) can be solved by elementary algebra and yields, together with (16.13), the contributions to the spin torque. In zeroth order, we find

$$\mathbf{T}^{(0)} = -\frac{\Delta_{\text{sd}}}{\hbar v} I_s \alpha_R \Big[\mathbf{n} \times \hat{\mathbf{y}} - \beta \mathbf{n} \times (\mathbf{n} \times \hat{\mathbf{y}})\Big] + \mathcal{O}(\alpha_R^3) . \tag{16.20}$$

which represent the known field-like and anti-damping-like SOI torques. The first-order result yields the adiabatic and non-adiabatic STT terms

$$\mathbf{T}^{\mathrm{ad}} = -I_s \partial_x \mathbf{n} + \mathcal{O}(\alpha_R^2)\,, \tag{16.21}$$

$$\mathbf{T}^{\mathrm{non-ad}} = \beta I_s \mathbf{n} \times \partial_x \mathbf{n} + \mathcal{O}(\alpha_R^2)\,, \tag{16.22}$$

as well as a correction to the field-like torque

$$\mathbf{T}^{(1)} = -\frac{I_s}{\gamma_0} \alpha_R^2 \big[(\mathbf{n} \times \partial_x \mathbf{n}) \cdot \hat{\mathbf{y}} \big] \mathbf{n} \times \hat{\mathbf{y}} + \mathcal{O}(\alpha_R^3)\,. \tag{16.23}$$

The higher orders are in principle accessible and would yield further corrections, but are very tedious to calculate and we will find an exact solution in the next section anyway. Before we come to that, we next discuss the implications of the field-like torque on the chirality of the DW.

With the explicit expressions for the spin torque at hand, we can numerically solve the LLG and with it obtain the dynamics of the system. We choose model parameters which correspond to the estimated values of Pt/Co/AlO$_x$ [17]: $A_{ex} = 10^{-11}\,\mathrm{J/m}$, $M_s = 1090\,\mathrm{kA/m}$, $K_\parallel = 0.92\,\mathrm{T}$, $K_\perp = 0.03\,K_\parallel, 2\alpha = \beta = 0.12$. In addition, we set $\Delta_{\mathrm{sd}} = 0.5\,\mathrm{eV}$, $v_F = 10^6\,\mathrm{m/s}$ and the polarity of the spin current $P = 1$. The easy axis is in the z-direction and the hard axis in y-direction. Thus, in equilibrium we have Néel walls with neither chirality favored (cf. Fig. 16.1). When we calculate the average DW velocity, however, we find differences between the different chiralities as shown in Fig. 16.3. These differences appear particularly for low current densities. Intuitively, it is clear that field-like terms prefer a distinct chirality, as they prefer a certain direction of the magnetization. Thus, it might be more interesting why the differences disappear for larger current densities, since the SOI torques even increase with increasing charge current. The explanation for that can be found by inspecting Fig. 16.4.

At low current densities, the DWs with different chiralities both move with a different, but steady velocity. But as soon as the current density, and with it the SOI torque, reaches a critical value, it can force the DW with the non-preferred chirality to overcome the hard axis and switch to the other chirality. In Fig. 16.4, we use the component of the magnetization in x direction (the easy axis) as a simple quantifier for that. When the current density increases even further, both chiralities switch frequently from one to the other. This is nothing but the common Walker breakdown, which also yields a reduction of the velocity in general. Again, no chirality is preferred in view of the DW velocity. This is even more obvious in Fig. 16.5. Here, we can identify a relatively large parameter regime where we are able to switch the chirality of a DW by a charge current. This opens, in principle, the possibility to use the chirality as an information bit in storage devices. As the SOI torque is proportional to the current density, we can simply switch the chirality back and forth by reversing the current direction.

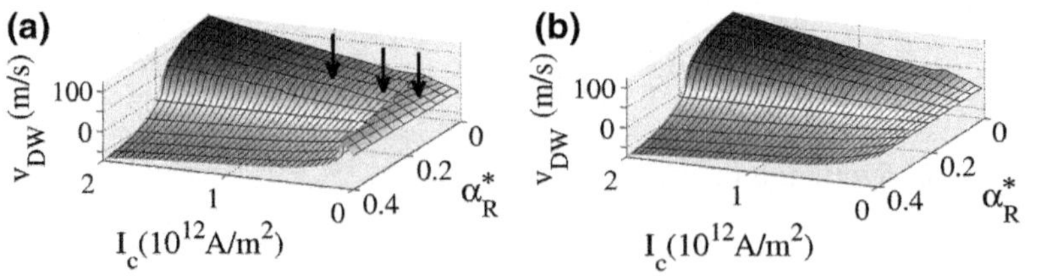

Fig. 16.3 Average DW velocities vs current density I_c and relative Rashba coupling strength α_R for a DW with **a** chirality $C = -1$ and **b** chirality $C = +1$. Differences between both cases appear, e.g., the vanishing of the Walker breakdown related velocity drop in (**b**) for low current densities. Black arrows in (**a**) indicate the parameter combinations of $\{I_c, \alpha_R\}$ shown in Fig. 16.4

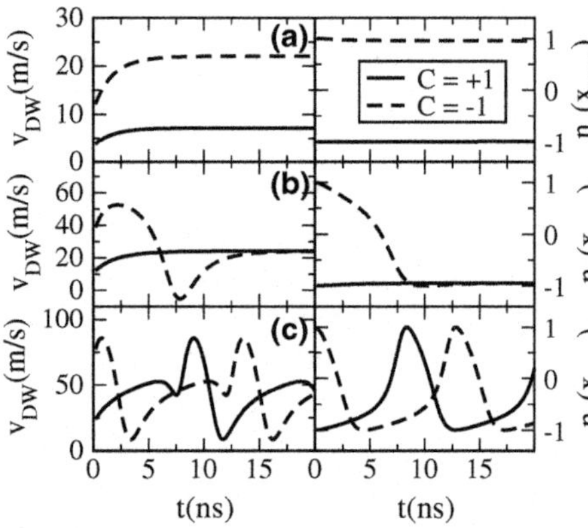

Fig. 16.4 DW velocities (left) and magnetization component in the x-direction at the DW center (right) for three different current densities: **a** $I_c = 0.125 \times 10^{12}\text{A/m}^2$, **b** $I_c = 0.4 \times 10^{12}\text{A/m}^2$ and **c** $I_c = 0.8 \times 10^{12}\text{A/m}^2$. The sign of $n_x(x_{DW})$ indicates the (negative) chirality of the DW. Three scenarios appear: **a** no switching of the chirality, **b** a single chirality switching for a positive initial chirality, and, **c** alternating chirality switching (and a Walker breakdown). We have set $\alpha_R = 0.1$

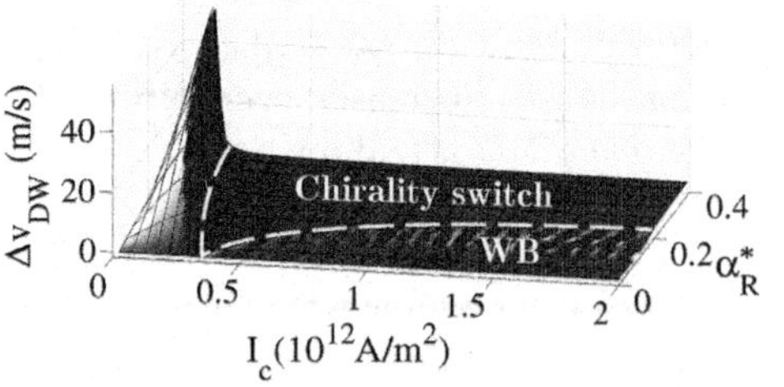

Fig. 16.5 Velocity difference between DWs of different chirality in dependence of the current density I_c and the relative Rashba coupling α_R. At intermediate current densities or large Rashba couplings an unpreferred chirality is switched to the preferred one. WB indicates the regime of the Walker breakdown

16.4 Steep Domain Walls

Solving (16.17) by the gradient expansion in Sect. 16.3 shows that higher-order corrections appear which are proportional to the magnetic gradient [18]. As is it not feasible to go to arbitrarily high orders in the gradient expansion, this approach is inappropriate for very steep structures with a large magnetic gradient. To avoid such an approximation, we follow a different line of reasoning and introduce a new Ansatz [19] for the spin density according to

$$\mathbf{J}_p = a_p(x,t)\overline{\mathbf{m}}_p + b_p(x,t)\overline{\partial_{(p)}\mathbf{m}_p} + c_p(c,t)\overline{\mathbf{m}}_p \times \overline{\partial_{(p)}\mathbf{m}_p} \,, \tag{16.24}$$

where the bars indicate a normalization of the vectors, i.e., $\overline{\mathbf{m}}_p = \mathbf{m}_p/|\mathbf{m}_p|$ and $\overline{\partial_{(p)}\mathbf{m}_p} = \partial_{(p)}\overline{\mathbf{m}}_p/|\partial_{(p)}\overline{\mathbf{m}}_p|$. Moreover, we have introduced for convenience $\partial_{(p)} = \partial_t + vp\partial_x$. In contrast to the gradient expansion, the prefactors a_p, b_p and c_p now depend on the spatial coordinate and on time. When we insert this Ansatz into (16.17) and arrange the resulting equation according to the linearly independent parts, we find a new differential equation in the form

$$\partial_{(p)}\begin{pmatrix} a_r \\ b_r \\ c_r \end{pmatrix} = A_r \begin{pmatrix} a_r \\ b_r \\ c_r \end{pmatrix} + \frac{\Delta_{\rm sd}\beta}{\hbar}\begin{pmatrix} a_r^{\rm relax} \\ b_r^{\rm relax} \\ c_r^{\rm relax} \end{pmatrix}, \tag{16.25}$$

which now has the prefactors a_p, b_p and c_p as independent variables. The coefficient matrix

$$A_p = \begin{pmatrix} -\Delta_{\rm sd}\beta/\hbar & |\partial_{(p)}\overline{\mathbf{m}}_p| & 0 \\ -|\partial_{(p)}\overline{\mathbf{m}}_p| & -\Delta_{\rm sd}\beta/\hbar & -\Delta_{\rm sd}|\mathbf{m}_p|/\hbar - g \\ 0 & \Delta_{\rm sd}|\mathbf{m}_p|/\hbar + g & -\Delta_{\rm sd}\beta/\hbar \end{pmatrix} \tag{16.26}$$

also depends on the spatial coordinate and on time via the magnetization and its gradient. Here, we have defined $g = [\partial_{(p)}\overline{\mathbf{m}}_p \cdot (\overline{\mathbf{m}}_p \times \partial^2_{(p)}\overline{\mathbf{m}}_p)]/(\partial_{(p)}\overline{\mathbf{m}}_p)^2$. This equation has to be solved numerically in the general case. However, we can still find some trends by analytical considerations. To simplify the discussion, we neglect the temporal derivatives from now on and set $\partial_{(p)} \to vp\partial_x$. This approach is not generally justified for all systems, but still is a good starting point since the Fermi velocity v is typically much larger than other velocities involved. First, in the limit of a vanishing gradient, the original results of the gradient expansion are recovered, which we have shown in [19]. Second, we can determine the eigenvalues of the coefficient matrix A_p as $\xi_1 = -\beta\Delta_{\rm sd}/(\hbar v)$ and $\xi_\pm = [-\beta\Delta_{\rm sd} \pm i\sqrt{(\Delta_{\rm sd}|\mathbf{m}_p| + g)^2 + (\hbar v\partial_x|\mathbf{m}_p|)^2}]/(\hbar v)$. When we assume this magnetic gradient to be small, i.e., $(\hbar v\partial_x|\mathbf{m}_p|)^2 \ll \Delta^2_{\rm sd}$, we can expect that the solutions of (16.25) do not differ so much from the solutions with $A_p = \text{const}$. Thus, we would expect a solution $\propto \exp(\xi_i x)$. As all eigenvalues ξ_i have a real part $\text{Re}\xi_i = -\beta\Delta_{\rm sd}/(\hbar v)$, we can further expect a universal damping length of $x_{\rm damp} = \hbar v/(\beta\Delta_{\rm sd})$. Moreover, two of the eigen-

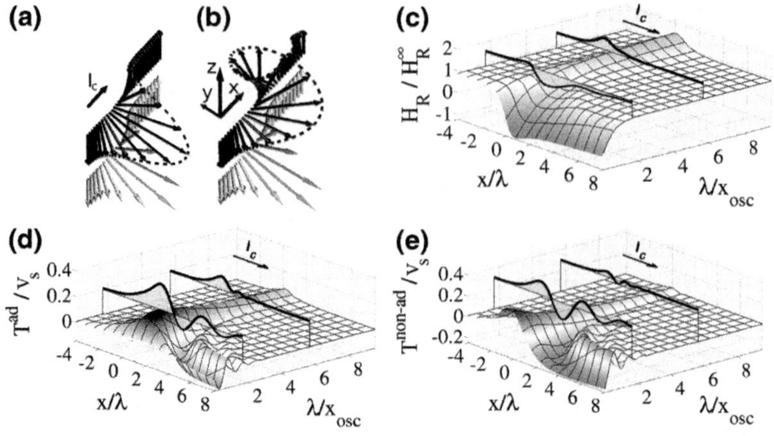

Fig. 16.6 Spin density of the current (dark arrows) flowing over a DW (light gray arrows) for (**a**) a smooth and (**b**) a steep DW. For steep DWs, the spin density cannot follow the DW's magnetization adiabatically. This results in a change of (**c**) the field-like torque $\mathbf{H}_R = H_R(x,t)\hat{\mathbf{y}}$, (**d**) the adiabatic STT $\mathbf{T}^{\rm ad} = T^{\rm ad}(x,t)\partial_x\mathbf{n}/|\partial_x\mathbf{n}|$, and the non-adiabatic STT $\mathbf{T}^{\rm non-ad} = T^{\rm non-ad}(x,t)\mathbf{n}\times\partial_x\mathbf{n}/|\partial_x\mathbf{n}|$ depending on the distance x from the DW center and the DW width λ. In each of the graphs (**c**) to (**e**), two curves (solid lines) are shown for the DW widths $\lambda/x_{\rm osc} = 1.7, 5.7$, which are vertically shifted to highlight the oscillatory behavior of the torques. We have used the relative Rashba coupling strength $\alpha_R = 0.4$ and choose $\beta = 0.2 = 2\alpha$

values have an imaginary part of $\operatorname{Im}\xi_\pm = \pm\sqrt{(\Delta_{\rm sd}|\mathbf{m}_p| + g)^2 + (\hbar v\partial_x|\mathbf{m}_p|)^2}]/(\hbar v)$ $\approx \pm\Delta_{\rm sd}/(\hbar v)$. Hence, we can expect oscillations of at least parts of the solutions with an oscillation length of around $2\pi x_{\rm osc}^\infty \equiv 2\pi\hbar v/\Delta_{\rm sd}$. However, this oscillation length will become large for very steep structures.

Using the same parameters as in the last section, we can confirm the previous qualitative arguments by the numerical results shown in Fig. 16.6. The DW width $\lambda = \sqrt{J/K_\parallel}$ is changed by setting $J = \lambda J^{(0)}$ and $K_\parallel = K_\parallel^{(0)}/\lambda$ with the ratio $\sqrt{J^{(0)}/K_\parallel^{(0)}} = 1$. The results of the spin torque can be directly calculated from the solutions of (16.25) by $\mathbf{T} = -\frac{\Delta_{\rm sd}}{\hbar}\mathbf{n}\times\mathbf{s}$ to

$$\mathbf{T} = \sum_{\nu=x,t}\left[T_\nu^{\rm nonad}(b_p)\overline{\partial_\nu\mathbf{n}} + T_\nu^{\rm ad}(c_p)\mathbf{n}\times\overline{\partial_\nu\mathbf{n}}\right] - H_{\rm R}(a_p,b_p)\mathbf{n}\times\hat{\mathbf{y}}\,. \tag{16.27}$$

Note that we use a slightly different definition of the STT which is proportional to the normalized derivative, e.g., $\mathbf{T}^{\rm ad} = T_x^{\rm ad}\overline{\partial_x\mathbf{n}}$ instead of the non-normalized one $\mathbf{T}^{\rm ad} = \tilde{T}_x^{\rm ad}\partial_x\mathbf{n}$. This is because the prefactors in the latter, conventional definition would diverge, even though the total STT would stay finite [19]. Another issue is that the anti-damping SOI torque seem to be missing. In fact, this is not the case, but these terms are hidden in the other contributions. The spin torque is a three-dimensional quantity in spin space, which means that it can be described by three basis vectors. We can choose as a complete basis set $\mathbf{n}$, $\partial_x\mathbf{n}$ and $\mathbf{n}\times\partial_x\mathbf{n}$. Since $\mathbf{T}$ is

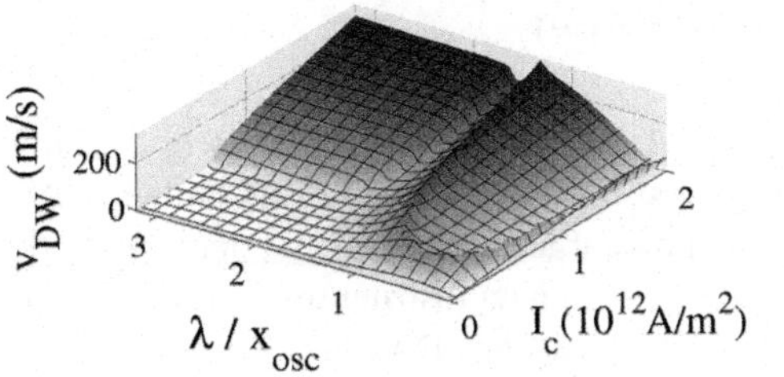

Fig. 16.7 Average velocity of a DW depending on the DW width λ and the current density I_c. Because we have chosen $\beta = 0$, we recover the typical intrinsic pinning of the DW at low current densities. However, this only occurs for smooth DWs. In contrast, we find for steep textures the typical velocity dependence connected to a Walker breakdown, i.e., a large increase of the velocity with increasing I_c and a sudden drop of the velocity at a certain critical current density. This can be traced back to a new contribution to the β parameter due to the precession of the spin current density, as shown in [19]. The Rashba interaction is set to zero for the case shown in this figure

perpendicular to $\mathbf{n}$, we actually only need $\partial_x \mathbf{n}$ and $\mathbf{n} \times \partial_x \mathbf{n}$. Thus, the expression in (16.27) is already overcomplete and the contribution of the Rashba field $H_R \mathbf{n} \times \hat{\mathbf{y}}$ is only present for a more convenient physical interpretation.

From the results of Fig. 16.6 we recover the analytical prediction. For steeper textures, i.e., for decreasing DW widths, we find spatial oscillations of the STT with a period of $\approx x_{osc}$. They are damped on a length scale of $x_{damp} = x_{osc}/\beta$. This is connected to a damped precession of the spin current density around the magnetization $\mathbf{n}$, which becomes only apparent when the DW width is smaller than the damping length. Obviously, this nonlocal spin torque changes the dynamics of steep DWs compared to smoother ones. This is perhaps best seen when considering the ratio $T^{nonad}/T^{ad} = \beta^*$ which is commonly used to determine the β parameter [20]. First, this effective β^* is now space dependent such as the torque itself. Second, we can define an averaged $\overline{\beta^*}$ which increases for decreasing DW width. This adds to the conventional nonadiabatic β parameter which we put into (16.17) in the beginning. Even for an originally vanishing $\beta = 0$, there would be an effective non-zero β^* parameter present. We will not discuss this in detail here, but the interested reader is referred to [19]. However, the effect of the DW width-dependent $\overline{\beta^*}$ parameter is visible in the dependence of the DW velocity on the current density, which is shown in Fig. 16.7. In this figure, we have chosen $\beta = 0$ which yields an intrinsic pinning at low current densities and for smooth DWs. For steep DWs, however, this dependence changes to the phenomenon of the Walker breakdown which only appears for $\beta > \alpha$. One can recognize this in the large increase of the DW velocity at low current densities and the sudden drop of the velocity at some critical value of the current. These effects have to be taken into consideration regarding a miniaturization of possible spintronic devices.

Notably, also the SOI torque is affected by the magnetic gradient. It can even change its sign for very steep structures (cf. Fig. 16.6) and may result in even further deviations of the dynamics of small magnetic structures.

16.5 Skyrmion Creation

Domain walls have been the first magnetic objects proposed to be used in a racetrack memory set-up, but they suffer from certain disadvantages. A major shortcoming is the necessity of comparably large applied current densities to depin DWs from impurities or notches in the supporting nanostructure. In this context, magnetic skyrmions can be more favorable [21–27] as they have the striking feature of topological protection. Magnetic skyrmions, which are essentially 2D magnetic vortices with a finite topological charge, cannot be destroyed in continuous systems at least from a principle point of view [28, 29]. Even though the theoretical concept of the topological protection cannot be rigorously transferred to real skyrmions on atomic lattices, there still exists a finite energy barrier which protects them against annihilation [30–32]. Thus, the concept of topology is still a useful tool to describe this protection, although it strictly applies to closed continuous systems without dissipation. In particular, it allows for the definition of quasi-particle-like properties as the skyrmion's topological charge density

$$q(x, y, t) = \mathbf{n} \cdot (\partial_x \mathbf{n} \times \partial_y \mathbf{n}) , \tag{16.28}$$

which we can use in the same way as, e.g., the electrical charge density of charged particles. The total topological charge of a skyrmion then is

$$Q = Q(t) = \frac{1}{4\pi} \int \mathrm{d}^2\mathbf{r}\, q(x, y, t) , \tag{16.29}$$

which always is an integer number $Q \in \mathbb{Z}$ [33, 34]. The aforementioned topological protection normally prevents any change of Q, but this is restricted to continuous systems. Actually, we describe a way to create skyrmions [35–39] in this section and thus a change of the skyrmion number over time. To understand this, it is necessary to discuss the real-time dynamics of q. However, the central equation describing the dynamics of the system, the LLG

$$\partial_t \mathbf{n} = -\mathbf{n} \times \mathbf{B}_{\text{eff}} + \alpha \mathbf{n} \times \partial_t \mathbf{n} + (\mathbf{v}_s \cdot \nabla)\mathbf{n} - \beta \mathbf{n} \times (\mathbf{v}_s \cdot \nabla)\mathbf{n} , \tag{16.30}$$

is an equation for the magnetization only. Yet, we can use the LLG to establish a new equation of motion for the topological charge density. To simplify the discussion, we restrict to the STT in lowest order in the magnetic gradient only which we have already used in (16.30). Moreover, we do not consider any SOI. Eventually, we have to calculate $\partial_t q$. Due to the product rule, several terms $\propto \partial_t \mathbf{n}$ appear which we can replace according to the LLG (16.30). For convenience, we can exploit rotational invariance in the definition of q and replace (16.28) by

$$q(x, y, t) = \mathbf{n} \cdot [(\hat{\mathbf{v}} \cdot \nabla)\mathbf{n} \times (\hat{\mathbf{v}}_\perp \cdot \nabla)\mathbf{n}] , \tag{16.31}$$

where $\hat{\mathbf{v}} = \mathbf{v}_s/|\mathbf{v}_s|$ and $\hat{\mathbf{v}}_\perp = \hat{\mathbf{z}} \times \hat{\mathbf{v}}$. Then, it is straightforward to find the skyrmion equation of motion

$$\partial_t q + \nabla \cdot \left(\mathbf{j}_{\mathrm{Sk}}^{(1)} + \mathbf{j}_{\mathrm{Sk}}^{(2)}\right) = 0\,, \tag{16.32}$$

which resembles a continuity equation when we introduce the skyrmion topological charge current densities

$$\mathbf{j}_{\mathrm{Sk}}^{(1)} = -\, j_1 q \mathbf{v}_s\,, \tag{16.33a}$$

$$\mathbf{j}_{\mathrm{Sk}}^{(2)} = j_2 \left\{[(\hat{\mathbf{v}} \cdot \nabla)\mathbf{n} \cdot (\hat{\mathbf{v}}_\perp \cdot \nabla)\mathbf{n}]\mathbf{v}_s - [(\hat{\mathbf{v}} \cdot \nabla)\mathbf{n}]^2 \mathbf{v}_\perp\right\}\,, \tag{16.33b}$$

with

$$j_1 = [1 + \alpha\beta + \alpha b_{\perp 1} + b_{\perp 2}]/(1 + \alpha^2)\,, \tag{16.34a}$$

$$j_2 = [\alpha - \beta - b_{\perp 1} + \alpha b_{\perp 2}]/(1 + \alpha^2)\,. \tag{16.34b}$$

The terms $b_{\perp 1}$ and $b_{\perp 2}$ are the contributions of the effective field in the LLG (16.30) represented according to

$$\mathbf{B}_{\mathrm{eff}} = b_\parallel \mathbf{n} + b_{\perp 1}(\mathbf{v}_s \cdot \nabla)\mathbf{n} + b_{\perp 2}\mathbf{n} \times (\mathbf{v}_s \cdot \nabla)\mathbf{n}\,. \tag{16.35}$$

Actually, these terms render a strict analytical analysis of the skyrmion equation of motion rather complicated. Fortunately, we can neglect these terms for current densities for which the skyrmions do not get too much deformed from their stationary state in their center-of-mass frame. Then it holds that $\partial_t \mathbf{n}(\mathbf{x}_{\mathrm{Sk}} - \mathbf{v}_{\mathrm{Sk}} t) \approx 0$, where $\mathbf{x}_{\mathrm{Sk}}$ is the skyrmion center moving with the current induced skyrmion velocity $\mathbf{v}_{\mathrm{Sk}}$. This requires the effective field $\mathbf{B}_{\mathrm{eff}}$ to be parallel to $\mathbf{n}$ which also means that $b_{\perp 1}$ and $b_{\perp 2}$ vanish. Equations (16.34a) and (16.34b) now simplify and we find $\mathbf{j}_{\mathrm{Sk}}^{(1)} \propto (1 + \alpha\beta)$ and $\mathbf{j}_{\mathrm{Sk}}^{(2)} \propto (\alpha - \beta)$. In particular, when $\alpha = \beta$ it holds that $\mathbf{j}_{\mathrm{Sk}}^{(2)} = 0$ and (16.32) is solved by $q(\mathbf{x}, t) = q(\mathbf{x} - \mathbf{v}_{\mathrm{Sk}} t)$ with $\mathbf{v}_{\mathrm{Sk}} = -\mathbf{v}_s$. Thus, we find that the topological charge density, and so the skyrmion, is just translationally moved against the charge current flow.

Probably more interesting is the skyrmion dynamics in the presence of dissipation with $\alpha \neq \beta$. In this case, $\mathbf{j}_{\mathrm{Sk}}^{(2)}$ does not vanish and a component of the skyrmion current density perpendicular to the charge current appears because of the part $\propto \mathbf{v}_\perp$ in (16.33b). When we write this part in the generic form of a current

$$\mathbf{j}_{\mathrm{Sk},\perp}^{(2)} = -j_2[(\hat{\mathbf{v}} \cdot \nabla)\mathbf{n}]^2 \mathbf{v}_\perp \equiv q\mathbf{v}^*\,, \tag{16.36}$$

we find for the velocity

$$\mathbf{v}^* = \frac{\beta - \alpha}{1 + \alpha^2} \frac{[(\hat{\mathbf{v}} \cdot \nabla)\mathbf{n}]^2}{q} \mathbf{v}_\perp\,, \tag{16.37}$$

where we have neglected the contributions of the effective field in (16.34b). As the sign of $[(\hat{\mathbf{v}} \cdot \nabla)\mathbf{n}]^2$ is always positive, the direction of $\mathbf{v}^*$ depends on the sign of q. Thus, we obtain the important conclusion that skyrmions and antiskyrmions move in different directions which means that they will be separated from each other.

We can use this result and set up a procedure to create skyrmion-antiskyrmion pairs. To do this we use the two-dimensional model Hamiltonian

$$\begin{aligned} H_d = & - J \sum_{\mathbf{r}} \mathbf{n}_{\mathbf{r}} \cdot \left(\mathbf{n}_{\mathbf{r}+\mathbf{e}_x} + \mathbf{n}_{\mathbf{r}+\mathbf{e}_y}\right) - \sum_{\mathbf{r}} \mathbf{B}_{\mathbf{r}} \cdot \mathbf{n}_{\mathbf{r}} \\ & - D \sum_{\mathbf{r}} \left[\left(\mathbf{n}_{\mathbf{r}} \times \mathbf{n}_{\mathbf{r}+\mathbf{e}_x}\right) \cdot \mathbf{e}_x + \left(\mathbf{n}_{\mathbf{r}} \times \mathbf{n}_{\mathbf{r}+\mathbf{e}_y}\right) \cdot \mathbf{e}_y\right] , \end{aligned} \tag{16.38}$$

where $\mathbf{r}$ denotes the site vector and which includes an exchange interaction $\propto J$, an external magnetic field $\mathbf{B}_{\mathbf{r}}$ and the Dzyaloshinskii-Moriya interaction (DMI) $\propto D$. We have chosen a bulk DMI which stabilizes Bloch skyrmions, but the results remain qualitatively unchanged for an interfacial DMI and Néel skyrmions [11, 40]. This is a standard model for skyrmions [41] and depending on the magnetic field, either the helical phase (low field), a skyrmion lattice (intermediate field), or the field-polarized/ferromagnetic phase (high field) is the ground state [42, 43]. For our study we choose typical parameter ratios as, for instance reported for MnSi [41] or for Mn layers on W [44, 45]. To be specific, we set $J = 1\,\text{meV}$, $D/J = 0.18$ and a magnetic field of $\mathbf{B} = (0, 0, B_z) = -0.03\,J\,\hat{\mathbf{z}}$. This value of $\mathbf{B}$ is just high enough to put the system in the field-polarized phase. However, when all moments are aligned parallel to each other we would have no finite topological density q and no skyrmion currents, which could create skyrmion-antiskyrmion pairs. Thus, to initialize the pair formation process at all, we add a tiny, inhomogeneous magnetic field pointing in the y-direction, i.e., $B_y = b_0[\sin(2\pi x/L_x) + \sin(2\pi y/L_y)]$ and $b_0 = B_z/100$. Note that it is necessary to have a gradient of the field in both x and y direction to create a finite q.

Even though the topological charge density created by this field is very small, we can use the dissipative skyrmion current $\mathbf{j}^{(2)}_{\text{Sk},\perp}$ to separate topological densities of different sign. This increases both the positive and negative topological charge densities and thus creates proper skyrmion-antiskyrmion pairs as shown in Fig. 16.8. This has also been investigated recently in [46]. Because we have set up a system with a particular form of the DMI, only the skyrmion is stable. After some time, when the antiskyrmion has reached some critical size, the antiskyrmion decays very fast on a timescale $\propto 1/\alpha$ and only the skyrmion survives. Thus, by the intermediate step of a skyrmion-antiskyrmion pair, we have eventually produced a stable skyrmion.

We can also evaluate the total skyrmion number, or the topological charge Q defined in (16.29). Indeed, Q changes over times in integer steps, as shown in Fig. 16.9. To check if each skyrmion creation is actually connected to an antiskyrmion decay, we define the total antiskyrmion charge

$$Q_{\text{ASk}} = \frac{1}{4\pi} \int_{q<0} \mathrm{d}^2\mathbf{r}\, q(x, y, t) \, , \tag{16.39}$$

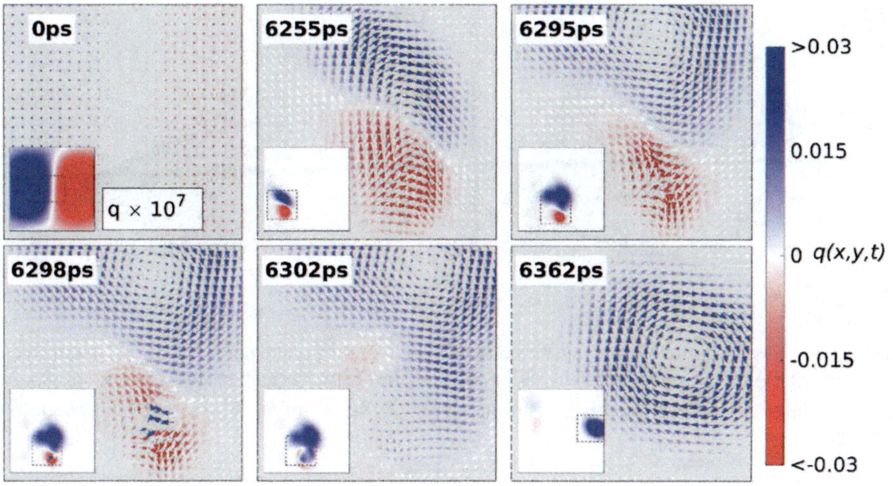

Fig. 16.8 Snapshots of the magnetic configuration $\mathbf{n}(x, y, t)$ in the xy plane at the indicated times. The color code refers to the topological charge density q. Starting from slight fluctuations of q (mind the multiplication of q by 10^7 at $t = 0$) a charge current produces skyrmion-antiskyrmion pairs. Around $t = 6300$ ps the antiskyrmion (red) decays and only the skyrmion remains. We have used $I_c = 7.7 \times 10^{11}$ A/m^2, $\alpha = 0.25$ and $\beta = 0$

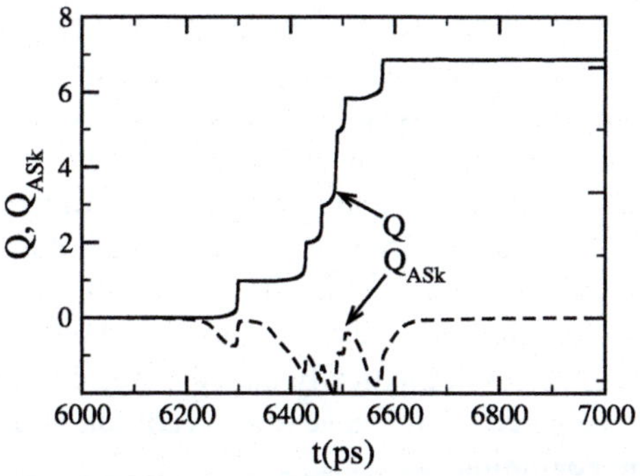

Fig. 16.9 Change of the total topological charge, or skyrmion number, Q over time. Parameters are chosen as in Fig. 16.8. We have defined the total antiskyrmion charge Q_{ASk} by summing over regions of negative topological charge $q < 0$ only. The increase of the magnitude of Q_{ASk}, connected to the creation of a skyrmion-antiskyrmion pair, occurs before the production of a (single) skyrmion and is directly visible, as well as the decay of the antiskyrmion right when the actual skyrmion is created

where we only sum over regions of negative q. This number does not have to be an integer as it is canceled out by the skyrmion part in each pair, keeping the total charge of a pair at zero. We find the magnitude of Q_{ASk} to increase before a skyrmion is created which we can connect to the creation of a pair. Exactly at the time when the skyrmion is created, Q_{ASk} is lost which indicates the decay of an antiskyrmion. Thus, every created skyrmion in fact stems from a skyrmion-antiskyrmion pair.

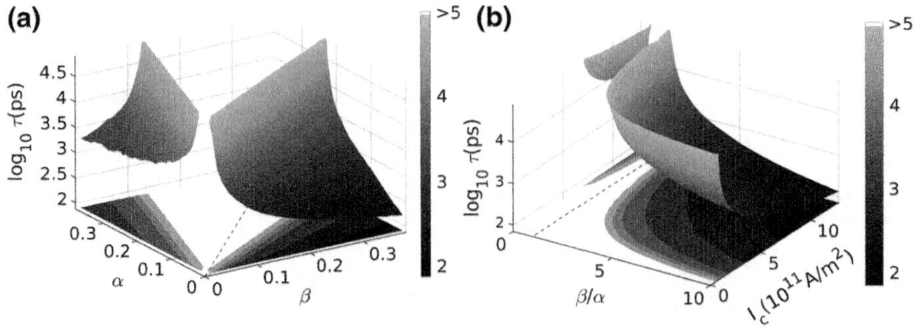

Fig. 16.10 Skyrmion creation time τ versus (**a**) Gilbert damping α and non-adiabaticity parameter β for $I_c = 10^{12}$ A/m^2 and (**b**) versus the ration β/α and the charge current density I_c for $\alpha = 0.05$. As the dissipative skyrmion current $\mathbf{j}^{(2)}_{\text{Sk},\perp}$ vanishes at $\alpha = \beta$, no skyrmion are created in this parameter region

Finally, we want to identify the general prerequisites for the production of skyrmions while we focus on (charge) current related parameters. For this, we use as a single quantifier for the efficiency of the skyrmion production process the skyrmion creation time τ which we define as the time between the onset of the current and the creation of the first skyrmion.

Because we trace back the skyrmion creation to a skyrmion-antiskyrmion pair production with additional antiskyrmion decay, we can roughly set $\tau = \tau_{\text{pair}} + \tau_{\text{ASk}}$ as the sum of the single duration of each individual process. However, since $\tau_{\text{pair}} \gg \tau_{\text{ASk}}$, the skyrmion creation time is almost entirely determined by the pair creation time. The origin of the pair creation is the dissipative skyrmion current $\mathbf{j}^{(2)}_{\text{Sk},\perp}$, which is proportional to $(\alpha - \beta)$ and $\mathbf{v}_s$. Thus, we would expect the skyrmion creation to happen faster the larger the current and the difference of α and β are. This analytical prediction is confirmed by numerical results which are shown in Fig. 16.10. We find the skyrmion creation times to dramatically increase the low $\mathbf{j}^{(2)}_{\text{Sk},\perp}$. In particular, for $\alpha = \beta$, no skyrmion is created at all, because no skyrmion-antiskyrmion pair can be formed right from the beginning.

The results of this section promise an efficient production of skyrmions by an electrical current in real physical systems. In principle, by any fluctuation of the topological charge density skyrmion-antiskyrmion pairs can be created, as long as the current and the difference of α and β is large enough. The production of skyrmions, however, is connected to further prerequisites, such as the existence of a sufficiently strong DMI and a certain external magnetic field. In real materials, there are obviously other relevant factors to be considered, for instance, the presence of pinning centers or the sample geometry [47]. These properties certainly influence details of the skyrmion production and should be taken into account in future studies.

16.6 Conclusions

In this chapter, we have presented a way to calculate the charge current induced spin torques, as the spin-transfer torque or the spin-orbit torques, which can be used to drive given magnetic textures in nanostructures. Starting from a 1D model, we have solved arising equations of motion first by a gradient expression. This approach is valid for magnetic textures with rather small gradients and already shows that corrections to the spin torque due to the magnetic gradient appear. In turn, for steep magnetic structures a gradient expansion usually becomes insufficient and we thus have introduced a way to determine an exact solution for the spin torque. The results from this nonperturbative approach indeed differ drastically from those on the basis of the conventional expressions for the spin torques especially in structures with a large magnetic gradient. Furthermore, we have shown how the chirality of a domain wall can be switched by a charge current on the basis of a spin-orbit torque. Finally, we have also discussed magnetic skyrmions which are vortex-like textures with a finite topological charge. We have established a skyrmion equation of motion which allowed us to interpret the general features of the dynamics of skyrmions and antiskyrmions particularly in the presence of dissipation. This technique has revealed that the skyrmion production occurs via the intermediate step of an skyrmion-antiskyrmion pair creation, with the subsequent decay of the unstable antiskyrmion. In addition, this skyrmion equation of motion allows us to identify the necessary prerequisites for an efficient production of skyrmions.

References

1. M.N. Baibich, J.M. Broto, A. Fert, F. Nguyen Van Dau, F. Petroff, P. Etienne, G. Creuzet, A. Friederich, J. Chazelas, Phys. Rev. Lett. **61**, 2472 (1988)
2. G. Binasch, P. Grünberg, F. Saurenbach, W. Zinn, Phys. Rev. B **39**, 4828 (1989)
3. I. Ennen, D. Kappe, T. Rempel, C. Glenske, A. Hütten, Sensors (Basel) **16**, 904 (2016)
4. S.S.P. Parkin, M. Hayashi, L. Thomas, Science **320**, 190 (2008)
5. S. Zhang, Z. Li, Phys. Rev. Lett. **93**, 127204 (2004)
6. A. Brataas, A.D. Kent, H. Ohno, Nat. Mater. **11**, 372 (2012)
7. E. van der Bijl, R.A. Duine, Phys. Rev. B **86**, 094406 (2012)
8. A. Qaiumzadeh, R.A. Duine, M. Titov, Phys. Rev. B **92**, 014402 (2015)
9. V. Risinggård, J. Linder, Phys. Rev. B **95**, 134423 (2017)
10. M. Stier, M. Creutzburg, M. Thorwart, Phys. Rev. B **90**, 014433 (2014)
11. M. Stier, W. Häusler, T. Posske, G. Gurski, M. Thorwart, Phys. Rev. Lett. **118**, 267203 (2017)
12. F.D.M. Haldane, J. Phys. C **14**, 2585 (1981)
13. A.O. Gogolin, A.A. Nersesyan, A.M. Tsvelik, Cambridge University Press (2004)
14. K. Obata, G. Tatara, Phys. Rev. B **77**, 214429 (2008)
15. M. Sayad, R. Rausch, M. Potthoff, Phys. Rev. Lett. **117**, 127201 (2016)
16. M. Stier, R. Egger, M. Thorwart, Phys. Rev. B **87**, 184415 (2013)
17. I.M. Miron, T. Moore, H. Szambolics, L.D. Buda-Prejbeanu, S. Auffret, B. Rodmacq, S. Pizzini, J. Vogel, M. Bonfim, A. Schuhl, G. Gaudin, Nat. Mater. **10**, 419 (2011)
18. M. Thorwart, R. Egger, Phys. Rev. B **76**, 214418 (2007)
19. M. Stier, M. Thorwart, Phys. Rev. B **92**, 220406 (2015)

20. S. Rößler, S. Hankemeier, B. Krüger, F. Balhorn, R. Frömter, H.P. Oepen, Phys. Rev. B **89**, 174426 (2014)
21. J. Sampaio, V. Cros, S. Rohart, A. Thiaville, A. Fert, Nat. Nanotechnol. **8**, 839 (2013)
22. S. Krause, R. Wiesendanger, Nat. Mater. **15**, 493 (2016)
23. S. Woo, K. Litzius, B. Kruger, M.-Y. Im, L. Caretta, K. Richter, M. Mann, A. Krone, R.M. Reeve, M. Weigand, P. Agrawal, I. Lemesh, M.-A. Mawass, P. Fischer, M. Kläui, G.S.D. Beach, Nat. Mater. **15**, 501 (2016)
24. X. Zhang, G.P. Zhao, H. Fangohr, J. Ping Liu, W.X. Xia, J. Xia, F.J. Morvan, Sci. Rep. **5**, 7643 (2015)
25. A. Fert, V. Cros, J. Sampaio, Nat. Nanotechnol. **8**, 152 (2013)
26. N. Nagaosa, Y. Tokura, Nat. Nanotechnol. **8**, 899 (2013)
27. R. Wiesendanger, Nat. Rev. Mater. **1**, 16044 (2016)
28. N.S. Kiselev, A.N. Bogdanov, R. Schäfer, U.K. Rößler, J. Phys. D Appl. Phys. **44**, 392001 (2011)
29. A. Bogdanov, A. Hubert, J. Magn. Magn. Mater. **138**, 255 (1994)
30. S. Mühlbauer, B. Binz, F. Jonietz, C. Pfleiderer, A. Rosch, A. Neubauer, R. Georgii, P. Böni, Science **323**, 915 (2009)
31. S. Heinze, K. von Bergmann, M. Menzel, J. Brede, A. Kubetzka, R. Wiesendanger, G. Bihlmayer, S. Blügel, Nat. Phys. **7**, 713 (2011)
32. J. Hagemeister, N. Romming, K. von Bergmann, E.Y. Vedmedenko, R. Wiesendanger, Nat. Commun. **6**, 8455 (2015)
33. X.-L. Qi, Y.-S. Wu, S-Ch. Zhang, Phys. Rev. B **74**, 085308 (2006)
34. M.W. Hirsch (Springer, New York, NY 1976), pp. 120–141
35. J. Iwasaki, M. Mochizuki, N. Nagaosa, Nat. Nanotechnol. **8**, 742 (2013)
36. Y. Tchoe, J.H. Han, Phys. Rev. B **85**, 174416 (2012)
37. W. Jiang, P. Upadhyaya, W. Zhang, G. Yu, M.B. Jungfleisch, F.Y. Fradin, J.E. Pearson, Y. Tserkovnyak, K.L. Wang, O. Heinonen, S.G.E. te Velthuis, A. Hoffmann, Science **349**, 283 (2015)
38. W. Koshibae, N. Nagaosa, Nat. Commun. **7**, 10542 (2016)
39. N. Romming, C. Hanneken, M. Menzel, J.E. Bickel, B. Wolter, K. von Bergmann, A. Kubetzka, R. Wiesendanger, Science **341**, 636 (2013)
40. E.C. Corredor, S. Kuhrau, F. Kloodt-Twesten, R. Frömter, H.P. Oepen, Phys. Rev. B **96**, 060410 (2017)
41. J. Iwasaki, M. Mochizuki, N. Nagaosa, Nat. Commun. **4**, 1463 (2013)
42. L. Schmidt, J. Hagemeister, P.-J. Hsu, A. Kubetzka, K. von Bergmann, R. Wiesendanger, New J. Phys. **18**, 075007 (2016)
43. J. Hagemeister, E.Y. Vedmedenko, R. Wiesendanger, Phys. Rev. B **94**, 104434 (2016)
44. P. Ferriani, K. von Bergmann, E.Y. Vedmedenko, S. Heinze, M. Bode, M. Heide, G. Bihlmayer, S. Blügel, R. Wiesendanger, Phys. Rev. Lett. **101**, 027201 (2008)
45. M. Bode, M. Heide, K. von Bergmann, P. Ferriani, S. Heinze, G. Bihlmayer, A. Kubetzka, O. Pietzsch, S. Blügel, R. Wiesendanger, Nature **447**, 190 (2007)
46. K. Everschor-Sitte, M. Sitte, T. Valet, A. Abanov, J. Sinova, New J. Phys. **19**(9), 092001 (2017)
47. J. Hagemeister, D. Iaia, E.Y. Vedmedenko, K. von Bergmann, A. Kubetzka, R. Wiesendanger, Phys. Rev. Lett. **117**, 207202 (2016)

Chapter 17
Imaging the Interaction of Electrical Currents with Magnetization Distributions

Robert Frömter, Edna C. Corredor, Sebastian Hankemeier, Fabian Kloodt-Twesten, Susanne Kuhrau, Fabian Lofink, Stefan Rößler and Hans Peter Oepen

17.1 Introduction

About a decade ago, the understanding of the interaction mechanisms between electrical currents and magnetic objects like domain walls or vortices was still in its infancy. Both, theory and experiments were moving on uncharted grounds, with new and many times contradicting results being published from different groups at increasing pace. Previously, only the Oersted field from a current was available to manipulate the magnetic state of a system. The latter was extensively used in magnetic storage devices, starting from the early core memory up to the write heads of the latest hard drive technologies based on perpendicular recording and tunnel-magnetoresistive (TMR) readout sensors, to alter the state of the magnetic bits. Even early prototypes of magnetic random-access memory (MRAM) used the Oersted field for writing.

17.1.1 Spin-Transfer Torque

With the experimental confirmation of the spin-transfer-torque (STT) effect in magnetic wires [1–4] a new method for manipulating magnetic bits became available. The STT arises, when an electrical current passes a boundary between regions (domains) of different magnetization orientation like, e.g. a domain wall in a magnetic wire with in-plane magnetization. In a simple picture for a 3d ferromagnet like Co

R. Frömter (✉) · E. C. Corredor · S. Hankemeier · F. Kloodt-Twesten ·
S. Kuhrau · F. Lofink · S. Rößler · H. P. Oepen
Institut für Nanostruktur- und Festkörperphysik,Universität Hamburg, Hamburg, Germany
e-mail: rfroemte@physik.uni-hamburg.de

H. P. Oepen
e-mail: oepen@physik.uni-hamburg.de

R. Wiesendanger (ed.), *Atomic- and Nanoscale Magnetism*, NanoScience and Technology, https://doi.org/10.1007/978-3-319-99558-8_17

or Fe, the s-like conduction electrons are spin polarized by the itinerant 3d electrons that carry the magnetization via s-d scattering processes. While crossing the domain wall the conduction electrons constantly adjust their polarization to the locally altered magnetization direction. The conservation of spin angular momentum in turn rotates the localized moments by a certain degree into the opposite direction, thus gradually moving the wall along the electron current direction. The efficiency of this process is characterized by the so-called spin-drift velocity. To manipulate a magnetization distribution in this way has two consequences that stirred the excitement in the magnetism community: First, the local action of a current in a magnetic wire or multilayer stack promised potential for miniaturization of devices like STT-MRAM, in contrast to the far-reaching Oersted field from separate current carrying wires, where cross-talk had to be circumvented by size. Also, the energy required for switching could be strongly reduced for an efficient material, making it interesting for mobile devices. Second, if a wire with alternating up-down or left-right domains is exposed to STT all domain walls are pushed into the same direction, whereas when applying an Oersted field the domains alternatingly shrink or grow. This led to the proposal of the so-called racetrack memory [5], a domain-wall shift register that, if successfully extended into the third dimension, could satisfy future storage-density needs.

Following early theoretical predictions by Berger [6] and Slonczewski [7], a micromagnetic model for describing the effect of the local current density on the local magnetization was developed in 2004 [8–10]. It consists of two additional current-dependent terms in the Landau-Lifshitz-Gilbert (LLG) equation, describing both, an adiabatic and a non-adiabatic contribution. The degree of nonadiabaticity is denoted by the parameter β. It was thought to originate mainly from spin relaxation, namely a miss-tracking of itinerant electrons with respect to the domain wall profile, leading to transverse spin accumulation. The importance of β is that the non-adiabatic torque acts like a negative damping term to the adiabatic torque, thus a large value will increase the domain wall velocity under current. Simulations predict that the initial domain wall velocity is strongly increased for $\beta > \alpha$, the latter being the Gilbert damping constant in LLG. Thus, a high-β material would have significant technological advantage. Unfortunately, it turned out to be experimentally very difficult to measure β and the reported values, even for the best-studied material at that time, permalloy, varied by orders of magnitude. In Sect. 17.2 an experiment is explained in more detail, where we measured the β parameter of permalloy with unprecedented accuracy, following a proposal from theory [11]. It is based on measuring the equilibrium displacement of a vortex in a permalloy square under direct current flow. The result of the analysis led to the unexpected conclusion, that the micromagnetic model used to describe the current-vortex interaction was not yet complete [12]. An additional, nonlocal contribution to the non-adiabatic STT has to be included to correctly describe the strong magnetization gradients occurring in tiny objects like magnetic vortices.

17.1.2 SEMPA as a Unique Tool for Magnetic Imaging

To carry out magnetization imaging we use the Scanning Electron Microscope with Polarization Analysis (SEMPA), which we have developed in our group and keep adding features, like recently the time resolution, which is discussed in more detail in Sect. 17.4. SEMPA is based on the effect that the secondary electrons that are emitted during image formation in a scanning electron microscope (SEM) carry with their spin the information about the magnetization at the surface area they originated from. Thus, by measuring the spin polarization of the secondary electrons a map of the surface magnetization is obtained. To measure the spin-polarization we use spin-polarized low-energy electron diffraction (SPLEED) at a tungsten single crystal, where at 104.5 eV the (2, 0) beams show an intensity asymmetry depending on the spin-polarization of the incoming electrons. A sketch of the setup is shown in Fig. 17.1. As there are four (2, 0) beams in the LEED pattern, two orthogonal components of the spin polarization can be measured simultaneously, which allows for the parallel acquisition of two orthogonal magnetization components in the sample surface. As the information depth for the initial process of spin polarization is very small (only about 0.5 nm in iron) and LEED is also very surface sensitive, SEMPA is necessarily an ultrahigh-vacuum (UHV) experiment. This allows to carry out clean surface-science experiments on uncapped epitaxial magnetic layers to study ideal material properties. But also ex-situ prepared samples can be analyzed following some surface treatment like Co- or Fe-atom decoration or mild Ar-ion sputtering in a Kaufman-type source. We have found that some cappings like graphene are even stable to atmosphere and still magnetically transparent [13].

Key features of our SEMPA setup are: A lateral resolution of down to 10 nm can be obtained using 5 nA primary beam current at 5 keV kinetic energy. Both in-plane components of the surface magnetization are imaged simultaneously. With standard image quality, for each pixel an angular accuracy $<4°$ for determining the magnetization orientation is obtained. A second setup allows for imaging the out-of-plane component together with one in-plane component by using a 90° electrostatic deflector (cylindrical mirror analyzer) in front of the spin detector. The sample temperature can be varied continuously from 40–400 K using a liquid-helium cryostat coupled to the sample holder. An in-situ coil on a ferrite core with gap allows to apply magnetic fields up to 60 mT at the sample. The working pressure of the microscope is below 5×10^{-11} mbar, which allows for several days of analysis by maintaining a sufficiently clean surface. A sample preparation chamber with surface science equipment like LEED and Auger electron spectroscopy is directly attached. In addition, ex-situ prepared samples can be transferred in and out of UHV by means of a fast-entry lock. Several thermal evaporators for in-situ contrast enhancement with clean ferromagnetic layers, as well as sputter sources for Ar-ion cleaning at different energies from 100–2000 eV are installed directly in the microscope chamber. Coaxial electrical wires and matched contacts to the sample allow for high-frequency stimulation up to several GHz or, alternatively, for in-situ four-point-probe measurements of sample resistivity. Acquisition times from 10 min (for regular images) to several hours (for

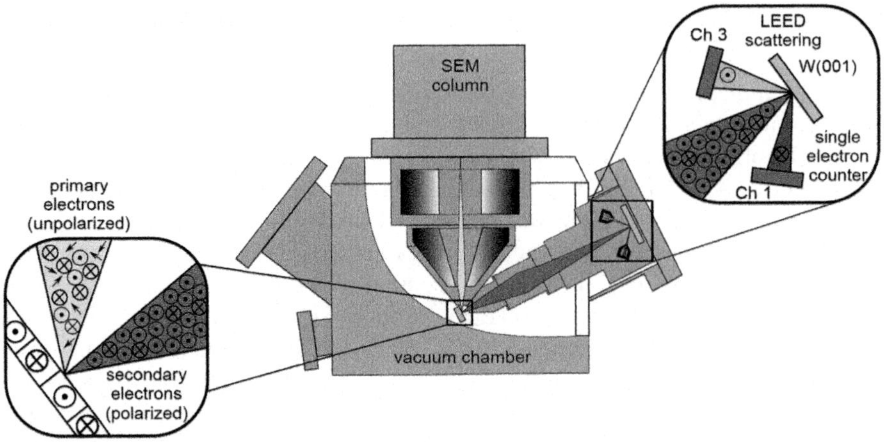

Fig. 17.1 Essentials of SEMPA. In the center a schematic of the setup is shown, consisting primarily of the SEM column and the spin detector in a UHV chamber. The two insets illustrate the main processes involved: To the left, the generation of spin-polarized secondary electrons at the surface of a magnetic sample is shown and to the right the tungsten-based SPLEED detector that yields an intensity asymmetry of opposed (2, 0) beams as function of the spin-polarization of the incoming beam. Of the four channels recorded, only the detector pair in the paper plane is shown. An electron optics between sample and tungsten crystal serves to efficiently pick up the secondary electrons, as well as to collimate and align the beam of secondary electrons with the correct energy onto the crystal (from [14])

time-resolved imaging) are feasible using digital drift correction. A time resolution of 700 ps has been demonstrated.

A more general review of the technique can be found in [15]. Details on our setup including a method to calculate and optimize the detection efficiency are given in [16]. The long-time stability of the SPLEED detector used for spin analysis against residual gas contamination is analyzed in [17]. The latest technical contribution is the introduction of nanosecond time resolution to SEMPA [18], which will be described in more detail in Sect. 17.4.

17.2 Determining the Nonadiabaticity Parameter from the Displacement of Magnetic Vortices

As described in the introduction, the parameter β that quantifies the amount of non-adiabaticity in STT turned out to be hard to measure experimentally. The reason for this was obvious, as most experiments attempted to derive β from the critical current density required to initiate domain wall motion in a magnetic wire. For a non-ideal system with possibly some amount of edge roughness and internal grain structure, however, this critical current density is more related to the local pinning landscape than to intrinsic material properties. In addition, the type of domain wall was not necessarily well-defined, as transitions under STT were observed. Thus published values varied over orders of magnitude.

17.2.1 Proposal from Theory

After developing a description for confined magnetic vortices under STT and external field using the Thiele model [19], a proposal to use the current-induced displacement of the magnetic vortex in the Landau state of a square magnetic element came up to determine β [11]. It was shown that the four degenerate ground states (polarity $p = \pm 1$ and curl $c = \pm 1$) of the Landau state behave differently under the various torques. While the non-adiabatic contribution and the Oersted field of the current shift the vortex in the axis of the current direction, the adiabatic contribution shifts the vortex orthogonal to the current. The amount of displacement against the parabolic confinement can be used like a balance to measure the strengths of the individual contributions from a comparison of displacements for the different ground states. As the vortex stays close to the center, an influence of edge roughness can be excluded. The extended period of time during which the vortex is exposed to a constant drive current will reduce pinning at grains, as the system can gradually creep into its stationary equilibrium. In this way, frequent problems of other methods can be avoided.

17.2.2 Sample Preparation

In order to realize such an experiment in SEMPA, several challenges had to be faced. To obtain well observable displacements of the order of 100 nm in a $5\,\mu\text{m} \times 5\,\mu\text{m} \times 12$ nm sized permalloy square, a current density of several 10^{11} A/m^2 is required. In contrast to most common experiments that require only current pulses from nanosecond to microsecond duration, this ultrahigh current density needs to be sustained during the whole time of acquiring an image, which can be several 10 min. During this period, no excessive heating of the structure is allowed. Under normal circumstances any metal wire would soon be destroyed by Joule heating or electromigration. To circumvent this problem we have developed a method to prepare metal wires on synthetic single-crystalline diamond substrates, held at cryogenic temperature. A comparison of low-temperature thermal conductances shows that single-crystalline diamond has the highest thermal conductivity (10^4 W/mK) of all materials at liquid-nitrogen temperature [20], about 20 times higher than copper at room temperature. Still, the purely phononic conductivity does not conduct electrical current, so no shunt is created. The available surface roughness can be below 0.3 nm RMS, as checked by AFM. In a test experiment a 22.5 nm high, 650 nm wide and 25 μm long permalloy wire on diamond was placed in vacuum on a cryostat held

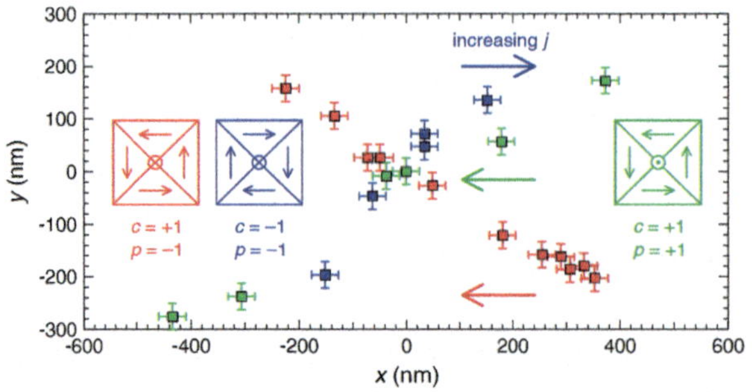

Fig. 17.2 Map of the current-dependent vortex positions in the permalloy square for the three different ground states illustrated. The current density is varied over a range of $\pm 3 \times 10^{11}$ A/m in the direction indicated for each case. All three states show a reasonably linear behavior, which allows to extract three slopes from the data and to calculate the non-adiabaticity. For details, see [12]

at 100 K. It could withstand a current density of 2×10^{12} A/m^2 for at least an hour without damage [21].

Our attempts to perform lithography on the diamond surface were unsuccessful. We therefore used a stencil mask deposition technique with focused-ion-beam (FIB) machined silicon nitride membranes and optical alignment to prepare the permalloy square and the gold contact pads.

17.2.3 Experimental Results

The experiment has been carried out without attaching the cryostat, so the state without current is at room temperature. Careful analysis of the resistance of the sample while applying the current allows to calculate the sample temperature. The maximum temperature recorded at the maximum current of $\pm 3 \times 10^{11}$ A/m applied was 350 K. As we found no temperature dependence of β over this range [22], thermal effects can be neglected. By using field pulses from an in-situ electromagnet we were able to seed three of the four possible ground states of the permalloy square. This turned out to be sufficient to perform the analysis, as only three unknown torque terms had to be separated. Fig. 17.2 shows the obtained positions of the vortex cores for the three states and for the current ramping from negative to positive values as indicated. Three different linear slopes can be extracted from the dataset, which allow for further analysis. As result we find for the parameter of non-abiabaticity $\beta = 0.119 \pm 0.022$ for the permalloy vortex. In addition, the spin drift velocity $v_j = (4.79 \pm 0.26)$ m/s can be directly determined form the slopes. As we also performed ferromagnetic resonance measurements on the material to determine the saturation magnetization, we can give as third material parameter the spin polarization, which is $(65 \pm 4)\%$, and the damping constant $\alpha = 0.0085 \pm 0.0006$.

When comparing this data to results from other groups, one observation was stringent: Two other experiments that also used vortices as test objects showed similarly high values for β [23, 24], albeit with less precision. In contrast, another carefully performed study using spin-wave dispersion instead of vortices resulted in a 5 times smaller value of $\beta = 0.022$ [25]. In fact this value is only about twice as large as α - similar values for both constants have indeed been predicted. As all the other material parameters of our system mentioned above were very similar in this study, the huge difference can only be appointed to the different object investigated. The spin waves analyzed in [25] are of the order of 10 μm in period, whereas vortices have a very narrow core of the order of 10 nm, so the spatial variation of the magnetization takes place on a much smaller length scale. Indeed theory predicts a second [26] truly non-adiabatic contribution to β that is due to spin scattering and that is essentially nonlocal [8]. It is oscillatory in space due to the precession of conduction electron spins [27]. Micromagnetic calculations with a more complex code that accounts for these nonlocal effects have already shown a threefold increase of the effective β for a vortex wall compared to a transverse wall [28]. Our experimental results prove that for small magnetic structures like vortices or skyrmions, which are small on the length scale of the Larmor precession, the local description using the LLG equation with two additional spin torque terms is not sufficient.

17.3 Applications of Vectorial Magnetic Imaging

Magnetic imaging can be performed to solve a variety of different tasks. A standard task is the question if and under which conditions a certain magnetic structure exists or ceases to exist. High-resolution vectorial magnetization imaging, which is possible directly by using SEMPA or following phase retrieval in Lorentz transmission electron microscopy (TEM) [29] or TEM holography [30], can supply a lot more information. The whole magnetization map of a structure becomes available and can be compared to micromagnetic simulations or analytical predictions. Micromagnetic simulations have developed over the last years into a very powerful and easy to use toolkit for predicting micromagnetic structures and devices, even including dynamic effects like STT or "defects" like grain boundaries [31, 32]. In reality, however, magnetic structures often look or behave quite differently. This can be most sensitively tested in a vectorial imaging experiment, as already small angular deviations from the predicted pattern become directly evident. Reasons for such discrepancies can be systematic deviations of the real structure from the one modeled, like the often non-ideal, inclined edges of magnetic wires. There can be defects like kinks resulting from problems in the lithography process of a real structure, that will pin domain wall motion. Stochastic variations of magnetic properties like grain boundaries in a polycrystalline material or surface/interface roughness can act as internal pinning sites. Finite temperature can have a strong impact on the depinning mechanism from those sites. However, for more ideal structures it is often possible to extract numbers for unknown magnetic properties by comparison with simulation. In some cases even

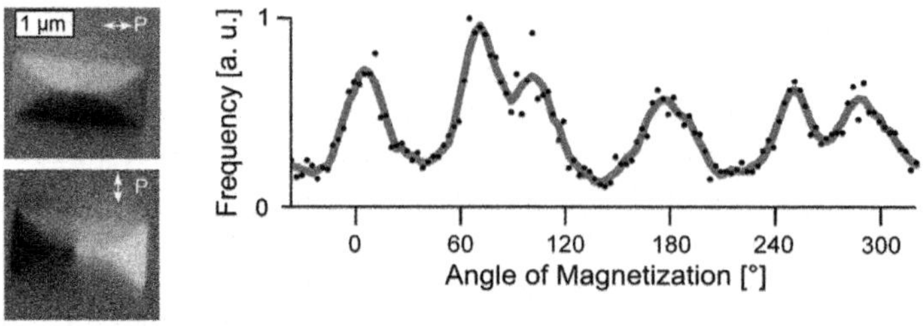

Fig. 17.3 Magnetic fine structure of a permalloy rectangle of size 2 μm×1 μm×20 nm observed in a vectorial SEMPA polarization map. The angular magnetization histogram to the right reveals that not four but six domains are clearly separated. The splitting of the larger two domains leads to the emergence of stray field and thus enables coupling between such rectangles. For details, see [36]

the magnetic theory behind the simulation may be incomplete and simply predict wrong results.

In several cases we have used the vectorial imaging capability to clarify magnetic ground state properties of various systems. The internal structure of the canted magnetization (also known as cone state) in a continuous Co/Pt multilayer film close to the spin-reorientation transition has been imaged for the first time in [33]. We found a temporally stable isotropic and gradual in-plane modulation, superimposed onto an alternating up/down domain pattern, both modulated on a similar length scale.

A similar study has focussed on the temperature-driven spin-reorientation transition between two different in-plane easy axes in $NdCo_5$, where the temperature-dependent anisotropy could be extracted and the microstructure analyzed [34, 35].

A third study deals with the internal magnetic structure of 2 μm×1 μm×20 nm sized rectangles made of permalloy [37] - the so-called standard problem #1 of micromagnetism. In contrast to simple reasoning, the angular histogram of the SEMPA dataset reveals not only the four domains of the expected Landau state, instead six domains are clearly separated. It turns out that the two larger domains along the long side of the rectangle are actually split along the short symmetry axis by some degrees, as shown in Fig. 17.3. This is a consequence of total energy minimization of the magnetic volume charge of the 180° wall along the center, which is compensated by surface charges along the edges of the structure. This tiny effect had previously been overlooked in the simulation, even though present, and it is responsible for the coupling between adjacent rectangles placed in a row, as the structures are no longer stray-field free. As the experimentally obtained splitting angle of 35° is much larger than the simulated 15° we adjusted the edge inclination and added edge roughness to obtain agreement. The additional effect of the contrast-enhancing iron layer on the coupling between rectangles was studied in [36].

Two studies deal with the ground state of domain walls in nanowires with a bend as function of bend angle, thickness and width [38], as well as a method to nucleate walls of fixed rotational sense in such bends [39]. Figure 17.4 shows as example the asymmetric transverse wall that is found in certain regions of parameter space instead

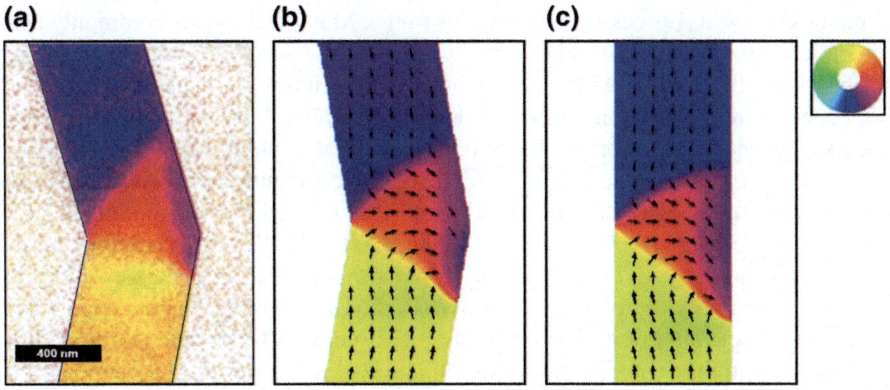

Fig. 17.4 Magnetic fine structure of an asymmetric transverse wall (ATW) in a soft-magnetic $Co_{39}Fe_{54}Si_7$ nanowire of width $w = 600$ nm and thickness $t = 10$ nm. **a** SEMPA measurement of the ATW in a 30° bend. **b** OOMMF simulation of the exact geometry as in (**a**). **c** OOMMF simulation of an ATW in a straight section of the same wire. The local orientation of the magnetization is color coded according to the given color wheel. For details, see [38]

of symmetric transverse walls or vortex walls. These experiments form the basis of current and field driven domain wall displacement studies, as they allow to control and restore a selected starting configuration. Such an experiment was carried out with a permalloy nanowire on a liquid-nitrogen-cooled diamond substrate using the total magnetoresistance as probe for the magnetic state, while sweeping the external field as function of current density [22]. We found, that for permalloy there is no temperature dependence of the spin-torque efficiency over a temperature range from 77 to 327 K.

Furthermore, we analyzed the magnetic structure of the clean, single-crystalline Ni(111) surface and the effect of a graphene monolayer deposited on top [13]. In this multiscale, multitechnique approach, theory provided first-principles calculations of the spin-resolved band structure and spin-polarized STM analyzed low-temperature domain walls with close-to-atomic resolution. Using SEMPA, we performed the first vectorial imaging on this peculiar Ni surface, where no easy axis of the system is located. We could in principle confirm a multi-level branching of closure domains, that was predicted in [40]. The graphene layer in fact protected the surface magnetization of nickel against oxidation, so even after transfer at ambient conditions the domain pattern could be observed. We found a reduction of contrast, but no spin inversion for the secondary electrons after passing through the graphene.

17.4 Development of Time-Resolved SEMPA

Time-resolved magnetic microscopy is an essential tool to understand magnetization dynamics or to analyze switching processes in magnetic devices. Relevant timescales for LLG dynamics are in the sub-ns range, while ultrafast magnetization dynamics concerns processes in the sub-ps range. Many techniques commonly used for static magnetic imaging can be empowered for time-resolved imaging by using a stroboscopic illumination with a source that can be periodically blanked, like a pulsed laser for scanning [41] or full-field [42] Kerr microscopy. Very recent developments have succeeded in transferring the time structure of a pump laser to an electron beam in a transmission electron microscope (TEM). The ultrafast Lorenz TEM uses a 150-femtosecond-laser-driven Schottky emitter [43] and the spin-polarized TEM [44] uses a picosecond-laser-pumped GaAs/GaAsP cathode. By synchronizing the illumination pulses with the periodically driven magnetization dynamics under investigation and varying the phase, a movie of the time evolution is obtained frame by frame. In scanning transmission X-ray microscopy (STXM), the internal bunch structure at a storage ring is used for illumination [45]. However, as here the illumination frequency is fixed, a more sophisticated measurement scheme is used, where the intensity information from each consecutive X-ray pulse is sorted by real-time electronics and accumulated into the according time slices. The synchronization here runs over an integer number of loops, so each revolution of the magnetic system is hit at different phases. In this way, all of the pulses can be used and maximum efficiency obtained.

17.4.1 Concept

To use time resolution in SEM, a stroboscopic approach with a blanked source or detector is in principle possible [46] and has already been used to record magnetic movies at microsecond time resolution [47]. However, due to the already long acquisition times per image this is not very attractive for SEMPA, as the time required to record a stroboscopic movie of n frames scales with n^2. Inspired by the STXM detection scheme, we came up with a much more efficient solution that does not even require in-vacuum modifications of the SEMPA (as long as high-frequency connections to the sample are already present) [18]. It is based on the fact that the SPLEED detector operates in single electron counting mode. Each electron of the four recorded LEED beams is amplified by a microchannel plate stack to give a 2 ns wide voltage pulse, that is normally only used for counting. Under the assumption of constant transit times for all electrons through the detector, the arrival time of each voltage pulse is just delayed by a constant with respect to the time the secondary electron was emitted at the sample surface. In this way the voltage pulses do not only carry the information about the local magnetization but also the information about the time at which this magnetic state was present. By using a periodic excitation and electronically recording the phase of all arriving voltage pulses with respect to this excitation, the counts can be grouped into time slices. In this way the whole time evolution at one pixel is recorded in parallel and with maximum efficiency, as (except for some timing overhead) all pulses are processed. A movie of the time evolution of a whole magnetic object is then recorded by scanning pixel after pixel.

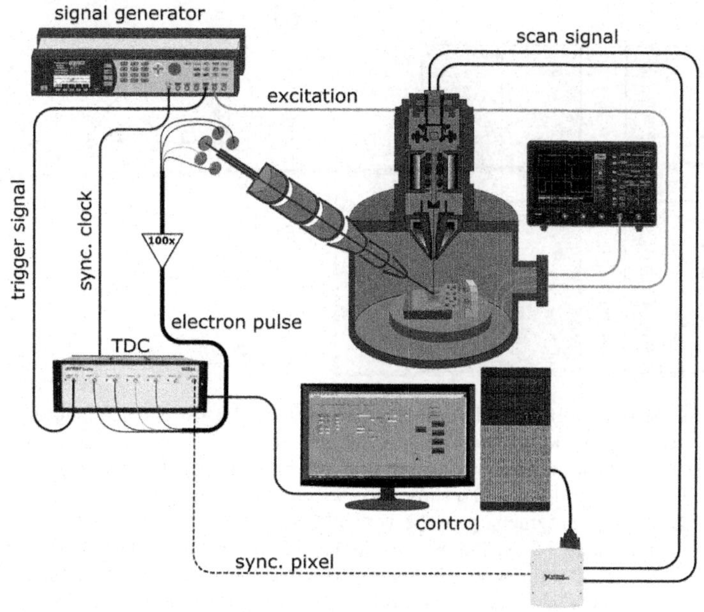

Fig. 17.5 Sketch of the setup used for TR-SEMPA. The time-to digital converter (TDC) receives the amplified voltage pulses from the SPLEED detector. A synchronized signal generator drives a current through the sample and the transmitted voltage is monitored using a scope. For details, see [18]

For this detection scheme the time to record n frames scales linearly with n, which is normally of the order of hours and thus feasible.

17.4.2 Experimental Setup

A sketch of the experimental setup for TR-SEMPA is given in Fig. 17.5. The only new hardware required is a time-to-digital converter (TDC) that measures the arrival time of each voltage pulse with respect to a trigger signal that is generated from the signal generator driving the magnetization dynamics. We use a four-channel TDC with 100 ps time resolution that transmits the digital data for all the pulses in real-time to a computer for on-line processing.

The sample used in the demonstration experiment is a 60 nm thick and 3.5 μm wide square made from soft-magnetic amorphous $Fe_{70}Co_8Si_{12}B_{10}$ [48]. The square is in its magnetic Landau ground state. It is deposited onto a gold (150 nm)/chromium (24 nm) microstrip of 5 μm width on a sapphire substrate. An electrical current running along the microstrip gives rise to an Oersted field across the strip and thus shifts the magnetic vortex of the Landau structure to the left or right.

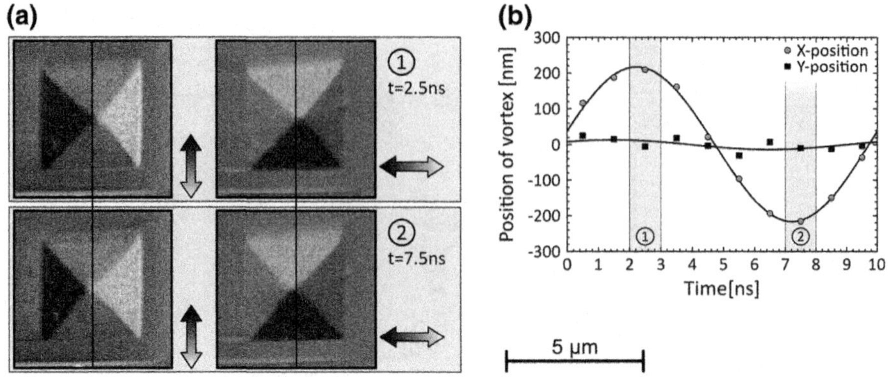

Fig. 17.6 Field-driven vortex oscillation in the Landau state of a soft-magnetic square element. The 60 nm thick and 3.5 μm wide square has been prepared from $Fe_{70}Co_{8}Si_{12}B_{10}$ on a gold microstrip. The Oersted field from a sine-modulated current at 100 MHz running horizontally in the microstrip is used to drive the dynamics. Panel **a** shows two selected frames of 1 ns duration from a 10-frame movie. The SEMPA micrographs give the vertical and horizontal magnetization components at maximum positive (delay 2.5 ns) and negative (delay 7.5 ns) displacements. Dashed lines indicate the horizontal zero-position around which the vortex oscillates. In **b** the horizontal and vertical coordinates of the vortex have been extracted from the movie. For details, see [18]

17.4.3 Results and Analysis

Figure 17.6 shows two frames at the positions of maximum displacement from a 10-frame movie acquired with TR-SEMPA. The sample was driven with a 100 MHz sine current. A sine displacement along the horizontal axis with 440 μm peak-to-peak amplitude is observed in the movie. The corresponding vortex positions have been extracted and are plotted in panel (b). Surprisingly, the vertical displacement is only at the noise limit, so the gyration that would be expected not too far from nominal resonance is strongly suppressed. An analysis of the dynamics in this structure, when driven at different frequencies, reveals that the suppression of gyration corresponds to the behavior of an oscillator with strongly damped dynamics. Cross-sections of the structure suggest that the damping is of extrinsic nature and can be attributed to a very rough surface of the gold microstrip, that was copied during deposition into the magnetic film.

From the high-quality image, not only the position of the vortex can be extracted but the full vectorial map, which shows the same image quality as a static SEMPA image at this relative acquisition time, is accessible in a time-resolved way. For instance, in this example the motion of the 90° bends can be studied in detail.

The time resolution available in this experiment has been determined from the motion blur effect of a sample edge, that is rapidly shifting when a square-wave modulation of the sample potential is performed. The sample edge gets widened (blurred) as a function of apparent velocity, as can be extracted from line profiles. Quantitative analysis yields a time resolution of (720 ± 80) ps [18]. Electron-optical simulation reveals that this resolution is to be expected with the design parameters of the currently used spin detector. The energy and angular distributions of the secondary electrons at the time of emission from the sample surface lead to a jitter in transit time through the spin detector, that smears out the correlation between starting time and pulse detection.

17.5 Conclusion and Outlook

The application of high-resolution vectorial magnetic imaging has been very successful in resolving a large number of open questions. Especially, we could confirm the importance of non-local effects when describing the current-driven dynamics of vortices in contrast to more extended objects. With the invention of TR-SEMPA we have opened a new experimental access to study magnetization dynamics that preserves all the advantages of stationary SEMPA, especially the ability to work on ideal epitaxial samples on single-crystalline substrates. This will allow to study dynamical processes in pure elements that are free from pinning at grain boundaries and show the unaltered behavior that can be directly compared to first principles calculations. Technologically more relevant are, however, glassy alloys containing, for instance, boron, where grain growth is suppressed.

Already for some time the interest in the community has shifted away from in-plane systems for spintronic devices towards out-of-plane magnetized systems. Especially if these are prepared in the form of artificially layered antiferromagnets [49], such that the stray field interaction is fully compensated, a much closer packing of structures and therefore of information can be obtained. In addition to the STT discussed here, new and more efficient torque terms appear in current-driven dynamics that can sustain higher domain wall velocities up to 750 m/s [50], which are required for application. These spin-orbit torques can result from Rashba effects or the inverse spin-Hall effect and open the promising field of spin-orbitronics. With the new SEMPA setup with out-of-plane sensitivity this class of materials can be analysed. The high surface sensitivity warrants, that only the topmost layer of an antiferromagnetically coupled stack contributes to the contrast, so that the full magnetic signal is available.

In addition to the geometry that has changed over the course of years, also a new class of magnetic particles has today strongly moved into the focus of research. Instead of domain walls, magnetic interface skyrmions, are envisioned for use in magnetic wires instead of domain walls to carry and store information [51]. They are topologically protected against spontaneous decay and get repelled from wire boundaries and by each other. Interface skyrmions were observed for the first time by low-temperature SP-STM on epitaxial Fe/Ir(111) [52]. The key to stabilizing such skyrmions at room temperature is a strong interfacial Dzyaloshinskii-Moriya interaction (DMI) that can typically be found at non-symmetric interfaces between ferromagnets and materials with strong spin-orbit coupling.

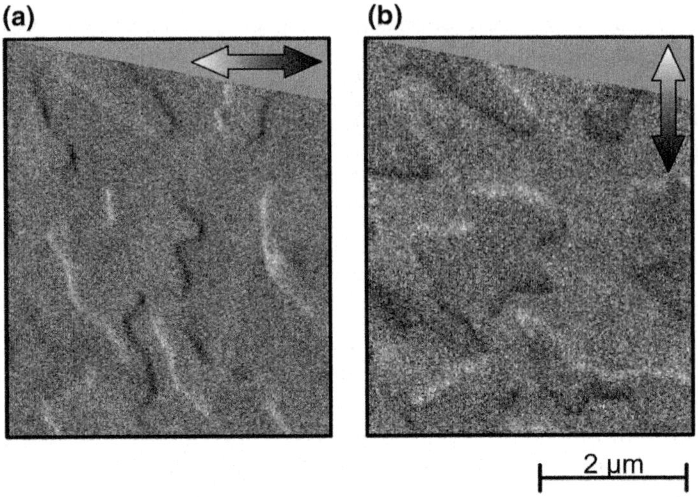

Fig. 17.7 SEMPA measurement of an epitaxial Co/Pt(111) layer at 1.4 nm thickness. All domain walls show a pure Néel character with fixed anticlockwise rotational sense, indicative of a positive DMI. From a quantitative analysis upper and lower bounds for the DMI strength have been derived. For details, see [53]

Recently we have obtained results on SEMPA imaging of such DMI systems by analysing the domain wall angles in epitaxially grown, uncapped cobalt on a Pt(111) single crystal surface [53]. The single layer Co films were grown as a wedge, so that thicknesses from 0 to 1.5 nm were accessible. From LEED data it is clear that the DMI-generating interface layer of Co grows pseudomophically; at higher coverage the growth becomes less ideal. The analysis of SEMPA data shows that over the whole out-of-plane magnetized range all domain walls are of pure Néel character. In addition, all domain walls show a fixed counterclockwise rotational sense, indicating a positive sign of the DMI (see Fig. 17.7). Even though the DMI is an interface effect it still dominates the energy balance of the domain walls, where otherwise uncharged Bloch walls would be favored. From the domain wall energy balance at 1.4 nm coverage a lower bound of 0.8 eV per interface Co atom for the DMI strength was calculated. The observation of a multi-domain state even at the onset of ferromagnetism at 0.3 nm allows to deduce an upper bound of 4.3 meV per interface Co atom. These findings confirm previous first-principles calculations.

Another study on DMI materials focussed on ex-situ-prepared multilayer films [54]. Multilayers consisting of several repeats of Co/Pt/Ir are considered as good candidates for room-temperature skyrmions because the DMI of both interfaces is of opposite sign and thus the total DMI adds up. We have shown, that 150 eV argon ion sputtering can be used to mill away the topmost Pt layer and to image the rotational sense of domain walls in such systems. A second method introduced is the deposition of a thin iron layer onto the topmost Pt layer. The iron layer shunts the magnetic flux of the out-of-plane magnetized domains underneath thus yielding an in-plane contrast.

By calculating the divergence of the in-plane SEMPA map the domain structure of the out-of-plane domains in the Co/Pt/Ir system becomes visible. We confirmed that both methods did not alter the magnetic state of the multilayer.

References

1. J. Grollier, D. Lacour, V. Cros, A. Hamzic, A. Vaurès, A. Fert, D. Adam, G. Faini, J. Appl. Phys. **92**(8), 4825 (2002)
2. M. Kläui, C.A.F. Vaz, J.A.C. Bland, W. Wernsdorfer, G. Faini, E. Cambril, L.J. Heyderman, Appl. Phys. Lett. **83**(1), 105 (2003)
3. M. Tsoi, R. Fontana, S. Parkin, Appl. Phys. Lett. **83**(13), 2617 (2003)
4. A. Yamaguchi, T. Ono, S. Nasu, K. Miyake, K. Mibu, T. Shinjo, Phys. Rev. Lett. **92**(7), 077205 (2004)
5. S.S.P. Parkin, M. Hayashi, L. Thomas, Science **320**(5873), 190 (2008)
6. L. Berger, J. Appl. Phys. **55**, 1954 (1984)
7. J.C. Slonczewski, J. Magn. Magn. Mater. **159**, L1 (1996)
8. S. Zhang, Z. Li, Phys. Rev. Lett. **93**(12), 127204 (2004)
9. G. Tatara, H. Kohno, Phys. Rev. Lett. **92**(8), 086601 (2004)
10. A. Thiaville, Y. Nakatani, J. Miltat, Y. Suzuki, Europhys. Lett. **69**(6), 990 (2005)
11. B. Krüger, M. Najafi, S. Bohlens, R. Frömter, D.P.F. Möller, D. Pfannkuche, Phys. Rev. Lett. **104**(7), 077201 (2010)
12. S. Rößler, S. Hankemeier, B. Krüger, F. Balhorn, R. Frömter, H.P. Oepen, Phys. Rev. B **89**, 174426 (2014)
13. L.V. Dzemiantsova, M. Karolak, F. Lofink, A. Kubetzka, B. Sachs, K. von Bergmann, S. Hankemeier, T.O. Wehling, R. Frömter, H.P. Oepen, A.I. Lichtenstein, R. Wiesendanger, Phys. Rev. B **84**, 205431 (2011)
14. S. Rößler, Magnetic imaging of static and dynamic states in systems of reduced dimensions. (Shaker, Aachen, 2015)
15. H.P. Oepen, R. Frömter, *Handbook of Magnetism and Advanced Magnetic Materials* (Wiley, New York, 2007). Chap. Scanning Electron Microscopy with Polarization Analysis
16. R. Frömter, S. Hankemeier, H.P. Oepen, J. Kirschner, Rev. Sci. Instrum. **82**(3), 033704 (2011)
17. F. Lofink, S. Hankemeier, R. Frömter, J. Kirschner, H.P. Oepen, Rev. Sci. Instrum. **83**, 023708 (2012)
18. R. Frömter, F. Kloodt, S. Rößler, A. Frauen, P. Staeck, D.R. Cavicchia, L. Bocklage, V. Röbisch, E. Quandt, H.P. Oepen, Appl. Phys. Lett. **108**, 142401 (2016)
19. B. Krüger, A. Drews, M. Bolte, U. Merkt, D. Pfannkuche, G. Meier, Phys. Rev. B **76**(22), 224426 (2007)
20. S. Barman, G.P. Srivastava, J. Appl. Phys. **101**(12), 123507 (2007)
21. S. Hankemeier, K. Sachse, Y. Stark, R. Frömter, H.P. Oepen, Appl. Phys. Lett. **92**(24), 242503 (2008)
22. B. Beyersdorff, S. Hankemeier, S. Rößler, Y. Stark, G. Hoffmann, R. Frömter, H.P. Oepen, B. Krüger, Phys. Rev. B **86**(18), 184427 (2012)
23. L. Heyne, J. Rhensius, D. Ilgaz, A. Bisig, U. Rüdiger, M. Kläui, L. Joly, F. Nolting, L.J. Heyderman, J.U. Thiele, F. Kronast, Phys. Rev. Lett. **105**, 187203 (2010)
24. S. Pollard, L. Huang, K. Buchanan, D. Arena, Y. Zhu, Nat. Commun. **3**, 1028 (2012)
25. K. Sekiguchi, K. Yamada, S.M. Seo, K.J. Lee, D. Chiba, K. Kobayashi, T. Ono, Phys. Rev. Lett. **108**, 017203 (2012)
26. G. Tatara, P. Entel, Phys. Rev. B **78**, 064429 (2008)
27. K.J. Lee, M. Stiles, H.W. Lee, J.H. Moon, K.W. Kim, S.W. Lee, Phys. Rep. **531**(2), 89 (2013)
28. D. Claudio-Gonzalez, A. Thiaville, J. Miltat, Phys. Rev. Lett. **108**, 227208 (2012)

29. V. Volkov, Y. Zhu, M.D. Graef, Micron **33**(5), 411 (2002)
30. R.E. Dunin-Borkowski, T. Kasama, A. Wei, S.L. Tripp, M.J. Hÿtch, E. Snoeck, R.J. Harrison, A. Putnis, Microsc. Res. Techn. **64**(5–6), 390 (2004)
31. A. Drews, Fast Micromagnetic Simulator for Computations on Cpu and Graphics Processing Unit (gpu) (2014)
32. M.J. Donahue, D.G. Porter, *Interagency Report NISTIR 6376* (National Institute of Standards and Technology, MD, 1999)
33. R. Frömter, H. Stillrich, C. Menk, H.P. Oepen, Phys. Rev. Lett. **100**, 207202 (2008)
34. M. Seifert, L. Schultz, R. Schäfer, S. Hankemeier, R. Frömter, H.P. Oepen, V. Neu, New J. Phys. **19**(3), 033002 (2017)
35. M. Seifert, L. Schultz, R. Schäfer, V. Neu, S. Hankemeier, S. Rößler, R. Frömter, H.P. Oepen, New J. Phys. **15**(1), 013019 (2013)
36. J. Jelli, K.M. Lebecki, S. Hankemeier, R. Frömter, H.P. Oepen, U. Nowak, IEEE Trans. Magn. **49**(3), 1077 (2013)
37. S. Hankemeier, R. Frömter, N. Mikuszeit, D. Stickler, H. Stillrich, S. Pütter, E.Y. Vedmedenko, H.P. Oepen, Phys. Rev. Lett. **103**(14), 147204 (2009)
38. F. Lofink, A. Philippi-Kobs, M.R. Rahbar Azad, S. Hankemeier, G. Hoffmann, R. Frömter, H.P. Oepen, Phys. Rev. Appl. **8**, 024008 (2017)
39. S. Hankemeier, A. Kobs, R. Frömter, H.P. Oepen, Phys. Rev. B **82**, 064414 (2010)
40. A. Hubert, R. Schäfer, *Magnetic Domains: The Analysis of Magnetic Microstructures* (Springer, Berlin, 2009)
41. W.K. Hiebert, A. Stankiewicz, M.R. Freeman, Phys. Rev. Lett. **79**(6), 1134 (1997)
42. N.O. Urs, B. Mozooni, P. Mazalski, M. Kustov, P. Hayes, S. Deldar, E. Quandt, J. McCord, AIP Adv. **6**(5), 055605 (2016)
43. T. Eggebrecht, M. Möller, J.G. Gatzmann, N. Rubiano da Silva, A. Feist, U. Martens, H. Ulrichs, M. Münzenberg, C. Ropers, S. Schäfer, Phys. Rev. Lett. **118**, 097203 (2017)
44. M. Kuwahara, Y. Nambo, K. Aoki, K. Sameshima, X. Jin, T. Ujihara, H. Asano, K. Saitoh, Y. Takeda, N. Tanaka, Appl. Phys. Lett. **109**(1), 013108 (2016)
45. Y. Acremann, V. Chembrolu, J.P. Strachan, T. Tyliszczak, J. Stöhr, Rev. Sci. Instrum. **78**, 014702 (2007)
46. G.S. Plows, W.C. Nixon, J. Phys. E **1**(6), 595 (1968)
47. T. Ikuta, R. Shimizu, J. Phys. E **9**(9), 721 (1976)
48. V. Hrkac, E. Lage, G. Köppel, J. Strobel, J. McCord, E. Quandt, D. Meyners, L. Kienle, J. Appl. Phys. **116**(134302) (2014)
49. S. Parkin, S.H. Yang, Nat. Nano **10**(3), 195 (2015)
50. S.H. Yang, K.S. Ryu, S. Parkin, Nat. Nano **10**(3), 221 (2015)
51. A. Fert, V. Cros, J. Sampaio, Nat. Nanotechnol. **8**, 152 (2013)
52. S. Heinze, K. von Bergmann, M. Menzel, J. Brede, A. Kubetzka, R. Wiesendanger, G. Bihlmayer, S. Blügel, Nat. Phys. **7**, 713 (2011)
53. E.C. Corredor, S. Kuhrau, F. Kloodt-Twesten, R. Frömter, H.P. Oepen, Phys. Rev. B **96**, 060410(R) (2017)
54. J. Lucassen, F. Kloodt-Twesten, R. Frömter, H.P. Oepen, R.A. Duine, H.J.M. Swagten, B. Koopmans, R. Lavrijsen, Appl. Phys. Lett. **111**(13), 132403 (2017)

Chapter 18
Electron Transport in Ferromagnetic Nanostructures

Falk-Ulrich Stein and Guido Meier

Abstract The proposal of logic- and memory devices based on magnetic domain-wall motion in nanostructures created a great demand on the understanding of the dynamics of domain walls. We describe the controlled creation and annihilation of domain walls by Oersted-field pulses as well as their internal dynamics during motion. Electric measurements of the magnetoresistance are utilized to identify permanent- or temporal creation and continuous motion of domain walls initiated by nanosecond short field pulses in external magnetic fields. The injection of domain walls into nanowires with control of their magnetic pattern (transverse or vortex), their type (head-to-head or tail-to-tail magnetization orientation) and their sense of magnetization rotation (clockwise or counter clockwise chirality) is reliably achieved. Influencing the creation process of consecutively created domain walls to obtain multiple walls inside one wire or to mutually annihilate the walls is found to be possible by changes of magnetic field parameters. The time structure of the creation process is analysed by time-resolved transmission X-ray microscopy. After complete formation wall transformations are observed above a critical driving field known as the Walker breakdown. Internal excitations of vortex domain walls are also found in low field motion. A strong interplay between internal dynamics and the macroscopic motion is identified.

18.1 Introduction

To overcome increasingly complex memory setups fast and large-scale non-volatile technologies are needed. From various proposed techniques [1] two have been most feasible: The magnetoresistive random access memory (MRAM) and the racetrack

F.-U. Stein · G. Meier (✉)
Max-Planck Institute for the Structure and Dynamics of Matter, Luruper Chaussee 149, 22761 Hamburg, Germany
e-mail: gmeier@mpsd.mpg.de

R. Wiesendanger (ed.), *Atomic- and Nanoscale Magnetism*, NanoScience and Technology, https://doi.org/10.1007/978-3-319-99558-8_18

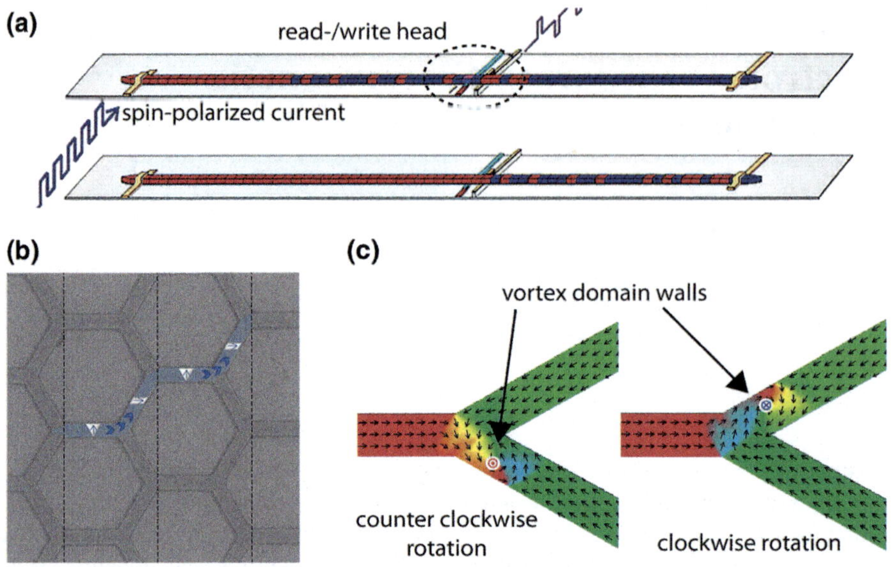

Fig. 18.1 **a** Horizontal racetrack memory of a ferromagnetic nanowire with alternating magnetization (red/blue), from [3]. Data writing and shifting is done by applying current pulses across or along the nanowire, respectively. **b** Path of a domain wall through an artificial spin-ice lattice, from [9]. **c** Motion of a vortex domain wall through a Y-shaped junction in dependence of its magnetization structure, from [10]

memory [2, 3]. The former is already commercially available and is fast and robust. However, it is too expensive for broad mass storage since every data bit requires individual wiring. The racetrack promises to be as fast but cheaper by using one read-/write head for multiple bits as it is done in conventional hard drives. But instead of physically moving the ferromagnetic data carrier only the magnetic bit pattern consisting of domains and domain walls is shifted through the ferromagnet. Concepts of data shifting by magnetic fields had early been proposed by Smith with a domain-wall based shift register [4] and realised in the bubble memory that was discontinued being uncompetitive [5]. The modern racetrack concept uses nanowires and replaces the magnetic field as a driving force by a spin-polarized electric current passing through the alternating magnetization of a data pattern as shown in Fig. 18.1a. Recently also an out-of-plane field was proposed as a synchronous driving mechanism for multiple domain walls with the same magnetic structure [6]. Together with the proposal of logic devices based on domain-wall motion by Allwood et al. [7] the racetrack created a demand for an extensive understanding of domain-wall dynamics. Today the interest extends to sensor application [8] and is of particular concern for artificial spin-ice systems. Spin-ice lattices from ferromagnetic material as shown in Fig. 18.1b contain degenerated ground states similar to water ice resulting in magnetic quasi monopoles that can travel through the system. Connected lattices from nanowires allow to study this effect at room temperature by motion of domain walls [9]. The path taken through the lattice depends heavily on the magnetic structure of the wall shown by Pushp et al. and presented in Fig. 18.1c [10]. For both, the racetrack- and the spin-ice application, control of all properties of the utilized domain walls as well as their dynamic behaviour is crucial. While many aspects of domain wall motion like velocity [11–16], dynamic properties [17–19] and depinning processes [16, 20–24] have been studied [25] the fast injection via Oersted-field pulses itself has not been in focus. Most dynamic parameters have to be gained indirectly by electric measurements or static imaging techniques. Here a detailed analysis of the controlled injection of domain wall dynamics is presented by a combination of quasi-static electric measurements and time-resolved transmission X-ray microscopy. The latter enables the observation of the magnetic structure of domain walls with a spatial resolution of 25 nm and a temporal resolution of less than 1 ns. A reliable control of all domain wall properties is achieved and the interplay between internal excitations and the macroscopic motion is experimentally revealed.

18.2 Domain Walls

One of the consequences from the competing energy contributions in micromagnetics is the existence of domains that had been proposed by Weiss [26]. Due to the exchange energy the magnetic moments tend to align parallel. In order to minimize the total energy it can be favourable for the magnetization to break up into domains that are not parallel [27] to each other. Inside the walls the magnetization continuously rotates from one orientation to the other. This rotation increases the exchange energy in the region of the wall but can reduce the stray field and thereby the total energy. In a homogeneously magnetized flat rectangular ferromagnet the magnetic surface charges create a large stray field and thus the magnetostatic energy is high. This energy is reduced if the magnetization splits up in two domains that are separated by a 180° domain wall and the stray field is decreased. The energy of these walls depends on their internal structure.

The magnetization in domain walls in bulk ferromagnets usually rotates parallel to the wall plane as shown in Fig. 18.2b, which is known as Bloch wall. The formation of these walls is dominated by crystalline anisotropy and exchange energy. In thin films the magnetostatic energy plays an increasingly important role. The shape anisotropy results in an in-plane magnetization parallel to the surface of the film and a Bloch wall creates magnetic charges on the film surface. The formation of Néel walls where the magnetization rotates perpendicular to the wall plane can reduce the magnetic charges but also increases the exchange energy inside the wall (Fig. 18.2a). While the magnetization in the wall centre is perpendicular to the wall plane for Néel walls it is parallel for Bloch walls. The energy of a 180° Bloch wall can be calculated by

$$\sigma_B = A\left(\frac{\pi}{a}\right)^2 a + \frac{1}{2}aK + -\mu_0\frac{a^2}{4(a+D)}M_S^2 \tag{18.1}$$

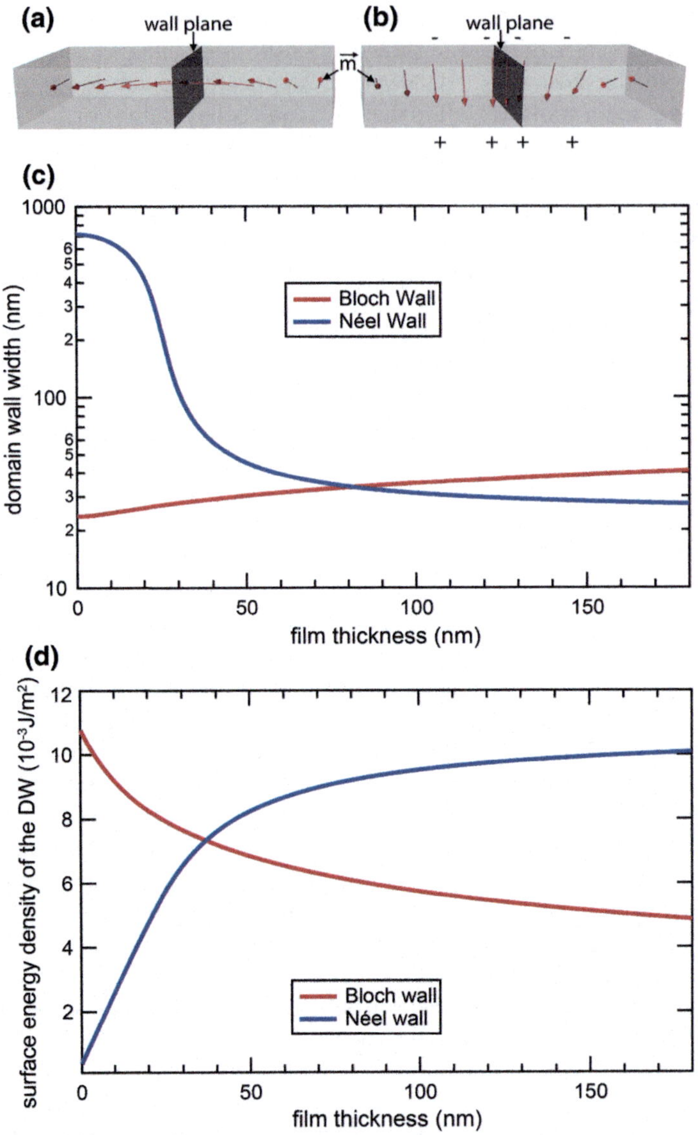

Fig. 18.2 **a** Néel and **b** Bloch domain walls with a rotation of the magnetization perpendicular and parallel to the wall plane, respectively. **c** Numerically calculated domain-wall widths for Néel and Bloch walls in thin $Ni_{80}Fe_{20}$ films. **d** Energy per unit area for both domain wall types in dependence of the film thickness

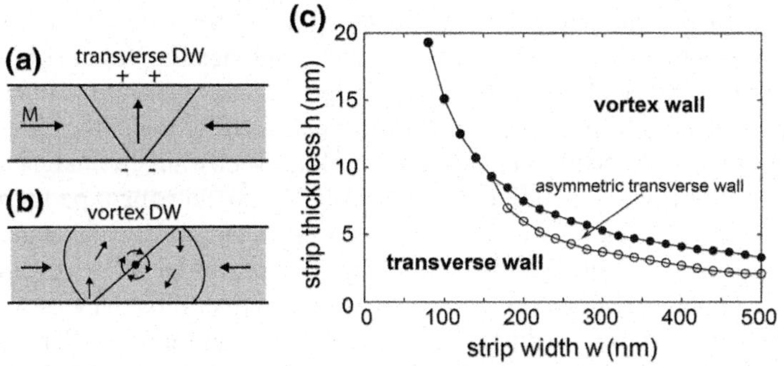

Fig. 18.3 Magnetic structure of **a** transverse and **b** vortex domain walls. **c** Phase diagram of domain-wall classes in dependence of the wire geometry (altered from [30]). The asymmetric transverse wall is a slanted transverse wall

with the domain wall width a, the thickness of the film D [28], the saturation magnetization M_S, the exchange constant A, and the anisotropy constant K. For Néel walls the last term resulting from the magnetostatic energy is different and the wall energy is

$$\sigma_N = A\left(\frac{\pi}{a}\right)^2 a + \frac{1}{2}aK + -\mu_0\frac{aD}{4(a+D)}M_S^2. \tag{18.2}$$

By minimizing the domain wall energy for a given film thickness with respect to the wall width the latter can be computed. A numerical evaluation of (18.1) and (18.2) with $M_S = 8\times10^5$ A/m, $K = 500$ J/m^3 and $A = 13$ pJ/m is shown in Fig. 18.2c, d. For film thicknesses up to 180 nm the width of a Bloch wall increases from 23 to 41 nm. Néel walls can be much wider with up to 700 nm in films thinner than 30 nm. In thicker films the wall width would decrease rapidly to about 30 nm and continue to decline afterwards. But as shown in Fig. 18.2d the surface energy for Néel walls is only lower than that of a Bloch wall in films thinner than 37 nm. In thicker films Bloch walls are energetically favoured. The energy of the wall is also proportional to the length of the wall. Therefore short domain walls are energetically favoured. For intermediate thicknesses a third type of domain wall with a more complex structure, the so called cross-tie wall, has been observed in thin films [29].

In flat nanowires with low crystalline anisotropy special types of domain walls occur that are modifications of the above mentioned wall types. Due to the shape anisotropy resulting from the magnetostatic energy the magnetization in such wires in favoured to be aligned along the wire direction. As a consequence neighbouring domains point towards- or away from each other in contrast to the 180° bulk domain walls shown in Fig. 18.2. The former are called head-to-head and the latter tail-to-tail domain wall. Depending on the wire dimensions two varying wall classes with distinct geometries occur [30]. Figure 18.3 shows the magnetic structure of both classes, the vortex- and the transverse domain wall. In a vortex wall the magnetization curls around a magnetic singularity in its centre, which is pointing out of plane. This configuration avoids magnetic charges on the sides of the wire. On the other hand the transverse domain wall has no out-of-plane component but induces a large stray field due to the surface charges on the wire edges. These walls are neither Néel nor Bloch walls but are referred to as Néel-like because the rotation of the magnetization is perpendicular to the wall plane although unlike Néel walls the magnetization in the centre of the wall is parallel to the wall plane. The occurring type for a range of wire thicknesses and widths for $Ni_{80}Fe_{20}$ has been calculated by Nakatani et al. and is shown in Fig. 18.3c. In general vortex walls are preferred for wider and thicker wires. As we focus on vortex domain walls in this chapter most wires studied range from 150 nm to 300 nm in width and from 10 to 30 nm in thickness. The energy for a transverse wall E_T can be approximated similar to equations (18.1) and (18.2) by integrating the energy contributions in the volume V of the domain wall

$$E_T = \int dV A \left(\frac{\partial\theta}{\partial x}\right)^2 + A\sin^2\theta\left(\frac{\partial\phi}{\partial x}\right)^2 + K_{\parallel}\sin^2\theta + K_{\perp}\sin^2\theta\sin^2\phi \quad (18.3)$$

with the angle around the wire axis ϕ and the angle θ between the magnetization and the wire axis [31]. The third and fourth term describe the anisotropy resulting from the magnetostatic energy along the wire direction ($K_{\parallel}$) and perpendicular ($K_{\perp}$) to it if the cross section of the wire is not quadratic. Since the length of the wall is proportional to its energy, a narrowing of the nanowire acts as an attractive potential where the wall gets pinned. Pinning occurs either at natural width variations or at defined positions via notches with an intentionally decreased wire width [22, 23, 32].

18.3 Domain-Wall Dynamics

To account for spin-transfer torques in the micromagnetic model the Landau-Lifshitz-Gilbert equation has been extended by two terms to incorporate the influence of the current density $\boldsymbol{j}$ on the magnetization $\boldsymbol{M}$. Zhang and Li derived the extended description

$$\begin{aligned}\frac{\mathrm{d}\boldsymbol{M}}{\mathrm{d}t} &= -\gamma\boldsymbol{M}\times\boldsymbol{H}_{eff} + \frac{\alpha}{M_s}\boldsymbol{M}\times\frac{\mathrm{d}\boldsymbol{M}}{\mathrm{d}t}\\ &\quad -\frac{b_j}{M_S^2}\boldsymbol{M}\times[\boldsymbol{M}\times\boldsymbol{j}\cdot\nabla)\boldsymbol{M}] - \xi\frac{b_j}{M_S}\boldsymbol{M}\times(\boldsymbol{j}\cdot\nabla)\boldsymbol{M}\end{aligned} \quad (18.4)$$

with the magnetization $\boldsymbol{M}$, the current density $\boldsymbol{j}$, the effective magnetic field $\boldsymbol{H}_{eff}$, the Gilbert damping parameter α, and the gyromagnetic ratio γ. The nonadiabacity constant ξ is the ratio between the exchange-relaxation- and the spin-flip-relaxation time $\xi = \frac{\tau_{ex}}{\tau_{sf}}$ [33]. The strength of the coupling between spatial inhomogeneities and the current is expressed by the coupling constant $b_j = \frac{P\mu_B}{eM_S(1+\xi^2)}$ with the spin polar-

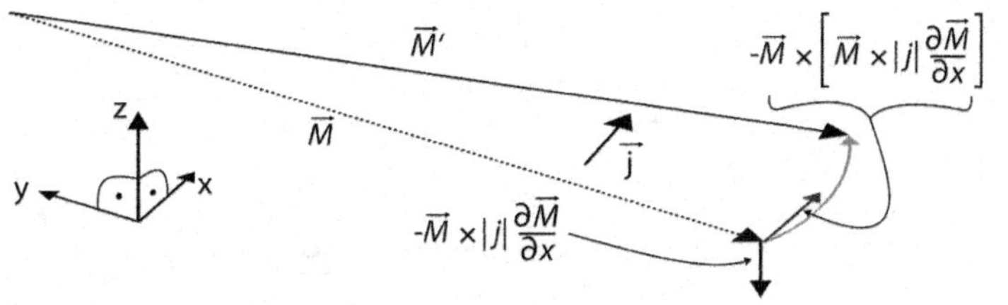

Fig. 18.4 Directions of the torques on a magnetization $\boldsymbol{M}$ resulting from a current density j along the x-axis from (18.5)

ization of the current P and the Bohr magneton μ_B. The third term describes the adiabatic rotation of the magnetization while the second takes account of a nonadiabatic torque.

At a given temperature the saturation magnetization M_S is constant and $\frac{d\boldsymbol{M}}{dt}$ is perpendicular to $\boldsymbol{M}$. For a current along the x-axis (18.4) can then be written as an explicit equation of motion for the magnetization [34]

$$\frac{d\boldsymbol{M}}{dt} = -\gamma'\boldsymbol{M} \times \boldsymbol{H}_{eff} + \frac{\alpha\gamma'}{M_s}\boldsymbol{M} \times (\boldsymbol{M} \times \boldsymbol{H}_{eff})$$
$$-\frac{b'_j}{M_S^2}(1+\alpha\xi)\boldsymbol{M} \times \left[\boldsymbol{M} \times |j|\frac{\partial \boldsymbol{M}}{\partial x}\right] - \frac{b'_j}{M_S}(\xi - \alpha)\boldsymbol{M} \times \left(|j|\frac{\partial \boldsymbol{M}}{\partial x}\right) \tag{18.5}$$

with the abbreviations $b'_j = \frac{b_j}{1+\alpha^2}$ and $\gamma' = \frac{\gamma}{1+\alpha^2}$. The directions of the torques from the spin-polarized current are depicted in Fig. 18.4. The dynamic behaviour of a magnetization in a magnetic field is well described by the first two terms of (18.5) and results in a parallel alignment of the magnetization along the field direction. If the magnetization is initially not homogeneous but is broken up into domains a magnetic field will increase domains with a parallel orientation. By increasing the size of a domain the walls in between have to move through the ferromagnet. The motion of 180° domain walls driven by magnetic fields has been examined by Thiele, Slonczewski as well as Schryer and Walker [35–37]. The velocity v of a wall is proportional to the applied field H but only up to a critical field value. Below the critical field the magnetic structure of the wall remains unchanged but above this critical Walker breakdown[1] the wall is no longer stable. The transformations during motion in fields just above the breakdown reduce the wall velocity. For much higher driving fields the relationship between the field and the velocity becomes linear again but with a smaller mobility ($\frac{\partial v}{\partial H}$) than below the Walker breakdown. This behaviour has been observed in films as well as in nanowires [12, 35] as shown in Fig. 18.5a, b, respectively. As mentioned the applied field enlarges domains with parallel orientation and thereby moves head-to-head and tail-to-tail domain walls in a nanowire into opposite directions (compare Fig. 18.5c–e). In contrast the spin-

[1] Named after L. R. Walker based on unpublished calculations on domain walls that have been redone and published in [38].

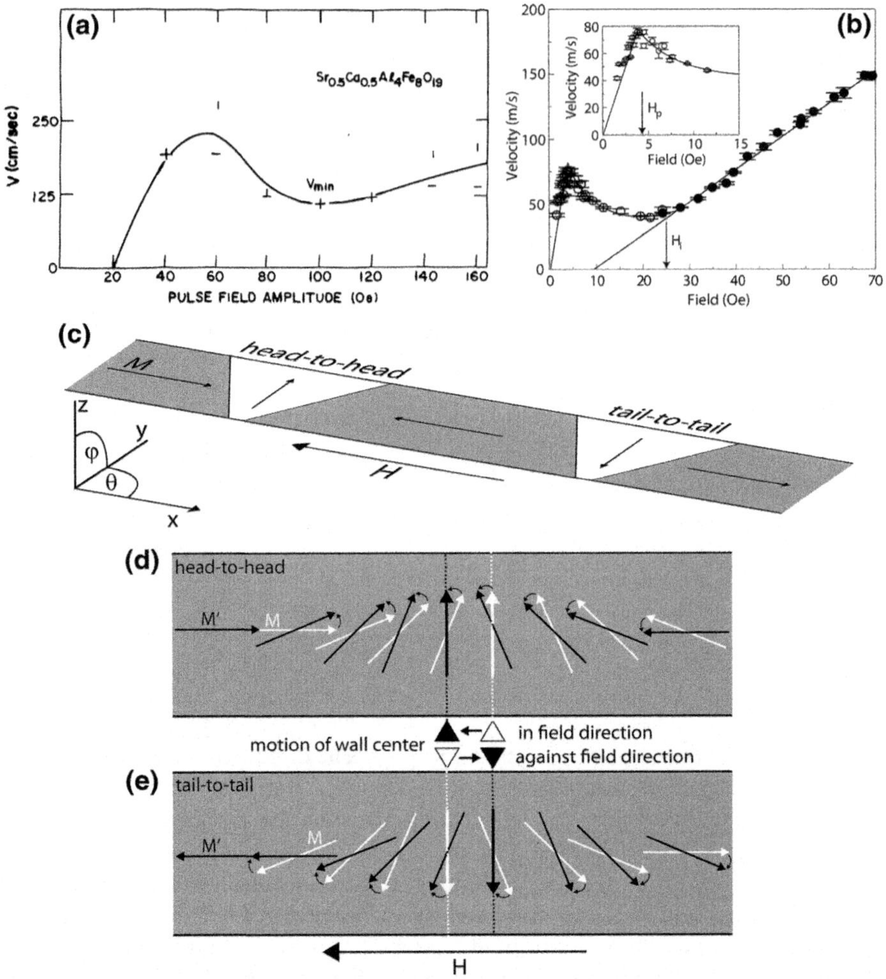

Fig. 18.5 Velocity of a magnetic field driven domain wall in **a** a $Sr_{0.5}Ca_{0.5}Al_4Fe_8O_{19}$ film and **b** a permalloy nanowire (graphs from [35] and [12]). The Walker breakdown is observed at (a) 60 Oe (6 mT) and **b** just below 5 Oe (0.5 mT). **c** A head-to-head and a tail-to-tail transverse domain wall. The field driven motion of both types is depicted in **d** and **e**. In the former the motion is along the field direction and in the latter it moves in the opposite direction

transfer torque from a spin-polarized current moves all wall types in the direction of the current.

To describe the motion of 180° (Néel-like) domain walls a one-dimensional model using only two dynamic parameters, the position of the wall centre along the wire X and the averaged angle of the magnetization around the wire axis Φ, is useful [37]. The domain wall is then handled as a quasi particle making it possible to calculate domain wall behaviour and apply the model to experimental data instead of performing full micromagnetic simulations. To calculate equations of motion for the two dynamic parameters (18.4) is written in the polar spin basis where $\boldsymbol{M} = M_S(\cos\theta, \sin\theta\cos\phi, \sin\theta\cos\phi)$ with the angle θ between the wire axis and the magnetization as well as ϕ the local angle around the wire axis as shown in Fig. 18.5c. In this coordinate system a transverse domain wall in a wire along the x direction has the form

$$\phi = 2\arctan\left(e^{\frac{x-X}{\lambda}}\right) \tag{18.6}$$

with the domain wall width parameter[2] $\lambda \approx \sqrt{\frac{A}{K}}$. The extended Landau-Lifshitz-Gilbert equation (18.5) including effects from magnetic fields and the spin-transfer torque then reads ($\boldsymbol{M}$ no longer handled in total, $|\boldsymbol{M}| = M_S = const.$)

$$\begin{aligned}\frac{d\theta}{dt} = &-\frac{\gamma'}{\mu_0 M_S \sin(\theta)}\frac{\delta E_T}{\delta\phi} - \frac{\gamma'\alpha}{\mu_0 M_S}\frac{\delta E_T}{\delta\theta}\\ &+b_j'(1+\alpha\xi)\boldsymbol{j}\cdot\nabla\theta + b_j'(\xi-\alpha)\sin(\theta)\boldsymbol{j}\cdot\nabla\phi\end{aligned} \tag{18.7}$$

and

$$\begin{aligned}\frac{d\phi}{dt}\sin\theta = &\frac{\gamma'}{\mu_0 M_S}\frac{\delta E_T}{\delta\theta} - \frac{\gamma'\alpha}{\mu_0 M_S\sin(\theta)}\frac{\delta E_T}{\delta\phi}\\ &+\sin(\theta)b_j'(1+\alpha\xi)\boldsymbol{j}\cdot\nabla\phi - b_j'(\xi-\alpha)\boldsymbol{j}\cdot\nabla\theta.\end{aligned} \tag{18.8}$$

with the variational derivatives of the energy of a transverse domain wall E_T from (18.3) that is extended by an energy $E_p(X)$ from a potential and an energy resulting from an external magnetic field $H_{ext} = -\mu_0 M_S H_{ext}\cos(\theta)$ [34, 40]. If the domain wall is a stable Néel-like wall (as for transverse walls) below the Walker breakdown an approximate for $\frac{dX}{dt}$ and $\frac{d\Phi}{dt}$ can be written using the domain wall width parameter. The angle ϕ is no longer considered for every position, instead the averaged angle Φ for the whole domain wall is used. Assuming that the external field varies slowly compared to the domain wall width parameter and the out-of-plane angle ϕ is smaller than 10° the equations of motion following the 1-D model of Schryer and Walker [37] for the domain wall are

$$\frac{dX}{dt} = -\lambda 2\gamma' K_\perp \Phi\frac{1}{\mu_0 M_S} - \frac{\lambda\gamma'\alpha}{2S\mu_0 M_S}\left(2S\mu_0 M_S H_{ext}(X) + \frac{dE_p}{dX}\right) - b_j'(1+\alpha\xi)j \tag{18.9}$$

and

$$\frac{d\Phi}{dt} = -\frac{\gamma'}{2S\mu_0 M_S}\left(2S\mu_0 M_S H_{ext}(X) + \frac{dE_p}{dX}\right) - 2\gamma'\alpha K_\perp\Phi\frac{1}{\mu_0 M_S} + \frac{b_j'(\xi-\alpha)j}{\lambda} \tag{18.10}$$

for a current j along the x direction. In this calculation the cross section of the wire is S and detailed derivations of the above can be found in [31] and in a slightly different

[2] Actual domain wall widths in nanowires are about 3λ as calculated in [39].

notation in [41]. Strictly, these equations are only valid for transverse domain walls with their Néel-like structure, and the domain wall width parameter λ is the real width of the wall. However, it was shown that (18.9) and (18.10) can also be applied to Bloch domain walls as well as vortex walls [31, 40]. In the latter case the domain wall width parameter is no longer of comparable magnitude to the real wall width but smaller.

18.4 Domain-Wall Mass

With the equations of motion from the previous section the domain wall is handled in a one-dimensional model as a quasiparticle with two dynamic variables. Compared to second order differentials for classical particles, the equations (18.9) and (18.10) are first order differential equations [34]. But to a certain extent a comparison between them is useful. Even before the development of the 1-D model the idea of an inertial domain wall mass was proposed [42]. Since the magnetization inside the wall gets distorted by a moving force due to the nature of the torques from (18.4) the energy associated with the domain wall increases. This increase in internal energy is reflected as an inertia of the domain wall and has been observed for transverse domain walls in nanowires [16, 17]. This inertial behaviour is expressed by an effective mass of the domain wall [17, 18, 42] that only exists for walls in motion. The time derivative of (18.9), the acceleration of a wall a, is

$$\begin{aligned}\frac{\mathrm{d}^2X}{\mathrm{d}t^2} = &-\frac{2K_\perp\gamma'\alpha}{\mu_0 M_\mathrm{S}}\frac{\mathrm{d}X}{\mathrm{d}t} - \frac{\gamma'\alpha\lambda}{2S\mu_0 M_\mathrm{S}}\frac{\mathrm{d}}{dt}\frac{\mathrm{d}E_p}{\mathrm{d}X} - \frac{K_\perp\gamma'\gamma\lambda}{S\mu_0^2 M_S^2}\frac{\mathrm{d}E_p}{\mathrm{d}X} \\ &-\frac{2K_\perp\gamma'\gamma\lambda}{S\mu_0^2 M_S^2}H_{ext}(X) - \gamma'\alpha\lambda\frac{\mathrm{d}H_{ext}}{\mathrm{d}t} - \frac{2K_\perp\gamma'\xi}{\mu_0 M_\mathrm{S}}b_j j - (1+\alpha\xi)b_j' j.\end{aligned} \tag{18.11}$$

Here, (18.9) solved for $-\frac{2K_\perp\gamma'\alpha}{\mu_0 M_\mathrm{S}}\phi$ is used [31]. For the freely moving domain wall, where $E_p = 0$, $H_\mathrm{ext} = 0$ and $j = 0$, the wall slows down to $1/e$ from its initial velocity in the damping time

$$\tau_d = \frac{\mu_0 M_\mathrm{S}}{2\gamma' K_\perp\alpha}. \tag{18.12}$$

The equation for a classical particle under the influence of a force F from a linear potential E reads

$$a = \frac{F}{m_c} = \frac{1}{m_c}\frac{\mathrm{d}E}{\mathrm{d}X} \tag{18.13}$$

with the classical mass m_c. In comparison, for a domain wall inside a linear potential E_p with no current or external magnetic field applied, (18.11) is much simplified to

$$\frac{\mathrm{d}^2X}{\mathrm{d}t^2} = \frac{1}{\tau_d}\frac{\mathrm{d}X}{\mathrm{d}t} - \frac{K_\perp \gamma' \gamma \lambda}{S\mu_0^2 M_S^2}\frac{\mathrm{d}E_p}{\mathrm{d}X} = \frac{1}{\tau_d}\frac{\mathrm{d}X}{\mathrm{d}t} - \frac{1}{m'}\frac{\mathrm{d}E_p}{\mathrm{d}X}. \tag{18.14}$$

Due to the resemblance of equations (18.14) and (18.13) the proportionality factor between the acceleration of a wall and the force from a linear potential can be interpreted as the inertial mass m of a domain wall in the 1-D model with

$$m' = \frac{m}{1+\alpha^2} = -\frac{\mathrm{d}E}{\mathrm{d}x}\frac{1}{\frac{v}{\tau_d}+a} \tag{18.15}$$

with the acceleration $a = \frac{\mathrm{d}^2X}{\mathrm{d}t^2}$, the velocity $v = \frac{\mathrm{d}X}{\mathrm{d}t}$ and the damping time τ_d. For transverse domain walls in nanowires as well as for an extended 180° domain wall in an infinite structure an effective mass on the order of 10^{-23} kg has been measured [17, 18, 43]. For vortex domain walls Clark et al. have proposed a more complex analytical model than the 1-D model [44]. For a nanowire similar to wires studied in this work that is 200 nm wide and 20 nm thick they calculated a mass of 10^{-22} kg. This is in agreement with the measurements that are discussed in Sect. 18.6.

18.5 Fast Generation of Domain Walls with Defined Chirality in Nanowires

Oersted fields can be used for a controlled creation of transverse and vortex domain walls [11, 13, 16, 20, 22, 40, 45–47]. An important property is the sense of rotation of the magnetization inside the wall called chirality. This can have great impact on the motion of domain walls especially through y-shaped junctions in spin-ice systems [9, 10] and for synchronous motion of multiple walls driven by out-of-plane fields [6]. The chirality can be probed by breaking the symmetry of the wire via a notch and utilizing the anisotropic magnetoresistance for detection [13, 20, 48]. The pinned domain walls at the notch show slightly different resistances that can be used to probe the wall structure [22]. Notches with a depth of 130 nm are placed on one side of the 8 μm long, 20 nm thick and 300 nm wide nanowires. Samples with six differently aligned 500 nm wide and 100 nm thick copper contact lines are prepared [48]. Figure 18.6a shows a scanning-electron micrograph of one sample as well as a scheme of the setup. The angle between the right stripline and the perpendicular direction to the nanowire is varied between ±15° as shown in the top left of Fig. 18.6a. The electron propagation direction is shown within the right stripline and downwards corresponds to negative voltage pulses. The 10 ns long voltage pulses, with amplitudes varied from 1–4 V with negative and positive polarity, result in Oersted fields from 20–75 mT. Prior to every measurement the wire is saturated in an external field of 100 mT so that the magnetization of the wire and the Oersted field beneath the wire are aligned antiparallel. In addition to the voltage pulse an external field is applied in x-direction (10 mT to −10 mT, see coordinate system Fig. 18.6a) to investigate

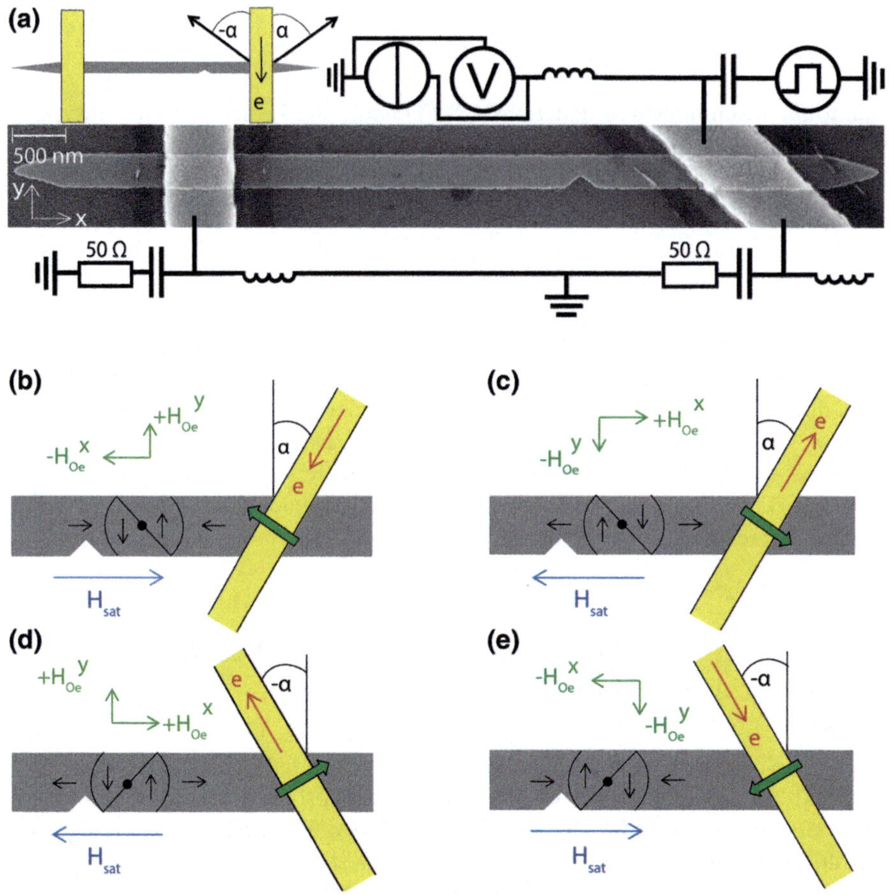

Fig. 18.6 **a** Schematic measurement setup and scanning electron micrograph of the nanowire with an artificial notch and two crossing striplines. The current for Oersted field generation runs through the right stripline. **b** Depending on the stripline angle, the external field (blue) and the Oersted field (green) beneath the stripline (yellow), domain walls with different chiralities can be generated. Shown are also the x- and y- components of the Oersted field. In **b** and **e** the situation for head-to-head walls and in **c** and **d** the situation for tail-to-tail walls is shown

its influence on the domain-wall creation. Every combination of pulse- and external field parameters is measured ten times. The resistance changes are used to detect the domain wall type (transverse or vortex) and their chirality [22]. The sample dimensions are again chosen to preferably create vortex walls (see Fig. 18.3). The resistance of the wire is measured by a small direct current of 100 μA and all resistance changes are referred to the resistance of the wire without a domain wall. Whether a head-to-head (hh) wall or a tail-to-tail (tt) wall is created depends on the initial magnetization of the wire. Head-to-head walls and tail-to-tail walls with opposite chiralities have the same magnetic structure and therefore yield the same resistance change when

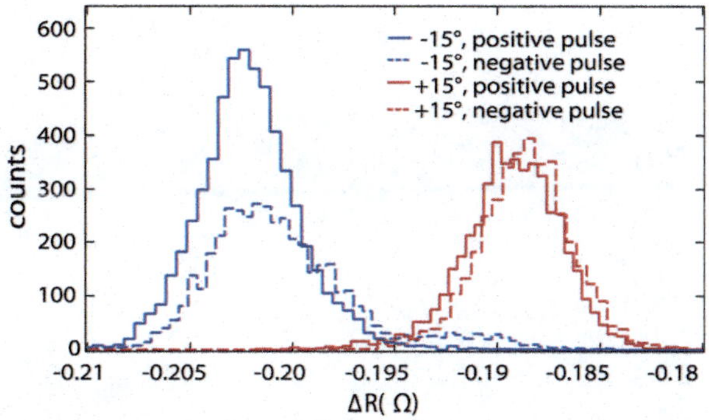

Fig. 18.7 Measurement of resistance changes due to domain wall generation for two angles between stripline and nanowire (±15°) as well as positive and negative pulse polarity

pinned at the notch. If both wall types have identical chiralities they have a different magnetic structure and therefore result in dissimilar resistance changes. The domain walls pin either inside (Fig. 18.6d, e) or at the right side of the notch (Fig. 18.6b, c) [23]. For example the hh wall with a clockwise chirality pins inside the notch (see Fig. 18.6e) whilst the hh wall with a counter-clockwise chirality pins at the right side of the notch (see Fig. 18.6a). The difference in the measured resistance change is larger if a hh wall with a clockwise chirality is present than in the case of a hh wall with a counter-clockwise chirality [22]. Because of the inversion symmetry the behaviour is the opposite for tt walls (Fig. 18.6b, c). The following analysis focuses on vortex walls which are predominantly generated in the present experiments. Figure 18.7 shows results for a stripline angle of $\pm 15°$, which yields the best control of chirality when compared to the results for $\pm 30°$ and $\pm 45°$ (not shown). In Fig. 18.7 the number of generated domain walls for every resistance change level depending on the angle and pulse polarity are plotted. The diagram indicates that a negative stripline angle generates vortex walls with the larger resistance change ΔR_2 irrespective of the pulse polarity. In concordance a positive stripline angle generates vortex walls with the smaller resistance change ΔR_1 again irrespective of the pulse polarity. To quantitatively determine the various resistance changes a Gaussian fit for every maximum is performed with amplitude, width and resistance change $\Delta R_{1/2}$ of the particular maximum resulting in $\Delta R_1 = (-0,1891 \pm 0,0004)\,\Omega$ and $\Delta R_2 = (-0,2021 \pm 0,0006)\,\Omega$. When the angle of the stripline is negative there are only hh walls with a clockwise chirality and tt walls with counter-clockwise chirality and vice versa if it is positive. This correlation is validated by imaging the domain walls in the nanowires with full field soft X-ray transmission microscopy at the Advanced Light Source in Berkeley, CA, USA. The images were recorded at the iron L3 edge (707 eV) with a spatial resolution of about 25 nm. Figure 18.8a shows a direct image with two 300 nm wide and 20 nm thick permalloy nanowires without notches and a tilted copper stripline. The electron flow of the 10 ns long pulses is downwards. The situation on the left side of the stripline is the same as in Fig. 18.6e with a negative tilting of the stripline and an electron flow downwards. When rotated by 180° the situation on the right side is comparable to the one in Fig. 18.6d. To enhance the magnetic contrast an image of the saturated state is subtracted from an

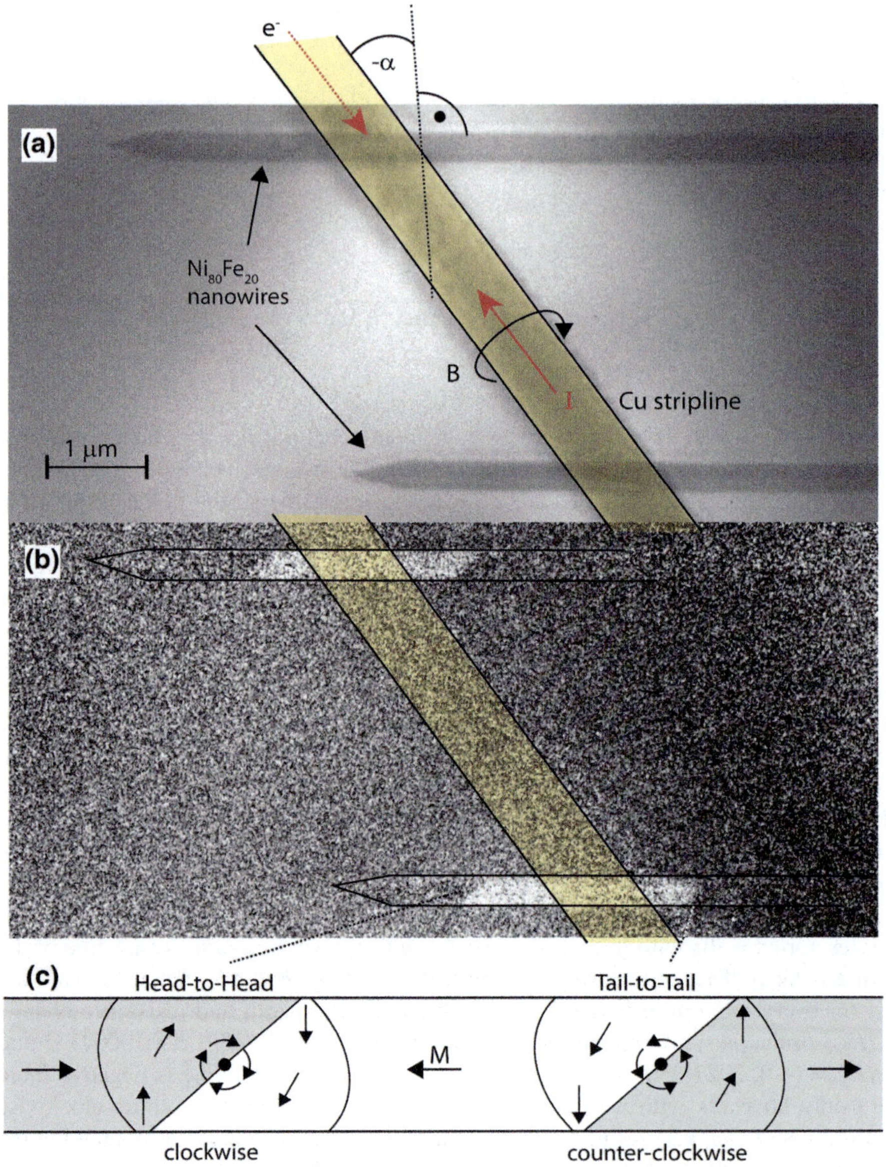

Fig. 18.8 **a** Transmission X-ray micrograph of two permalloy nanowires with a crossing copper stripline recorded at the XM-1 microscope at the Advanced Light Source. **b** Differential image of the saturated- and the post current-pulse state. Domain walls are created on both sides of the stripline (transitions form grey to white). The occurring magnetization configuration is shown in **c**

Table 18.1 Summary of the domain wall structure depending on the angle and pulse polarity. Tail-to-tail wall (tt) or head-to-head wall (hh) with a clockwise (cw) or counter-clockwise (ccw) chirality. Indicated are the resistance changes which arise from the particular wall, whereby $\Delta R_2 > \Delta R_1$

	Positive angle	Negative angle
Positive pulse	tt, cw $\Rightarrow \Delta R_1$	tt, ccw $\Rightarrow \Delta R_2$
Negative pulse	hh, ccw $\Rightarrow \Delta R_1$	hh, cw $\Rightarrow \Delta R_2$

image after the pulse has been applied. The differential image is shown in Fig. 18.8b. On both sides of the stripline domain walls are created as indicated by the transitions from grey to white. The occurring domain-wall structure is determined by the angle of the transitions and is displayed in Fig. 18.8c. On the left side a clockwise hh wall and on the right side a counter clockwise tt wall is created. This is in agreement with the electric measurements shown in Fig. 18.7. These results can be understood by looking at the x- and y-components of the Oersted field of the current pulses as depicted in Fig. 18.6b–e. The type (hh or tt) of the domain wall is determined by the initial magnetization. For domain wall creation an antiparallel x-component of the Oersted field, compared to the initial magnetization, is necessary, which corresponds to a certain pulse polarity. On the other hand the chirality of the wall is controlled by the y-component which depends on the stripline angle and the pulse polarity. For example a negative angle and a negative pulse (Fig. 18.6e) result in a hh wall with clockwise chirality. This wall has the same structure as the tt wall with counter-clockwise chirality in Fig. 18.6d resulting in the same resistance change ΔR_2 for both cases. The four possible configurations of the angle and the pulse polarity (see Table 18.1) allow to inject a domain wall with defined chirality and type into the nanowire. To show the dependence of the wall creation on the pulse amplitude and the applied external magnetic field a statistical analysis of the individual pulse polarity/angle configuration is performed. Figure 18.9 shows the probabilities that a vortex domain wall is present after a pulse depending on the external field and pulse amplitude. The regime in which a wall generation is most likely ranges from ∓ 4 to $\pm 2\,mT$ and the pulse amplitude must be at least about $\mp 1.9\,V$ which corresponds to an Oersted field of 38 mT. The results in Fig. 18.9a, d give asymmetric domain-wall generation probabilities. For example Fig. 18.9a shows that the probability decreases with higher negative background fields, while the probability is zero for positive external fields higher than +3 mT. The reason for this asymmetry is due to the notch. Up to field values of −8 mT the wall pins at the notch, whilst at a field value of +3 mT the two generated walls annihilate each other. No change in the resistance level is then measured. Neither the external field nor the amplitude have shown an impact on the chirality of the created domain walls.

These results show that using a tilted contact line and short current pulses a generation of domain walls with defined type and chirality can be achieved. The measurements are performed with 10 ns long current pulses for domain wall creation. Lower pulse lengths can result in a less reliable creation [49]. Therefore the evolution of the creation process is studied in detail in the following section.

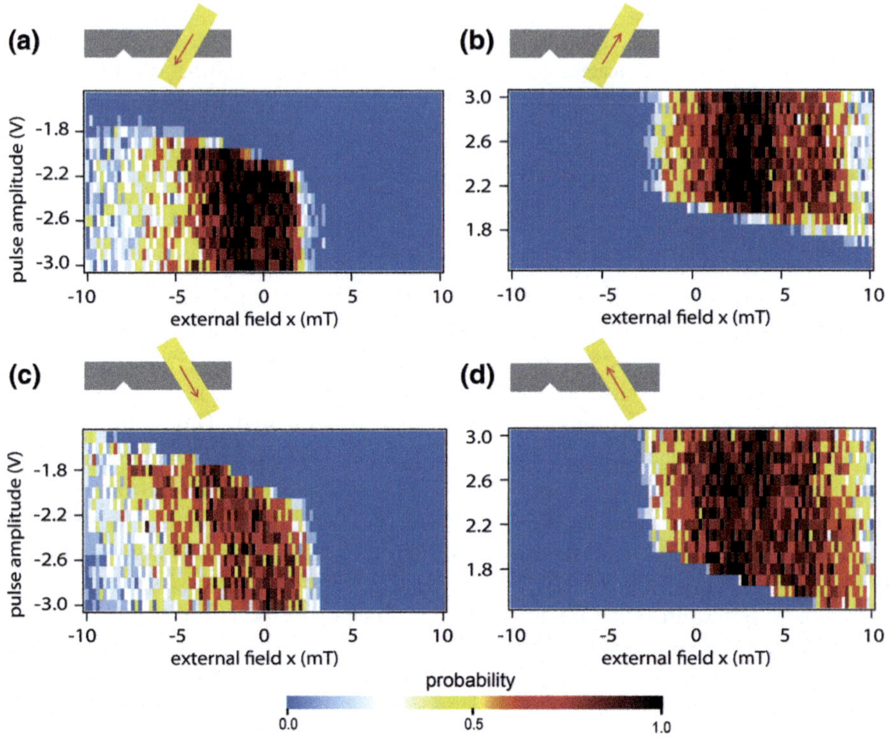

Fig. 18.9 Statistical analysis of domain wall generation. Shown are the probabilities that a domain wall is present depending on the pulse and external field for $+15°$ with negative pulse (**a**) and positive pulse (**b**) and for $-15°$ with negative pulse (**c**) and positive pulse (**d**)

18.6 Time-Resolved Imaging of Nonlinear Magnetic Domain-Wall Dynamics in Ferromagnetic Nanowires[3]

Time-resolved scanning transmission X-ray microscopy (STXM) is used to observe the creation and the motion of domain walls with sub-nanosecond time and 25 nm spatial resolution [50–52]. Domain walls are created by current pulses through a stripline crossing a permalloy nanowire as shown in Fig. 18.10a. The pulse is accompanied by an Oersted field of 66 mT creating two domain walls underneath the stripline. Both walls move away from the stripline into opposite directions predominantly driven by the in-plane component of the Oersted field (Fig. 18.10b, c). Vortex domain walls lead to a black/white contrast in the X-ray micrographs as seen in Fig. 18.10b. The initial magnetization of the wire is achieved by sending a second inverted current pulse through the stripline to again create two domain walls. These collide with

[3]Figures 18.10–18.13 and parts of the text are reproduced from F.-U. Stein, L. Bocklage, M. Weigand and G. Meier, Sci. Rep. 3, 1737 (2013) with the permission of SpringerNature Publishing.

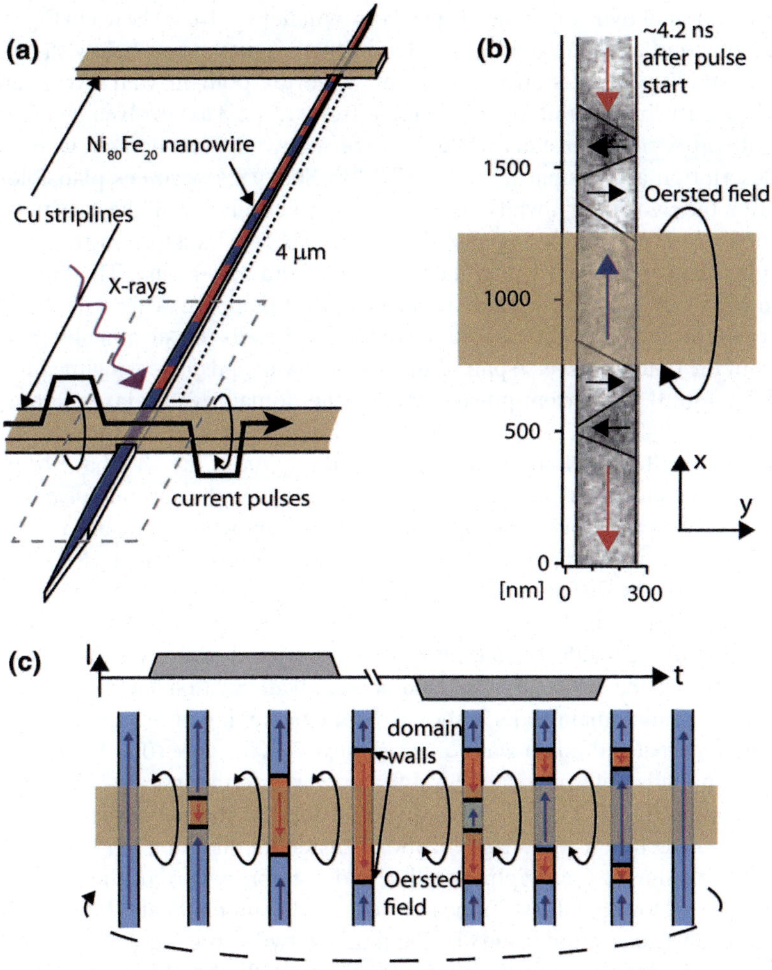

Fig. 18.10 Schematic views of experimental setup and measurement procedure. **a** Alternatingly magnetized nanowire with two crossing striplines to create and detect domain walls. **b** Scanning transmission X-ray microscopy superimposed with a scheme of the nanowire with two vortex domain walls in the vicinity of the lower stripline. **c** Illustration of the repeated process for time-resolved imaging. A field pulse creates one domain wall at the upper and one at the lower edge of the stripline. The succeeding domain walls from the second pulse annihilate the ones created by the first pulse therefore resetting the nanowire to its initial state

the primarily created domain walls from the first pulse and annihilate each other as schematically shown in Fig. 18.10c. As the Oersted field of the second pulse points in the direction of the initial magnetization the initial state of the wire is restored. It has been shown that this process is reliable for a wide range of pulse amplitudes and external magnetic fields [49].

The creation of two domain walls by a 6 ns long field pulse is shown in Fig. 18.11a. One nanosecond after the field pulse begins the magnetization below the stripline becomes significantly distorted and forms a vortex domain wall. An alternating domain pattern (black-white-black contrast from 1.7–2.3 ns) evolves by creation of a second vortex core above the white domain. As two distinct domain walls emerge their description as quasiparticles described in Sect. 18.4 becomes plausible. After separation the two domain walls take on a more transverse wall like structure where the vortex core is not visible any more (2.6–2.9 ns) before transforming back into vortex walls (black/white and white/black contrast from 3.4–4.6 ns). The macroscopic motion of the domain wall is almost linear up to this point. After 4 ns the domain wall decelerates and the vortex core moves back towards the stripline by increasing the size of the outer transverse part of the domain-wall and decreasing the inner ones (4.6–5.7 ns) until the current pulse ends and the domain wall relaxes into a stable vortex wall (5.7–8.0 ns). This oscillation of the domain-wall width as well as the vortex core oscillation along the propagation direction of the wall have been theoretically predicted above the Walker breakdown [44]. As the current pulse is applied a fraction of the current is transmitted through the nanowire. The current causes an additional force on the upper wall via spin-transfer torque (see Sect. 18.3) thus most likely leading to the difference in velocity of the upper and the lower domain wall. The time scale of 2.7 ns for the domain wall generation is found to be in very good agreement with the results from time-resolved electrical measurements [49].

In Fig. 18.11b the consecutive second domain wall creation leading to the mutual annihilation of all domain walls is shown. At the beginning of the second pulse only the primarily created domain walls are present within the wire (0.3–1.1 ns) until two new domain walls with opposite chirality compared to the initial walls are created (1.4–2.3 ns, see Fig. 18.10c). The magnetization between the new domain walls points in the same direction as the magnetization outside the first domain walls. The two walls on each side of the stripline collide and the vortex core of the outer domain wall gets pulled towards the stripline before the walls mutually annihilate each other (3.1–5.7 ns). The attraction occurs by the time the two vortices are 500 nm apart and the transverse parts of the walls merge. By the time the 6 ns long pulse is over the initial magnetization is restored. Although the domain walls collide about 3 ns after the Oersted field is present the final relaxation of the magnetization takes another 3 ns (light black and white contrast until 6.3 ns). The almost constant velocity of the domain wall within the first nanoseconds after its creation is also visible in the averaged domain-wall positions in Fig. 18.12. The motion directly correlates with the spatially varying in-plane component of the Oersted field shown on the right side of the graph. The domain wall moves with high velocities until the edge of the stripline is reached and the magnetic field rapidly drops. The 66 mT field underneath the stripline is far above the Walker breakdown field and therefore leads to the observed transformations from transverse to vortex walls and domain wall width oscillations in Fig. 18.11. Next to the stripline where the field decreases the domain wall slows down and finally stops when the in-plane component of the Oersted field subsides below 3 mT which is the propagation threshold field for domain walls in our samples. The theoretical description of the vortex wall-dynamics for high fields is rather complex

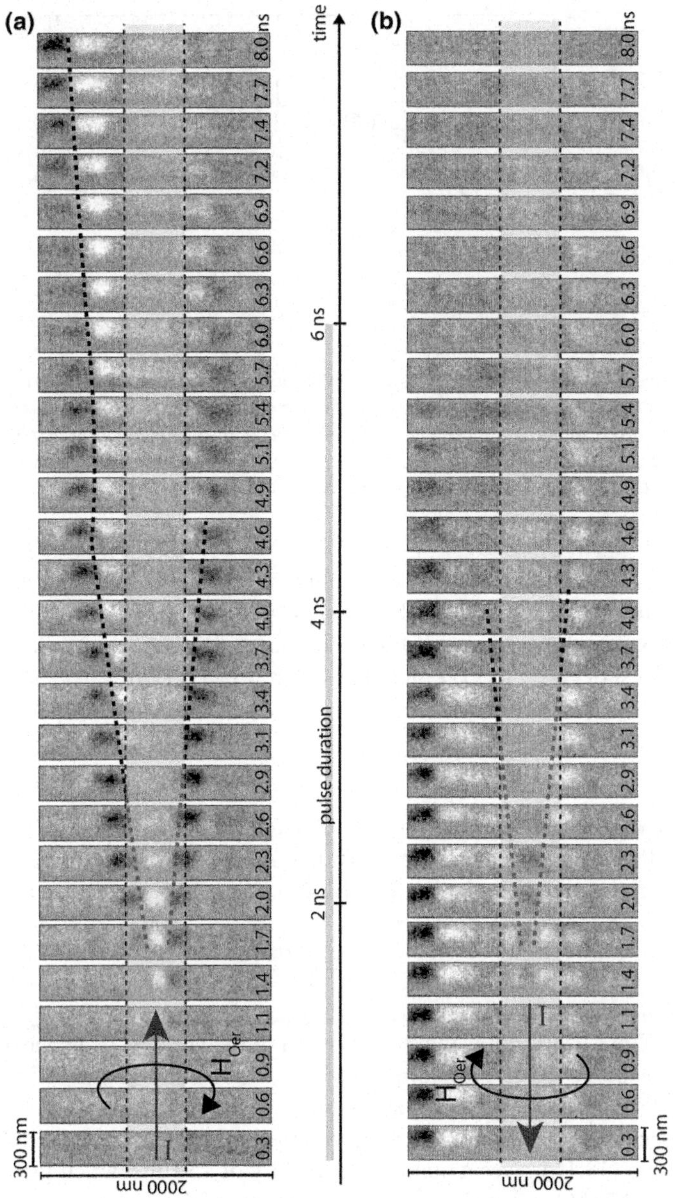

Fig. 18.11 Time-resolved STX micrographs in time steps of 286 ps. **a** Creation of two domain walls by the field of a current pulse through the perpendicularly aligned stripline indicated by the horizontal dotted lines. A guide to the eye for the domain wall motion is given by the dashed lines. **b** Collision of two further domain walls created by an inverted current pulse. The collision leads to an annihilation of all four walls. The chirality is opposite for the first and second creation because of the opposite current flow direction through the stripline

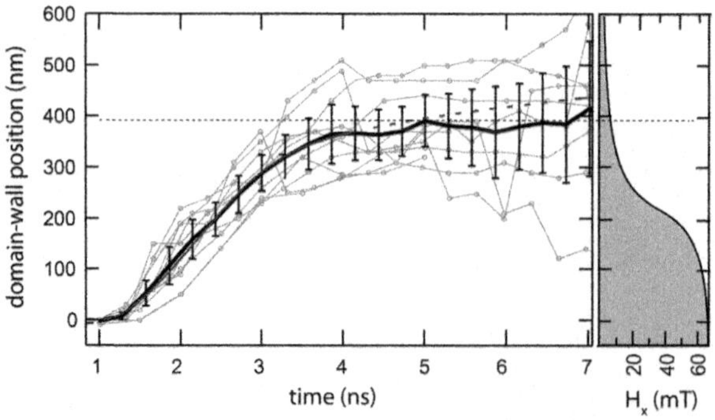

Fig. 18.12 Averaged (black solid line) domain-wall position with concomitant error bars relative to the centre of the stripline in dependence of the time from the beginning of the current pulse. Data of ten time-resolved image sequences is shown in grey. A fit of the equation of motion from the 1D model is extended as dahed line in the region where the domain wall is pinned. The in-plane component of the Oersted field along the wire is plotted on the right

and has been carried out in detail and compared to numerical simulations for the presented wire dimensions by Clarke et al. [44]. Although the detailed dynamics is very complex the total displacement of the domain wall above the Walker breakdown can still be described by the 1D model as derived in Sect. 18.3. The acceleration of the wall is

$$\ddot{X} = \frac{\lambda\gamma}{\alpha\tau_D} H_{Oer} - \frac{\dot{X}}{\tau_D} + \frac{\lambda\gamma\alpha}{1+\alpha^2}\dot{H}_{Oer}. \tag{18.16}$$

H_{Oer} in our case is the spatially varying Oersted field and $\alpha = 0.0076$ the Gilbert damping. The damping time of the domain wall τ_D and the effective domain-wall width λ can be used to calculate a domain wall mass equivalent by using (18.14):

$$m = \frac{2\alpha S \mu_0 M_S \tau_D}{\lambda\gamma} \tag{18.17}$$

where $S = 20\,\text{nm} \times 200\,\text{nm}$ is the cross section of the wire, μ_0 is the vacuum permeability, $M_S = 8\times10^5$ A/m the saturation magnetization and $\gamma = \mu_0 \cdot 1.76\times10^{11}$ C/kg $= 2.211\times10^5$ m/C is the gyromagnetic ratio.

A fit of the displacement in Fig. 18.12 with the numerically integrated acceleration of a wall from (18.16) gives a maximum velocity of 159 m/s which is in agreement with other studies [14, 45, 53] and the damping time of the domain wall $\tau_D = 0.44$ ns is the same order of magnitude as in other experiments on transverse walls [18, 40] but is 1/10th of that found for non-transforming walls [54]. The acceleration of the wall is proportional to the Oersted field scaled by a domain wall width parameter λ. This parameter is only comparable in magnitude to the real wall width for transverse walls and is about one order of magnitude smaller than the real width for vortex domain walls in low fields as the vortex core magnetization strongly influences the dynamics

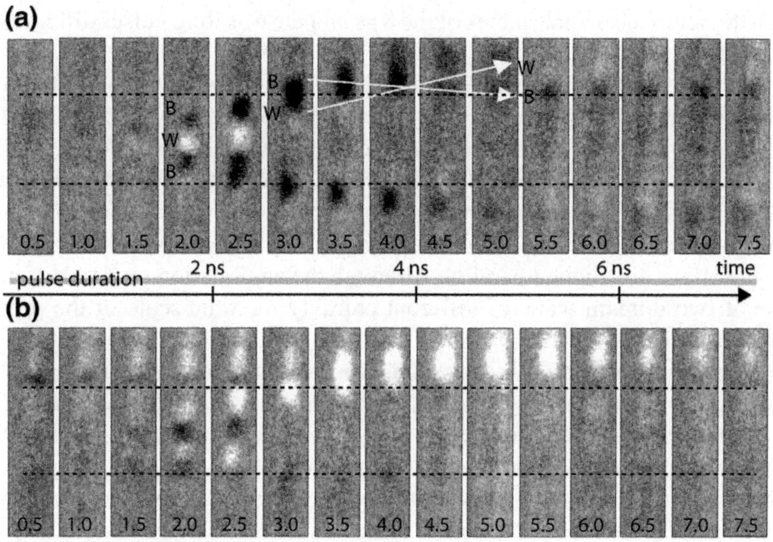

Fig. 18.13 STX micrographs in time steps of 500 ps. **a** Domain-wall creation by a 8 ns long field pulse with a chirality change at the end of the creation process. The domain wall stops moving before the pulse has ended. **b** Collision of two domain walls of the same chirality due to the chirality change of the first domain wall after creation (compare Fig. 18.11b)

[30, 55]. The observed transformations in the spin structure for high driving fields in Fig. 18.11 also influence the macroscopic motion of the wall leading to a lower but also constant mobility [12]. Therefore a reduced effective domain wall width of $\lambda = 0.27$ nm is obtained.[4] Hence the calculated mass of the vortex domain wall is $m = 4.5\times10^{-22}$ kg in agreement with the estimations for vortex walls in this geometry [44].

In Fig. 18.13a the creation of domain walls by a longer pulse of 8 ns with a lower time resolution compared to Fig. 18.11 is imaged. Directly after creation the upper domain wall has a black/white contrast being a vortex wall. After transformation to a transverse wall (3.5 ns) a vortex wall is again observed 2 ns later (5.5 ns) with the opposite white/black contrast showing that the chirality has changed. Note that the domain wall stops after the transformation even though the Oersted field is still applied because the driving field dropped below the propagation field (compare Fig. 18.12). In contrast to the continued motion of the non-transformed wall under its own inertia (Fig. 18.11a) the transformation seems to suppress such motion and the wall stops in the low field beyond the edge of the stripline (Fig. 18.13a). Although the transformation of the wall occurs after 5 ns it is not observed in the creation process with 6 ns long pulses in Fig. 18.11. Due to the non-ideal transmission of square pulses

[4]In [56] a wrong numerical value $\lambda = 0.13$ nm was written on page 2 line 78. Still the calculation had been carried out with the correct value of $\lambda = 0.27$ nm in the publication.

through the setup the actual shapes of the 8 ns and the 6 ns long pulses differ resulting in a not fully comparable creation process.

Since the domain wall has changed its chirality the mutual annihilation of two domain walls with the same chirality is observed in Fig. 18.13b where in Fig. 18.11b the collision of two domain walls with opposite chirality is shown. Unlike in Fig. 18.11 the second emerging domain wall does not attract the existing domain wall (2.0–3.0 ns) before merging into one horizontally oriented domain. This merged domain moves away from the stripline and then vanishes as the surrounding domains below and above are aligned parallel. Although the annihilation process differs from the one of two domain walls of different chirality the time scale of the process is identical.

In summary the temporally resolved microscopy of non-linear motion of domain walls in high magnetic fields has been used for the observation of transformations of the wall between vortex and transverse type. The creation process of vortex domain walls as well as the mutual annihilation of domain walls have been directly observed. Despite the complex processes during motion the displacement of transforming domain walls on the observed time scale can still be described by the equations of motion from the 1D model using a reduced dynamic domain wall width.

18.7 Conclusion

The dynamics of domain walls in nanowires has been studied by anisotropic resistance measurements and transmission X-ray microscopy. A static transport measurement is used to statistically evaluate occurring wall structures for a wide range of parameters while time-resolved microscopy enabled analysis of spin-structural changes during dynamical processes on a nanometre length- and sub-nanosecond time scale. The combination of anisotropic magnetoresistance measurements and transmission X-ray microscopy has been shown to be a comprehensive method for the study of domain walls.

The fast domain wall creation by nanosecond long Oersted-field pulses is found to be a highly robust technique that allows full control of the domain wall class (transverse or vortex), its type (head-to-head or tail-to-tail) and the chirality (clockwise or counter clockwise). In the presence of small assisting external magnetic fields ($\leq$1.5 mT) the class is predefined by the wire dimensions chosen to favour vortex domain walls. External fields during the Oersted-field pulse up to the depinning threshold can be used to inject transverse walls instead. The type is controlled by the pulse polarity and the direction of the sample saturation. A tilting of the contact line used for pulse transmission results in an Oersted-field component perpendicular to the wire and governs the magnetic sense of rotation inside the wall. The consecutive injection of domain walls by pulses with alternating polarity can either result in mutual annihilation and thereby deletion of domain walls or can lead to multiple walls inside the nanowire depending on the external magnetic field. Without an external field a reliable annihilation is found enabling a fully electrical fast creation and annihilation. High repetition rates of both processes have been utilized for time-resolved imaging at the MAXYMUS transmission X-ray microscope to study their evolution. Domain-wall transformations in high driving fields above the Walker breakdown are observed. The creation process by successive vortex-core formation underneath the contact line within 4 ns is in agreement with the results from AMR measurements and micromagnetic simulations.The inertial mass of a vortex domain wall is found to be 5×10^{-22} kg in agreement with theoretical predictions and is one order of magnitude smaller than the mass equivalent of a transverse domain wall.

The measurement techniques established in this work yield a unique access to the evolution of the domain-wall spin structure in combination with statistical analysis of occurring processes. The detailed knowledge of the domain wall dynamics and especially their controlled injection will be relevant for efficient implementation into devices and studies on connected artificial spin ice lattices.

Acknowledgements We are grateful to Ulrich Merkt for continuous support and fruitful discussions over many years. We thank MarkusWeigand, Hermann Stoll, and Gisela Schütz, Max-Planck-Institute for Intelligent Systems, Stuttgart, Germany as well as Mi-Young Im and Peter Fischer, LBNL Berkeley, CA,USA for excellent, long-standing cooperation. We acknowledge financial support from the Deutsche Forschungsgemeinschaft via SFB 668 'Magnetism from the Single Atom to the Nanostructure', via Graduiertenkolleg 1286 'Functional Metal-Semiconductor Hybrid Systems', and via excellence cluster 'The Hamburg Centre for Ultrafast Imaging - Structure, Dynamics and Control of Matter on the Atomic Scale'.

References

1. M.H. Kryder, C.S. Kim, IEEE Trans. Magn. **45**, 3406 (2009)
2. Everspin Technolgies - The MRAM company, http://www.everspin.com as of (2014)
3. S.S.P. Parkin, M. Hayashi, L. Thomas, Science **320**, 190 (2008)
4. D.H. Smith, IEEE Trans. Magn. **1**, 281 (1965)
5. H. Banks, The New York Times, September 20th (1981)
6. J.-S. Kim, M.-A. Mawass, A. Bisig, B. Krüger, R.M. Reeve, T. Schulz, F. Büttner, J. Yoon, C.-Y. You, M. Weigand, H. Stoll, G. Schütz, H.J.M. Swagten, B. Koopmans, S. Eisebitt, M. Kläui, Nat. Commun. **5**, 3429 (2014)
7. D.A. Allwood, G. Xiong, C.C. Faulkner, D. Atkinson, D. Petit, R.P. Cowburn, Science **309**, 1688 (2005)
8. R. Mattheis, S. Glathe, M. Diegel, U. Hübner, J. Appl. Phys. **111**, 113920 (2012)
9. K. Zeissler, S.K. Walton, S. Ladak, D.E. Read, T. Tyliszczak, L.F. Cohen, W.R. Branford, Sci. Rep. **3**, 1252 (2013)
10. A. Pushp, T. Phung, C. Rettner, B.P. Hughes, S.-H. Yang, L. Thomas, S.P.P. Parkin, Nat. Phys. **9**, 505 (2013)
11. A. Yamaguchi, T. Ono, S. Nasu, K. Miyake, K. Mibu, T. Shinjo, Phys. Rev. Lett. **92**, 077205 (2004)
12. G.S.D. Beach, C. Nistor, C. Knutson, M. Tsoi, J.L. Erskine, Nanowires Nat. Mater. **4**, 741 (2005)
13. M. Hayashi, L. Thomas, Y.B. Bazaliy, C. Rettner, R. Moriya, X. Jiang, S.S.P. Parkin, Phys. Rev. Lett. **96**, 197207 (2006)
14. G. Meier, M. Bolte, R. Eiselt, B. Krüger, D.-H. Kim, P. Fischer, Phys. Rev. Lett. **98**, 187202 (2007)

15. M. Hayashi, L. Thomas, C. Rettner, R. Moriya, Y.B. Bazaliy, S.S.P. Parkin, Phys. Rev. Lett. **98**, 037204 (2007)
16. L. Thomas, R. Moriya, C. Rettner, S.S.P. Parkin, Science **330**, 1810 (2010)
17. E. Saitoh, H. Miyajima, T. Yamaoka, G. Tatara, Nature **432**, 203 (2004)
18. L. Bocklage, B. Krüger, R. Eiselt, M. Bolte, P. Fischer, G. Meier, Phys. Rev. B **78**, 180405(R) (2008)
19. A. Bisig, J. Rhensius, M. Kammerer, M. Curcic, H. Stoll, G. Schütz, B. Van Waeyenberge, K.W. Chou, T. Tyliszczak, L.J. Heyderman, S. Krzyk, A. von Bieren, M. Kläui, Appl. Phys. Lett. **96**, 152506 (2010)
20. M. Kläui, P.-O. Jubert, R. Allensbach, A. Bischof, J.A.C. Bland, G. Faini, U. Rüdiger, C.A.F. Vaz, L. Vila, C. Vouille, Phys. Rev. Lett. **95**, 026601 (2005)
21. M. Kläui, H. Ehrke, U. Rüdiger, T. Kasama, R.E. Dunin-Borkowski, D. Backes, L.J. Heyderman, C.A.F. Vaz, J.A.C. Bland, G. Faini, E. Cambril, W. Wernsdorfer, Appl. Phys. Lett. **87**, 102509 (2005)
22. M. Hayashi, L. Thomas, C. Rettner, R. Moriya, X. Jiang, S.S.P. Parkin, Phys. Rev. Lett. **97**, 207205 (2006)
23. P. Lendecke, Verlag Dr. Hut, München (2010)
24. C. Wuth, L. Kolbe, G. Meier, J. Appl. Phys. **114**, 103901 (2013)
25. C.H. Marrows, G. Meier, J. Phys.: Cond. Mat. **24**, 020301 (2012)
26. P.-E. Weiss, Comptes rendus de l'Académie des sciences **143**, 1136 (1906)
27. L. Landau, E. Lifshits, Phys. Zeitsch. der Sow. **8**, 153 (1935)
28. S. Middlehoek, J. Appl. Phys. **34**, 1054 (1963)
29. E.E. Huber Jr., D.O. Smith, J.B. Goodenough, J. Appl. Phys. **29**, 294 (1958)
30. Y. Nakatani, A. Thiaville, J. Milat, J. Magn. Magn. Mater. **290**, 750 (2005)
31. B. Krüger, *Current-Driven Magnetization Dynamics: Analytical Modeling and Numerical Simulation* (Universität Hamburg, 2011)
32. D. Petit, A.-V. Jausovec, D. Read, R.P. Cowburn, J. Appl. Phys. **103**, 114307 (2008)
33. S. Zhang, Z. Li, Phys. Rev. Lett. **93**, 127204 (2004)
34. B. Krüger, D. Pfannkuche, M. Bolte, G. Meier, U. Merkt, Phys. Rev. B **75**, 054421 (2007)
35. J.C. Slonczewski, J. Appl. Phys. **44**, 1759 (1973)
36. A.A. Thiele, J. Appl. Phys. **45**, 377 (1974)
37. N.L. Schryer, L.R. Walker, J. Appl. Phys. **45**, 5406 (1974)
38. J.F. Dillon Jr., published in *Treatise on Magnetism, III, 450* (Academic, New York, 1963)
39. K. Ramstöck, W. Hartung, A. Hubert, Phys. Stat. Sol. (a) **155**, 505 (1996)
40. L. Bocklage, B. Krüger, T. Matsuyama, M. Bolte, U. Merkt, D. Pfannkuche, G. Meier, Phys. Rev. Lett. **103**, 197204 (2009)
41. M. Hayashi, *Current Driven Dynamics of Magnetic Domain Walls in Permalloy Nanowires* (Stanford University, 2006)
42. W. Döring, Z. Naturforsch. **3a**, 373 (1948)
43. L. Bocklage, B. Krüger, P. Fischer, G. Meier, Phys. Rev. B **81**, 054404 (2010)
44. D.J. Clarke, O.A. Tretiakov, G.-W. Chern, Ya B. Bazaliy, O. Tchernyshyov, Phys. Rev. B **78**, 134412 (2008)
45. M. Hayashi, L. Thomas, C. Rettner, R. Moriya, S.S.P. Parkin, Nat. Phys. **3**, 21 (2007)
46. J.L. Prieto, M. Muñoz, E. Martinez, Phys. Rev. B **83**, 104425 (2011)
47. L. O'Brien, D. Read, D. Petit, R.B. Cowburn, J. Phys.: Cond. Mat. **24**, 024222 (2012)
48. K. Sentker, F.-U. Stein, L. Bocklage, T. Matsuyama, M.-Y. Im, P. Fischer, G. Meier, Appl. Phys. Lett. **104**, 172404 (2014)
49. L. Bocklage, F.-U. Stein, M. Martens, T. Matsuyama, G. Meier, Appl. Phys. Lett. **103**, 092406 (2013)
50. H. Ade, H. Stoll, Nat. Mater. **8**, 281 (2009)
51. H. Stoll, A. Puzic, B. van Waeyenberge, P. Fischer, J. Raabe, M. Buess, T. Haug, R. Höllinger, C. Back, D. Weiss, G. Denbeaux, Appl. Phys. Lett. **84**, 3328 (2004)
52. H. Stoll, M. Noske, M. Weigand, K. Richter, B. Krüger, R.M. Reeve, M. Hänze, C.F. Adolff, F.-U. Stein, G. Meier, M. Kläui, G. Schütz, Front. Phys. **3**, 26 (2015)

53. X. Jiang, L. Thomas, R. Moriya, S.S.P. Parkin, Nano Lett. **11**, 96 (2011)
54. F.-U. Stein, L. Bocklage, M. Weigand, G. Meier, Phys. Rev. B **89**, 024423 (2014)
55. L. Thomas, M. Hayashi, X. Jiang, R. Moriya, C. Rettner, S.S.P. Parkin, Nature **443**, 197 (2006)
56. F.-U. Stein, L. Bocklage, M. Weigand, G. Meier, Sci. Rep. **3**, 1737 (2013)

Index

R. Wiesendanger (ed.), *Atomic- and Nanoscale Magnetism*, NanoScience and Technology, https://doi.org/10.1007/978-3-319-99558-8

Printed by Printforce, the Netherlands